ADVANCES IN
Applied Microbiology

VOLUME 13

CONTRIBUTORS TO THIS VOLUME

Richard Bartha

Robert G. Benedict

Jnanendra K. Bhattacharjee

A. C. Blackwood

Robert K. Blackwood

R. Elsworth

Arthur R. English

A. G. Fredrickson

P. A. J. Gorin

J. M. Ingram

E. G. Jeffreys

Arthur Lewis

R. D. Megee, III

J. K. Peterson

David Pramer

P. J. Radlett

R. C. Righelato

K. Sargeant

M. L. Sassiver

J. F. T. Spencer

R. C. Telling

H. M. Tsuchiya

George E. Ward

John V. Wittenburg

ADVANCES IN

Applied Microbiology

Edited by D. PERLMAN

School of Pharmacy
The University of Wisconsin
Madison, Wisconsin

VOLUME 13

 1970

ACADEMIC PRESS, New York and London

Copyright © 1970, by Academic Press, Inc.
ALL RIGHTS RESERVED.
NO PART OF THIS BOOK MAY BE REPRODUCED IN ANY FORM,
BY PHOTOSTAT, MICROFILM, RETRIEVAL SYSTEM, OR ANY
OTHER MEANS, WITHOUT WRITTEN PERMISSION FROM
THE PUBLISHERS.

ACADEMIC PRESS, INC.
111 Fifth Avenue, New York, New York 10003

United Kingdom Edition published by
ACADEMIC PRESS, INC. (LONDON) LTD.
Berkeley Square House, London W1X 6BA

LIBRARY OF CONGRESS CATALOG CARD NUMBER: 59-13823

PRINTED IN THE UNITED STATES OF AMERICA

CONTENTS

LIST OF CONTRIBUTORS ix
PREFACE xi
CONTENTS OF PREVIOUS VOLUMES xiii

Chemotaxonomic Relationships Among the Basidiomycetes

ROBERT G. BENEDICT

I. Introduction 1
II. Macrochemical Color Reactions 2
III. Other Chemotaxonomic Aids in Fleshy Fungi 4
References 21

Proton Magnetic Resonance Spectroscopy—An Aid in Identification and Chemotaxonomy of Yeasts

P. A. J. GORIN AND J. F. T. SPENCER

I. Introduction 25
II. Chemotaxonomy of Yeasts According to Their Mannose-Containing Polysaccharides 33
III. Classification of the Yeasts according to the Pmr Spectra of Their Mannose-Containing Polysaccharides 54
IV. Summary and Conclusions 84
References 87

Large-Scale Cultivation of Mammalian Cells

R. C. TELLING AND P. J. RADLETT

I. Introduction 91
II. Selection, Storage, and Maintenance of Cells 92
III. Production of Culture Media 99
IV. Mass Cultivation of Cells 104
V. Conclusions 115
References 116

Large-Scale Bacteriophage Production

K. SARGEANT

I. Introduction 121
II. General Considerations 122
III. DNA Bacteriophages 125
IV. RNA Bacteriophages 128
V. Bacteriophage-Infected Cells 133
References 136

Microorganisms as Potential Sources of Food

Jnanendra K. Bhattacharjee

I.	Introduction	139
II.	Inadequacy of Cereal and Vegetable Protein	141
III.	Desirable Characteristics of Microorganisms	142
IV.	Algae	142
V.	Fungi	149
VI.	Yeast	152
VII.	Lichens	155
VIII.	Bacteria	156
IX.	Chemosynthesis	158
X.	Palatability	158
XI.	Conclusion	158
	References	159

Structure–Activity Relationships Among Semisynthetic Cephalosporins

M. L. Sassiver and Arthur Lewis

I.	Introduction	163
II.	Semisynthetic Cephalosporins	165
III.	Resistance to Cephalosporinases	219
IV.	Pharmacology, Metabolism, and Mode of Action	225
	References	230

Structure–Activity Relationships in the Tetracycline Series

Robert K. Blackwood and Arthur R. English

I.	Introduction	237
II.	Basic Structural Requirement for Tetracycline Activity	239
III.	Electronic and Lipophilic Effects on Tetracycline Activity	258
	References	265

Microbial Production of Phenazines

J. M. Ingram and A. C. Blackwood

I.	Introduction	267
II.	Pyocyanine	267
III.	Chlororaphine	270
IV.	Phenazine-1-Carboxylic Acid	272
V.	Iodinin	274
VI.	Other Phenazines	275
VII.	Discussion and Possible Future Studies	279
	References	280

The Gibberellin Fermentation

E. G. Jefferys

I.	Introduction and Scope of Review	283
II.	Gibberellins — Structure and Terminology	284
III.	Other Compounds Isolated from Gibberellin Fermentations	284
IV.	Organisms, Nomenclature, and Strains	284
V.	Assay, Extraction, and Purification Methods	291
VI.	The Biosynthesis of Gibberellins	294
VII.	The Course of Fermentation	296
VIII.	The Effects of Environmental Changes on the Course of Fermentation	299
IX.	The Production of Gibberellins	307
X.	Present and Potential Applications	309
	References	310

Metabolism of Acylanilide Herbicides

Richard Bartha and David Pramer

I.	Introduction	317
II.	Metabolism of Acylanilides	319
III.	Biological Activity of Metabolites	336
IV.	Concluding Comments	338
	References	339

Therapeutic Dentifrices

J. K. Peterson

I.	Introduction	343
II.	Plaque- and Calculus-Inhibiting Dentifrices	344
III.	Desensitizing Dentifrices	344
IV.	Caries-Inhibitory Dentifrices	345
V.	Sodium Fluoride Dentifrices	350
VI.	An Amine Fluoride Dentifrice	350
VII.	Stannous Fluoride Dentifrices	350
VIII.	Phosphate Fluoride Dentifrices	355
IX.	Summary	358
	References	359

Some Contributions of the U.S. Department of Agriculture to the Fermentation Industry

George E. Ward

I.	Introduction	363
II.	Citric Acid Research of Dr. J. N. Currie	364
III.	Arlington Farm Research Period	364
IV.	Regional Research Laboratory Period: 1940–1945	371
V.	Regional Research Laboratory Period: 1945–1969	377
VI.	Vitamin B_{12} Fermentation	379
	References	379

Microbiological Patents in International Litigation

John V. Whittenburg

Text	383
References	398

Industrial Applications of Continuous Culture: Pharmaceutical Products and Other Products and Processes

R. C. Righelato and R. Elsworth

I.	Introduction	399
II.	Pharmaceutical Products	401
III.	Organic Chemicals	405
IV.	Drink	407
V.	Sewage and Trade Wastes	412
VI.	General Conclusions	415
	References	416

Mathematical Models for Fermentation Processes

A. G. Fredrickson, R. D. Megee, III, and H. M. Tsuchiya

I.	Introduction	419
II.	Formulation of Models for Fermentation Processes	423
III.	Some Basic Simplifying Assumptions	427
IV.	Fermentors: Some Engineering Considerations	431
V.	Some Models for the Growth of a Pure Culture of Unicellular Microorganisms	438
VI.	Models for the Growth of Filamentous Organisms	459
VII.	Concluding Remarks	463
	References	464

Author Index	467
Subject Index	485

LIST OF CONTRIBUTORS

Numbers in parentheses indicate the pages on which the authors' contributions begin.

RICHARD BARTHA, *Department of Biochemistry and Microbiology, College of Agriculture and Environmental Science, Rutgers–The State University, New Brunswick, New Jersey* (317)

ROBERT G. BENEDICT, *Drug Plant Laboratory, College of Pharmacy, University of Washington, Seattle, Washington* (1)

JNANENDRA K. BHATTACHARJEE, *Department of Microbiology, Miami University, Oxford, Ohio* (139)

A. C. BLACKWOOD, *Department of Microbiology, Macdonald College of McGill University, Quebec, Canada* (267)

ROBERT K. BLACKWOOD, *Medical Research Laboratories, Chas. Pfizer & Co., Inc., Groton, Connecticut* (237)

R. ELSWORTH,° *New Brunswick Scientific Co., Inc., New Brunswick, New Jersey* (399)

ARTHUR R. ENGLISH, *Medical Research Laboratories, Chas. Pfizer & Co., Inc., Groton, Connecticut* (237)

A. G. FREDRICKSON, *Chemical Engineering Department, University of Minnesota, Minneapolis, Minnesota* (419)

P. A. J. GORIN, *National Research Council of Canada, Prairie Regional Laboratory, Saskatoon, Saskatchewan, Canada* (25)

J. M. INGRAM, *Department of Microbiology, Macdonald College of McGill University, Quebec, Canada* (267)

E. G. JEFFERYS, *Imperial Chemical Industries Limited, Pharmaceutical Division, Alderley Park, Macclesfield, Cheshire, Great Britain* (283)

ARTHUR LEWIS, *Lederle Laboratories, American Cyanamid Company, Pearl River, New York* (163)

R. D. MEGEE, III, *Chemical Engineering Department, University of Minnesota, Minneapolis, Minnesota* (419)

J. K. PETERSON, *Divison of Dental Health, North Dakota State Department of Health, Bismarck, North Dakota* (343)

°Present address: 25, Potters Way, Laverstock, Salisbury, Wilts., England.

DAVID PRAMER, *Department of Biochemistry and Microbiology, College of Agriculture and Environmental Science, Rutgers–The State University, New Brunswick, New Jersey* (317)

P. J. RADLETT, *Animal Virus Research Institute, Pirbright, Surrey, England* (91)

R. C. RIGHELATO, *Glaxo Laboratories, Ltd., Ulverston, Lancashire, England* (399)

K. SARGEANT, *Microbiological Research Establishment, Porton, Salisbury, Wiltshire, England* (121)

M. L. SASSIVER, *Lederle Laboratories, American Cyanamid Company, Pearl River, New York* (163)

J. F. T. SPENCER, *National Research Council of Canada, Prairie Regional Laboratory, Saskatoon, Saskatchewan, Canada* (25)

R. C. TELLING, *Animal Virus Research Institute, Pirbright, Surrey, England* (91)

H. M. TSUCHIYA, *Chemical Engineering Department, University of Minnesota, Minneapolis, Minnesota* (419)

GEORGE E. WARD, *Dawe's Laboratories, Inc., Chicago, Illinois* (363)

JOHN V. WHITTENBURG, *American Cyanamid Co., Stanford, Connecticut* (383)

PREFACE

In this volume as in previous volumes in this series we have attempted to focus attention on those facets of applied microbiology that are currently receiving or will receive attention in the immediate future. Many microbiologists have shifted their interests from the traditional pathways of applied microbiology including food microbiology, soil microbiology, and industrial (fermentation) microbiology, to the charms of molecular biology. We expect that they will return to the applied areas and hope that their renewed interest is attributable in part to some of the articles in this volume.

The topics covered in this volume are centered around microbial chemistry, antimicrobial agents, and fermentations. The use of microbial products and components in cells in chemotaxonomy is considered by Benedict for basidiomycetes and by Gorin and Spencer for bacteria. Telling and Radlett, in discussing large-scale growth of animal cells in fermentation equipment, Sargeant doing the same for 'phage, and Righelato and Elsworth, in surveying continuous fermentations, all contribute to our understanding of the problems of maintaining microbial cultures in a vigorous condition. Fredrickson *et al.* use data collected in both batch and continuous processes to formulate mathematical expressions to predict bacterial growth under a variety of conditions. If microbial cells are to serve as a major protein source for human diets (Bhattacharjee), we will have to understand more about the economic aspects of growth.

Although 1970 marks 30 years of intensive effort on the use of fermentation processes for antibiotics, the future is still bright for finding useful compounds in fermentations for a variety of purposes. Ingram and Blackwood summarize reports on phenazine production by microorganisms, and Jefferys discusses the plant growth substances, the gibberellins. Ward reviews the history of the United States Department of Agriculture's research in microbial chemistry, and Whittenburg presents some aspects of trying to obtain patent protection on fermentation processes and products. The combination of microbiology and chemistry to produce new antibiotic substances is reviewed by Sassiver and Lewis for the cephalosporins and by Blackwood and English for the tetracyclines. Peterson indicates some of the requirements for antimicrobial agents if they are to be useful in dental practice, and Bartha and Pramer carry antimicrobial activities one step further in their discussion of the limitations of microbial systems in degradation of organic compounds.

The scope of the areas covered in this volume indicates the expanding horizons of applied microbiology. We hope that these essays will encourage many to join the groups examining the potential of microbial processes. There is opportunity for all.

CONTENTS OF PREVIOUS VOLUMES

Volume 1

Protected Fermentation
Miloš Herold and Jan Nečásek

The Mechanism of Penicillin Biosynthesis
Arnold L. Demain

Preservation of Foods and Drugs by Ionizing Radiations
W. Dexter Bellamy

The State of Antibiotics in Plant Disease Control
David Pramer

Microbial Synthesis of Cobamides
D. Perlman

Factors Affecting the Antimicrobial Activity of Phenols
E. O. Bennett

Germfree Animal Techniques and Their Applications
Arthur W. Phillips and James E. Smith

Insect Microbiology
S. R. Dutky

The Production of Amino Acids by Fermentation Processes
Shukuo Kinoshita

Continuous Industrial Fermentations
Philip Gerhardt and M. C. Bartlett

The Large-Scale Growth of Higher Fungi
Radcliffe F. Robinson and R. S. Davidson

AUTHOR INDEX—SUBJECT INDEX

Volume 2

Newer Aspects of Waste Treatment
Nandor Porges

Aerosol Samplers
Harold W. Batchelor

A Commentary on Microbiological Assaying
F. Kavanagh

Application of Membrane Filters
Richard Ehrlich

Microbial Control Methods in the Brewery
Gerhard J. Hass

Newer Development in Vinegar Manufactures
Rudolph J. Allgeier and Frank M. Hildebrandt

The Microbiological Transformation of Steroids
T. H. Stoudt

Biological Transformation of Solar Energy
William J. Oswald and Clarence G. Golueke

SYMPOSIUM ON ENGINEERING ADVANCES IN FERMENTATION PRACTICE

Rheological Properties of Fermentation Broths
Fred H. Deindoerfer and John M. West

Fluid Mixing in Fermentation Processes
J. Y. Oldshue

Scale-up of Submerged Fermentations
W. H. Bartholemew

Air Sterilization
Arthur E. Humphrey

Sterilization of Media for Biochemical Processes
Lloyd L. Kempe

Fermentation Kinetics and Model Processes
Fred H. Deindoerfer

Continuous Fermentation
W. D. Maxon

Control Applications in Fermentation
George J. Fuld

AUTHOR INDEX – SUBJECT INDEX

Volume 3

Preservation of Bacteria by Lyophilization
Robert J. Heckly

Sphaerotilus, Its Nature and Economic Significance
Norman C. Dondero

Large-Scale Use of Animal Cell Cultures
Donald J. Merchant and C. Richard Eidam

Protection Against Infection in the Microbiological Laboratory: Devices and Procedures
Mark A. Chatigny

Oxidation of Aromatic Compounds by Bacteria
Martin H. Rogoff

Screening for and Biological Characterizations of Antitumor Agents Using Microorganisms
Frank M. Schabel, Jr., and Robert F. Pittillo

The Classification of Actinomycetes in Relation to Their Antibiotic Activity
Elio Baldacci

The Metabolism of Cardiac Lactones by Microorganisms
Elwood Titus

Intermediary Metabolism and Antibiotic Synthesis
J. D. Bu'Lock

Methods for the Determination of Organic Acids
A. C. Hulme

AUTHOR INDEX – SUBJECT INDEX

Volume 4

Induced Mutagenesis in the Selection of Microorganisms
S. I. Alikhanian

The Importance of Bacterial Viruses in Industrial Processes, Especially in the Dairy Industry
F. J. Babel

Applied Microbiology in Animal Nutrition
Harlow H. Hall

Biological Aspects of Continuous Cultivation of Microorganisms
T. Holme

Maintenance and Loss in Tissue Culture of Specific Cell Characteristics
Charles C. Morris

Submerged Growth of Plant Cells
L. G. Nickell

AUTHOR INDEX – SUBJECT INDEX

Volume 5

Correlations between Microbiological Morphology and the Chemistry of Biocides
Adrien Albert

Generation of Electricity by Microbial Action
J. B. Davis

Microorganisms and the Molecular Biology of Cancer
G. F. Gause

Rapid Microbiological Determinations with Radioisotopes
Gilbert V. Levin

The Present Status of the 2,3-Butylene Glycol Fermentation
Sterling K. Long and Roger Patrick

Aeration in the Laboratory
W. R. Lockhart and R. W. Squires

Stability and Degeneration of Microbial Cultures on Repeated Transfer
Fritz Reusser

Microbiology of Paint Films
Richard T. Ross

The Actinomycetes and Their Antibiotics
Selman A. Waksman

Fusel Oil
A. Dinsmoor Webb and John L. Ingraham

AUTHOR INDEX — SUBJECT INDEX

Volume 6

Global Impacts of Applied Microbiology: An Appraisal
Carl-Göran Hedén and Mortimer P. Starr

Microbial Processes for Preparation of Radioactive Compounds
D. Perlman, Aris P. Bayan, and Nancy A. Giuffre

Secondary Factors in Fermentation Processes
P. Margalith

Nonmedical Uses of Antibiotics
Herbert S. Goldberg

Microbial Aspects of Water Pollution Control
K. Wuhrmann

Microbial Formation and Degradation of Minerals
Melvin P. Silverman and Henry L. Ehrlich

Enzymes and Their Applications
Irwin W. Sizer

A Discussion of the Training of Applied Microbiologists
B. W. Koft and Wayne W. Umbreit

AUTHOR INDEX — SUBJECT INDEX

Volume 7

Microbial Carotenogenesis
Alex Ciegler

Biodegradation: Problems of Molecular Recalcitrance and Microbial Fallibility
M. Alexander

Cold Sterilization Techniques
John B. Opfell and Curtis E. Miller

Microbial Production of Metal-Organic Compounds and Complexes
D. Perlman

Development of Coding Schemes for Microbial Taxonomy
S. T. Cowan

Effects of Microbes on Germfree Animals
Thomas D. Luckey

Uses and Products of Yeasts and Yeastlike Fungi
Walter J. Nickerson and Robert G. Brown

Microbial Amylases
Walter W. Windish and Nagesh S. Mhatre

The Microbiology of Freeze-Dried Foods
Gerald J. Silverman and Samuel A. Goldblith

Low-Temperature Microbiology
Judith Farrell and A. H. Rose

AUTHOR INDEX — SUBJECT INDEX

Volume 8

Industrial Fermentations and Their Relations to Regulatory Mechanisms
Arnold L. Demain

Genetics in Applied Microbiology
S. G. Bradley

Micotoxins
Alex Ciegler and Eivind B. Lillehoj

Microbial Ecology and Applied Microbiology
Thomas D. Brock

The Ecological Approach to the Study of Activated Sludge
Wesley O. Pipes

Control of Bacteria in Nondomestic Water Supplies
Cecil W. Chambers and Norman A. Clarke

The Presence of Human Enteric Viruses in Sewage and Their Removal by Conventional Sewage Treatment Methods
Stephen Alan Kollins

Oral Microbiology
Heiner Hoffman

Media and Methods for Isolation and Enumeration of the Enterococci
Paul A. Hartman, George W. Reinbold, and Devi S. Saraswat

Crystal-Forming Bacteria as Insect Pathogens
Martin H. Rogoff

Mycotoxins in Feeds and Foods
Emanuel Borker, Nino F. Insalata, Colette P. Levi, and John S. Witzeman

AUTHOR INDEX – SUBJECT INDEX

Volume 9

The Inclusion of Antimicrobial Agents in Pharmaceutical Products
A. D. Russell, June Jenkins, and I. H. Harrison

Antiserum Production in Experimental Animals
Richard M. Hyde

Microbial Models of Tumor Metabolism
G. F. Gause

Cellulose and Cellulolysis
Brigitta Norkrans

Microbiological Aspects of the Formation and Degradation of Cellulosic Fibers
L. Jurášek, J. Ross Colvin, and D. R. Whitaker

The Biotransformation of Lignin to Humus – Facts and Postulates
R. T. Oglesby, R. F. Christman, and C. H. Driver

Bulking of Activated Sludge
Wesley O. Pipes

Malo-lactic Fermentation
Ralph E. Kunkee

AUTHOR INDEX – SUBJECT INDEX

Volume 10

Detection of Life in Soil on Earth and Other Planets. Introductory Remarks
Robert L. Starkey

For What Shall We Search?
Allan H. Brown

Relevance of Soil Microbiology to Search for Life on Other Planets
G. Stotzky

Experiments and Instrumentation for Extraterrestrial Life Detection
Gilbert V. Levin

Halophilic Bacteria
D. J. Kushner

Applied Significance of Polyvalent Bacteriophages
S. G. Bradley

Proteins and Enzymes as Taxonomic Tools
Edward D. Garber and John W. Rippon

Transformation of Organic Compounds by Fungal Spores
Claude Vézina, S. N. Sehgal, and Kartar Singh

Microbial Interactions in Continuous Culture
Henry R. Bungay, III and Mary Lou Bungay

Chemical Sterilizers (Chemosterilizers)
Paul M. Borick

Antibiotics in the Control of Plant Pathogens
M. J. Thirumalachar

AUTHOR INDEX – SUBJECT INDEX

CUMULATIVE AUTHOR INDEX – CUMULATIVE TITLE INDEX

Volume 11

Successes and Failures in the Search for Antibiotics
Selman A. Waksman

Structure-Activity Relationships of Semisynthetic Penicillins
K. E. Price

Resistance to Antimicrobial Agents
J. S. Kiser, G. O. Gale, and G. A. Kemp

Micromonospora Taxonomy
George Luedemann

Dental Caries and Periodontal Disease Considered as Infectious Diseases
William Gold

The Recovery and Purification of Biochemicals
Victor H. Edwards

Ergot Alkaloid Fermentations
William J. Kelleher

The Microbiology of the Hen's Egg
R. G. Board

Training for the Biochemical Industries
I. L. Hepner

AUTHOR INDEX – SUBJECT INDEX

Volume 12

History of the Development of a School of Biochemistry in the Faculty of Technology, University of Manchester
Thomas Kennedy Walker

Fermentation Processes Employed in Vitamin C Synthesis
Miloš Kulhánek

Flavor and Microorganisms
P. Margalith and Y. Schwartz

Mechanisms of Thermal Injury in Non-sporulating Bacteria
M. C. Allwood and A. D. Russell

Collection of Microbial Cells
Daniel I. C. Wang and Anthony J. Sinskey

Fermentor Design
R. Steel and T. L. Miller

The Occurrence, Chemistry, and Toxicology of the Microbial Peptide-Lactones
A. Taylor

Microbial Metabolites as Potentially Useful Pharmacologically Active Agents
D. Perlman and G. P. Peruzzotti

AUTHOR INDEX – SUBJECT INDEX

ADVANCES IN
Applied Microbiology

VOLUME 13

Chemotaxonomic Relationships Among the Basidiomycetes

ROBERT G. BENEDICT

Drug Plant Laboratory, College of Pharmacy,
University of Washington,
Seattle, Washington

I.	Introduction	1
II.	Macrochemical Color Reactions	2
III.	Other Chemotaxonomic Aids in Fleshy Fungi	4
	A. Indolyl Derivatives	4
	B. Urea	6
	C. Studies on Mycorrhizal-Associated Fungi	9
	D. Tetronic Acids	13
	E. Terphenylquinones	15
	F. Styrylpyrones in *Gymnopilus* and *Polyporus*	18
	G. Polyacetylenic Compounds	19
	References	21

I. Introduction

The classification of basidiomycetes is a difficult task and the disagreements that often occur between expert classical taxonomists in matters of this sort do not simplify the task of interpreting the chemotaxonomic data that have appeared in the literature in the past 10 years. The writer knows one mycology professor who hands out eleven different taxonomic keys to students to ensure that no paucity of diagnostic characteristics prevails for the identification of field specimens.

Differentiation of tissue structures on the surface and within the basidiocarps has permitted the classical taxonomist to separate many species within related genera. Botanists working their way along a perhaps never-ending road toward "full truth" must in Singer's words (1962) be willing to "accept reshifts in classification, the application of more and more time-absorbing methods of investigation, the opposition of some of our colleagues working in other fields who may denounce our inability to state shortly and simply the characters on which the groups of fungi are separated."

The writer agrees with Cronquist (1962) that a "backcross of biosystematics to classical taxonomy should give us the modified classical taxonomy of the future, receptive to data of all sorts." In the latter category, the feverish activity that has taken place in the "new chromatographic era" includes a start on some potentially useful chemotaxonomic studies among the basidiomycetes. If the current review stimulates a few young systematists to continue along these lines, the reviewer will consider worthwhile the time spent in preparing it.

The numerous references throughout this review to Singer (1962) are in support of the reviewer's choice of this eminent agaricologist as his taxonomic authority. In this paper, author citations are given on all basidiomycete genera and species except in those cases where the authors of cited papers chose to omit them.

II. Macrochemical Color Reactions

Basidiomycetes collected at random (in season) and broken apart will sometimes reveal the presence of some leuco chemical which changes color upon oxidation by air. A simple example is that of some types of white latex in the genus *Lactarius* (lacking in older, dry specimens). The presence of latex readily distinguishes this mushroom from all others in the order Agaricales. Without doubt, these natural changes stimulated taxonomists to spot various chemical reagents on cap and stipe tissues of fleshy fungi and to record color changes that occurred.

The reviewer is indebted to Professor R. Watling for a prepublication copy of his excellent review on "Chemical Tests in Agaricology" (Watling, 1969). One section contains a detailed table with eleven categories of macrochemical test reagents commonly used by agaricologists along with the color changes that occur in representatives of several families of the order Agaricales. The diagnostic value of some reactions of alkalis, iodine complexes, and simple iron salts with compounds in the mushrooms is discussed in detail. Instructions for the preparation of the various reagents are carefully explained.

When Watling's data are applied to the families in Singer's (1962) classification system, one finds that members of several families respond little or not at all to macrochemical tests. These are the Polyporaceae (9), Hygrophoraceae (8), Coprinaceae (7), Bolbitiaceae (5), Crepidotaceae (7), Rhodophyllaceae (3), and Strobilomycetaceae (4). (The numbers in parentheses represent genera in each of these families.) Families with representatives that respond to the tests are shown in Table I.

It appears that chemical categories 9 (nitrogen-containing aromatics, e.g., aniline, benzidine, etc.), 8 (aromatic alcohols and acids, e.g., phenol, resorcinol, etc.), and 4 (simple complexes containing halogens, e.g., Melzer's iodine complex, calcium hypochlorite, etc.) are the most useful, although these totals are less significant for 8 and 9 when one

TABLE I
COMPARATIVE UTILITIES OF CHEMICAL CATEGORIES IN MACROCHEMICAL TESTS

Family (genera)	1 (5)[a]	2 (4)	3 (2)	4 (6)	5 (3)	6 (5)	7 (3)	8 (12)	9 (10)	10 (4)
Russulaceae (2)	0	2[b]	1	0	1	5	1	10	10	1
Boletaceae (14)	2	2	1	3	1	1	2	1	4	1
Tricholomataceae (84)	0	1	1	2	0	1	1	3	2	1
Cortinariaceae (15)	2	2	1	4	0	0	0	0	0	0
Amanitaceae (7)	1	0	0	1	1	0	0	2	2	0
Agaricaceae (17)	2	0	0	0	0	0	0	0	1	1
Paxillaceae (5)	0	0	0	0	1	0	1	0	1	0
Gomphidiaceae (2)	1	0	0	0	0	0	2	0	0	0
Totals	8	7	4	10	4	7	7	16	20	4

[a]Number of test reagents in this category.
[b]Equals number of tests from category 2 reported as positive in members of the Russulaceae.

reflects that at least three reagents in these categories are repetitious, i.e., in 8, guaiacol, pyrogallol, and phlorglucinol are each oxidized by the same enzymes (laccase or tyrosinase), and in 9, aniline, o-aminophenol, and benzidine by laccase or tyrosinase, laccase, and laccase, respectively. Enzyme data and substrate color changes are those noted in an extensive study of cultures of wood-rotting basidiomycetes by Käärik (1965).

The importance of the iodine complexes in Agaricales classification should not be underestimated. Reagents such as Melzer's (a mixture of chloral hydrate plus KI and metallic I_2) often react with constituents on fungal cell walls, and the terms "amyloid" and "nonamyloid" (with respect to the darkening or nondarkening by iodine) of spores or parts thereof, have had a strong impact on basidiomycete classification. An outstanding example of this (Watling, 1969) is the darkening effect of Melzer's solution on the spore ornamentation of members of the Russulaceae, helping to set this family apart from all other agarics. The genus *Russula* itself probably contains more than 500 species, and in Singer's opinion (1962) "our knowledge of *Russula* is very good, the difficulties one often encounters in the identification of specimens arise from the multitude of species all very similar to each other and the sparsity of specialists to assist in the identification of specimens."

One of the main objections to the macrochemical tests is lack of specificity, e.g., the mushroom compound that reacts with the test agent is often unknown. Watling (1969) believes that chemical tests will have more value when we know more about the substrates we are testing for, their variation and distribution in other organisms.

III. Other Chemotaxonomic Aids in Fleshy Fungi

A. INDOLYL DERIVATIVES

1. Panaeolus *Species*

Singer (1962) characterizes *Panaeolus* (Fr.) Quél. as having no cystidia on the sides of the gills, spores smooth, and spore print black. In the subfamily Panaeoloideae Sing., one also finds the genera *Panaeolina* R. Maire, *Copelandia* Bres., and *Anellaria* Karst. Tyler (1958) prepared extracts of *Panaeolus campanulatus* (Fr.) Quél. and found serotonin (5-hydroxytryptamine) plus other Ehrlich-positive indolyl compounds. Weir and Tyler (1963) reported that *Panaeolus campanulatus* and *Panaeolina foenisecii* (Pers. ex Fr.) R. Maire are excellent sources of serotonin, averaging about 1234 and 3728 µg/gm dry wt, respectively.

In 1938, a specimen of *Paneolus campanulatus* Fr. ex L. var. *sphinctrinus* (Fr.) Bres. was brought in from the jungle to R. E. Schultes as a presumed sample of "teonanacatl," the divine mushroom of Mexico (Heim and Hofmann, 1958). Both *P. sphinctrinus* and *P. papilionaceus* (Bull.) Quél. were reported by Singer (1949) to be sources of intoxicating drugs used by Central American Indians. Tyler and Smith (1963b) examined the following *Panaeolus* species for their protoalkaloid content: *P. campanulatus* (2), *P. foenisecii* (2), *P. acuminatus* (Schaeff. ex Fr.) Quél., *P. acuminatus* f. *gracilis* f. nov., *P. fontinalis* Smith, *P. semiovatus* (Fr.) Lundell, *P. solidipes* Peck., *P. subbalteatus* (Berk. & Br.) Sacc., and *P. texensis* sp. nov. All of these extracts contained serotonin and 5-hydroxytryptophan except *P. solidipes*, a 55-year-old herbarium specimen, which may have lost its protoalkaloids during the long storage period. The authors found no chromatographic indication of the presence of psilocybin or psilocin, although some strains of *P. subbalteatus* are definitely psychotomimetic, according to Stein (1958) who, several years earlier, had eaten this mushroom mixed with *Agaricus bisporus* (Lange) Imbach. Later on, Stein *et al.* (1959) found a 4-phosphoryl indolyl derivative in *P. subbalteatus* similar to psilocybin, but melting above 250°C. The structure of this compound has not been completely elucidated, but Leung and Paul (1968) agree that it may be identical with baeocystin, the monomethyl analog of psilocybin.

The controversy surrounding the status of *Panaeolus sphinctrinus* as a hallucinogenic mushroom still exists. The reported isolation of psilocybin from *P. sphinctrinus* carpophores by Heim and Hofmann (1958) could not be substantiated with later specimens of the same fungus (Hofmann *et al.*, 1963). Specimens of *P. sphinctrinus* authenticated by Singer, and reputedly the same species as the Mexican material originally collected by Schultes (1939) were carefully analyzed for protoalkaloids by Tyler and Gröger (1964a). The extracts contained relatively large amounts of serotonin, 5-hydroxytryptophan, and urea (see Section III,B), but no psilocybin or psilocin. In a report by Ola'h (1968) one finds *P. sphinctrinus* listed in a group under "Psilocybiennes latentes," i.e., specimens that "contain with certainty, indolic compounds with psychodisleptic capacity but do not offer a constant presence." These are *P. africanus, P. castaneifolius, P. fimicola, P. microsporus,* and *P. foenisecii*. One must select one of two possibilities with respect to *P. sphinctrinus*, either that different chemical races exist, or that the experts disagree on what should constitute an authentic specimen of this species. Ola'h's list of psychotomimetic Panaeoli alleged to contain psilocybin includes *P. ater* (Lange) Kühner and Romagnesi, *P. cambodginiensis, P. cyanescens* [listed in Singer as *Copelandia cyanescens* (Berk. & Br.) Singer], *P. subbalteatus*, and *P. tropicalis*. The UV absorption spectra of compounds isolated from these fungi as shown by Ola'h are indeed similar to those of psilocybin, baeocystin, and norbaeocystin, but unequivocal proof of identity has not been provided.

From a chemotaxonomic view, we may say that many species of *Panaeolus* contain the indolyl compounds 5-hydroxytryptophan and serotonin, and, in addition, a few have psychotomimetic compounds related to psilocybin.

2. Psilocybe *and* Conocybe

The genus *Psilocybe* (Fr.) Quél., as arranged by Singer (1962), contains 38 species in five sections. Section Caerulescentes (23 species) is characterized in part by specimens whose contexts blue on exposure, reaching "deep Medici blue" Ridgway, and react strongly with the macrochemical reagent monomethylparamidophenol.

Psilocybin and/or psilocin has been reported from a high percentage of these species, as shown by Heim and Hofmann (1958), Benedict *et al.* (1962, 1967), and Benedict and Brady (1969). In addition, *Psilocybe baeocystis* Singer and Smith, *Psilocybe quebecensis* Heim and Ola'h, and *Psilocybe serbica* spec. nov. may also be added to the list [Leung and Paul (1968), Ola'h (1967), and Moser and Horak (1968)].

Watling (in Benedict *et al.*, 1967) pointed out that psilocybin-containing *Psilocybe semilanceata* (Fr. ex Secr.) Kummer, Section Psilocybe, may or may not show blueing and in reality is synonymous with Section Caerulescentes. He suggests, therefore, that these two sections be combined.

Conocybe *Fayod*

The genus *Conocybe* is placed in the family Bolbitiaceae, and there are some doubts regarding early reports by Heim (1957) that *Conocybe siligeneoides* was probably used by Mexican Indians in religious ceremonies. However, later reports of psilocybin in *Conocybe cyanopus* (Atk.) Kühner by Benedict *et al.* (1962), and in *Conooybe smithii* Watling, formerly *Galera cyanopes* Kauffman (Benedict *et al.*, 1967), show that the drug is present in some species. The number of species is too small, however, to have any chemotaxonomic potential. Probable and postulated biosynthetic pathways to these compounds are summarized in Fig. 1.

B. UREA

1. Survey of Many Genera

Urea reacts with Ehrlich reagent to give a bright yellow color and is so characteristic and runs so well on chromatographs (Smith, 1960) that it is often added as a marker to solutions not containing it. The R_f values in commonly used solvent systems, e.g., *n*-butanol-acetic acid-water, *n*-butanol-pyridine-water, and isopropyl alcohol-ammonia-water are about 0.5±0.06. Other Ehrlich-positive compounds with yellow chromophores and similar R_f values are almost totally lacking in mushrooms. Biosynthetic origins of urea are not only complex but diverse, according to Tyler *et al.* (1965), and both of these factors are important with respect to its biosystematic utility. They collected and analyzed for their urea content 344 species of higher fungi, and most of these came from western Washington State. Shown in Table II are the results obtained on representative genera from all but one of the sixteen families placed in the Agaricales by Singer (1962), plus four other families of nonagarics.

Perhaps one should suspect heterogeneity in a family as large and diverse as that of the Tricholomataceae (see Section III,C,1,a). We found that *Armillaria, Clitocybe, Lepista, Lyophyllum, Marasmius, Melanoleuca,* and *Tricholomopsis* accumulate urea, whereas *Armillariella, Catathelasma, Laccaria, Leucopaxillus, Marasmiellus, Omphalotus, Oudemansiella, Panellus, Pleurocybella, Tricholoma,*

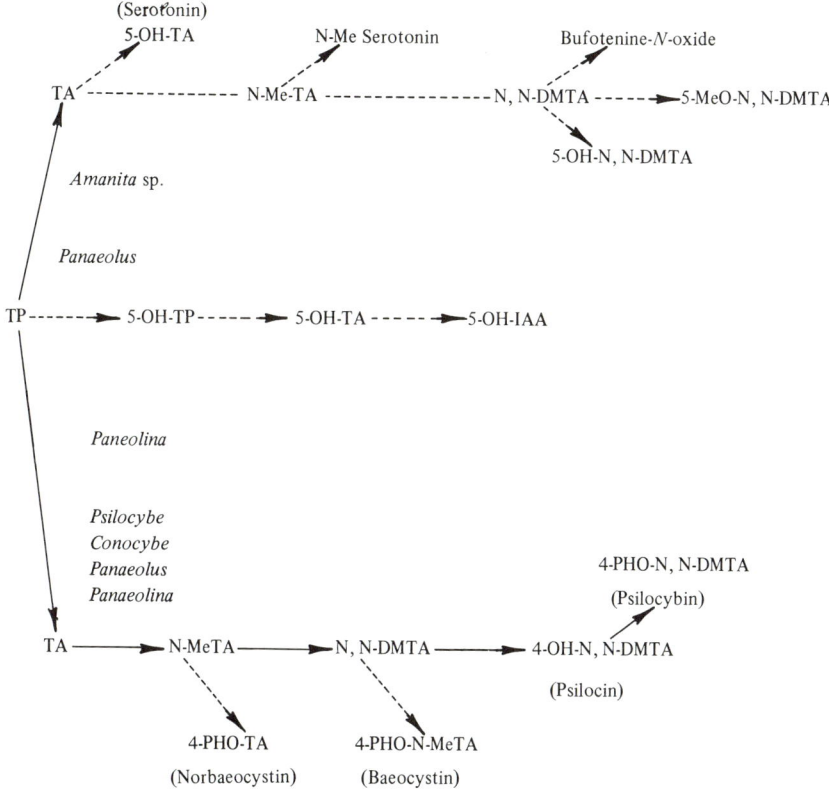

FIG. 1. Biosynthetic pathways to indolyl hallucinogens. TA = Tryptamine; DM = dimethyl-; MeO = methoxy-; OH = hydroxy-; TP = tryptophan; Me = methyl; PHO = phosphoryl-; IAA = indolylacetic acid.

and *Xeromphalina* do not. The apparent inconsistent results noted with *Collybia* (2+ and 2−) are less puzzling when one observes that *Collybia confluens* (Pers. ex Fr.) Kummer and *C. dryophila* (Bull. ex Fr.) Kummer were both formerly classified in the genus *Marasmius*. Despite Singer's judgment (1962) that "the characters of the epicutis clearly separate *Collybia* from nearly all species of *Marasmius*," the findings with respect to urea could further assist in settling taxonomic differences between the two genera. Examination of more species of both *Mycena* and *Xeromphalina* should help clarify the previously reported inconsistencies.

The broad range of urea concentrations in the Amanitaceae, (0–5+), is less startling when one recalls that subgenus AMANITA contains

TABLE II
UREA IN BASIDIOMYCETES[a]

Family	No. of genera tested	Species	Levels of urea
Hygrophoraceae	3	17	0-1+
Tricholomataceae	21	58	(7) = 3-5+
			(11) = 0-1+
			(3) = Variable
Amanitaceae	2	16	0-5+
Agaricaceae	5	17	3-5+
Coprinaceae	5	16	3-5+
Bolbitiaceae	1	1	2+
Strophariaceae	4	14	0-2+
Cortinariaceae	6	75	0-1+
Rhodophyllaceae	2	6	Variable
Paxillaceae	2	4	0-1+
Boletaceae	6	17	0
Russulaceae	2	25	0-1+
Lycoperdaceae	2	7	3-5+

[a] Members of the Cantharellaceae, Clavariaceae, Hydnaceae, and Strobilomycetaceae totaling 7 genera and 49 species, contained no urea.

nonamyloid spores and little or no urea, and subgenus EUAMANITA, amyloid spores and moderate to high urea levels (Section III,C,c). The results shown with *Rhodophyllus* species were both interesting and challenging, ranging from 0 to 5+. The two urea-negative species, *R. sericellus* (Fr.) Quél. and *R. serrulatus* (Fr.) Quél., are classified by Dennis *et al.* (1960) as species of *Leptonia*. *Rhodophyllus nidorosus* (Fr.) Quél. and *R. sinuatus* (Fr.) Quél., both with moderate amounts of urea, were placed in the genus *Entoloma*, and *R. sericeus* (Fr.) Quél., rich in urea, in the genus *Nolanea*. As part of an extensive chemotaxonomic study, D. Largent and R. G. Benedict (1969, unpublished) examined the ethanolic extractives from 289 collections of rhodophylloid fungi. We found all of our *Nolanea* species to be strongly urea-positive (3-5+), whereas species of *Leptonia* and *Entoloma* ranged between 0 and 2+ in 90-95% of all samples examined.

In contrast with the confused chemotaxonomic picture regarding certain indolyl compounds in *Panaeolus*, *Panaeolina*, and *Psilocybe*, the results on urea are quite clear, e.g., species of the former genera showing 3-5+ urea levels, and *Psilocybe* species, none.

In the family Boletaceae, it is surprising indeed that one species,

Suillus piperatus (Bull. ex Fr.) O. Kuntze, should have urea when all other species examined had none.

The apparent stability of urea in long-held herbarium specimens, plus the ease of detection on chromatograms and the improbability of finding false positives, undoubtedly strengthens its position as a valuable chemotaxonomic tool in the fleshy fungi.

C. Studies on Mycorrhizal-Associated Fungi

1. Miscellaneous Compounds

The submerged mycelia of mycorrhizally associated basidiomycetes are intimately connected with the minute rootlets of certain trees and derive a portion of their nutritional requirements from the host.

a. Tricholoma *Species*

Benedict et al. (1964) made a preliminary study of twelve species of *Tricholoma* and three species of *Lepista* and *Tricholomopsis* (both formerly included in *Tricholoma*) and found that approximately thirty compounds could be detected on acid-washed paper, following one-dimensional chromatography and an Ehrlich reagent spray. Urea was present in both *Lepista* and *Tricholomopsis* species and in *Tricholoma sclerotoideum* Morse. The chemical profile of the latter pointed away from *Tricholoma*; in fact, Singer (1962) had already indicated the close relationship of *T. sclerotoideum* and *Clitocybe inornata* (Fr.) Gill. Several species of *Clitocybe* proved to be strongly urea-positive. The remaining *Tricholoma* species in our survey were urea-negative, and the chance is remote that one would find this compound in any mushroom, "with a tricholomatoid habit and with lamallae not free, or spores permanently without an internal pigment" (*Tricholoma*, in Singer's Key II).

The 80 species of *Tricholoma* listed by Singer are classified among four subgenera. Our survey contained too few species within any subgenus to permit evaluation of comparative tests for common chemotaxonomic markers. However, the occurrence of identical markers in extracts of *Tricholoma inamoenum* (Fr. ex Fr.) Quél. and *T. sulphureum* (Bull. ex Fr.) Kummer suggests that they are varieties of the same basic form rather than distinct species. The six identical chemotaxonomic markers present in both *T. aurantium* (Schaeff. ex Fr.) Ricken and *Armillaria zelleri* Stuntz and Smith suggests that the latter fungus should be transferred from *Armillaria* to Stirps Aurantium, in Subgenus TRICHOLOMA.

b. Inocybe *Species*

Some of the results of a chemical and chemotaxonomic evaluation of 39 species of *Inocybe* by Robbers *et al.* (1964) brought up some interesting taxonomic questions. In this survey, some *Inocybe* species in unrelated stirpes contained the same secondary metabolite (ergothioneine), raising the problem of how one should interpret both morphological and biochemical characteristics. The reviewer is in accord with Robbers and co-worker's statement that "Both characteristics are expressions of genetic differences, but, in systematics, there is no existing finite knowledge that permits greater emphasis to be placed on either." On the other hand, *Inocybe kauffmanii* A. H. Smith and *Inocybe picrosma* Stuntz, closely related according to morphological characteristics, have dissimilar chemical compositions, the latter species not showing muscarine, ergothioneine, tyrosine, and valine. Despite these apparent anomalies, the authors set up a key, based on the chemotaxonomic data, and then correctly identified coded specimens with the aid of that key. The genus *Inocybe* has by far the highest percentage of muscarine-synthesizing species. Twenty-six of the 39 collections examined were positive, far more than either *Amanita* or *Clitocybe* species (Section III,1,C) and (Section III,G). A few compounds were limited to single species, i.e., 5-hydroxytryptophan to *Inocybe* 1838 and tryptamine to *Inocybe hirsuta* var. *maxima* A. H. Smith. The use of two-dimensional chromatography, multisolvent systems, and a variety of sprays permitted an indication of the general chemical nature of some compounds not specifically identified, but which were useful in setting up the chemotaxonomic key.

c. Amanita *Species*

Singer (1962) states that the poisonous species in subgenus I, AMANITA, contain muscarine rather than amanitatoxins. Among species listed in Section 2 of the subgenus, one finds *A. muscaria* (Fr.) S. F. Gray, *A. pantherina* (Fr.) Secr., and *A. gemmata* (Fr.) Gill. The first two species have minute amounts of muscarine ($\sim 0.00025\%$ wet weight), usually not enough to cause symptoms of muscarine poisoning. Pacific Northwest specimens of *A. muscaria* and *A. pantherina* are reported to have the narcotic-intoxicant isoxazoles ibotenic acid and muscimol (the latter formerly known as pantherine) in levels from 0.18–0.45% on a dry solids basis (Benedict *et al.*, 1966). We found that "true" *A. gemmata* lacks these compounds and that apparent hybrids, taxonomically designated as *A. gemmata-pantherina* intergrades, can be differentiated from *A. gemmata* only by the presence of these isoxazoles. None of the remaining species in Singer's Subgenus I are known to have these compounds. *Amanita strobiliformis* (Vitt.) Quél. is the type species in a section of Subgenus EUAMANITA (wherein the spores are amyloid). Takemoto *et al.* (1964) isolated ibotenic acid from *A. strobiliformis* (Paul.) Quél. The latter contains an incorrect author citation and raises doubt regarding the identity of this Japanese mushroom. Specimens of *A. strobiliformis* (Vitt.) Quél. *sensu* W. C. Coker, collected in Tennessee and examined by Tyler *et al.* (1966), contained neither ibotenic acid nor muscimol. Examinations of other collections of both Japanese and American *A. strobiliformis* might settle the question.

Dennis *et al.* (1960) consider *A. strobiliformis* to be conspecific with *A. solitaria* (Fr.) Secr., but our collections of the latter species, *sensu* D. E. Stuntz, had no ibotenic acid or muscimol; in their place we found the related compounds solitaric acid and solitarine. Thus, a considerable amount of chemotaxonomic confusion exists in this group at present.

In subgenus II, EUAMANITA, section Euamanita, Singer (1962) says, "Several or all species in this section contain smaller or larger amounts of amanitatoxin (deadly poisonous or suspect)." The type species is *A. phalloides* (Vaill. ex Fr.) Secr. All specimens of this taxon examined in our laboratory have shown amatoxins, although the ratios of α- and β-amanitin vary in different collections, suggesting the existence of different chemical races. Some strains of *A. verna* (Fr.) Vitt. s. Boud. or *A. virosa* Secr. contain little or no detectable toxin (Tyler *et al.*, 1966). Probably *A. ocreata* Peck and certainly *A. bisporigera* Atk. should be transferred from section Amidellae to section Euamanita. The latter species, assaying from 3 to 5 mg/gm dry weight of α- and β-amanitin, is probably the most toxic mushroom yet reported. *Amanita tenuifolia* Murr., also known to contain amatoxins, is not treated in Singer.

In the same section one also finds stirps Citrina listing four species, at least two of which have the odor of sprouting potatoes—*A. citrina* (Schaeff. ex) S. F. Gray and *A. porphyria* (A. & S. ex Fr.) Secr. They lack the toxic cyclopeptides, but other secondary metabolites present include several indolyl protoalkaloids, e.g., bufotenine, N-methylserotonin, etc. (Tyler and Gröger, 1964b) (see Fig. 1). If chemotaxonomic leads are to be meaningful, the reviewer believes that these species merit a separate section.

Neither *Galerina* Earle (family Cortinariaceae) nor *Lepiota* (Pers. ex) S. F. Gray (family Agaricaceae) are strictly mycorrhizal, nor are they closely related to *Amanita,* but they are mentioned here because certain sections of each have species with the same complex metabolites (amatoxins, phallotoxins, or both). Singer's Section Naucoriopsis in *Galerina* has at least three toxic species—*G. autumnalis* (Peck) Sm. and Singer, *G. marginata* (Fr.) Kühner, and *G. venenata* A. H. Smith, according to Tyler and Smith (1963a) and Tyler *et al.* (1963). In genus *Lepiota*, Section Ovisporae, one finds the deadly *L. helveola sensu* Josserand and the related *L. brunneoincarnata* Chodat & Martin. *Lepiota fuscovinacea* Moeller and Lange is also suspect in the opinion of Mortara and Filipello (1967). The principal toxins are easily identified on thin-layer chromatograms, but mycologists are often reluctant to release the small samples of herbarium specimens needed to provide presumptive identification.

d. Cortinarius *Species*

The genus *Cortinarius* is a very large and taxonomically difficult one, including more than 500 species. Some of these are brilliantly colored with red, purple, orange, yellow, and brown pigments predominating. The majority of these pigments are probably anthraquinones, but relatively few have been positively identified (see Fig. 2). Most of the 29 sections treated by Singer (1962) contain species primarily from European collections. Gabriel (1960, 1965) extracted and studied the pigments of several colorful species authenticated by R. Kühner, an eminent French mycologist. Three species were examined from Section Sanguinei, whose representatives by definition "contain vacuolar red pigments (at times only in the gills)." Chromatographic separation in an amyl alcohol-pyridine-water mixture showed thirteen pigments, which she numbered according to mobility, the highest spot (emodin) being assigned number one.

Cortinarius sanguineus Wulf ex Fr. exhibited pattern 1–6 and 12 (seven pigments); *C. semisanguineus* Wulf ex Fr. 4–6, 9, 12, and 13; and *C. phoeniceus* Bull. ex Fr. 4–6 and 12. All three species contained purpurin. According to Steglich and Losel (unpublished reference 2) in Steglich *et al.* (1968), *C. sanguineus* extracts also yield endocrocin. Other specimens studied by Gabriel (1960) in Section Dermocybe included *Cortinarius cinnamomeus* L. ex Fr. var. *conformis* Fr. pattern 1, 3, 4', 5–9, and 12, and L. ex Fr. var. *lutescens* Gil. pattern 5–10. *Cortinarius sphagneti* Orton and *C. ulignosus* Berk. are similar to *C. sanguineus*, evidence offered by Gabriel that the two groups should be rejoined. The reviewer believes that more data should be obtained before chemotaxonomic reevaluations can be made.

e. Boletus *and Related Species*

A search for chemotaxonomic markers in the Boletaceae by Benedict and Tyler (1968) showed that mannitol, glucose, and trehalose were present in various species of *Boletinus, Suillus, Phylloporus, Boletus, Xerocomus, Leccinum,* and *Boletellus.* The alcohol arabitol was present in all extracts except those of *Xerocomus chrysenteron* (St. Am.) Quél. and *Boletellus zelleri* (Murr.) Snell, Sing. & Dick. Fructose occurred only in *Boletus miniato-olivaceus* var. *sensibilis* Peck and two species of *Leccinum*, *L. aurantiacum* (St. Am.) S. F. Gray and *L. scabrum* (Fr.) S. F. Gray, and heptulose only in *L. aurantiacum* and *Boletellus zelleri* (four separate collections). Although no clear-cut chemotaxonomic conclusions could be drawn from these studies, a survey of a larger number of species may prove of value at the species level.

f. Gomphus *Species*

Agaricic acid (α-hexadecylcitric acid), structure (**I**), was isolated many years ago from the wood-rot fungus *Fomes officinalis* Fr. (Faull.) by Thoms and Vogelsang (1907).

$$CH_3-(CH_2)_n-CH-COOH$$
$$HO-C-COOH$$
$$H_2-C-COOH$$

I, $n = 15$

II, $n = 13$

A second compound of this type was not observed in higher fungi until Miyata *et al.* (1966) found norcaperatic acid (**II**) (α-tetradecylcitric acid) in *Cantharellus floccosus* Schw., now known as *Gomphus floccosus* (Schw.) Sing. Other fleshy fungal sources reported by Sullivan (1968) and Henry and Sullivan (1969) include *Polyporus fibrillosus* Karst. (22%) and *Gomphus kauffmanii* (A. H. Smith) Corner (2%). None was found in *Cantharellus cibarius* Fr., *C. infundibuliformis* Fr., *C. subalbidans* Smith & Morse, or in *Gomphus clavatus* S. F. Gray. Should other species of *Gomphus* be tested, it will be interesting to see if they too possess this compound.

D. Tetronic Acids

1. Boletus *and Related Species*

Until recently, naturally occurring tetronic acids and related compounds had been found mainly in lichens or produced fermentatively by yeasts, molds, and a few actinomycetes.

The first compound common to several species of *Boletus* in Singer's Section Luridi was designated "boletol" (see Fig. 2). However, Steglich et al. (1968) claim that this compound, isolated by Kögl and Deijs (1934) from *B. satanus* Lenz, *B. luridus* Schaeff. ex Fr., and *B. badius* (Fr.) Kühner ex Gilbert, gives no blue color with oxidase, hence cannot be the specific compound that gives the blueing reaction in certain of the Boletaceae. Steglich and co-workers (1968) ex-

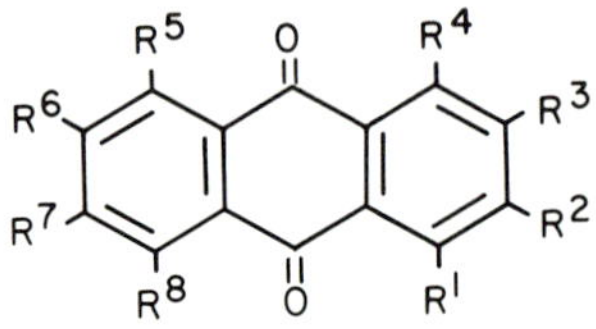

FIG. 2. Anthraquinone pigments in basidiomycetes.

	R^1	R^2	R^3	R^4	R^5	R^6	R^7	R^8
Emodin	OH	H	CH_3	H	H	OH	H	OH
Purpurin	OH	H	OH	OH	H	H	H	H
Endocrocin	OH	COOH	CH_3	H	H	OH	H	OH
Dermocybin[a]	—	—	—	—	—	—	—	—
Boletol	COOH[b]	H	H	COOH[b]	OH	OH	H	OH

[a]Four —OH groups and one —OCH_3 group on the outside rings, but localities not certified.

[b]COOH Group on R^1 or R^4, but not on both. In the opinion of Beaumont et al. (1968), this compound does not exist in boletes.

amined *Boletus erythropus* (Fr. ex Fr.) Pers., *B. calopus* Fr., and *Xerocomus chrysenteron* (Bull. ex St. Am.) Quél. and found that the blue pigments forming in air with the aid of an oxidase are substituted tetronic acids related to pulvinic acid from the lichen, *Sticta coronata*. The first one of these from the Boletaceae was variegatic acid (see Fig. 3), originally isolated from *Suillus variegatus* (Sow. ex Fr.) O. Kuntze by Edwards and Elsworthy (1967). The same compound occurred in *B. erythropus* (0.04%), *B. calopus* (0.005%), and *Xerocomus chrysenteron* (0.09%). The recovery percentages are those of Steglich et al., who also reported the latter species to have more xerocomic acid as well as small amounts of atromentic and chloroxerocomic acids. These investigators also believe that the R_f values and the UV spectrum of the so-called "boletol," detected chromatographically in 22 species of the genera *Boletus, Phylloporus,*

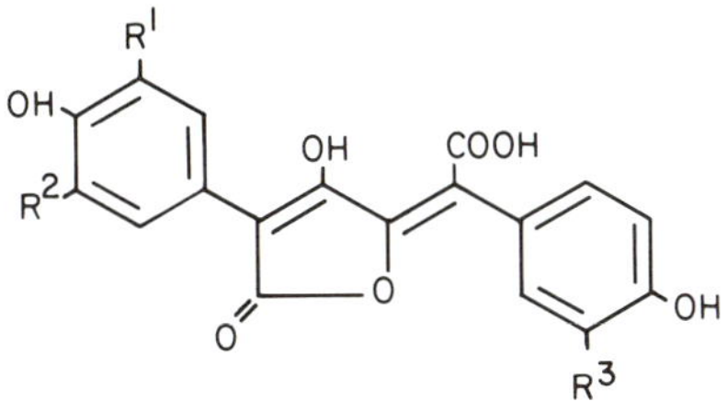

FIG. 3. Tetronic acid derivatives in *Boletus* and *Gomphidius* species.

$R^1 = R^3 = OH$; $R^2 = H$ — Variegatic acid
$R^1 = OH$; $R^2 = R^3 = H$ — Xerocomic acid
$R^1 = R^2 = R^3 = H$ — Atromentic acid
$R^1 = R^2 = OH$; $R^3 = H$ — Gomphidic acid

Xerocomus, Boletinus, Suillus, and *Gyrodon* by Gabriel (1965), point toward variegatic acid, and, on the same basis, "pseudobolitol" from *B. erythropus, X. parasiticus* (Bull. ex Fr.) Quél., *Suillus placidus* (Bon.) Sing., and *S. piperatus* (Bull. ex Fr.) O. Kuntze becomes xerocomic acid.

In his discussion of the close relationship of the Paxillaceae with the Boletaceae, Singer (1962) mentions that blueing does often occur in many species of the latter, but the reviewer finds that little or no taxonomic use of this characteristic is made in the key for the subfamilies in the Boletaceae.

The small genus *Gomphidius* Fr. is closely related to *Boletus*, and it is not surprising that similar secondary metabolites appear in both. Singer lists only eighteen species of the former. The stipe ends of *Gomphidius glutinosus* (Schaeff. ex Fr.) Fr., *G. oregonensis* Peck, and *G. smithii* Sing. are bright yellow in color. Steglich *et al.* (1969) isolated a new pulvinic acid derivative from *G. glutinosus* and named it gomphidic acid (see Fig. 3). No mention is made of stipe end colors in Singer's key, and it is difficult to see why this characteristic has been ignored in past considerations of species differentiation within the genus.

E. TERPHENYLQUINONES

1. Hydnellum *and Related Genera*

The natural occurrence of terphenylquinones, with one exception, is thus far restricted to certain basidiomycetes and lichens. Among the former, the family Hydnaceae, including the genera *Hydnellum, Hydnum, Lopharia,* and *Phellodon*, are good sources of these compounds. One of the most interesting compounds is atromentin (see Fig. 4) which is found in *Hydnellum peckii* Harrison, formerly *H.*

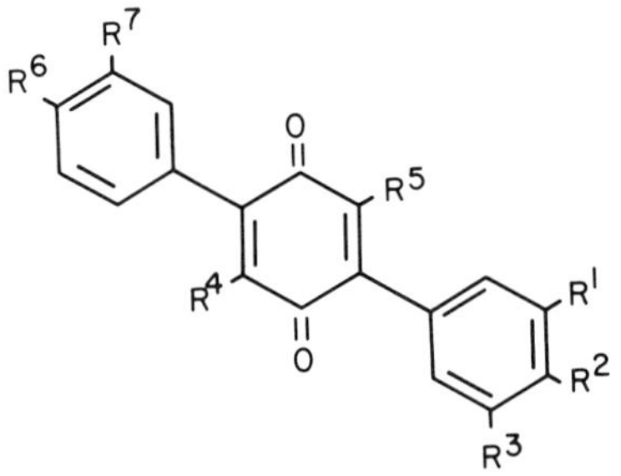

FIG. 4. Some terphenylquinones in basidiomycetes.

	R^1	R^2	R^3	R^4	R^5	R^6	R^7
Atromentin	H	OH	H	OH	OH	OH	H
Polyporic acid	H	H	H	OH	OH	H	H
Leucomelone	OH	OH	H	OH	OH	OH	H
Volucrisporin[a]	H	H	OH	H	H	H	OH
Aurantiacin	H	OH	H	BZO[b]	BZO[b]	OH	H

[a] From the hyphomycete *Volucrispora aurantiaca* Haskins. See Divekar et al. (1959).
[b] BZO = benzoyloxy

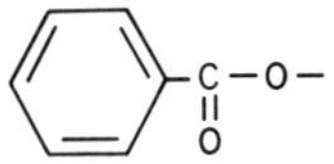

TABLE III
TERPHENYLQUINONES IN THE GENUS *Hydnellum*

Hydnellum species	Aurantiacin	Dihydro-aurantiacin dibenzoate	Atromentin
H. HA 202[a]	+	+	−
H. caeruleum (Pers.) Karst.	+[b]	+	−
H. aurantiacum Batsch.	+[b]	+[b]	N.R.[c]
H. scrobiculatum var. zonatum (Fr.) Harr.	+	−	−
H. scrobiculatum var. scrobiculatum (Secr.) Karst.	−	−	−
H. peckii Harrison (formerly H. diabolus Banker)	−	−	+[b]
H. suaveolens (Scop.) Karst.	−	−	−

[a] A University of Washington Mycological Herbarium number referring to a recognizable entity, not yet described in the literature.
[b] Reported by workers other than Sullivan et al. (1967).
[c] Not reported.

diabolus Banker (Euler *et al.*, 1965; and Khanna *et al.*, 1965), *Clitocybe subilludens* Murr. (Sullivan and Guess, 1969), and *Paxillus atrotomentosus* (Batsch. ex Fr.) Fr. (Kögl and Postowsky, 1924).

The distribution of various terphenylquinones in *Hydnellum* species collected in the Pacific Northwest was undertaken by Sullivan *et al.* (1967). All of the fungi listed in Table III contained a black pigment, later identified as thelephoric acid, the structure of which is not included in Fig. 4. It is 2,3,8,9-tetrahydroxybenzo-bis(1,2-*b*: 4,5-*b'*)benzofuran-6,12-quinone. Thelephoric acid is probably the terphenylquinone most frequently encountered in basidiomycetes; Sawada (1952, 1958) isolated and quantitated this compound in *Hydnum cyathiforme, H. nigrum, H. graveolens, H. amarescens, H. imbricatum, H. aspratum, H. scabrosum, Phlebia strigosozonata, Polystictus versicolor,* and *Cantharellus multiplex.* On a fresh weight basis, the amounts recovered ranged from 1.7 to 0.004%.

2. Paxillus *Species*

The presence of atromentin in *P. atrotomentosus* has already been mentioned. Although not a terphenylquinone, a new, naturally occurring relative of this type compound has recently been isolated from *Paxillus involutus* (Batsch. ex Fr.) Fr. by Edwards *et al.* (1967). It is a diphenylcyclopenteneone named "involutin" (see Fig. 5). *Paxillus* is a small genus with not more than a dozen species and is closely related to *Gyrodon*, in the Boletaceae (see Section III,D).

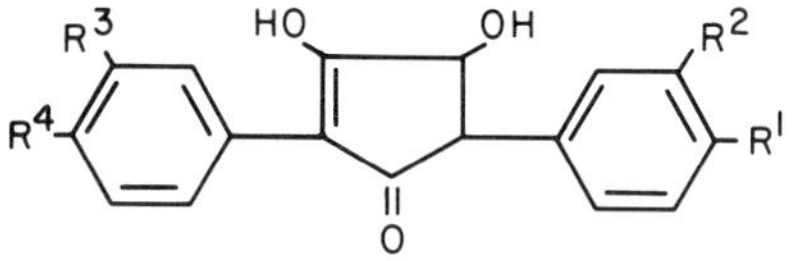

FIG. 5. Involutin.
$R^1 = R^3 = R^4 = OH; R^2 = H$
$R^1 = R^2 = R^4 = OH; R^3 = H$

3. Polyporus *Species*

Terphenylquinones in *Polyporus* and related genera are limited to too few species at present for any sort of chemotaxonomic evaluation. However, accumulations of some of these compounds reach high proportions. For example, polyporic acid (Fig. 4) occurs in *Polyporus rutilans* (P.) Fr. to the extent of 23% of the dry weight, according to Frank *et al.* (1950a). These same workers postulate (1950b) that the ability of polyporic acid to form highly insoluble complexes with most metallic ions may serve a useful function, i.e., to protect its spores against microorganisms and insects by precipitating and removing essential minerals from such invaders. Fruiting bodies of this fungus have an unusually low mineral content. According to Murray (1952), polyporic acid is also present in *Peniophora filamentosa* Burt.

F. Styrylpyrones in *Gymnopilus* and *Polyporus*

During a survey of various fleshy fungi for polyacetylenic antibiotics and other secondary metabolites of interest, Bu'Lock and Smith (1961) isolated and characterized hispidin from *Polyporus hispidus* (Bull.) Fr. (see Fig. 6). The assigned structure was confirmed

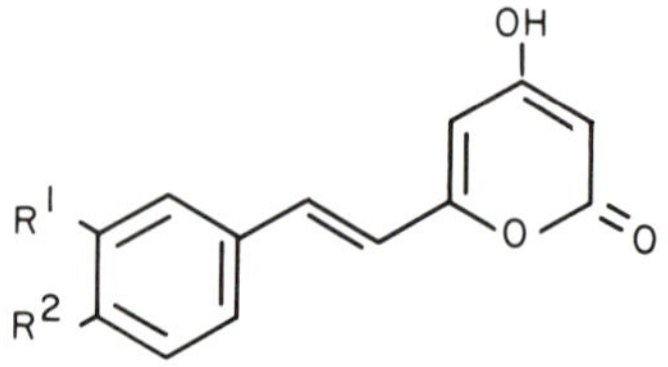

Fig. 6. Styrylpyrones in *Polyporus* and *Gymnopilus*.
$R^1 = R^2 = OH = $ Hispidin
$R^1 = H; R^2 = OH = $ bis-Noryangonin

by Edwards *et al.* (1961) by synthesis of this and many other similar styrylpyrones. Bu'Lock and coworkers suggested that other wood-rot species in the genus *Polyporus* might also have pigments of the hispidin type. They found such pigments in *Polyporus schweinitzii* (Fr.) Pat., grew the fungus on ordinary media, and produced hispidin as the main product (Bu'Lock, 1967). To date, these two species are the only polypores showing hispidin, but more species could be added later because the genus *Polyporus* is very large and versatile with respect to secondary metabolites found in mature carpophores.

During the urea survey (Section III,B,1), Tyler *et al.* (1965) noted that Ehrlich-positive compounds extracted from *Polyporus schweinitzii* closely resembled those from *Gymnopilus* Karsten. Species of the latter are primarily lignicolous basidiomycetes whose predominant colors are yellow to yellow-brown. A recent monograph on this genus by Hesler (1969) recognizes 73 species in North America, approximately half of which contain yellow, KOH-soluble pigments. One of these, *G. spectabilis* (Fr.) Singer, is especially interesting in that it has an unidentified hallucinogen exerting an effect in humans similar to that of psilocybin (Walters, 1965; Romagnesi, 1964; and Buck, 1967).

Investigation of the major Ehrlich-positive compound in *G. decurrens* Hesler by Hatfield and Brady (1968) showed it to be bis-noryangonin [4-hydroxy-6-(4-hydroxystyryl) pyrone] (Fig. 6), a compound closely related to some of those found in Kava root. Later, the same authors (1969) isolated bis-noryangonin from *G. spectabilis*. This finding enhances the possibility that the unidentified hallucinogen in *G. spectabilis* is a styrylpyrone or a close relative of compounds in this chemical class. Current studies in our laboratory on various Gymnopoli show their extracts to contain a variety of Ehrlich-positive compounds, some of which should be of chemotaxonomic value within the genus when their chemical identities and distributions become better known.

G. Polyacetylenic Compounds

During the feverish search for new antibiotics of medicinal importance in the late 1940's and early 1950's, many fermentation broths of various basidiomycete cultures were tested and some showed good antibacterial activity. Some factors were too unstable for isolation and purification. Analyses of the more stable molecules revealed both allenic and acetylenic groups, ranging in length from eight to ten carbons. Thus far, about fifty compounds have been characterized from species within fourteen genera of basidiomycetes scattered among several different families. For comprehensive reviews on the polyacetylenes from both basidiomycetes and higher plants, see Sørensen (1963) and Johnson (1965). To the reviewer's knowledge, none of these compounds has yet been isolated directly from the original carpophores from which the subcultures were obtained.

There appeared little chance that meaningful chemotaxonomic conclusions could be drawn from the over-all "scatter pattern" of polyacetylenic biosynthesis among the basidiomycetes. However, examinations of groups of these compounds proved more fruitful. One noteworthy contribution along these lines is that of Anchel *et al.* (1962) who had previously observed that certain basidiomycetes produced singles, pairs, or groups of polyacetylenes. To study possible chemotaxonomic implications, they selected the "diatretynes" formed by *Clitocybe diatretra* (Fr.) Quél. (family Tricholomataceae). The three diatretynes are shown below:

$$HOOC-CH=CH-C\equiv C-C\equiv C-CONH_2$$
I

$$HOOC-CH=CH-C\equiv C-C\equiv C-C\equiv N$$
II

$$HOOC-CH=CH-C\equiv C-C\equiv C-C\equiv C-CH_2OH$$
III

The chromophores of diatretynes **I, II,** and **III** exhibit very characteristic UV absorption spectra. Additional supporting evidence for their presence was obtained by paper chromatography and in some cases by use of their bacterial spectra or behavior on countercurrent distribution, or both.

A total of 254 species in 55 genera, representing 12 families in Singer's classification system (1949), were utilized in the study. After a preliminary screening of the mycelia on agar slant cultures, those showing evidence of diatretyne production were grown in liquid media. The culture broths were carefully processed to partially purify the diatretynes before assay. Only one species outside the family Tricholomataceae produced small amounts of diatretyne **III** [*Psathyrella (Drosophila) sarcocephala* (Fr.) Quél.]. The remaining twelve species occurred in the family Tricholomataceae; *Lepista nuda* (Fr.) Quél., *L. luscina* (Fr.) Singer [formerly *Tricholoma panaeolum* (Fr.) Quél.], *Pleurotus ulmarius* (Fr.) Quél., and nine species of *Clitocybe*. The reviewer shows in Table IV how the latter are distributed among the thirteen sections of *Clitocybe* in Singer (1962). Please note that almost half of these taxa occur in Section Candicantes. [Although having no direct connection, it is noteworthy that the parent carpophores of two of the diatretyne producers in this section *(C. rivulosa* and *C. cerrusata)* also synthesize muscarine (Section III,C,1,b)]. Each of the remaining species is allocated to a different section. Coproduction of

TABLE IV
TAXONOMIC POSITIONS OF DIATRETYNE PRODUCERS
IN *Clitocybe* SPECIES

Clitocybe	Section[a]	Status
C. rivulosa (Fr.) Quél.[b]	Candicantes	Type species
C. cerrusata (Fr.) Quél.	Candicantes	—
C. diatreta (Fr.) Quél.	Candicantes	—
C. fragrans (Fr.) Quél.	Candicantes	—
C. inversa (Fr.) Quél.	Eulepista	Type species
C. nebularis (Fr.) Quél.	Disciformis	Type species
C. umbilicata (Fr.) Quél.	Umbilicatae	Type species
C. robusta Peck[c]		—
C. gallinacea (Fr.) Gill.	Bulluliferae	*C. hydrogramma* (Bull. ex Fr.) Kummer or *C. adirondackensis* (Peck) Sacc.

[a]As revised by Singer (1962).
[b]All author citations are those of the culture suppliers.
[c]Not treated in Singer.

the diatretynes was the rule, with some species of *Clitocybe* producing all three types. The authors concluded that "diatretyne production is probably limited to the family Agaricaceae *sensu lato* and in support of Singer's classification (1949) it may be limited in this group to the genera segregated by him as Tricholomataceae."

ACKNOWLEDGMENTS

This survey was supported in part by National Institutes of Health research grant 5-R01-GM 07515-9. The writer wishes to thank Professor D. E. Stuntz, Department of Botany, University of Washington, for assistance on certain taxonomic aspects.

REFERENCES

Anchel, M., Silverman, W. B., Valanju, N., and Rogerson, C. T. (1962). *Mycologia* **54**, 249–257.
Beaumont, P. C., Edwards, R. L., and Elsworthy, G. C. (1968). *J. Chem. Soc.* (**C**) 2968–2974.
Benedict, R. G., and Brady, L. R. (1969). *In* "Fermentation Advances" (D. Perlman, ed.), pp. 63–88. Academic Press, New York.
Benedict, R. G., and Tyler, V. E. (1968). *Herba Hung.* **7**, 17–21.
Benedict, R. G., Brady, L. R., Smith, A. H., and Tyler, V. E., Jr. (1962). *Lloydia* **25**, 156–159.
Benedict, R. G., Tyler, V. E., Jr., Brady, L. R., and Stuntz, D. E. (1964). *Planta Med.* **12**, 100–106.
Benedict, R. G., Tyler, V. E., Jr., and Brady, L. R. (1966). *Lloydia* **29**, 333–342.
Benedict, R. G., Tyler, V. E., and Watling, R. (1967). *Lloydia* **30**, 150–157.
Buck, R. W. (1967). *New Engl. J. Med.* **276**, 391–392.
Bu'Lock, J. D. (1967). *In* "Essays in Biosynthesis and Microbial Development," E. R. Squibb Lectures on Chemistry of Microbial Products, pp. 2–18. Wiley, New York.
Bu'Lock, J. D., and Smith, H. G. (1961). *Experientia* **17**, 553–554.
Cronquist, A. (1962). *In* "Taxonomic Biochemistry and Serology" (C. A. Leone, ed.), p. 11. Ronald Press, New York.
Dennis, R. W. G., Orton, P. D., and Hora, F. B. (1960). *Trans. Brit. Mycol. Soc. Suppl.* pp. 1–225.
Divekar, P. V., Read, G., and Vining, L. C. (1959). *Chem. Ind. (London)* pp. 731–732.
Edwards, R. L., and Elsworthy, G. C. (1967). *Chem. Commun.* pp. 373–374.
Edwards, R. L., Lewis, D. G., and Wilson, D. V. (1961). *J. Chem. Soc.* pp. 4995–5002.
Edwards, R. L., Elsworthy, G. C., and Kale, N. (1967). *J. Chem. Soc.* (**C**), 405–409.
Euler, K. L., Tyler, V. E., Jr., Brady, L. R., and Malone, M. H. (1965). *Lloydia* **28**, 203–206.
Frank, R. L., Clark, G. R., and Coker, J. N. (1950a). *J. Am. Chem. Soc.* **72**, 1824–1826.
Frank, R. L., Clark, G. R., and Coker, J. N. (1950b). *J. Am. Chem. Soc.* **72**, 1827–1829.
Gabriel, M. (1960). *Bull. Trim. Soc. Mycol. Fr.* **76**, 208–215.
Gabriel, M. (1965). Thesis, Univ. de Lyon, Lyon, France.
Hatfield, G. M., and Brady, L. R. (1968). *Lloydia* **31**, 225–228.
Hatfield, G. M., and Brady, L. R. (1969). *J. Pharm. Sci.* **58**, 1298.
Heim, R. (1957). *Rev. Mycol.* **22**, 183–207.

Heim, R., and Hofmann, A. (1958). In "Les Champignons Hallucinogènes du Mexique" (R. Heim and R. G. Wasson, eds.), Vol. 6, pp. 123-272. Archives. Muséum National d'Historie Naturelle, Paris.
Henry, E. G., and Sullivan, G. (1969). Lloydia 32, 523.
Hesler, L. R. (1969). "North American Species of Gymnopilus," Mycologia Memoir No. 3. 117 pp. Hafner Publishing Co., New York.
Hofmann, A., Heim, R., and Tscherter, H. (1963). C. R. Acad. Sci. 257, 10-12.
Johnson, A. W. (1965). Endeavour 34, 126-130.
Käärik, A. (1965). Stud. Forest. Suec. No. 31, pp. 7-80.
Khanna, J. M., Malone, M. H., Euler, K. L., and Brady, L. R. (1965). J. Pharm. Sci. 54, 1016-1020.
Kögl, F., and Deijs, A. (1934). Ann. Chem. 515, 10-23.
Kögl, F., and Postowsky, J. J. (1924). Ann. Chem. 440, 19-35.
Leung, A., and Paul, A. G. (1968). J. Pharm. Sci. 57, 1667-1671.
Miyata, J. T., Tyler, V. E., Jr., and Brady, L. R. (1966). Lloydia 29, 43-49.
Mortara, M., and Filipello, S. (1967). Minerva Med. 58, 3628-3632.
Moser, M., and Horak, E. (1968). Z. Pilzkunde 34, 137-144.
Murray, J. (1952). J. Chem. Soc. 1345-1350.
Ola'h, G. M. (1967). Natur. Can. 94, 573-587.
Ola'h, G. M. (1968). C. R. Acad. Sci. 267, 1369-1372.
Robbers, J. E., Brady, L. R., and Tyler, V. E., Jr. (1964). Lloydia 27, 192-202.
Romagnesi, M. H. (1964). Bull. Trim. Soc. Mycol. Fr. 80, IV-V.
Sawada, M. (1952). Nippon Rin Gakkai Shi 34, 110-113.
Sawada, M. (1958). Nippon Rin Gakkai Shi 40, 195-197.
Schultes, R. E. (1939). Bot. Mus. Leafl. Harvard Univ. 7, 37-56.
Singer, R. (1949). Lilloa 22, 1-832.
Singer, R. (1962). "The Agaricales in Modern Taxonomy." 2nd ed., 915 pp. Cramer, Weinheim.
Smith, I. (1960). "Chromatographic and Electrophoretic Techniques." Vol. I: Chromatography, p. 194. Wiley (Interscience), New York.
Sørensen, N. A. (1963). In "Chemical Plant Taxonomy" (T. Swain, ed.), pp. 219-252. Academic Press, New York.
Steglich, W., Furtner, W., and Prox, A. (1968). Z. Naturforsch. 23B, 1044-1050.
Steglich, W., Furtner, W., and Prox, A. (1969). Z. Naturforsch. 24B, 941-942.
Stein, S. (1958). Mycopath. Mycol. Appl. 9, 263-267.
Stein, S., Closs, G. L., and Gabel, N. W. (1959). Mycopath. Mycol. Appl. 11, 205-216.
Sullivan, G. (1968). J. Pharm. Sci. 57, 1804-1806.
Sullivan, G., and Guess, W. L. (1969). Lloydia 32, 72-74.
Sullivan, G., Brady, L. R., and Tyler, V. E., Jr. (1967). Lloydia 30, 84-90.
Takemoto, T., Yokobe, T., and Nakajima, T. (1964). Yakugaku Zasshi 84, 1186-1188.
Thoms, H., and Vogelsang, J. (1907). Ann. Chem. 357, 145-170.
Tyler, V. E., Jr. (1958). Science 128, p. 718.
Tyler, V. E., Jr., and Gröger, D. (1964a). J. Pharm. Sci. 53, 462-463.
Tyler, V. E., Jr., and Gröger, D. (1964b). Planta Med. 12, 397-402.
Tyler, V. E., Jr., and Smith, A. H. (1963a). Mycologia 55, 358-359.
Tyler, V. E., Jr., and Smith, A. H. (1963b). In "Abhandlungen der Deutschen Akademie der Wissenschaften zu Berlin" (K. Mothes and H.-B. Schröter, eds.), pp. 45-54. Akademie-Verlag, Berlin.
Tyler, V. E., Jr., Brady, L. R., Benedict, R. G., Khanna, J. M., and Malone, M. H. (1963). Lloydia 26, 154-157.

Tyler, V. E., Jr., Benedict, R. G., and Stuntz, D. E. (1965). *Lloydia* **28**, 342–353.
Tyler, V. E., Jr., Benedict, R. G., Brady, L. R., and Robbers, J. E. (1966). *J. Pharm. Sci.* **55**, 590–593.
Walters, M. B. (1965). *Mycologia* **57**, 837–838.
Watling, R. (1969). *In* "Methods in Microbiology" (C. Booth, ed.), Vol. 4, Chemical Tests in Agaricology. in press. Academic Press, New York.
Weir, J. K., and Tyler, V. E., Jr. (1963). *J. Pharm. Sci.* **52**, 419–422.

Proton Magnetic Resonance Spectroscopy — An Aid in Identification and Chemotaxonomy of Yeasts[1]

P. A. J. GORIN AND J. F. T. SPENCER

National Research Council of Canada,
Prairie Regional Laboratory,
Saskatoon, Saskatchewan, Canada

I.	Introduction	25
	A. Definition of the Yeasts	25
	B. The Genera and Species of the Yeasts	26
	C. Methods of Classification of the Yeasts	29
	D. Chemotaxonomy of the Yeasts in Terms of Nucleic Acid, Protein, and Antigenic Structures	31
II.	Chemotaxonomy of Yeasts According to Their Mannose-Containing Polysaccharides	33
	A. Early Chemical Studies on Yeast Polysaccharides	33
	B. Application of the Proton Magnetic Resonance Spectroscopic Method to Yeast Polysaccharides	37
	C. Isolation of Mannose-Containing Polysaccharides of Yeasts and Preparation of Pmr Spectra	40
	D. Methods Used in Determination of Chemical Structure of Yeast Mannose-Containing Polysaccharides	43
	E. Interpretation of Pmr Spectra of Mannose-Containing Polysaccharides in Terms of Chemical Structure	48
	F. Interpretation of Pmr Spectra of Oligosaccharides in Terms of Chemical Structure	50
	G. Chemical Structures of Yeast Mannose-Containing Polysaccharides	53
III.	Classification of the Yeasts According to the Pmr Spectra of Their Mannose-Containing Polysaccharides	54
IV.	Summary and Conclusions	84
	References	87

I. Introduction

A. DEFINITION OF THE YEASTS

The term "yeast" according to Alexopoulos (1962) refers to Ascomycetes of the subclass Hemiascomycetidae, Order Endomycetales, Family Saccharomycetaceae, which possess a predominantly unicellular thallus; reproduce asexually by budding, transverse division

[1] Issued as NRCC No. 11349.

(fission) or both; and produce ascospores in a naked ascus, originating either from a zygote or parthenogenetically from a single somatic cell. Forms not known to produce ascospores, but which possess all other characteristics listed above, and are not obviously related to other groups of fungi, are also included generally under the term yeast, for it is believed that many yeasts have lost their ability to form ascospores or that they may actually form them under conditions as yet unknown to us.

This definition, like most other finished or semifinished products of scientific endeavor, gives no hint of the tortuous path that had to be followed to reach it. Nor does it hint of the blind alleys, which sometimes retard the development of yeast taxonomy, nor of the unexplored ones that are being investigated in the hope that they will lead to a more accurate map of the area occupied by these organisms.

The above definition includes those species placed by Phaff *et al.* (1966) in the family Cryptococcaceae as asporogenous forms, some of which are transferred at intervals to the family Saccharomycetaceae whenever strains that produce ascospores are discovered. It does not, however, include the family Sporobolomycetaceae, the members of which produce ballistospores that are discharged by drop-excretion. These species are classified by Alexopoulos as Basidiomycetes, though Phaff *et al.*, and Lodder *et al.* (1958) classify them as yeasts, along with the similar organisms in the genus *Rhodotorula*. Thus the question of a possible multiphyletic origin of the yeasts is not touched upon by Alexopoulos (1962) in an otherwise excellent definition of the yeasts. Finally, such definitions give no clue to the difficulty often involved in defining the relationship of an organism to other members of the group, nor its similarities to and differences from them. The statement by Lodder *et al.* (1958) that "the delimitation of the yeasts is subject to arbitrary decisions and as a group they are far from homogeneous" gives some warning of the problems which have been encountered and which still exist in the identification and classification of yeasts.

B. The Genera and Species of the Yeasts

Phaff *et al.* (1966) place the yeasts in three families, Saccharomycetaceae, Sporobolomycetaceae, and Cryptococcaceae. The first family contains the known ascosporogenous yeasts. Alexopoulos (1962) however, places three groups, containing the genera *Eremascus*, *Endomyces*, and *Endomycopsis*, in a separate family, Endomycetaceae. Phaff *et al.* (1966) place these genera in two of six subfamilies within the family Saccharomycetaceae, so that *Eremascus* and *Endomyces* are placed in the subfamilies Eremascoideae and Endomycetoideae, respectively. *Endomycopsis* is placed in the subfamily Saccharomycetoideae. The other subfamilies are Schizosaccharomycetoideae, Lipomycetoideae, and Nematosporoideae.

The family Sporobolomycetaceae contains two genera of yeasts which produce ballistospores, *Sporobolomyces* and *Bullera*. They are distinguished by the shape of the ballistospores, kidney- to sickle-shaped in the former and round to oval in the latter. Most *Sporobolomyces* species are red or pink but white or cream-colored species have been described. *Bullera* species are colorless to cream-colored.

The family Cryptococcaceae contains a heterogeneous collection of asporogenous forms grouped according to morphological and physiological criteria. At intervals individual species are found to produce spores or otherwise engage in some form of sexual activity, and are transferred to an existing ascosporogenous genus or a new one is created for them. The genera *Kloeckera, Torulopsis* and *Candida*, for instance, contain anascosporogenous forms of *Hanseniaspora, Debaryomyces, Pichia, Hansenula, Saccharomyces* and *Metschnikowia*. Recently, conjugation has been reported in strains of *Rhodotorula* by Banno (1967) and a new genus, *Rhodosporidium*, has been suggested to include them.

Most of the genera of this family form only budding cells and sometimes pseudomycelium. However, true mycelium is formed by members of the genus *Trichosporon* and a few *Candida* species. The shape of the cells is very variable, including round, oval, elongate, bottle- and flask-shaped, and triangular forms.

The family Saccharomycetaceae as envisaged by Phaff *et al.* (1966) includes forms producing true mycelium only, and which cannot properly be classed as yeasts. These include the genera *Eremascus* and *Endomyces*. *Endomycopsis* is an intermediate form in this respect, producing true mycelium as well as pseudomycelium and budding cells. The latter genus is placed in the family Endomycetaceae by Alexopoulos (1962).

The *Schizosaccharomyces* species, which multiply vegetatively by fission, are placed in a separate subfamily Schizosaccharomycetoideae, by Phaff *et al.* (1966). According to the same workers the subfamily Saccharomycetoideae is considerably larger than the others, containing sixteen genera. These include *Saccharomyces*, an old and rather heterogeneous genus, which contains the commercially important species *Saccharomyces cerevisiae* and *Saccharomyces carls-*

bergensis; the genus *Fabospora,* which contains species which form bean-shaped spores, and which were originally placed in the genus *Saccharomyces;* the genera *Hansenula* and *Pichia,* which include many species that live in association with bark beetles and their host trees, and are separated only by the ability to utilize nitrate or the lack of it; the genus *Debaryomyces,* which includes a number of salt-tolerant species that grow on the surface of preserved meats, and several minor genera.

Two minor subfamilies, Lipomycetoideae and Nematosporoideae, contain the genera *Lipomyces,* and *Nematospora, Coccidiascus* and *Metschnikowia,* respectively. The three latter genera consist of species that form needle-shaped ascospores.

All genera of the family Saccharomycetaceae are separated largely on the basis of the shape, number, and mode of production of their ascospores. Whether this is a valid criterion for determining the relationships among these genera will have to await the results of future research. However, there are instances in the work reported here that suggest that unsuspected similarities among some genera may make a regrouping of them desirable.

There exist, in addition, a number of genera of yeastlike organisms, whose relationship to the "true" yeasts is also yet to be determined. Phaff *et al.* (1966) mention the genera of filamentous fungi *Pullularia (Aureobasidium), Geotrichum, Eremothecium, Ashbya,* and *Taphrina* and the genus of colorless algae, *Prototheca,* which is mentioned because it is often found in the same habitat as some yeasts and may be mistaken for a yeast by inexperienced yeast ecologists.

Another genus, *Ceratocystis* (and its imperfect form, *Graphium*) is worthy of mention in connection with the yeastlike fungi. Like other such organisms, many members of the genus grow readily in a yeastlike phase under some conditions. Also, many species of this genus live in association with bark beetles and their host trees, as do yeasts of the genera *Hansenula* and *Pichia.* Some *Ceratocystis* species form hat- and Saturn-shaped ascospores, as do most species of the above yeasts.

Most of the *Ceratocystis* species, unlike some of the other filamentous fungi, form mannose-containing polysaccharides. The genus *Ceratocystis* is of economic importance since it includes the pathogen *Ceratocystis ulmi,* the causal agent of Dutch Elm disease, and a number of other wilt pathogens and blue-stain fungi affecting the quality of lumber. While this genus has not hitherto been considered among the yeastlike organisms, its possible relationships with the yeasts are worthy of study.

Species of the genus *Alternaria* and *Nectria* have likewise been reported (Wolf and Wolf, 1947) as microbial commensals of bark beetles and plant pathogens, respectively. Some of these, too, have been found to form extractable mannose-containing polysaccharides which give distinctive pmr spectra.

C. Methods of Classification of the Yeasts

Yeasts are at present classified on the basis of their morphology, for assignment to genera, and by their biochemical reactions, for assignment to species. The workers of the Dutch school were responsible for much of the pioneering work on the classification of the yeast species known up to about 1950. This work culminated in the standard text on yeast taxonomy by Lodder and Kreger-van Rij (1952). These workers classified all the yeasts available to them on the basis of cellular morphology, spore shape and number, and the nature of the conjugation process, and, at the specific level, on the ability to ferment and assimilate six sugars, to use ethanol and nitrate, and to hydrolyze arbutin. The distinction between some species was rather fine, as judged by these criteria. The greatest of all their contributions, perhaps, was the introduction of order into the chaos of synonymous genera and species of yeasts, which had existed until that time.

Wickerham and Burton (1948), and Wickerham (1951) at about the same time, introduced a number of refinements of the Dutch system, especially the use of a much larger number of carbon compounds. These included additional hexoses, di-, tri- and tetrasaccharides, two polysaccharides, and a number of pentoses, polyhydric alcohols, and organic acids. They also introduced tests for vitamin requirements.

Current practice is to use approximately thirty carbon compounds and to test for fermentation of at least eleven of these, including including inulin. The ability to utilize nitrite as well as nitrate, to hydrolyze gelatin, to produce acid, and to grow at various elevated or depressed temperatures and on media of high sugar or salt content, are also used. The type and number of additional reactions tested vary with the interests and preferences of the individual investigator.

A certain number of difficulties, major and minor, accompany the use of these methods. One of these is the question of the stability of the reaction and its significance. This is true of morphological as well as biochemical criteria. For instance, the genera *Candida* and *Torulopsis* are separated solely on the ability of the former to produce pseudohyphae. At one time, great attention was paid to the form of the pseudohyphae themselves for differentiation of species, until it was observed that the same species might produce two or more forms, simultaneously or at different stages of growth. It has now become evident that different strains of the same species may differ in their ability to produce pseudomycelium, and the value of this criterion in distinguishing the two genera approaches the vanishing point.

Likewise, the genera *Hansenula* and *Pichia* are separated solely by the ability or lack of ability to utilize nitrate. If this should be no longer a valid criterion for separating the two genera, then the distinction between them would disappear, and one generic name or the other would have to be discarded.

Another problem concerns the instability of the physiological characters. Scheda and Yarrow (1966) observed enough variability in the fermentation and carbon assimilation patterns of a number of *Saccharomyces* species to cause difficulties in the assignment of their yeast strains to definite species.

A different type of difficulty lies in the relationship of the biochemical tests to the metabolism of the organisms. It was not originally sufficiently appreciated that the various carbon compounds are not necessarily assimilated independently, but may be metabolized by common pathways. Thus most yeasts that use one substrate can also use a structurally related one by the same metabolic pathway. Barnett (1968) gives several examples of such linked mechanisms: (a) a common initial enzyme, for instance, that which is involved in the metabolism of a number of β-D-glucopyranoside derivatives, the metabolism of raffinose and sucrose, and the metabolism of L-sorbose and D-mannitol; (b) interconvertible substrates, such as D-ribose and ribitol; and (c) common route to a central pathway, as in the utilization of L-arabinose and D-xylose. Barnett noted that there was a small percentage of yeasts that were exceptions to these rules, but in general the conclusion was valid, that the effective number of criteria for distinguishing yeast species was reduced by the number of substrates metabolized by such linked mechanisms. The metabolism of most or all of the compounds used involves a few distinct central pathways, and depends on the cells' ability to convert the substrates into intermediary metabolites of one of these pathways. From this it logically follows that probably only minor differences in carbohydrate metabolism are concerned in these tests. Pasteur's comment [quoted by Barnett (1968) from "Etudes sur la Bière"] is thus most significant, concerning his studies of different yeasts. He did not give names to his cultures, saying that "Many a time I have found that forms different in appearance, often belong to the same species and that similar forms can hide profound differences."

If, as has been shown, gross morphological differences, such as they are in relatively undifferentiated organisms such as yeasts, are unreliable as taxonomic criteria, and the biochemical and physiological criteria discussed by Barnett (1968) and by Scheda and Yarrow (1966) are not much better, then new criteria, which are hopefully more stable, must be sought. The chemical structures of the macromolecules of the yeast cell, such as the nucleic acids, proteins, and polysaccharides are obvious choices.

D. Chemotaxonomy of the Yeasts in Terms of Nucleic Acid, Protein, and Antigenic Structures

Recently, with the advent of more sophisticated chemical and physiochemical techniques, it has been found that the chemical structures of components of the yeast cell vary from species to species. This has led to greater interest in chemotaxonomy of the yeast cell using as criteria the chemical structures and physical and immunological properties of macromolecules. A brief description of the methods in use at present is given below and the polysaccharide structure–proton magnetic resonance method is described in detail later (Section II).

1. Yeast Nucleic Acids

From a phylogenetic point of view, it is advantageous to base yeast classification schemes on the chemical nature of informational macromolecules, i.e., DNA, RNA, and enzyme proteins. This approach has been covered in a comprehensive review on bacterial taxonomy by Marmur et al. (1963). Since DNA base compositions vary widely in microorganisms, this property is useful as a taxonomic aid, a fact first recognized by Lee et al. (1956). Conventional chemical methods of determining base ratios have now been largely replaced by convenient physiochemical methods. These methods include determination of the thermal denaturation temperature of DNA (Marmur and Doty, 1962; Falkow et al. 1962) and to a lesser extent the buoyant density in a cesium salt gradient (Rolfe and Meselson, 1959; Sueoka et al., 1959; Belozersky and Spirin, 1960). The base compositions of DNA of many bacteria have been determined but it was not until more recently that Storck (1966) examined a number of true fungi, yeastlike fungi, and yeasts for guanine+cytosine (G+C) content. He concluded that their G+C contents had a taxonomic and phylogenetic significance. Using the G+C content technique, Nakase and Komagata (1968) have analyzed 62 strains of yeast and yeastlike fungi and have also shown (Nakase and Komagata, 1969) that the genus *Hansenula* can be divided into two large groups, one with a G+C content of 32–34%, and another with 40–42%. In a similar manner Stenderup and Bak (1968) have examined nuclear and mitochondral DNA of some *Candida* species for G+C content. In general the G+C method agrees with taxonomic assignments based on other criteria and appears to have significance in yeast taxonomy.

The G+C base ratio method, which only clearly demonstrates differences in nucleic acid structures, is complemented by the molecular hybridization method. Nucleic acids, which form molecular hybrids on denaturation followed by annealing in solution (Schildkraut *et al.*, 1961), are related to an extent in their base sequences. [Factors governing the hybridization reaction have been summarized by Walker (1969) in a review on the nucleic acids of higher organisms.] In their studies on yeast Bak and Stenderup (1969) obtained cell DNA and radioactive RNA, prepared by the action of a synthetase using DNA as a template. Using the DNA–RNA hybridization technique of Gillespie and Spiegelman (1965) they measured the genetic relatedness of ten species of *Candida*, a genus of considerable heterogeneity. Very close relationships were found between three pairs of imperfect *Candida* species and their perfect counterparts. On the other hand, some *Candida* species with similar G+C base ratios, show very low homology figures.

Marmur *et al.* (1963) have pointed out that base sequence analysis of sRNA, a polymer of approximately 70 nucleoside units is feasible from a chemotaxonomic point of view. However, sRNA's of yeast have not yet been investigated in this manner.

2. Yeast Proteins

Enzymes have not yet been used to any extent in taxonomic studies despite the correlation of base sequence of RNA with enzyme structure. In the study of proteins Clare (1963) has shown that the protein profile of *Saccharomyces cerevisiae* extracts, obtained on zone electrophoresis, differs from those from *Pullularia pullulans* and *Fusarium oxysporum* [for other examples see the review of Garber and Rippon (1968)]. However, protein profiles can vary according to the stage of growth of the cell (Dawson, unpublished data). Claisse (1969) has reported that the types of cytochrome *c* vary depending on the species of yeast examined. The differences were detected using low-temperature spectrophotometry (Estabrook, 1966) of the α-region. The variations of amino acid composition and other properties of cytochrome *c* from yeasts and other fungi have been reviewed by Margoliash and Schejter (1966). This type of investigation, as applied to chemotaxonomy of fungi, may prove to have considerable value.

3. *Serological Properties of Yeasts as a Taxonomic Acid*

The classification and differentiation of yeasts according to their serological properties have been investigated in detail and summarized by Tsuchiya and co-workers (1965) following the report of Benham (1931) that yeasts can be differentiated in this way. The antigenic properties of the cell depend mainly on the mannans or mannose-containing polysaccharides which occur as layers in the cell wall (Fukazawa and Tsuchiya, 1969). Many of the yeast antigens have been analyzed serologically and subdivided into components termed "antigenic structures." As would be expected the serological properties of the yeast antigens are related (although not identical) to the chemical structures of the mannose-containing polysaccharides obtained by alkaline degradation of the cell wall. A few comparisons of serological properties and polysaccharide structures of individual yeasts will be made later in Section III.

II. Chemotaxonomy of Yeasts According to Their Mannose-Containing Polysaccharides

In the classification of yeasts according to their component mannose-containing polysaccharides, two distinct although closely related approaches are open to the investigator. The first is the determination of the polysaccharide structure in terms of its main chain and side chains (differences in component monosaccharides can occasionally be used to distinguish species). These structures are the ultimate criterion for determining similarities and differences between yeast species. The second approach is far less time consuming. The proton magnetic resonance (pmr) spectrum of the mannose-containing polysaccharide can be used directly in yeast taxonomy since similarities in the spectra are suggestive of similarities in chemical structure. The spectra can be used to confirm that one species is the asporogenous form of another, to differentiate between otherwise similar species, and to determine the homogeneity of ascosporogenous genera and suggest relationships between such genera.

A. Early Chemical Studies on Yeast Polysaccharides

The chemical studies on yeast polysaccharides directed toward classification of yeasts, prior to the advent of the pmr method, are described below.

Cell wall and capsular polysaccharides of yeasts have only been used relatively recently for identification of yeasts. This is due to the unavailability, prior to 1950, of adequate techniques for chemical and physiochemical investigations on complex polysaccharides. Early studies on yeast polysaccharides were based on monosaccharide composition. For example, Slodki and Wickerham (1966) showed that the exocellular polysaccharides of *Lipomyces lipoferus* and *Lipomyces starkeyi* can be distinguished by the monosaccharides formed on hydrolysis (Table I), whereas these two species are difficult to distinguish by their morphological characteristics and sugar-assimilation patterns. Gorin *et al.* (1965) found that *Rhodotorula* spp. form a mannan containing alternate β-(1 $\rightarrow$ 3)- and β-(1 $\rightarrow$ 4)-D-mannopyranose units whereas *Cryptococcus* spp. form starchlike material[1] (Aschner *et al.*, 1945) and a heteropolysaccharide containing xylose, mannose, and glucuronic acid units (Einbinder *et al.*, 1954; Abercrombie *et al.*, 1960; Miyazaki, 1961).

The formation of such related heteropolymers by some red yeasts has necessitated the reclassification of *R. aureae*, *R. peneaus*, *R. macerans*, and *R. infirmominiata* to the genus *Cryptococcus* (Gorin *et al.*, 1966). These reclassifications agree with those of Phaff and Spencer (1966) who have shown that, whereas *Cryptococcus* species assimilate inositol, *Rhodotorula* species do not. One of these, *Rhodotorula peneaus*, has now been reclassified by Foda and Phaff (1969) as *Cryptococcus laurentii*.

Many yeasts of the genus *Trichosporon* can be distinguished according to their polysaccharide components. Whereas α-D-linked D-galacto-D-mannans occur as cell components of *T. fermentans*, *T. hellenicum*, and *T. penicillatum*, pentosyl-D-mannans are formed by *T. cutaneum*, *T. inkin*, *T. pullulans*, and *T. sericeum* (Gorin and Spencer, 1967; Gorin and Spencer, 1968a). *Trichosporon aculeatum*, on the other hand, contains a cellular α-D-linked mannan (Gorin *et al.*, 1968) (Table II). Similarly the *Candida* spp. can be divided according to whether extraction of cells provides galactomannan or heteropolymer (Gorin and Spencer, 1968b). The group containing heteropolymers is only a minor one, including only 13 of the 81 *Candida* species examined (Tables II, III, and IV). Both *Trichosporon cutaneum* pentosylmannan (Gorin and Spencer, 1967) and *Candida bogoriensis*

---

[1]The formation of starchlike material is considered by Phaff and Spencer (1966) to have limited diagnostic value in differentiation of *Rhodotorula* and *Cryptococcus* spp. Furthermore iodine and starch can give a brown instead of a blue color in the presence of lipid (H. J. Phaff, private communication).

TABLE I
PROPERTIES OF POLYSACCHARIDES OF *Lipomyces* SPECIES

Species	Source	NRRL number	Lactose assimilation	$[\alpha]_D^{25}$, degrees	Sugars present		
					Mannose	Galactose	Glucuronic acid
Lipomyces starkeyi	Starkey 74	Y-1388	+	0	+	+	+
Lipomyces starkeyi	Starkey 74b	Y-2543	−	− 4	+	+	+
Lipomyces starkeyi	David Jones	Y-6333	−	+54	+	−	+
Lipomyces lipoferus	A. C. Thaysen	Y-1351	+	+42	+	−	+
Lipomyces lipoferus	Type strain CBS	Y-2542	+	+40	+	−	+

TABLE II
The Monosaccharide Components of Crude *Candida* and *Trichosporon* Polysaccharides, Purified via Their Copper Complexes

Species of *Trichosporon* or *Candida*	Monosaccharide components[a]
Trichosporon	
inkin IGC 3727	Man, Xyl
sericeum CBS 2545	Man,[b] Xyl
undulatum CBS 2546	Man,[b] Xyl
cutaneum PRL RS1	Man, Xyl, Ara
pullulans NCYC 477	Man, Xyl, tr. Fuc, Gal
fermentans PRL 2263	Man, Gal
hellenicum CBS 4099	Man, Gal
penicillatum IGC 3716	Man, Gal
aculeatum IGC 3551	Man
Candida	
buffonii CBS 2838	Man, Rha, Gal, G
curvata CBS 570	Man, Xyl
foliarum CBS 5234	Man,[b] Rha, tr. Gal, Fuc
diffluens CBS 5233	Gal, Man, Rha
marina CBS 5235	Man, Xyl
scottii CBS 614	Gal, Man, tr. G
humicola IGC 3391	Man, Xyl, uronic acid
nivalis 3AH2[b]	Man,[c] Xyl, Fuc
frigida 5A1[b]	Man,[c] Xyl, Fuc
gelida 2AH10[b]	Man,[c] Xyl, Fuc

[a] Gal, galactose; Rha, rhamnose; Man, mannose; G, glucose; Fuc, fucose; Xyl, xylose; GA, glucuronic acid; Ara, arabinose (tr, trace).

[b] The copper complex of the polysaccharide is relatively water-soluble resulting in low yield.

[c] From Dr. M. di Menna, D.S.I.R., Lower Hutt, New Zealand.

heteropolymer (Gorin and Spencer, 1968b) contain $\alpha\text{-}(1 \to 3)$-D-mannopyranose main chains and side chains of different chemical structure. (The method of classification of chemical structures of yeast mannose-containing polysaccharides according to the nature of their main chains and side chains is discussed on p. 43.

Certain *Hansenula, Pichia,* and *Pachysolen* species produce exocellular phosphonomannans, which have been structurally investigated by Jeanes and co-workers (1961, 1962; Jeanes and Watson, 1962) and Slodki (1962, 1963). The phylogenetic significance of phosphonomannans of different structure has been considered by Wickerham and Burton (1961) and by Slodki *et al.* (1961). The occurrence of phosphonomannans is probably widespread since they occur in the cell walls of *Saccharomyces cerevisiae* and *Candida pseudotropicalis* [Mill (1966) and Elinov and Vitovskaya (1965), respectively]. Also Jones and Ballou (1968b) reported the presence of phosphorus in mannans from *Candida tropicalis, Candida stellatoidea, Kloeckera brevis,* and two strains of *Candida albicans.*

B. APPLICATION OF THE PROTON MAGNETIC RESONANCE SPECTROSCOPIC METHOD TO YEAST POLYSACCHARIDES

Pmr spectroscopy was first used as a taxonomic tool by Gorin *et al.* (1968) to distinguish the chemical structures of mannans of *Trichosporon aculeatum* and *Saccharomyces cerevisiae*. A number of pmr spectra of mannans were found to be typical of the parent yeast, an observation consistent with that of Tsuchiya and co-workers (1965) who demonstrated serological differences between yeasts. These differences are due to the various structures of the mannan cell-wall components. The pmr spectra of mannose-containing polysaccharides of most of the available yeasts have now been obtained. The cells of of 450 yeast species have been extracted with hot aqueous alkali and attempts made to obtain the mannose-containing components of the extracts by purification via the insoluble copper complexes formed with Fehling solution. In 410 cases a purified polysaccharide was obtained and of these pmr spectra could be obtained from all but 23. These spectra, which include 150 distinguishable types, are presented in Section III. The sugar compositions of the polysaccharides from yeasts not amenable to the pmr technique are listed in Tables III and IV.

Some failures occurred since precipitates could not be obtained with Fehling solution (Table III). In others, some of the regenerated polysaccharides (Table IV) gave viscous solutions, from which pmr spectra could not be obtained. This and other limitations in the pmr technique, as applied to macromolecules, have been summarized by Bradbury and Crane-Robinson (1968).

The portion of the pmr spectrum used in taxonomic and identification studies is from $\tau 3.8$ to $\tau 4.8$. The signals in this region are from H-1 (for numbering of the mannose molecule see Fig. 1). Other well-defined signals $(C-CH_3)$ are occasionally found at $\tau 8.1$–8.2 (70°) in spectra of polysaccharides containing methylpentose units and at $\tau 7.2$ (25°) which corresponds to N-acetyl groups. This signal is particularly prominent in the polysaccharide from *Saccharomyces phaseolosporus* (Gorin and Spencer, unpublished results). A typical H-1 pmr spectrum, that of *Saccharomyces cerevisiae* mannan, is

TABLE III
Monosaccharide Components of Yeast Polysaccharides not Giving a Precipitate with Fehling Solution

Source of polysaccharide	Hydrolysis products[a]
Alternaria dauci PRL 504	G, Man
Bullera alba PRL 2125	G, Man, Xyl, tr. Gal
Candida	
aquatica CBS 5443	G, Man, tr. Gal, Xyl, Fuc
bogoriensis CBS 4101	Gal, Man, Fuc, Rha, GA, tr. G
ingens CBS 4603	G, Gal, Man
javanica CBS 5236	G, tr. Gal
mesenterica NCYC 390	G, Man, Xyl, uronic acid
muscorum CBS 2740	G, Gal, Man
Eremascus fertilis CBS 103.09	Gal, Man, tr. Fuc
Pityrosporum orbiculare UCD 62-16	G, Man
Prototheca wickerhamii UCD 60-47	Gal, Rha, uronic acid, Ara, G
Rhodotorula texensis UCD 48-23	G, Gal, Man
Sporobolomyces	
gracilis PRL 2123	G, Gal, Man
holstaticus PRL 1993	G, Gal, Man
johnsonii PRL 2111	G, Man, Fuc
odorus PRL 2122	G, Gal, Man, Fuc
pararoseus PRL 2124	G, Gal, Man, Fuc
salmonicolor PRL 2116	G, Gal, Man, Fuc
Trichosporonoides oedocephalus PRL 62-334	G, tr. Man

[a] Meanings of abbreviations are given in Table II.

depicted in Fig. 1. Each H-1 signal has a chemical shift, expressed as a τ value based on a value of τ 10 for the tetramethylsilane standard. It is possible to assign each signal to the H-1 of a particular mannopyranose unit, based on the known chemical structure 1 (Lee and Ballou, 1965) of *S. cerevisiae* mannan.

The results of several experiments (Gorin and Spencer, unpublished results) indicate that the structure of the cell-wall mannose-containing polysaccharides of a yeast are not readily susceptible to change. The pmr spectra of mannans of amino acid auxotrophs of *Saccharomyces fragilis, Saccharomyces dobzhanskii, Saccharomyces phaseolosporus* and four strains of *S. cerevisiae* are identical to each of the parent strains. Also the pmr spectra of galactomannans of colonial variants of *Trichosporon fermentans* induced by ultraviolet irradiation are also unchanged. Pmr spectra appear to be

TABLE IV
MONOSACCHARIDE COMPONENTS OF PURIFIED POLYSACCHARIDES
GIVING D_2O SOLUTIONS TOO VISCOUS FOR PMR DETERMINATIONS

Source of polysaccharide	Hydrolysis products[a]
Candida	
curvata CBS 570	Man, Xyl
humicola IGC 3391	Man, Xyl, uronic acid
marina CBS 5235	Man, Xyl
scottii PRL Y-124	Gal, Man, tr. G
Cryptococcus	
laurentii PRL 316-63	Gal, Man, Xyl, uronic acid
melibiosum PRL Y-125	Man, uronic acid, tr. Gal
neoformans	Man, Xyl, GA
skinneri PRL Y-126	G, Man, Xyl
terreus PRL 62-507	Man, Gal, Xyl
uniguttulatus PRL 62-518	G, Man, Xyl, tr. G
Dipodascus uninucleatus CBS 190.37	Man, uronic acid
Geotrichum vanrijii CBS 439.64	Man, Xyl, tr. G
Lipomyces lipofera CBS 944	Man, uronic acid
Prototheca ciferri UCD 60-49	G, Gal, Man
Sporobolomyces albidus PRL 1985	Gal, Man, Xyl, uronic acid
Taphrina deformans CBS 355.35	Man, Xyl, Fuc, uronic acid
Trichosporon	
cutaneum PRL RS-1	Man, Ara, Xyl
inkin IGC 3727	Man, Xyl
pullulans NCYC 477	Man, Xyl, tr. Fuc, Gal
sericeum CBS-2544	Man, Xyl
undulatum CBS 2546	Man, Xyl

[a] Meanings of abbreviations are given in Table II.

reproducible even when they contain very minor peaks, as in those prepared from the various rhamnomannans of *Ceratocystis* species (for examples see spectral Group 22, Fig. 10). Also, it is significant that the chemical structure of the mannan of *Candida utilis* does not vary with the stage of the cell cycle when the cells were grown in synchronous culture on nitrogen- and carbon-limiting media (Gorin, Spencer, and Dawson, unpublished results). Nevertheless, to avoid possible variations of chemical structure of a polysaccharide from a given yeast, the same cultural conditions (Gorin and Spencer, 1968a) were used as often as possible. In occasional cases when the cell contains two or more mannose-containing polysaccharides, as with *Torulopis magnoliae* (Gorin et al., 1969b), the proportions of the polysaccharides could vary with the growth conditions.

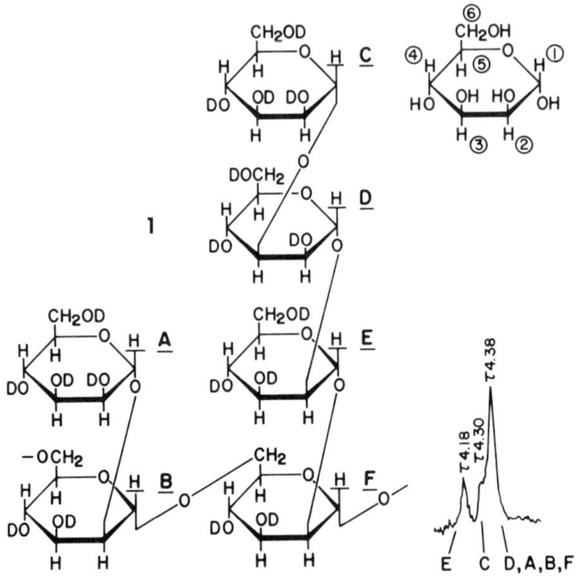

FIG. 1. Numbering of ring protons in α-D-mannose and correlation of H-1 pmr spectrum with structure of *S. cerevisiae* mannan.

C. Isolation of Mannose-Containing Polysaccharides of Yeasts and Preparation of Pmr Spectra

The majority of yeasts contain, in their cell walls, either mannan or galactomannan, which are generally in an insoluble, chemically bound form. In order to liberate and solubilize these polysaccharides the cells must be heated in aqueous alkali. In the case of bakers' yeast the mannan is partially phosphorylated (Northcote and Horne, 1952) and attached, most likely, to a peptide through a nitrogen glycosyl bond involving aspartamide and 2-acetamido-2-deoxy-D-glucose. In addition small oligosaccharide units appear to be joined to the hydroxyl groups of serine and threonine units in the peptide (Sentandreu and Northcote, 1968, 1969). The alkali cleaves the phosphate groups and the carbohydrate-protein linkages, and the resulting polysaccharide can be freed of residual mannose oligosaccharides and glucan (some yeasts contain soluble glucan) using Fehling solution (Fig. 2) by the following method (based on Gorin and Spencer, 1968a).

Dried yeast cells (ca. 10 gm) are suspended in 2% aqueous potassium hydroxide (200 ml) and heated for 2 hours at 100°C (cells containing an alkali-labile mannan with β-(1 → 3)- and β-(1 → 4)-linkages, such as obtained from *Rhodotorula* species, should be heated for only

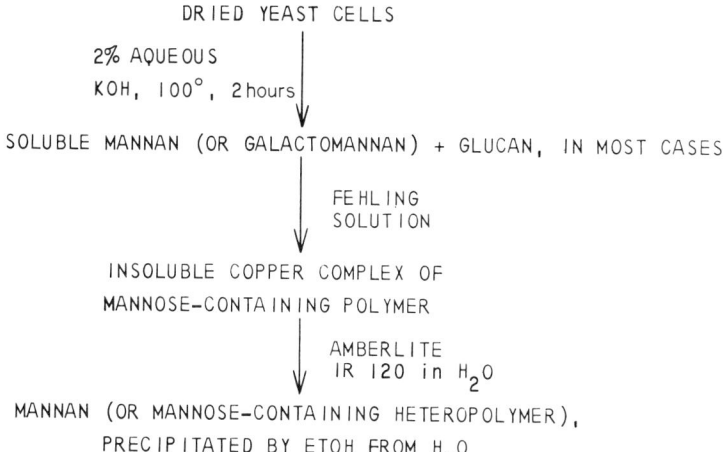

FIG. 2. Procedure for isolation of pure mannose-containing polysaccharides from yeasts.

1 hour). The solution is neutralized with acetic acid, centrifuged at 9000 rpm and the supernatant evaporated to 50 ml. Methanol (200 ml) is added and the precipitate filtered or centrifuged off, washed with methanol, and dried. At this stage a portion of the total polysaccharide fraction is hydrolyzed and its monosaccharide components analyzed on a paper chromatogram. The polysaccharide is dissolved in water (50 ml) by heating at 100°C for 2 hours, insoluble material centrifuged off, and Fehling solution (50 ml) added. The solution is stored overnight at 5°C and the insoluble copper complex filtered off. (In general, galactomannans and β-(1 → 3)-, β-(1 → 4)-linked mannans give copper complexes, which are relatively water soluble, so that a longer storage time is necessary.) The complex is filtered off, washed with 2% aqueous potassium hydroxide and then methanol. Remaining alkali is removed by homogenizing with methanol followed by washing with methanol on a filter. The copper complex is decomposed by shaking in water with Amberlite IR 120 (H^+ form) and after 1 hour the resin is filtered off. The filtrate is evaporated to approximately 10 ml and methanol or ethanol (100 ml) added followed by the addition of 3 drops of concentrated hydrochloric acid. The precipitate is filtered off, washed with methanol, and dried in vacuum (yields 1–10%). A portion of the purified polysaccharide is then hydrolyzed and its component sugars analyzed on a paper chromatogram.

The H-1 pmr spectra in D_2O of the phosphate-containing mannan (kindly supplied by I. Campbell) obtained by autoclaving *Saccharomyces cerevisiae* cells (Peat et al., 1961) and glycoprotein extracted with cold aqueous alkali (Falcone and Nickerson, 1956) are identical with that of mannan obtained by the above method (Gorin and Spencer, unpublished data).

The majority of mannose-containing polysaccharides from yeast cells give pmr spectra containing a distinctive fingerprint region in the τ3.8–4.8 region arising from H-1 signals. Optimum resolution is obtained using a 100-MHz nuclear magnetic resonance (nmr) spectrometer and a 5–20% polysaccharide solution in D_2O^2 (previously clarified by centrifugation) at 70°C with a suitable external standard such as tetramethylsilane. The elevated temperature is used to shift the temperature-dependant DOH signal upfield since at room temperature it can be superimposed with or interfere with the resolution of the H-1 signals because of its proximity to them. (see Fig. 3). Increasing the temperature to 110°C using a sealed pmr tube shifts the DOH signal further upfield, but little improvement of the resolution of the pmr H-1 spectrum occurs.

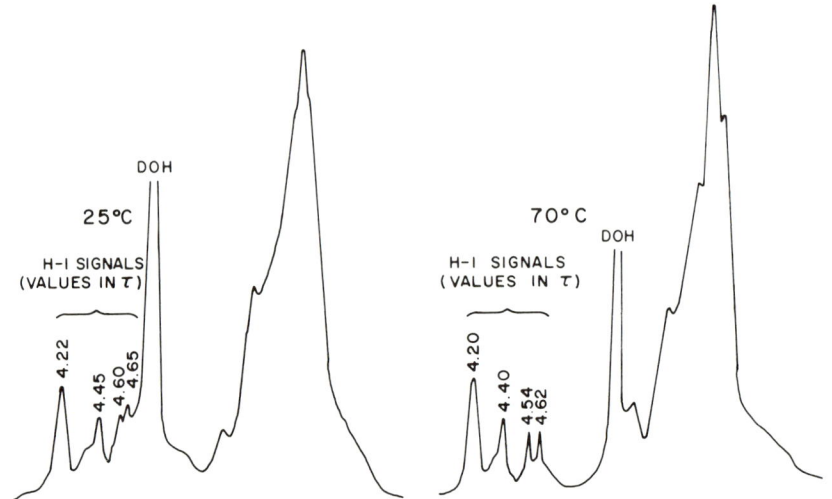

FIG. 3. Pmr spectrum of *Trichosporon aculeatum* mannan in D_2O at 25° and 70°C (tetramethylsilane external standard).

---

[2] Dimethyl sulfoxide-d_6 can also be used as a solvent (Frechet and Schuerch, 1969), but the resulting spectra are less well resolved (Gorin, unpublished results).

D. METHODS USED IN DETERMINATION OF CHEMICAL STRUCTURE OF YEAST MANNOSE-CONTAINING POLYSACCHARIDES

The mannose-containing polysaccharides of yeast cells can be defined structurally according to their main chains and side chains. To date, four different types of main chain are known. Of these the $(1 \rightarrow 6)$-linked α-D-mannopyranosyl main-chain appears to occur most frequently as a component of a branched polymer (as in baker's yeast mannan, **5**), (Jones and Ballou, 1968a; Gorin and Spencer, 1968a; Gorin, *et al.*, 1969a, b, c). The great variety of pmr spectra is probably mainly due to the many types of side-chain structures with this main chain. Some branched mannose-containing heteropolymers have $(1 \rightarrow 3)$-linked α-D-mannopyranose main-chains (Gorin and Spencer,

$$\left[-\beta\text{-}D-\text{Man}p\text{-}(1 \rightarrow 4)-\beta\text{-}D-\text{Man}p\text{-}(1 \rightarrow 3)- \right] \quad \mathbf{2}$$

$$\begin{bmatrix}
& \alpha\text{-}L\text{-}Ara p - 1 & \\
& \downarrow & \\
& 4 & \\
D-Xyl p - 1 & D-Xyl p - 1 & \\
\downarrow & \downarrow & \\
6 & 6 & \\
-\alpha\text{-}D\text{-}Man p\text{-}(1 \rightarrow 3)-\alpha\text{-}D\text{-}Man p\text{-}(1 \rightarrow 3)- &
\end{bmatrix} \quad \mathbf{3}$$

1967, 1968b) (as in *T. cutaneum* pentosylmannan, **3**) and one branched-chain mannan has been reported to have a main chain with $(1 \rightarrow 3)$-linked α-D-mannopyranosyl and $(1 \rightarrow 6)$-linked D-mannopyranosyl units arranged in a "block type" structure (Gorin and Spencer, 1970a), in which the predominant $(1 \rightarrow 3)$-links are grouped consecutively rather than dispersed throughout the chain. Straight-chain polysaccharides consisting of alternate $(1 \rightarrow 3)$- and $(1 \rightarrow 4)$-linked β-D-mannopyranosyl units (**2**) occur in cells of *Rhodotorula* spp. (Gorin *et al.*, 1965) and some other yeasts. Clearly other main-chain types are possible since the structures of only relatively few yeast mannose-containing polysaccharides are known. However, the report of Masler *et al.* (1968) that *Torulopsis ingeniosa* di Menna produces a branched-chain mannan with a β-$(1 \rightarrow 4)$-linked main-chain has not been substantiated. Instead, it has been found that the mannan is linear containing $(1 \rightarrow 3)$- and $(1 \rightarrow 4)$-linked D-mannopyranosyl units (Gorin and Spencer, unpublished results).

The methods available to the investigator for determination of main-chain and side-chain structures are summarized below.

1. The Methylation-Fragmentation Gas–Liquid Chromatography Technique

To show the structure of individual hexose-containing side-chain and main-chain units in the polysaccharides, the methylation-fragmentation gas–liquid chromatography technique is used. The polysaccharide is partially methylated by sodium hydroxide-dimethyl sulfate (Haworth, 1915) and methylation completed with silver oxide in a mixture of methyl iodide and N, N-dimethylformamide (Kuhn et al., 1955) giving a product that does not show hydroxyl absorption in the infrared, as a chloroform solution. This is then converted under acidic conditions to a mixture of partially methylated monosaccharides, which in the case of a mannan is fractionated by cellulose chromatography giving tetra-O-, tri-O-, and di-O-methylmannoses (Gorin et al., 1968). The fractions can be individually analyzed by gas–liquid chromatography (GLC) (Bhattacharjee and Gorin, 1969). The two possible tetra-O-methylmannoses are distinguishable as their trimethylsilyl (tms) derivatives. The 7 tri-O-methyl- and 9 possible di-O-methyl-D-mannoses may each be distinguished by the retention times of their methyl glycosides and the tms derivatives of the methyl glycosides.

2. Partial Acetolysis

The partial acetolysis technique is particularly useful in structural investigations of mannose-containing polysaccharides. It cleaves $(1 \rightarrow 6)$- in preference to $(1 \rightarrow 2)$- and $(1 \rightarrow 3)$-mannopyranoside linkages (Gorin and Perlin, 1956), which renders it ideal for investigation of mannans and galactomannans with α-$(1 \rightarrow 6)$-linked mannan main-chains. For example, with *Saccharomyces cerevisiae* mannan (**5**) the α-$(1 \rightarrow 6)$-linked main-chain is cleaved giving oligosaccharide fragments containing $(1 \rightarrow 2)$- and $(1 \rightarrow 3)$-linked D-mannopyranose units (**6** and **7**) (Lee and Ballou, 1965), which contain the side-chain linkages (Fig. 4).

The partial acetolysis fragments can be fractionated by cellulose column chromatography using the eluants described by Gorin et al. (1969b) and their structures partly elucidated by the methylation-fragmentation GLC method. Pmr spectroscopic examination of the H-1 signals of the oligosaccharide can show whether it is homogeneous and, if so, give information on its molecular size and structure (see pages 50–53).

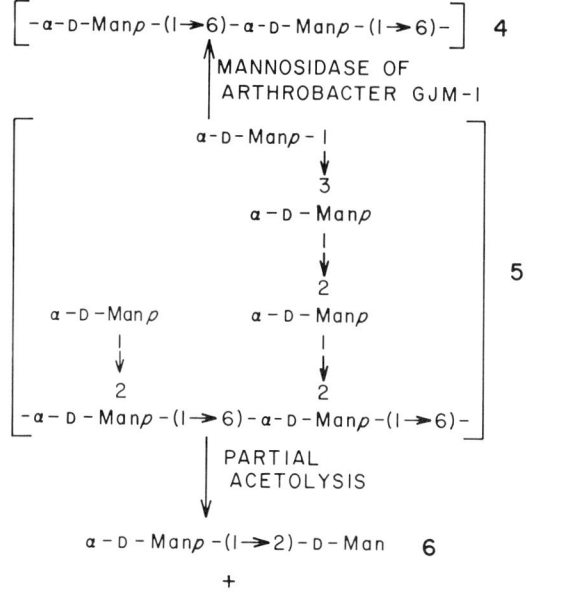

FIG. 4. Degradation of bakers' yeast mannan by partial acetolysis and enzymolysis.

3. Smith Degradation

The Smith degradation (Goldstein et al., 1959) can sometimes be used to obtain α-(1 → 6)-linked D-mannan main-chains from yeast cell-wall heteropolymers containing this structure. *Ceratocystis ulmi* L-rhamno-D-mannan consists of a (1 → 6)-linked α-D-mannopyranosyl main-chain substituted in each of its 3 positions with O-α-L-rhamnopyranosyl and a few O-L-rhamnopyranosyl-(1 → 4)-O-L-rhamnopyranosyl side-chains. Using a Smith degradation (Fig. 5), the rhamnomannan (**8**) was oxidized with sodium periodate to a polyaldehyde (**9**), which was reduced with sodium borohydride to a polyalcohol (**10**). Mild acidic hydrolysis removed the modified side chains giving the (1 → 6)-linked D-mannopyranosyl main-chain (**11**) (Gorin and Spencer, 1970b).

This method can be applied more generally to heteropolymers with a (1 → 3)-linked α-D-mannopyranosyl main-chain. For example *Trichosporon cutaneum* produces a L-arabino-D-xylo-D-mannan (**12**) which, on Smith degradation, gives a (1 → 3)-linked α-D-mannopyranosyl main-chain (**13**) by removal of the side chains (Fig. 6; Gorin and Spencer, 1967). In certain cases, such as with the complex hetero-

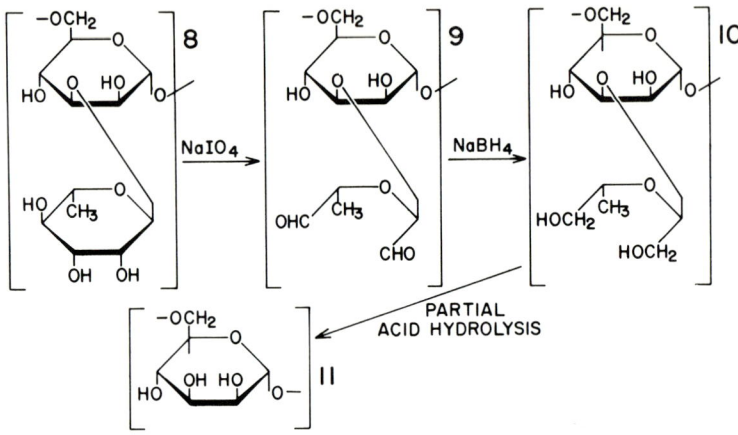

FIG. 5. Smith degradation of *Ceratocystis ulmi* L-rhamno-D-mannan.

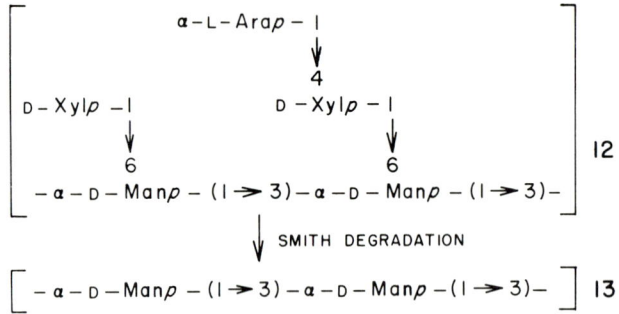

FIG. 6. Smith degradation of *Trichosporon cutaneum* pentosylmannan.

polymer from *Candida bogoriensis,* only the terminal units in some of the long side-chains were oxidizable with sodium periodate. In this instance it was necessary to use four successive Smith degradations to remove the side chains and obtain the α-(1 → 3)-linked mannan backbone (Gorin and Spencer, 1968b).

A Smith degradation of the mannan of *Rhodotorula* spp., which contains alternate (1 → 3)- and (1 → 4)-β-D-mannopyranosyl units gives 2-O-β-D-mannopyranosyl-D-erythritol only, characteristic of this type of structure (Gorin et al., 1965).

Smith degradations can be used to determine the structure of fragments formed on acetolysis. The pentasaccharide, α-D-Man$p$-(1 → 2)-β-D-Man$p$-(1 → 2)-β-D-Man$p$-(1 → 2)-α-D-Man$p$-(1 → 2)-D-Man from *Pichia pastoris* mannan was degraded using insufficient sodium periodate for complete oxidation. Consecutive comparison of the specific rotations of α-D-Man p-(1→2)-MT, β-D-Man p-(1→2)-α-D-Man p-(1→2)-MT, β-D-Man p-(1→2)-β-D-Man p-(1→2)-α-D-Man p-(1→2)-MT and α-D-Man p-(1→2)-β-D-Man p-(1→2)-β-D-Man p-(1→2)-α-D-Man p-(1→2)-MT (MT=D-mannitol), which were formed by the degradation, showed the configurations of each of the (1→2)-linkages in the pentasaccharide (Gorin et al., 1969a).

4. Partial Hydrolysis

This method is useful in the few polysaccharides that contain acid-labile side-chains. The galactomannan of *Schizosaccharomyces octosporus*, when partially hydrolyzed with acid, gives a (1→6)-linked α-D-mannopyranosyl main-chain by removal of its galactose-containing side-chains (Gorin et al., 1969a). However, the side chains of branched-chain mannans can be only partially removed by partial hydrolysis (Gorin et al., 1968).

5. Enzymolysis

The α-D-mannosidase isolated by Jones and Ballou (1968a) is particularly useful in removing mannan side chains, yielding (1→6)-linked α-D-mannopyranosyl main-chains. This has been carried out with the mannans of *Saccharomyces cerevisiae* (Fig. 3), *Candida stellatoidea* and *Kloeckera magna* (Jones and Ballou, 1968b), and *Endomycopsis fibuliger*, *Saccharomyces rouxii*, *Candida parapsilosis*, and *Torulopsis bombicola* (Gorin et al., 1969c). The main chains of the latter group were identified by their pmr signals at $\tau 4.57$, which were shifted downfield 10 Hz by addition of borate.

Mannans from *Citeromyces matritensis*, *Pichia pastoris*, *Saccharomyces rosei*, and *Saccharomyces lodderi* contain β-D-mannopyranosyl side-chain units (Gorin et al., 1969a) which are enzyme resistant. These, however, can be removed by partial acid hydrolysis giving an enzyme-vulnerable polysaccharide which is degraded to the (1→6)-linked α-D-mannopyranosyl main-chains (Gorin et al., 1969c). Similarly enzyme-resistant galactomannans from *Trichosporon fermentans*, *Torulopsis magnoliae*, *Torulopsis gropengiesseri*, *Torulopsis lactis-condensi*, and *Candida lipolytica* can be degraded to this type of main chain using successive acidic and enzymatic hydrolysis.

6. Degradations via 3,6-Anhydro Derivatives of the Side-Chain Units

Certain polysaccharides, such as the D-gluco-D-mannan of *Ceratocystis brunnea*, are not susceptible to the approaches described above for preparation of its main chain. It contains a (1→6)-linked α-D-mannopyranosyl main chain substituted in most of the 2-positions by α-D-glucopyranosyl- (**14**) and a few O-α-D-glucopyranosyl-(1 → 2)-O-α-D-mannopyranosyl (**16**) and O-α-D-mannopyranosyl side-chains (Gorin and Spencer, 1970b). Each of the side-chain units were converted to their 3,6-anhydro derivatives (**15** and **17**, respectively) by successive (1) tritylation, (2) acetylation, (3) detritylation, (4) tosylation, and (5) treatment with alkali (Ingle and Whistler, 1965). The 3,6-anhydro units were labile to partial hydrolysis leading to preparation of the main chain (**18**) (Fig. 7).

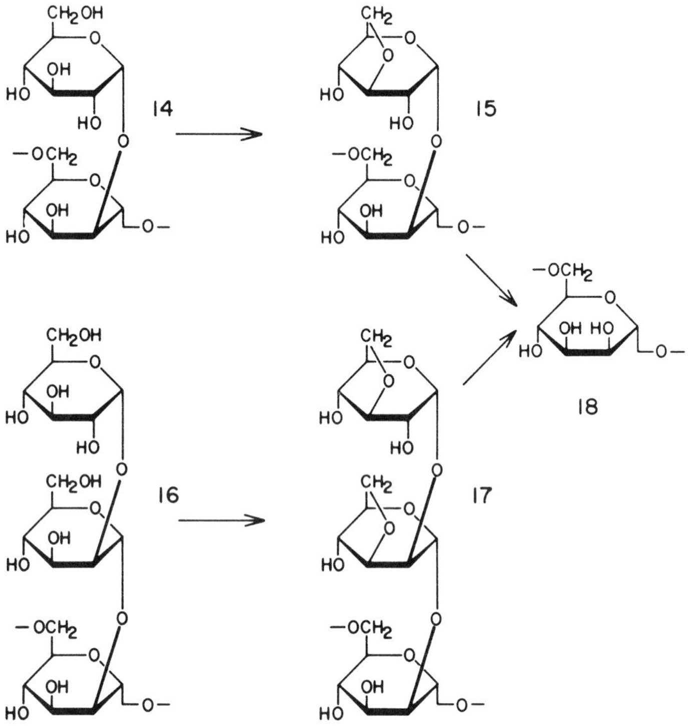

FIG. 7. Degradation of *C. brunnea* glucomannan to its α-(1 → 6)-linked α-D-mannopyranose main chain via the 3,6-anhydro intermediate.

E. Interpretation of PMR Spectra of Mannose-Containing Polysaccharides in Terms of Chemical Structure

The H-1 signals of mannose-containing polysaccharides, and the oligosaccharides that can be derived from them, occur in the $\tau 3.8$ to $\tau 4.8$ region, although the majority are at $\tau 4.1$ to $\tau 4.8$. In some cases certain signals can be interpreted in terms of chemical structure with good chance of success. For example, signals at τ4.20, τ4.57, and signals of greater chemical shift than τ4.57 are particularly meaningful and the significance of these and other signals are discussed below.

1. H-1 Signals of Mannans and Derived Oligosaccharides at τ4.20

The majority of yeast cell-wall mannans contain $(1 \to 6)$-linked α-D-mannopyranosyl main-chains (see page 54) and a high proportion of these contain side chains with consecutive α-$(1 \to 2)$-D-mannopyranosyl units, which gives an H-1 signal at τ4.20 whose chemical shift is unaffected by borate (Gorin et al., 1968). Mannans that give signals of this type should generally contain α-$(1 \to 2)$-linked side-chains of two units or longer (Gorin et al., 1969b).

2. H-1 Signals of Mannans at τ4.57

A τ4.57 signal is obtained from $(1 \to 6)$-linked α-D-mannopyranosyl polysaccharides representing the main chain of mannans, galactomannans (Gorin et al., 1969a,c), rhamnomannans and glucomannans (Gorin and Spencer, 1970b). This signal shifts downfield 10 Hz on addition of sodium borate, a property that distinguished the spectrum from *Torolopsis stellata* mannan, a branched-chain polymer, whose H-1 signal is unaffected by borate (Gorin and Spencer, unpublished results).

Since a large number of pmr spectra of mannans and galactomannans contain the τ4.57 signal it appears likely that in some cases it may arise from α-$(1 \to 6)$-linked D-mannopyranosyl units in the main chain that are unsubstituted with side chains.

3. H-1 Signals of Mannans at Higher Field than τ4.57

Of a representative group of 100 pmr spectra of mannans, 28 of them (Table V) contained signals at higher field than τ4.57 (Gorin et al., 1969a). These signals suggested that β-D-linkages are present since methyl β-D-mannopyranoside has an H-1 signal at higher field than those of H-1's of methyl α-D-mannopyranoside and the α- and β-anomers of methyl-D-mannofuranoside. This possibility was confirmed with the mannans of *Brettanomyces anomolus*, *Pichia pastoris*, *Saccharomyces rosei*, and *Citeromyces matritensis*. Their specific rotations were intermediate between +88° for the α-linked D-mannan of *Saccharomyces cerevisiae* (Haworth et al., 1941) and −78° for the β-linked D-mannan of *Rhodotorula glutinis* (Gorin et al., 1965). Hudson's definition (1938) states that in the D-series, sugars with α-linkages have a higher specific rotation than their β-linked counterparts.

TABLE V
YEASTS PRODUCING MANNANS WITH H-1 SIGNALS AT HIGHER
FIELD THAN τ4.57

Brettanomyces
  anomalus NCYC 449
  bruxellensis NCYC 362
  dublinensis NCYC 615
  lambicus NCYC 395
Candida
  obtusa UCD 60-17
  lusitaniae CBS 4413
Citeromyces matritensis CBS 2764
Debaryomyces
  castelli CBS 2923
  hansenii CBS 767
  kloeckeri NCYC 8
  phaffii PRL 199-64
  subglobosus PRL RS4
Hanseniaspora osmophila JPV 72
Kloeckera
  africana CBS 277
  magna CBS 105
Pichia
  farinosa HO E3b
  ohmeri CBS 2557
  pastoris PRL 63-208
  toletana UCD 66-1015
  vanrijii JPV 143
Saccharomyces
  lodderi JPV 192
  microellipsodes NRRL Y-1549
  pretoriensis JPV 111
  rosei PRL 16B5
  vafer NCYC 678
Schwanniomyces alluvius UCD 54-83
Schwanniomyces castelli CBS 2863
Torulopsis colliculosa NCYC 608
Trichosporon aculeatum IGC 3551

F. INTERPRETATION OF PMR SPECTRA OF OLIGOSACCHARIDES IN TERMS OF CHEMICAL STRUCTURE

Pmr spectroscopy is particularly useful in determining structures of oligosaccharides formed on fragmentation of yeast mannose-containing polysaccharides.

Disaccharides, such as 2-*O*-α-, 3-*O*-α-, and 6-*O*-α-D-mannopyranosyl-

D-mannose, can be identified by their H-1 pmr spectra, whether they are pure or mixed (Table VI). Interpretation of the complex H-1 pmr spectra of oligosaccharides in structural terms is more difficult and

TABLE VI
CHEMICAL SHIFTS OF H-1'S IN FOUR POSSIBLE
α-D-MANNOPYRANOSYL-D-MANNOSES

Linkage of α-D-mannopyranosyl-D-mannose	Chemical shift of H-1's (τ)	
	Nonreducing end	Reducing end
1→2	4.40	4.10
1→3	4.34	4.34
1→4	4.37	4.37
1→6	4.57	4.30

use must be made of chemical reagents. First, reducing-end signals can be recognized by their downfield shift on reaction of the oligosaccharide with methoxyamine at pH 9 (A. S. Perlin, unpublished results). Second, the H-1 signals of α-D-mannopyranosyl units that can form 2,3-borate complexes with sodium borate are shifted downfield approximately 10 Hz on addition of this salt to the oligosaccharide (Gorin et al., 1968). This is true of nonreducing end- and 6-O-linked units, but 2-O- and 3-O-substituted derivatives undergo H-1 downfield shifts of 2 and 5 Hz respectively (Table VII). A high proportion of the total of the latter shifts are due to a "salt effect" which could be avoided by use of an internal standard. With methyl α-D-mannopyranoside and methyl 4,6-O-ethylidene-α-D-mannopyranoside standards their H-1 signals shifted downfield on borate addition and collapsed from doublets to broad singlets, presumably due to long-range coupling of H-1's with the boron atoms (Fig. 8). This broadening is troublesome in determining polysaccharide spectra after treatment with borate since polysaccharide spectra themselves are often poorly resolved.

β-D-Mannopyranoside units capable of forming 2,3-borates undergo little downfield shift of their H-1 signals (Table VII) since the complexes are more sterically removed from H-1, compared with the α-anomers.

The unit size of an oligosaccharide can be ascertained by comparison of the intensities of the H-1 pmr signals. Oligosaccharides that have the reducing unit substituted in the 2-position by an α-D-manno-

TABLE VII
H-1 Proton Signals in Mannose Derivatives; Chemical Shifts
and Downfield Shifts on Borate Addition

Location of H-1 proton	Chemical shift (τ)	Downfield shift on borate addition (Hz)
Methyl α-D-mannopyranoside	4.71	14
Methyl 4,6-O-ethylidene-α-D-mannopyranoside	4.71	11
α-D-(1 → 3)-Linked mannopyranosyl units in linear polymer	4.35	5
Nonreducing end of α-(1 → 2)-linked D-mannopyranosyl oligosaccharides	4.40	10
Internal units of above	4.20	2
2,6-Di-O-substituted D-mannopyranose units in mannans from S. cerevisiae, S. rouxii, and T. aculeatum	4.38, 4.40	< 5
Methyl β-D-mannopyranoside	4.92	6
β-(1 → 3)-Linked units in Rhodotorula glutinis mannan	4.58	4
β-(1 → 4)-Linked units in above	4.69	7

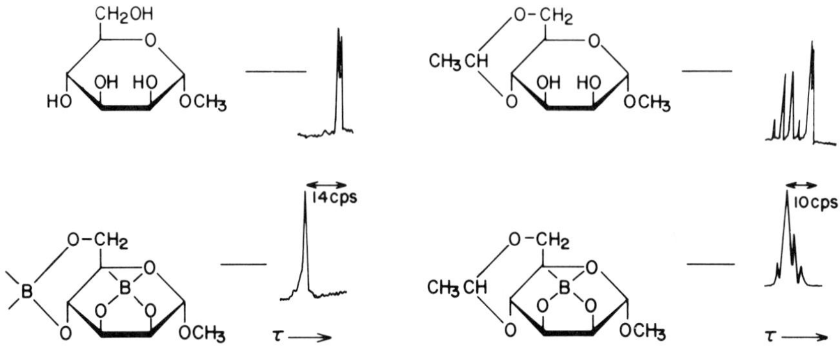

FIG. 8. Effect of sodium tetraborate on H-1 signals of methyl-α-D-mannopyranoside and its 4,6-O-ethylidene derivative.

pyranose unit have a low field signal at approximately $\tau 4.10$, corresponding to a reducing end which is virtually all in the α-form. The relative sizes of the H-1 signals can therefore be expressed as integers. On the other hand, oligosaccharides with their reducing ends substituted in the 2-position by β-D-mannopyranose units contain reducing ends with a higher proportion of the β-form. This gives a more complex signal and to obtain a simple H-1 pmr spectrum it is necessary to remove the reducing-end signal by reduction with sodium borohydride to the alditol.

In certain cases structural assignments to oligosaccharides can be made merely on the basis of pmr data. For example the pentasaccharide obtained following partial acetolysis of *Hansenula subpelliculosa* mannan has an H-1 pmr spectrum identical to that of α-D-Man$p$-(1 → 3)-α-D-Man$p$-(1 → 2)-α-D-Man$p$-(1 → 2)-D-Man (Lee and Ballou, 1965) except that the signal at τ4.20 is twice as large (Fig. 9). In each oligo-

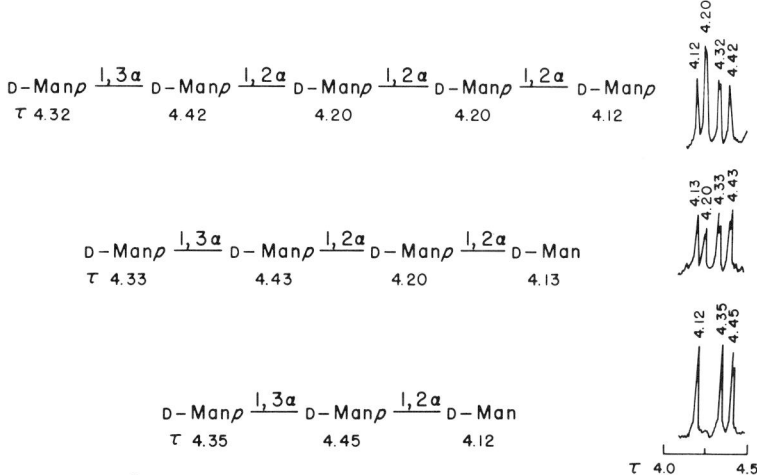

FIG. 9. H-1 pmr spectra of a homologous series of oligosaccharides.

saccharide the signal at τ4.12 corresponds to H-1 of the reducing end (downfield shift with methoxyamine) and the τ4.32 signal to the non-reducing end (downfield shift with borate). The 2-proton signal at τ4.20 is unaffected by borate and this corresponds to an additional (1 → 2)-linked D-mannopyranosyl unit leading to an assignment of an α-D-Man$p$-(1 → 3)-α-D-Man$p$-(1 → 2)-α-D-Man$p$-(1 → 2)-α-D-Man$p$-(1 → 2)-D-Man structure. This was confirmed by the spectrum of α-D-Man$p$-(1 → 3)-α-D-Man$p$-(1 → 2)-D-Man, which does not contain a τ4.20 signal (Gorin *et al.*, 1969b).

G. Chemical Structures of Yeast Mannose-Containing Polysaccharides

The structures of mannose-containing yeast polysaccharides having main chains other than the (1 → 6)-linked α-D-mannopyranosyl type are presented in Table VIII in terms of their main and side chains. Polysaccharides with $(1 \to 6)$-linked α-D-mannopyranosyl main-chains

TABLE VIII
STRUCTURES OF YEAST MANNOSE-CONTAINING POLYSACCHARIDES HAVING MAIN CHAINS OTHER THAN α-$(1 \to 6)$-D-MANNOPYRANOSE

Source of mannose-containing polysaccharide	Structure of polysaccharide	
	Main Chain	Side Chain
Trichosporon cutaneum PRL RS-1	α-$(1 \to 3)$-D-Man$p$	D-Xyl$p$ & α-L-Ara$p$-$(1 \to 4)$-D-Xyl
Candida bogoriensis CBS 4101	α-$(1 \to 3)$-D-Man$p$	Contains α-L-Rha$p$-$(1 \to 3)$-L-Rha, α-L-Rha$p$-$(1 \to 2)$-L-Fuc, α-L-Fuc$p$-$(1 \to 3)$-L-Fuc, β-D-GpA-$(1 \to 4)$-L-Fuc, α-L-Rha$p$-$(1 \to 4)$-β-D-GpA-$(1 \to 4)$-L-Fuc
Rhodotorula glutinis PRL RS-33	Alternate β-$(1 \to 3)$-D-Man$p$ and β-$(1 \to 4)$-D-Man$p$	
Rhodotorula mucilaginosa PRL 2S1	Alternate β-$(1 \to 3)$-D-Man$p$ and β-$(1 \to 4)$-D-Man$p$	
Rhodotorula minuta PRL RS55	Alternate β-$(1 \to 3)$-D-Man$p$ and β-$(1 \to 4)$-D-Man$p$	
Sporobolomyces roseus PRL 345-63	Alternate β-$(1 \to 3)$-D-Man$p$ and β-$(1 \to 4)$-D-Man$p$	
Candida sp. PRL 1S-20	Mixed α-$(1 \to 3)$-D-Man$p$ and $(1 \to 6)$-D-Man$p$	?

have their structures expressed in Table IX in terms of their partial acetolysis products. Of these fragments the largest oligosaccharide from a polysaccharide must arise from the largest side-chain. The reducing end of the fragment represents a former part of the mannan main-chain.

III. Classification of the Yeasts According to the Pmr Spectra of Their Mannose-Containing Polysaccharides

The following section consists of a discussion of yeasts which are grouped according to resemblances in the pmr spectra of their mannose-containing polysaccharides (Fig. 10, Groups 1–24). It is

TABLE IX

Oligosaccharide Fragments Formed Following Partial Acetolysis of Mannose-Containing Polysaccharides with $(1 \rightarrow 6)$-Linked α-D-Mannopyranose Main-Chains[a]

Source of mannose-containing polysaccharide	Oligosaccharide formed on partial acetolysis of polysaccharide
1. *Torulopsis bombicola* (*T. bombi*) PRL 319-67	$[\alpha\text{-}(1 \rightarrow 2)\text{-D-Man}p]_{2-8}$
2. *Torulopsis magnoliae* CBS 166	$[\alpha\text{-}(1 \rightarrow 2)\text{-D-Man}p]_{2-10}$
3. *Torulopsis apicola* (Hajsig strain) CBS 2868	$[\alpha\text{-}(1 \rightarrow 2)\text{-D-Man}p]_{2-6}$
4. *Torulopsis gropengiesseri* NRRL Y1445	$[\alpha\text{-}(1 \rightarrow 2)\text{-D-Man}p]_{2-3}$
5. *Saccharomyces rouxii* PRL 411-64	$[\alpha\text{-}(1 \rightarrow 2)\text{-D-Man}p]_{2-3}$
6. *Trichosporon fermentans* PRL 2263	α-D-Man$p$-$(1 \rightarrow 2)$-D-Man, α-D-Gal$p$-$(1 \rightarrow 2)$-α-D-Man$p$-$(1 \rightarrow 2)$-D-Man
7. *Candida lipolytica* CBS 599	α-D-Man$p$-$(1 \rightarrow 2)$-D-Man, α-D-Gal$p$-$(1 \rightarrow 2)$-α-D-Man$p$-$(1 \rightarrow 2)$-D-Man
8. *Endomycopsis fibuliger* NCYC 13	α-D-Man$p$-$(1 \rightarrow 2)$-D-Man, α-D-Man$p$-$(1 \rightarrow 3)$-α-D-Man$p$-$(1 \rightarrow 2)$-D-Man
9. *Torulopsis lactis-condensi* CBS 52	α-D-Man$p$-$(1 \rightarrow 2)$-D-Man, α-D-Gal$p$-$(1 \rightarrow 6)$-α-D-Man$p$-$(1 \rightarrow 2)$-D-Man
10. *Schizosaccharomyces octosporus* PRL F2	α-D-Gal$p$-$(1 \rightarrow 2)$-D-Man
11. *Trichosporon aculeatum* IGC 3551	$[\alpha\text{-}(1 \rightarrow 2)\text{-D-Man}p]_{2-6}$
12. *Saccharomyces cerevisiae*	α-D-Man$p$-$(1 \rightarrow 3)$-α-D-Man$p$-$(1 \rightarrow 2)$-α-D-Man$p$-$(1 \rightarrow 2)$-D-Man, α-D-Man$p$-$(1 \rightarrow 2)$-D-Man
13. *Pichia pastoris* PRL 63-208	α-D-Man$p$-$(1 \rightarrow 2)$-β-D-Man$p$-$(1 \rightarrow 2)$-β-D-Man$p$-$(1 \rightarrow 2)$-α-D-Man$p$-$(1 \rightarrow 2)$-D-Man
14. *Ceratocystis ulmi* CBS 374.67	α-L-Rha$p$-$(1 \rightarrow 3)$-D-Man
15. *Ceratocystis brunnea* CBS 161.61	α-D-Gp-$(1 \rightarrow 2)$-D-Man
16. *Citeromyces matritensis* CBS 2764	β-D-Man$p$-$(1 \rightarrow 2)$-β-D-Man$p$-$(1 \rightarrow 2)$-β-D-Man$p$-$(1 \rightarrow 2)$-D-Man, β-D-Man$p$-$(1 \rightarrow 2)$-β-D-Man$p$-$(1 \rightarrow 2)$-D-Man

[a] Gorin and Perlin, 1956; Lee and Ballou, 1965; Gorin and Spencer, 1968a; Gorin et al., 1969a,b,c; Gorin and Spencer, 1967; Gorin and Spencer, 1970b.

assumed that these similarities are due to similarities in chemical structures which, in turn, may mean that the parent yeasts are related in a phylogenetic sense. The conclusions presented below summarizes work carried out by Spencer and Gorin on microorganisms

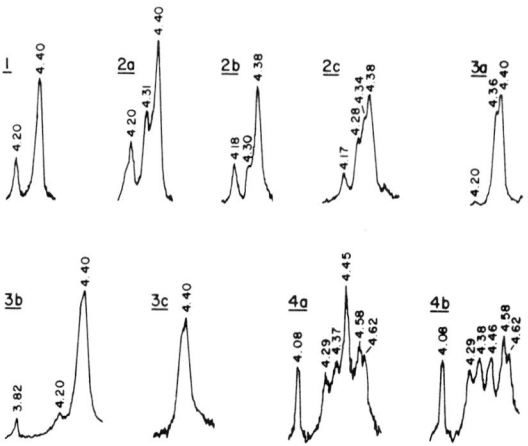

GROUP 1. *Saccharomyces rouxii* group
  Saccharomyces
    rouxii   CBS 732 (Type)
    acidifaciens   NRRL Y-1011
    bailii   CBS 680 (Type)
    carlsbergensis   CBS 1486
    cerevisiae   CBS 1171 (Type)
    cerevisiae (Syn *S. batatae*)   CBS 1199
    cidri   CBS 4575
    elegans   NRRL Y-2228
    exiguus   CBS 5648 (Type)
    hienipiensis   CBS 4903
    italicus var. *melibiosi*   CBS 2909
    mellis   NRRL Y-58
    montanus   NRRL Y-1559
    norbensis   CBS 5378
    oleaginosus   CBS 3081
    pastorianus   NCYC 392
    tellustris   CBS 2685
  Hanseniaspora
    valbyensis   CBS 479
    uvarum   CBS 2570
  Kluyveromyces osmophilus   CBS 5499
  Kloeckera apiculata   NCYC 466
  Candida
    kefyr   CBS 834
    slooffii   CBS 2419

FIG 10. [pages 56–70] The groups of yeasts and yeastlike organisms according to the pmr spectra of their mannose-containing polysaccharides. (Most of the yeasts form mannans. Those yeasts that form other mannose-containing polysaccharides are listed along with the component sugars of the polymer.)

    Torulopsis
      bovina — CBS 2760
      gropengiesseri (Gal, Man) — NRRL Y-1445
      holmii — CBS 135
      sphaerica — CBS 141

GROUP 2. *Saccharomyces carlsbergensis* group

a. *Saccharomyces*
    carlsbergensis — CBS 1485
    aceti — CBS 4054
    bayanus (strain from which *Saccharomyces inusitatus* is derived) — CBS 1546-1
    bayanus — NCYC 387
    capensis — CBS 2247
    carbajali — CBS 5313
    cerevisiae (syn S. *anamensis*) — CBS 1200
    cerevisiae (syn S. *elongatus*) — CBS 439
    cerevisiae (syn S. *sake*) — CBS 1198
    cerevisiae (haploid a) — CBS 5494
    cerevisiae (haploid α) — CBS 5495
    chevalieri — CBS 2992, CBS 5298, CBS 3077, CBS 3078, CBS 1544
    coreanus — CBS 5635
    diastaticus — CBS 1782
    ellipsoides — CBS 1395
    eupagycus — CBS 748
    fermentati — NCYC 572
    florentinus — NCYC 172 (also Type NRRL Y-1560)
    fructuum — NCYC 609
    globosus — CBS 424
    heterogenicus — CBS 5755
    hispanica — CBS 5835
    inusitatus — CBS 1546
    italicus — NCYC 108
    logos — NCYC 72
    onubensis (syn S. *capensis*) — CBS 5112
    oviformis — CBS 429
    prostoserdovii — CBS 5155
    steineri — NCYC 406
    transvaalensis — JPV 112
    uvarum — PRL 62-186
    veronae — NCYC 412
    wickerhamii — NCYC 546
    willianus — NCYC 122
  Pichia chambardii — CBS 1900
  Candida
    catenulata — CBS 565
    natalensis — JPV 184
    sake (Type, C. *natalensis*) — CBS 2935
    santamariae — CBS 4515
  Torulopsis dattila — CBS 137

b. *Saccharomyces*
  cerevisiae — NCYC 324
  cerevisiae (syn *S. muntzii*) — CBS 1195
  delphensis
c. *Torulopsis ernobii* — CBS 1737
  Pachysolen tannophilus — CBS 4044

GROUP 3. The *Saccharomyces fragilis* group
a. *Saccharomyces*
  fragilis — ATCC 12424
  chevalieri — CBS 1084 G, Man, CBS 400 (Type)
    CBS 403, CBS 405, NCYC 123
  exiguus — CBS 3019
  Candida
    macedoniensis — CBS 600
    pseudotropicalis — PRL 702-67
b. *Saccharomyces marxianus* — NRRL Y-1052
  Candida beechii — IGC 3423
c. *Torulopsis lactis-condensi* (Gal, Man) — CBS 52

GROUP 4. The *Saccharomyces rosei* group
a. *Saccharomyces*
  rosei — CBS 817 (Type) PRL 16B5
  delbrueckii — CBS 1146 (Type)
  fermentati — CBS 818 (Type)
  inconspicuus — NCYC 677
  vafer — NCYC 678
  Torulopsis
    cambresieri — CBS 158
    colliculosa — CBS 133 (Type) NCYC 608
b. *Saccharomyces rosei* — CBS 404 (Antigens 2,3,10,14,31)

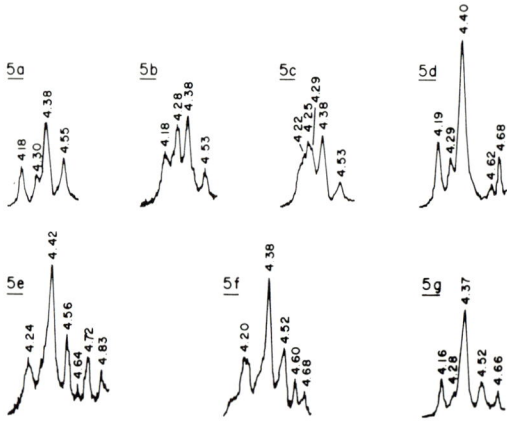

GROUP 5. *Saccharomyces aestuarii* group
a. *Saccharomyces aestuarii* — PRL Y-117
  Kluyveromyces polysporus — CBS 2163

Nematospora coryli	CBS 2608
Pichia	
pseudopolymorpha	UCD 57-3
quercuum	UCD 54-K-41
Wickerhamia fluorescens	CBS 4565
b. *Saccharomyces dobzhanskii*	NRRL Y-1974
c. *Saccharomyces*	
lactis	NRRL Y-1140
sociasi	CBS 4574
drosophilarum	PRL Y-116
d. *Saccharomyces exiguus*	CBS 4660, CBS 4661
	CBS 379, CBS 2141
e. *Saccharomyces kluyveri*	CBS 3082 (Type)
f. *Candida glaebosa*	CBS 5691
g. *Kloeckera javanica*	CBS 282

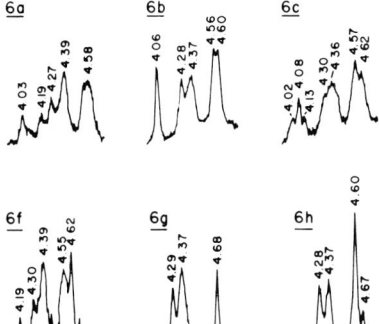
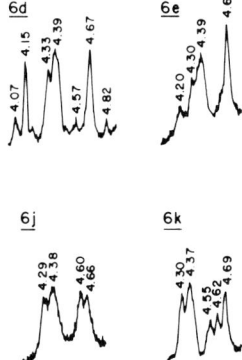

GROUP 6. *Saccharomyces microellipsodes* group
  a. *Saccharomyces microellipsodes*   NRRL Y-1549
  b. *Pichia*
    pastoris   UCD 64-1
    toletana   UCD 66-1015
  c. *Saccharomyces pretoriensis*   JPV 111
  d. *Saccharomyces kloeckerianus*   CBS 765 (Type)
  e. *Saccharomyces delbrueckii*   NCYC 147
  f. *Saccharomyces dairensis*   CBS 421
  g. *Saccharomyces lodderi*   JPV 193
    Hanseniaspora osmophila   JPV 72
    Kloeckera
      magna   CBS 105
      africana   CBS 277
  h. *Brettanomyces*
    anomalus   NCYC 499
    dublinensis   NCYC 615
  j. *Brettanomyces bruxellensis*   NCYC 362
    Torulopsis cylindrica   CBS 1947
  k. *Brettanomyces lambicus*   NCYC 395

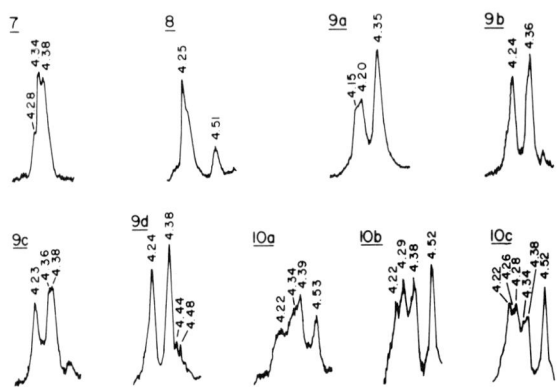

GROUP 7. *Endomycopsis capsularis* group
    Endomycopsis
        capsularis   PRL R_2a
        fibuliger   NCYC 13
        selenospora   CBS 2562
GROUP 8. *Schizosaccharomyces* group (galactomannans)
    Schizosaccharomyces
        malidevorans   NCYC 683
        octosporus   PRL F2
        pombe   PRL Y-30
        versatilis   CBS 103
GROUP 9. *Pichia membranaefaciens* group
  a. *Pichia*
        membranaefaciens   NCYC 326
        terricola   CBS 2617
    Candida
        mycoderma   NCYC 327
        valida   CBS 638
    Torulopsis pinus (G, Man)   CBS 970
  b. *Pichia*
        fluxuum   CBS 2237
        kluyveri   UCD C-117
        angophorae   UCD 65-106
        pinus   CBS 5096
    Hansenula
        glucozyma   NRRL YB-2185
        henricii   NRRL YB-2194
        minuta   NRRL Y-411
        nonfermentans   NRRL YB-2203
        polymorpha   NRRL Y-1798
        wickerhamii   NRRL Y-4943
    Candida diversa   CBS 4074
    Candida (*Pichia*)
        krusei   CBS 573
        silvae   CBS 5497
        sorbosa   CBS 1910

vini	CBS 639
Torulopsis	
maris	CBS 5151
nitratophila	CBS 2027
c. *Pichia*	
fermentans	UCD 56-5
orientalis	CBS 5147
trehalophila	UCD 62-509
Saccharomyces	
bisporus	NRRL Y-408
scandinavicus	CBS 4611
Candida lambica	CBS 1786
Torulopsis inconspicua	CBS 180
d. *Hansenula platypodis*	NRRL Y-6106

GROUP 10. The *Hansenula anomala* group

a. *Hansenula*	
anomala	NRRL Y-366
bimundalis	NRRL Y-5343
ciferri	NRRL Y-1031
fabianii	NRRL Y-1871
petersonii	NRRL Y-3808
saturnus	NRRL Y-1304
silvicola	NRRL Y-1678
subpelliculosa	NRRL Y-1683
wingei	NRRL Y-2340
Pichia	
bovis	UCD 60-20
salictaria	UCD 61-344
strassburgensis	CBS 2939
Candida	
berthetii	CBS 5452
freyschussii	CBS 2162
melinii	CBS 601
b. *Candida maritima*	CBS 5107
c. *Torulopsis norvegica*	CBS 4239

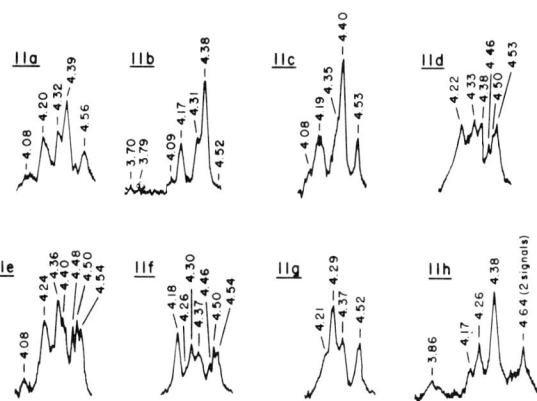

GROUP 11. Other *Hansenula* and *Pichia* species
  a. *Hansenula*
      californica   NRRL Y-1680
      dimennae   NRRL YB-3239
  b. *Hansenula holstii*   NRRL Y-2155
     Candida silvicola
  c. *Pichia polymorpha*   CBS 186
  d. *Hansenula*
      beijerinckii   JPV 182
      mrakii (G, Man)   NRRL Y-1364
  e. *Hansenula saturnus* var. *subsufficiens*   NRRL YB-1657
  f. *Hansenula*
      jadinii   NRRL Y-1542
     Candida utilis   NRRL Y-900
  g. *Hansenula*
      beckii   NRRL Y-1482
      canadensis   NRRL Y-1888
  h. *Hansenula capsulata*   NRRL Y-1842

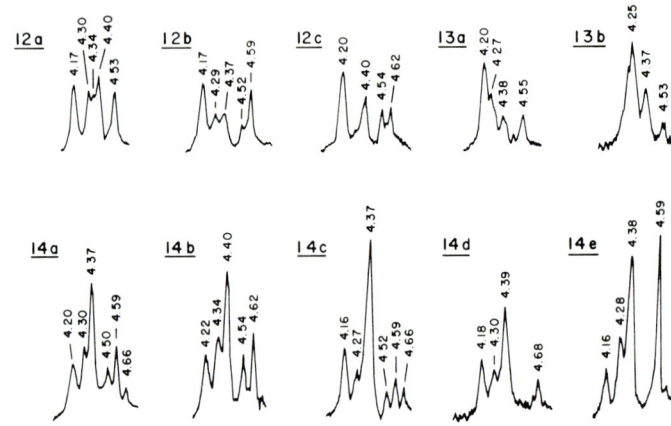

GROUP 12
  a. *Pichia rhodanensis*   CBS 2518
  b. *Pichia ohmeri*   CBS 2557
  c. *Trichosporon aculeatum*   IGC 3551
GROUP 13
  A. *Pichia pijperi*   JPV 101
  b. *Saccharomyces phaseolosporus*   NRRL Y-1975
GROUP 14
  a. *Pichia farinosa*   HO Ecb
  b. *Candida cacaoi*   IGC 3422
  c. *Torulopsis castellii*   CBS 4332
  d. *Torulopsis*
      glabrata   PRL ACB1

humilis　　　　　　　　　　　CBS 5658
e. Torulopsis pintolopesii　　　CBS 1787

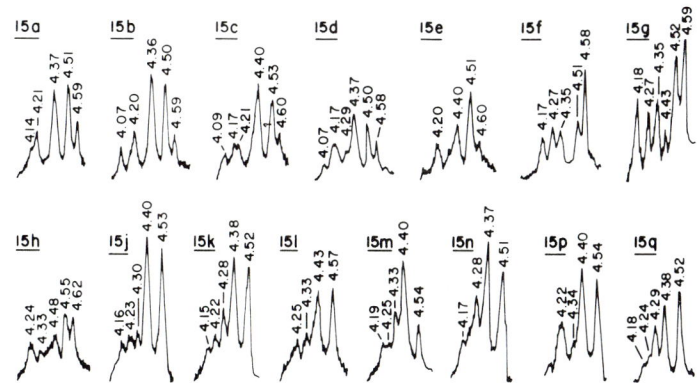

GROUP 15. Pichia robertsii–Candida tropicalis group
  a. Pichia
      robertsii　　　　　　　　　　　　UCD 60-22
      vanrijii　　　　　　　　　　　　JPV 143
      Candida guilliermondii　　　　　PRL RS3
      Schwanniomyces
        alluvius　　　　　　　　　　　UCD 54-83
        castellii　　　　　　　　　　UCD 58-3
  b. Schwanniomyces occidentalis　　NCYC 133
  c. Candida
      diddensii　　　　　　　　　　　CBS 2214
      rugosa　　　　　　　　　　　　CBS 613
      Endomycopsis chodatii　　　　　NCYC 437
  d. Candida tropicalis　　　　　　　NRRL Y-1410
  e. Candida zeylanoides　　　　　　CBS 619
      Selenotila intestinalis　　　　CBS 5946
  f. Candida
      lusitaniae　　　　　　　　　　CBS 4413
      obtusa　　　　　　　　　　　　UCD 60-17
  g. Candida
      sake　　　　　　　　　　　　　CBS 159
      salmonicola　　　　　　　　　　CBS 5690
  h. Candida albicans, serotype A
      (mannan from C. Bishop; NRCC, Ottawa)
  j. Candida tenuis　　　　　　　　　CBS 615
  k. Candida shehatae　　　　　　　　CBS 5813
  l. Candida friedrichii　　　　　　IGC 3570
  m. Candida brumptii
  n. Candida ravautii　　　　　　　　CBS 1904
  p. Candida viswanathii　　　　　　CBS 4024
  q. Endomycopsis javanensis　　　　NCYC 40

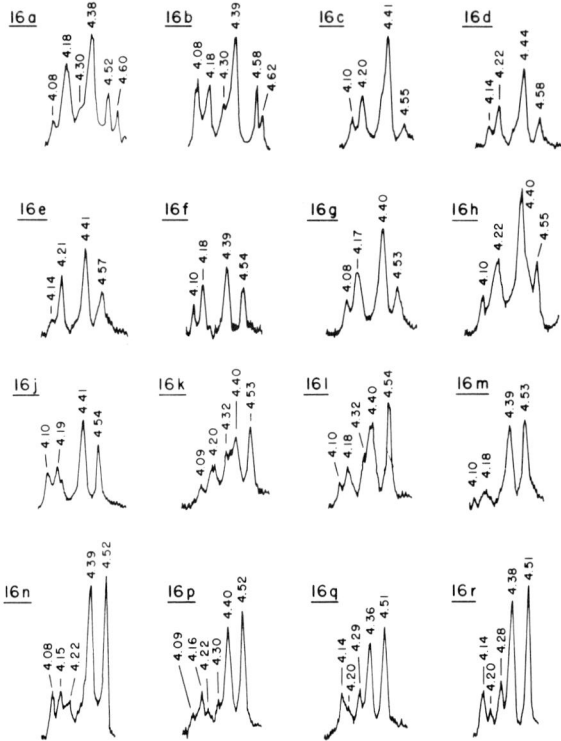

GROUP 16. The *Debaryomyces–Candida parapsilosis–Metschnikowia* group
  a. *Candida*
      claussenii   CBS 1949
      mogii   IGC 3688
      oregonensis   UCD 60-73
      Torulopsis haemulonii   CBS 5149
  b. *Candida intermedia*   CBS 572
  c. *Debaryomyces*
      castellii   CBS 2923
      hansenii (Syn *D. subglobosus*,   CBS 767
          D. kloeckeri)
      phaffii   PRL 199-64
  d. *Metschnikowia zobellii*   UCD 61-33
  e. *Metschnikowia kamienskii*   PRL-S9
  f. *Metschnikowia krissii*   UCD 61-31
  g. *Candida parapsilosis*   PRL-BMC1
     Torulopsis
       burgeffiana   CBS 4872
       torresii   CBS 5152
      candida (Syn *T. famata*)   HO 3569, CBS 843
     Candida melibiosica   IGC 2515
  h. *Candida albicans* (serotype B)   PRL H5958

j. *Metschnikowia (Candida) pulcherrima*   PRL 4R10
   Candida stellatoidea
k. *Lodderomyces elongasporus*   CBS 2605
l. *Metschnikowia (Candida) reukaufii*   PRL S-5 B-2
m. *Schwanniomyces persoonii*   UCD 61-9
n. *Candida conglobata*   CBS 2018
p. *Pichia vini*   CBS 810
q. *Pichia haplophila*   UCD 52-12
r. *Candida membranaefaciens*   CBS 1952

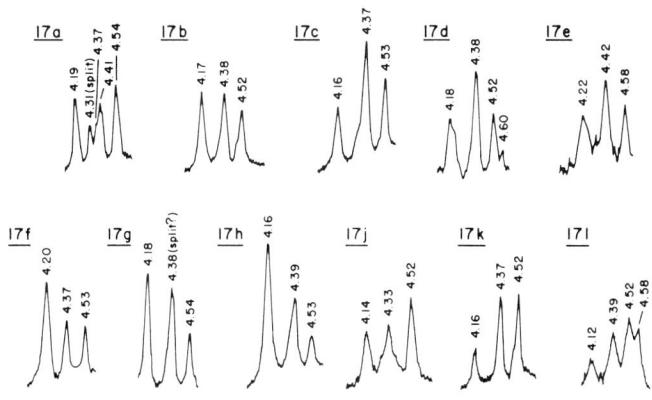

GROUP 17. The *Torulopsis apicola–Torulopsis bombicola* group
  a. *Torulopsis apicola*   CBS 2868
    Candida bombi   CBS 5836
  b. *Torulopsis*
    bombicola n. sp.   PRL 319-67
    halonitratophila   CBS 5240
    Candida vartiovaarii (Gal, Man)   CBS 4289
    Torulopsis
      nodaensis (Gal, Man)   HO (no number)
      apis (Gal, Man)   CBS 2674
      magnoliae (Gal, Man)   CBS 166
  c. *Candida ciferrii*   CBS 4856
  d. *Candida langeronii*   CBS 1912
  e. *Eremothecium ashbyi*   CBS 106.43
    Ashbya gossypii   CBS 109.51
    Candida pelliculosa (Gal, Man)   NCYC 471
    Kluyveromyces africanus   CBS 2517
  f. *Torulopsis etchellsii* (Gal, Man)   CBS 1751
  g. *Cephaloascus fragrans*   CBS 118.57
  h. *Candida suécica*   CBS 5724
  j. *Endomyces ovetensis* (Gal, Man)   CBS 192.55
  k. *Pichia etchellsii*   UCD 66-23
    Torulopsis sphaerica   NCYC 469
  l. *Candida rhagii*   CBS 4237

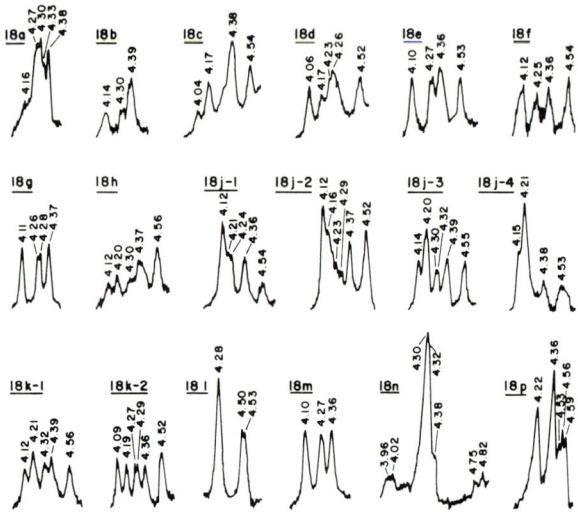

GROUP 18. Species producing galactomannans with spectra of varying degrees of complexity.

a.	*Trigonopsis variabilis*	CBS 1040
b.	*Trichosporon penicillatum*	IGC 3716
c.	*Candida edax*	CBS 5657
d.	*Candida steatolytica*	CBS 5839
e.	*Candida lipolytica*	CBS 599
f.	*Candida ingens* (Man)	CBS 4603
g.	*Trichosporon fermentans*	PRL 2263
	Geotrichum candidum	CBS 178.53
	Endomyces decipiens	CBS 165.29
	Endomyces reesii	CBS 179.60
	Torulopsis vanderwaltii	CBS 5524
h.	*Torulopsis halophila*	HO (no number)
	mannitofaciens	
	versatilis	CBS 1752
j. 1.	*Candida tepae*	(Grinbergs)
2.	*Candida salmanticensis*	CBS 5121
3.	*Candida blankii*	CBS 1898
4.	*Candida incommunis*	CBS 5604
k. 1.	*Nadsonia elongata*	CBS 2593
2.	*Nadsonia fulvescens*	CBS 2596
	Torulopsis domercqii	CBS 4351
	Trichosporon hellenicum	CBS 4099
l.	*Endomyces magnusii*	CBS 107.12
m.	*Dipodascus albidus*	CBS 152.57
n.	*Torulopsis cantarellii* (Gal, Man, G)	CBS 4878
p.	*Candida boidinii* (Man, tr. G)	CBS 2428

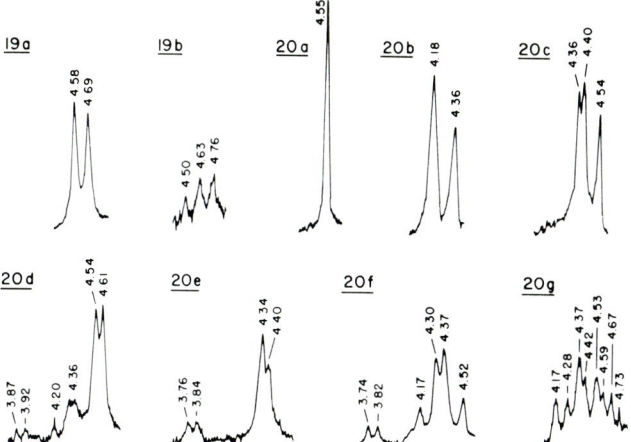

GROUP 19. The *Rhodotorula*—*Sporobolomyces* group
  a. *Rhodotorula*
      glutinis   PRL RS33
      minuta   PRL RS55
      pilamanea   CBS 4479
      Bullera tsugae   UCD 60-71
      Sporobolomyces roseus   PRL 345-63
      Torulopsis ingeniosa   CBS 4240
      Gymnosperma virginiae (Juniper rust)
      Puccinia graminis tritici (Wheat rust)
  b. *Sporobolomyces rubicundulus*   PRL 1992

GROUP 20. Miscellaneous species producing mannans with unusual pmr spectra
  a. *Torulopsis stellata*   NCYC 486
  b. *Candida norvegensis*   CBS 1922
  c. *Candida aaseri*   CBS 1913
  d. *Citeromyces matritensis*   CBS 2764
     Torulopsis globosa
  e. *Torulopsis (Selenotila) peltata*   CBS 5576
  f. *Torulopsis wickerhamii*   CBS 2928
  g. *Saccharomycodes ludwigii*   CBS 821

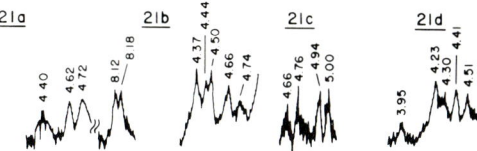

GROUP 21. Species producing heteropolymers
  a. *Candida buffonii* (Gal, G, Man, Rha)   CBS 2838
  b. *Candida diffluens* (Gal, Man, Rha)   CBS 5233
  c. *Ustilago maydis* (Man, G)   PRL 1092
  d. *Taphrina deformans* (Rha, Man)   CBS 355-35

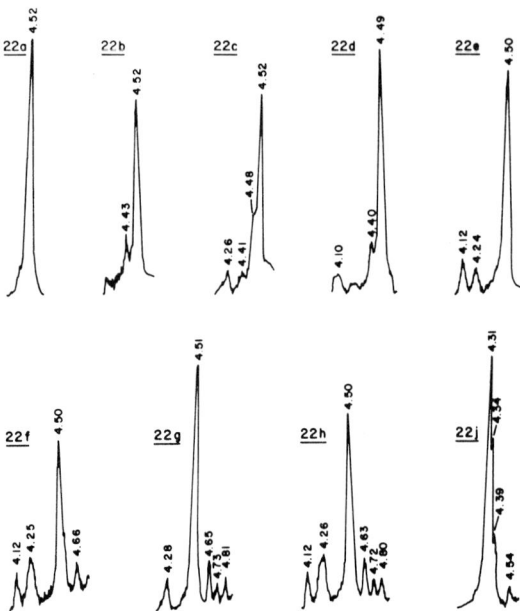

GROUP 22. *Ceratocystis* and *Graphium* species which produce polysaccharides containing mannose and glucose or mannose and rhamnose

  a. Ceratocystis dryocoetidus group
     dryocoetidus (Rha, Man, tr. G)   CBS 366.66
     multiannulata (Rha, Man, G)   CBS 124.39
     penicillata (Rha, Man)   CBS 140.36
    Graphium
     album (Rha, Man, tr. G, Gal)   CBS 276.54
     aureum (Rha, Man, tr. Gal)   CBS 266.54
     erubescens (Rha, Man, Gal)   CBS 278.54
  b. Ceratocystis europhioides group
     europhioides (Man, Rha, Gal)   CBS 275.65
     huntii (Man, Rha, tr. Gal)   CBS 153.65
     serpens (Man, Rha)   CBS 141.36
     seticollis (Man, Rha)   CBS 634.66
  c. Ceratocystis olivacea group
     olivacea (Man, Rha, tr. Gal)   CBS 138.51
    Graphium
     fragrans (Man, Rha, tr. Gal)   CBS 279.54
     silanum (Rha, Man, tr. G, Gal)   CBS 206.37
  d. Ceratocystis minuta group
     minuta (Rha, Man, Gal)   CBS 145.59
     minuta-bicolor (Rha, Man, tr. Gal, G)   CBS 635.66
  e. Ceratocystis ambrosiae group
     ambrosiae (Rha, Man, tr. Gal)   CBS 210.64
     cana (Rha, Man)   CBS 133.51

megalobrunnea (Rha, Man, G, tr. Gal)	CBS 360.65
nigrocarpa (Rha, Man, tr. Gal)	CBS 637.66
piceae (Man, Rha, G)	CBS 108.21
tetropii (Man, Rha, G)	CBS 140.51

f. *Ceratocystis minor* group

capillifera (Man, Rha, tr. Gal)	CBS 134.51
minor (Man, Rha, tr. Gal)	CBS 138.36
narcissi (Man, Rha)	CBS 138.50
perfecta (Man, Rha)	CBS 636.66
Graphium rigidum (Man, Rha, tr. Gal)	CBS 209.34

g. *Ceratocystis ulmi* group

auracariae (Man, Rha, tr. Gal)	CBS 114.68
ips (Man, Rha)	CBS 137.36
leptographioides (Man, Rha, tr. Gal)	CBS 144.59
populina (Man, Rha, tr. Gal)	CBS 212.67
tremulo-aurea (Man, Rha)	CBS 361.65
ulmi (Man, Rha, tr. Gal)	CBS 374.67
fagi (Man, Rha, tr. Gal)	CBS 236.32
catonianum (Man, Rha)	CBS 263.35

h. *Ceratocystis clavata* group

clavata (Rha, Man)	CBS 135.51
pilifera (Rha, Man, tr. G)	CBS 125.29
pluriannulata (Rha, Man, tr. Gal)	CBS 136.56
stenoceras (Rha, Man, tr. Gal)	CBS 237.32

j. *Ceratocystis adiposa* group

adiposa (Man, G)	CBS 127.27
brunnea (Man, G)	CBS 161.61
coerulescens (Man > G, Gal)	CBS 142.53
major (Man, G)	CBS 138.34
paradoxa (Man, G)	CBS 128.32 and CBS 453.66
polonica (Man > G, Gal)	CBS 133.38
radicicola (Man, G)	CBS 114.47

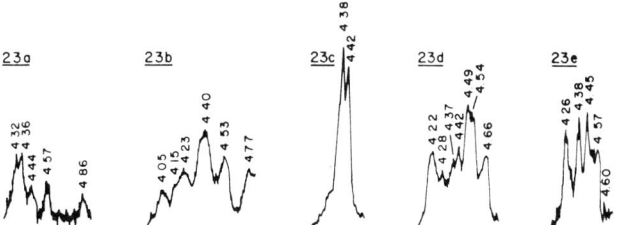

GROUP 23. Spectra of mannose-containing polysaccharides formed by *Ceratocystis* species not belonging in any previous group

a. *Ceratocystis fimbriata* (Gal, Man)	CBS 141.37
b. *Ceratocystis galeiformis* (Gal, G, Rha, Man)	CBS 137.51
c. *Ceratocystis moniliformis* (Man, tr. G)	CBS 155.62
d. *Ceratocystis nigra* (Rha, Man, tr. Gal)	CBS 163.61
e. *Ceratocystis obscura* (Rha, Man, Gal)	CBS 125.39

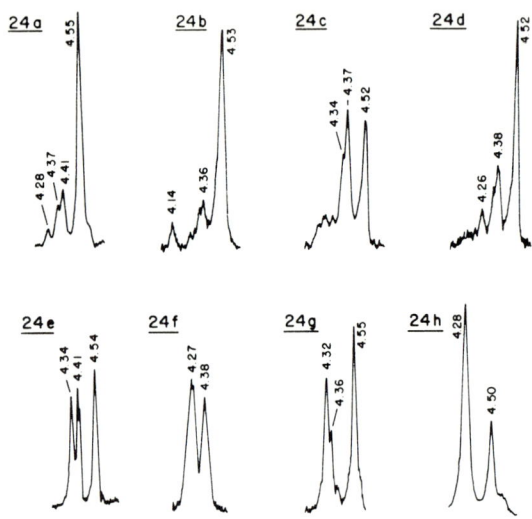

GROUP 24. Miscellaneous organisms
  a. *Alternaria brassicola* (Man, Gal, tr. G)   PRL 503
  b. *Alternaria linicola* (Man, G)   PRL 510
  c. *Alternaria raplani* (Man, G, tr. Gal)   PRL 505
  d. *Alternaria dianthicola* (Man, Gal, tr. G)   PRL 509
  e. *Microsporum quinckeanum* I (Gal, Man)
  f. *Microsporum quinckeanum* II (Gal, Man)
  g. *Mortierella renispora* (Man, G, tr. Gal)   PRL 26
  h. *Nectria* sp. (Man, G)   PRL 2242

of the genera *Nadsonia, Hanseniaspora, Kloeckera, Saccharomycodes* (1968), *Candida* (1969a), *Hansenula* and *Pichia* (1969b), *Saccharomyces, Schizosaccharomyces, Schwanniomyces, Endomycopsis, Kluyveromyces,* and *Brettanomyces* (1969c), *Torulopsis* (1970b), *Metschnikowia* and *Debaryomyces* (1970d), *Ceratocystis* (1970c), and further species of *Candida* (1970a).

In the figures and elsewhere the sources of the yeasts are abbreviated as follows: NCYC – National Collection of Yeast Cultures, Brewing Industry Research Foundation, Nutfield, Surrey, England; CBS – Centraalbureau voor Schimmelcultures, Delft and Baarn, The Netherlands; UCD – Collection of Dr. H. J. Phaff, University of California, Davis, California (United States); JPV – Collection of Dr. J. P. van der Walt, Pretoria, South Africa; HO – Collection of Dr. H. Onishi, Noda Institute of Scientific Research, Japan; PRL – Collection of the Prairie Regional Laboratory, National Research Council, Saskatoon, Saskatchewan, Canada; NRRL – Collection of the Northern Utilization Research and Development Division, United States Department of

Agriculture, Peoria, Illinois (United States); IGC—Collection of the Laboratorio de Microbiologia, Institute Gulbenkian de Ciencia, Oeiras, Portugal; ATCC—American Type Culture Collection, Washington, D.C. (United States).

Groups 1, 2, and 3 (Fig. 10). The Saccharomyces rouxii, Saccharomyces carlsbergensis, and Saccharomyces fragilis Groups

These groups contain the majority of the *Saccharomyces* species, including some of those recently placed in the genus *Kluyveromyces* or *Fabospora* (van der Walt, 1965b; Phaff *et al.*, 1966), some of the apiculate yeasts of the genera *Hanseniaspora* and *Kloeckera*, a number of species of *Candida* and *Torulopsis*, some of which may prove to be imperfect forms of known *Saccharomyces* species, and one species of *Pachysolen* and one of *Pichia*. *Torulopsis dattila*, for instance, has carbon and nitrogen assimilation patterns identical with those of *Saccharomyces veronae*, as well as forming a mannan whose pmr spectrum is the same as that of *S. veronae* mannan. The chief difference in the two species is the formation of pseudomycelium by the latter. This characteristic, however, may not be sufficient reason for maintaining *T. dattila* as a separate species.

Torulopsis bovina and *Torulopsis holmii* are the imperfect forms of *Saccharomyces telluris* (Lodder *et al.*, 1958), and *Saccharomyces exiguus* (Lodder and Kreger-van Rij, 1952), respectively, which is in accordance with the similarity in the spectra of their mannans.

Group 4. The Saccharomyces rosei Group

Group 4 constitutes a small group of species forming mannans whose pmr spectra are nearly all identical. *Saccharomyces rosei* forms a mannan whose spectrum is characteristic of the group. *Saccharomyces inconspicuus* and *Saccharomyces vafer* are considered to be derived from *Saccharomyces delbrueckii* (van der Walt, 1965a). *Torulopsis colliculosa* and *Torulopsis cambresieri* are the imperfect forms of *Saccharomyces fermentati* (Lodder and Kreger-van Rij, 1952) and *Saccharomyces rosei* (van Uden and Vidal-Leiria, 1970, respectively). The group includes one strain of *S. rosei* (Fig. 10, Group 4b) with an atypical antigenic structure, and the spectrum of its mannan is somewhat different.

Group 5. The Saccharomyces aestuarii Group

Group 5 is composed of yeasts forming mannans having spectra of increasing complexity. Additional signals in the high-field region appear, at chemical shifts from $\tau 4.53$, in the spectrum of *Saccharo-*

myces aestuarii mannan, to τ4.83, in the spectrum of *Saccharomyces kluyveri* mannan. A strain of *Saccharomyces exiguus*, CBS 379, is placed in this group. The spectrum of its mannan has signals at τ4.62 and τ4.68, which are not present in the spectrum of the mannan of *S. exiguus*, CBS 5648 (Fig. 10, Group 1).

The group also includes representatives of the genera *Kluyveromyces*, *Nematospora*, *Pichia*, *Wickerhamia*, and the asporogenous genera *Candida* and *Kloeckera*. *Pichia pseudopolymorpha* forms round spores and would be placed in the genus *Saccharomyces* except for its slow fermentation of sugars (Kreger-van Rij, 1964). *Pichia quercuum* forms hat-shaped spores, which is characteristic of numerous species in this genus.

Kloeckera javanica, one of the apiculate yeasts, forms a mannan whose spectrum (Fig. 10, Group 5f) is almost indistinguishable from that of *S. aestuarii* mannan.

Group 6. The Saccharomyces microellipsodes Group

This group includes a number of species of *Saccharomyces*, *Pichia*, *Hanseniaspora*, *Kloeckera*, and *Brettanomyces* which form mannans whose spectra are characterized by very prominent signals in the high-field region. These probably arise from β-D-mannopyranose units (Gorin *et al.*, 1969a). A few species such as *Pichia pastoris*, *Pichia toletana*, and *Saccharomyces pretoriensis*, form mannans which have prominent signals in the low-field region as well.

Hanseniaspora osmophila, *Kloeckera magna*, and *Kloeckera africana* form mannans whose spectra are the same as that of *Saccharomyces lodderi* mannan. *Brettanomyces anomalus* mannan has a spectrum very much like that of *S. lodderi* mannan except that the former spectrum has a high-field signal at τ4.62 instead of τ4.68. The spectra of all the mannans of the *Brettanomyces* species differ only slightly. *Torulopsis cylindrica* forms a mannan whose spectrum is identical with that of *Brettanomyces bruxellensis* mannan. *Brettanomyces bruxellensis* has been shown to form hat-shaped ascospores (van der Walt and van Kerken, 1960), which is consistent with the relationship to the yeasts of the genus *Hanseniaspora* suggested by the spectra of their mannans.

Group 7. The Endomycopsis capsularis Group

The spectra of the mannans of these three yeasts have points of similarity to those of the mannans of *Saccharomyces fragilis* and some other members of Group 3.

The mannans of *Endomycopsis fibuliger* (Gorin et al., 1969b) and *Saccharomyces fragilis* (Gorin and Spencer, unpublished results) resemble each other closely since they both contain $(1 \to 6)$-linked α-D-mannopyranosyl units in the main chains substituted in the 2-positions by α-D-mannopyranosyl side-chains. The only difference is that the smaller amounts of 2-unit side-chains are O-α-D-mannopyranosyl-$(1 \to 3)$-O-α-D-mannopyranosyl in the former yeast mannan and O-α-D-mannopyranosyl-$(1 \to 2)$-O-α-D-mannopyranosyl in the latter.

Group 8. The Schizosaccharomyces Group

These four species produce galactomannans having a spectrum characterized by a major signal at τ4.26. *Schizosaccharomyces octosporus* has a galactomannan with a $(1 \to 6)$-linked α-D-mannopyranosyl main-chain substituted in the 2-positions by α-D-galactopyranosyl and galactobiosyl units (Gorin et al., 1969a,b).

Group 9. The Pichia membranaefaciens Group

This group includes a large number of *Pichia, Hansenula, Saccharomyces, Candida,* and *Torulopsis* species, which form mannans characterized by spectra having major signals at approximately τ4.24 and τ4.38. Some of the spectra (Group 9a) have a broad or doubled signal at τ4.20, and some have a similar broad signal at τ4.38 (Group 9c). In the largest group (9b), both signals are sharp.

Many of the *Candida* and *Torulopsis* species are probably imperfect forms of *Pichia* and *Hansenula* species. *Torulopsis inconspicua* and *Candida lambica*, which differ only very slightly in their carbon assimilation patterns, have been suggested (Spencer and Gorin, 1969b; van Uden and Buckley, 1970) as imperfect forms of *Pichia fermentans*. The carbon and nitrogen assimilation patterns of *C. lambica* are most like those of *P. fermentans*, though the differences among these species and others in this group are slight. Likewise, the carbon assimilation patterns of *Candida valida* and *Candida vini* differ very little from that of *Pichia membranaefaciens*. They are probably imperfect forms of the latter. *Candida valida* is synonymous with *Candida mycoderma* (van Uden and Buckley, 1970), which was classified by Lodder and Kreger-van Rij (1952) as the imperfect form of *P. membranaefaciens*. At present a perfect form of *Torulopsis nitratophila* would be classified in the genus *Hansenula*. It may eventually prove to be one of several species intermediate between the genera *Hansenula* and *Pichia*.

Saccharomyces bisporus is probably wrongly placed in the genus *Saccharomyces* (Yarrow, private communication), and may therefore belong in the genus *Pichia*. On the basis of the pmr spectrum of the mannan, *Saccharomyces scandinavicus* should be transferred to the genus *Pichia* also.

Hansenula wickerhamii, *Hansenula nonfermentans*, *Hansenula henricii*, and *Hansenula glucozyma* are new species recently described by Wickerham (1968). They form mannans whose spectra place them in the *Pichia membranaefaciens* group. The spectrum of *H. platypodis* mannan, however, which also has major signals at approximately $\tau 4.24$ and $\tau 4.38$, has two additional small signals at $\tau 4.44$ and $\tau 4.48$. The composition of the DNA of this species (Nakase and Komagata, 1969) likewise indicates that it is not as closely related to the rest of the other new species as these are to each other.

Group 10. The Hansenula anomala Group

Species of this group form mannans characterized by a broad, undifferentiated signal from approximately $\tau 4.20$ to $\tau 4.39$, and a single signal at $\tau 4.53$. Many of the species in this group occur on Lines 2 and 4 of the phylogenetic scheme proposed by Wickerham (1968).

The group includes three *Pichia* species, which differ from the *Hansenula* species by being unable to utilize nitrate. *Candida freyschussii* and *Candida maritima* likewise do not utilize nitrate, and would be considered species of *Pichia* if they could be induced to sporulate.

Candida berthetii and *Candida melinii* do use nitrate and may prove to be nonsporulating forms of *Hansenula* species. *Candida melinii* is morphologically and physiologically similar to *Hansenula wingei*, though Wickerham (1968) was unable to demonstrate hybridization between the two forms.

Torulopsis norvegica likewise assimilates nitrate, and may prove to be the imperfect form of a *Hansenula* species.

Group 11. Other Hansenula and Pichia Species

The first three subdivisions in this group include species which produce mannans with spectra having fewer and more clearly defined signals in the region from $\tau 4.00$ to $\tau 4.40$. The signal at approximately $\tau 4.52$ is often reduced in size. *Hansenula dimennae*, which produces a mannan whose spectrum is like that of *Hansenula californica*, is a new species recently described by Wickerham, and placed on Line 1 of his phylogenetic scheme (Wickerham, 1968).

Hansenula beijerinckii, Hansenula saturnus var. *subsufficiens* (syn. *Hansenula suaveolens*), and *Hansenula jadinii* (imperfect form, *Candida utilis*) form mannans. *Hansenula mrakii* forms a glucomannan. Their spectra differ from that of *H. anomala* mannan in having three signals at approximately $\tau 4.46$, $\tau 4.50$, and $\tau 4.53$. The spectra of the mannans of *H. saturnus* var. *subsufficiens* and *H. jadinii* differ in some minor respects from that of *H. beijerinckii* mannan.

Hansenula beckii and *Hansenula canadensis* are the only yeasts now placed on Line 3 of Wickerham's phylogenetic scheme. The spectra of their mannans differ considerably from those of the mannans of their nearest neighbors, *H. henricii* and *H. nonfermentans*, as do the base compositions of their DNA's (Nakase and Komagata, 1968). *Hansenula beckii* is classified by Kreger-van Rij (1964) as *Endomycopsis bispora*. The spectra of the mannans are unusual in possessing a major signal at $\tau 4.29$.

Hansenula capsulata forms a mannan with signals at $\tau 3.86$ and $\tau 4.64$, which are absent from the spectra of the mannans of most other *Hansenula* and *Pichia* species.

The taxonomic regions occupied by the genera *Hansenula* and *Pichia* overlap considerably, according to the spectra of the mannans formed by the different species. At least three spectral types (Fig. 10, Groups 9a, 10a, 11a,b,c) are shared by members of both genera. Eventually the two genera may prove to be indistinguishable.

Group 12. *Pichia rhodanensis, Pichia ohmeri, and Trichosporon aculeatum*

The spectra of the mannans formed by these three species are characterized by a prominent signal at approximately $\tau 4.17$, and various high-field signals. The spectra of the mannans of *P. ohmeri* and *T. aculeatum* have a signal at approximately $\tau 4.59$ corresponding to β-linked D-mannopyranose units (Gorin *et al.*, 1969a), which the spectrum of *P. rhodanensis* mannan lacks.

Group 13. *Pichia pijperi and Saccharomyces phaseolosporus*

These two species form mannans whose spectra are superficially similar, each having a prominent signal in the low-field region. The signal at $\tau 4.20$ in the spectrum of *P. pijperi* mannan is somewhat farther downfield than the corresponding signal in that of *S. phaseolosporus* mannan. Also this mannan gives an *N*-acetyl signal at $\tau 7.2$ one half the size of the H-1 signal (Gorin and Spencer, unpublished results).

Group 14. The Pichia farinosa Group

The first three members of the group form mannans whose spectra have many points of similarity. The carbon and nitrogen assimilation patterns of *P. farinosa* and *Candida cacaoi* are almost identical. Other evidence of a relationship between these two species should be sought.

The spectra of the mannans of *Torulopsis glabrata* and *Torulopsis pintolopesii* lack some of the signals in the high-field region. The spectra of the mannans of *T. glabrata* and *Torulopsis humilis* are identical. Their carbon and nitrogen assimilation patterns likewise are almost alike. *Torulopsis pintolopesii* forms a mannan with a distinctive spectrum, having a very prominent signal at $\tau 4.59$.

Group 15. The Pichia robertsii–Candida tropicalis Group

The species of this group form mannans and galactomannans having numerous signals ranging in chemical shift from $\tau 4.09$ to $\tau 4.62$. The first few members of the group (Fig. 10, Groups 15a to 15f) form mannans with spectra that have a signal at $\tau 4.09$, which is lacking in the spectra of the mannans of the other species. Some of the remainder (Fig. 10, Groups 15j to 15q) lack the signal at $\tau 4.60$.

The group includes species of *Pichia, Schwanniomyces, Endomycopsis, Candida, Selenotila,* and *Torulopsis. Pichia robertsii, Pichia vanrijii, Candida (Pichia) guilliermondii, Schwanniomyces alluvius,* and *Schwanniomyces castellii* form mannans whose spectra are identical. The spectrum of *Schwanniomyces occidentalis* (Fig. 10, Group 15b) mannan differs in having more clearly defined signals at $\tau 4.07$ and $\tau 4.20$. *Candida diddensii, Candida rugosa,* and *Endomycopsis chodatii* (Fig. 10, Group 15c) form mannans whose spectra are alike and differ only slightly from the preceding ones.

Candida tropicalis forms a mannan whose spectrum differs slightly, in the small signal at $\tau 4.29$, from the preceding spectra.

Selenotila intestinalis is a member of a new genus characterized by the formation of lunate cells. The spectrum of *S. intestinalis* mannan is almost identical with that of *Torulopsis haemulonii* mannan (Fig. 10, Group 15e).

Candida zeylanoides forms a mannan whose spectrum has a prominent signal at $\tau 4.51$. The spectrum has many points of similarity to the others in the group, but is sufficiently distinctive to be useful in confirming the identity of isolates of this species (Spencer and Gorin, unpublished data).

Candida lusitaniae and *Candida obtusa* are two species earlier classified by Phaff and do Carmo-Sousa (1962) in the "*Candida parapsilosis* group." The spectra of the mannans of *C. lusitaniae* and *C. obtusa* are identical, but differ somewhat from the spectra of the mannans of *C. oregonensis* and *C. parapsilosis*, the other two members of the original group (Fig. 10, Groups 16a and g).

Candida sake, CBS 159, and *Candida salmonicola* form mannans whose spectra are alike and have many points of similarity to those of the mannans of *C. lusitaniae* and *C. tropicalis*. The carbon and nitrogen assimilation patterns of *C. sake* and *C. salmonicola* are very much alike.

Candida albicans, serotype A, forms a mannan whose spectrum resembles the others in the early part of this group. The spectrum of the mannan of serotype B differs somewhat, and is placed in Group 16 (h) (Fig. 10).

The remaining species in this group form mannans whose spectra have many points of similarity. They differ mainly in the size of corresponding signals. They are distinguishable from the spectra of the mannans formed by some of the yeasts in Group 16 since they lack a small but characteristic signal at $\tau 4.08$.

The spectrum of *Candida ravautii* mannan differs only slightly from that of *Candida brumptii* mannan. The carbon and nitrogen assimiliation patterns are likewise much alike. However, van Uden and Buckley (1970) maintained them as separate species.

Endomycopsis javanensis, also, forms a mannan whose spectrum is like the others in this group. Kreger-van Rij (1964) rejected a proposal to place *E. javanensis* in the genus *Schwanniomyces*. The spectra of the mannans formed by *Schwanniomyces alluvius* and *Schwanniomyces castellii*, differ in signals at $\tau 4.29$ and $\tau 4.59$, but are otherwise alike.

Group 16. The Debaryomyces–Candida parapsilosis–Metschnikowia Group

This group includes such diverse forms as all known species of *Metschnikowia* and a number of species of *Debaryomyces* as well as species of *Pichia, Lodderomyces, Schwanniomyces, Candida,* and *Torulopsis*. *Metschnikowia* species produce long, needle-shaped spores; *Debaryomyces*, round, warty ones. Nevertheless, the spectra of the mannans formed by these species are strikingly similar.

A few *Candida* and *Torulopsis* species (Fig. 10, Groups 16a and b) were placed in this group, even though they formed mannans whose spectra included a small signal at $\tau 4.60$ which is absent from most of the spectra in this group. *Candida oregonensis* (Fig. 10, Group 16a) was originally placed in the "*Candida parapsilosis* group" by Phaff and do Carmo-Sousa (1962), and is retained in Group 16 for that reason. The spectra of the mannans of *Candida mogii*, *Candida claussenii*, *Candida intermedia*, and *Torulopsis haemulonii* are almost identical with that of *C. oregonensis* mannan. The carbon assimilation patterns of these species differ somewhat, but show a number of points of similarity.

The spectra of the mannans of *Debaryomyces castellii*, *Debaryomyces hansenii*, *Debaryomyces phaffii*, *Metschnikowia kamienskii*, *Metschnikowia krissii*, *Metschnikowia zobellii*, and *Candida parapsilosis* are almost identical. Several other *Candida* and *Torulopsis* species, including *T. famata*, the imperfect form of *Debaryomyces hansenii*, form mannans whose spectra are identical with that of *C. parapsilosis* mannan.

The remaining species of *Metschnikowia*, *M. reukaufii* and *M. pulcherrima*, form mannans whose spectra differ slightly from those of the preceding group, but which have many points of similarity to the spectra of the mannans of *Pichia vini* and *Pichia haplophila*. *Pichia vini* was originally classed as an aberrant species of *Debaryomyces*. The taxonomic significance of the similarities in the spectra of the mannans of these species is not known.

Group 17. The Torulopsis apicola–Torulopsis bombicola Group

This group produces mannans and galactomannans characterized by spectra with signals at approximately $\tau 4.17$, $\tau 4.38$, and $\tau 4.52$. *Torulopsis apicola* and *Candida bombi*, which may be synonymous, form mannans whose spectra have additional signals at approximately $\tau 4.30$ and $\tau 4.35$. These two species, like *T. bombicola*, *T. magnoliae*, *T. apis*, and *T. nodaensis*, are found in similar habitats, and use very few carbon compounds. These species are not easy to distinguish by their carbon assimilation patterns, but *T. apicola* and *C. bombi* can be readily separated from the other members of the group by the differences in the spectra of the mannans.

The mannan of *T. bombicola* contains $(1 \rightarrow 6)$-linked α-D-mannopyranose main-chains partially substituted in the 2-positions by long side chains containing $(1 \rightarrow 2)$-linked α-D-mannopyranosyl units. Its spectrum fortuitously resembles that of the galactomannan of *T. magnoliae* which has a related structure except that the side chains are terminated by β-D-galactopyranosyl units (Gorin et al., 1969b).

The other members of this group include species of *Kluyveromyces*, *Endomycopsis* (*Endomyces*), *Pichia*, and such yeastlike fungi as

Ashbya, Eremothecium, and *Cephaloascus.* The relationships among most of these species, if any, are not known. *Cephaloascus fragrans* may be related to some species of *Hansenula* (Wickerham, 1968) but the spectrum of its mannan is unlike any of those of the mannans formed by known species of *Hansenula.*

Candida pelliculosa was suggested as the imperfect stage of *Hansenula anomala,* but it forms a galactomannan rather than a mannan, and the spectrum fits better into this group.

The spectrum of *Candida rhagii* mannan differs slightly from the others in this group, but resembles them more than the spectra of most other groups. It has several points of similarity to the spectrum of *Endomycopsis ovetensis* galactomannan. *Candida rhagii* was once considered synonymous with *C. tropicalis,* but the spectra of the mannans differ sufficiently that van Uden and Buckley (1970) are probably justified in maintaining it as a separate species. *Endomycopsis (Endomyces) ovetensis* has been suggested as the perfect stage of *Trichosporon sericeum. Trichosporon sericeum,* CBS 2545, however, formed a pentosylmannan (Gorin and Spencer, 1967).

Group 18. Species Producing Galactomannans which Have Not Been Placed in Other Groups

The spectrum of the galactomannan formed by *Trigonopsis variabilis* (Fig. 10, Group 18a) is unlike any other so far observed. As far as is known, there is no other yeast like this species, either.

Trichosporon penicillatum forms a galactomannan whose spectrum resembles that of the mannan formed by *Saccharomyces cerevisiae* (Fig. 10, Group 2b). Details of its chemical structure have not been determined.

Candida edax galactomannan (Fig. 10, Group 18c) has a spectrum like that of *Debaryomyces hansenii* mannan. The chemical structures of these two mannose-containing polysaccharides differ, however (Gorin and Spencer, unpublished results).

Candida steatolytica and *Candida lipolytica* form galactomannans (Gorin *et al.,* 1969b) with spectra which have some points of similarity (Fig. 10, Groups 18d and e). Yarrow (1969), who described the former species, stated that it resembled *Candida guilliermondii,* but this species forms a mannan. *Candida steatolytica* also resembles *C. lipolytica* in the ability to hydrolyze fat, and probably is more closely related to this species than to *C. guilliermondii. Candida ingens* forms a mannan. The spectrum resembles that of *C. lipolytica* galactomannan, and is included in this group.

Trichosporon fermentans, Geotrichum candidum, Endomyces reessii, and *Torulopsis vanderwaltii* all form galactomannans having identical spectra (Fig. 10, Group 18g). The relationships among these species, if any, are not known. *Torulopsis vanderwaltii*, however, has a carbon and nitrogen assimilation pattern that is the same as that of *T. fermentans*. If these two species are in fact mycelial and nonmycelial forms of the same organism, it shows that the ability of a yeast to form true mycelium, arthrospores, or pseudomycelium may be totally irrelevant for its classification. The spectra of *T. fermentans* and *C. lipolytica* galactomannans are closely related.

Torulopsis versatilis, Torulopsis mannitofaciens, and *Torulopsis halophila* form galactomannans with identical spectra (Fig. 10, Group 18h). *Torulopsis mannitofaciens* and *T. halophila*, isolated in Japan from salty environments, may prove to be variants of *T. versatilis*.

Candida tepae, Candida salmanticensis, Candida blankii, and *Candida incommunis* form galactomannans. The spectra of the polysaccharides from the first- and second-named yeasts have some similarities as do the spectra typical of the galactomannans of the third and fourth yeasts. The spectra are unusual in having strong low-field signals, at $\tau 4.12$ in one case and $\tau 4.20$ in the other (Fig. 10, Group 18j, 1-4). *Candida blankii* and *C. incommunis* both use inositol as a sole carbon source, which is not a common characteristic of yeasts.

Nadsonia elongata and *Nadsonia fulvescens* constitute another group of apiculate yeasts, differing from the other apiculate species in forming a galactomannan. The spectrum of *N. elongata* galactomannan is slightly simpler than that of *N. fulvescens* galactomannan (Fig. 10, Group 18k, 1 and 2). *Torulopsis domercqii* and *Trichosporon hellenicum* form galactomannans whose spectra are identical with the spectrum of the *N. fulvescens* galactomannan (Gorin and Spencer, 1968a). No relationship among these species has been otherwise demonstrated.

Endomyces magnusii (Fig. 10, Group 18l) and *Dipodascus albidus* (Fig. 10, Group 18m) form galactomannans whose spectra are unique. *Torulopsis cantarelli* forms a polysaccharide or polysaccharides containing galactose, mannose, and glucose, whose spectrum (Fig. 10, Group 18n) is the same as that of the glucomannan formed by *Ceratocystis paradoxa* and some other *Ceratocystis* species (Fig. 10, Group 22j) except for the presence of some small high- and low-field signals. No other similarities between these two species have been observed. *Candida boidinii* forms a mannan whose spectrum most closely resembles that of *Hansenula platypodis* mannan, except for small signals at $\tau 4.53$, $\tau 4.56$, and $\tau 4.59$. The spectrum of *H. platypodis* mannan has two small signals at τ4.44 and τ4.48, in this region. *Candida boidinii* assimilates nitrate, and may prove to be the imperfect form of a *Hansenula* species.

Group 19. The Rhodotorula-Sporobolomyces Group

Most members of this group, such as *Rhodotorula glutinis, Rhodotorula minuta, Sporobolomyces roseus, Torulopsis ingeniosa,* and *Puccinia graminis,* form a mannan, with (1 → 3) and (1 → 4) linked-β-D-mannopyranose units. Its spectrum (Fig. 10, Group 19a) has signals at approximately τ4.58 and τ4.69. *Sporobolomyces rubicundulus* forms a mannan whose spectrum has an additional signal at τ4.50. The chemical structure of the polysaccharide has not yet been determined.

Group 20. Miscellaneous Species Producing Mannans with Unusual Pmr Spectra

Torulopsis stellata (syn. *Torulopsis bacillaris*) forms a mannan whose spectrum (Fig. 10, Group 20a) has a single signal at τ4.55. The chemical structure of the mannan is not yet known. The spectrum of *Candida norvegensis* mannan has signals like those of the spectrum of *S. rouxii* mannan (Fig. 10, Group 20a). The sizes of the signals, however, differ considerably from the corresponding ones in the latter spectrum.

Candida aaseri mannan has a spectrum like that of *Saccharomyces fragilis* mannan, with an additional large signal at τ4.54.

Citeromyces matritensis and its imperfect form, *Torulopsis globosa,* form mannans whose spectra are identical and have unusual high-field signals at τ4.54 and τ4.61 due to a (1 → 6)-linked α-D-mannopyranose main-chain with (1 → 2)-linked β-mannopyranose side-chains (Gorin *et al.,* 1969a,c). There are also lower field signals at τ4.36, τ4.20, τ3.92, and τ3.87.

Torulopsis (Selenotila) peltata mannan gives rise to major signals at τ4.34 and τ4.40, with minor signals at τ3.76 and τ3.84. The spectrum of *Torulopsis wickerhamii* mannan has many points of similarity to it, but has additional signals at τ4.17 and τ4.52. The relationship between the two species has not been elucidated.

Saccharomycodes ludwigii is the only representative of the third group of apiculate yeasts. It forms a mannan with a very complex spectrum unlike any other examined.

Group 21. Species Producing Heteropolymers with Many Component Sugars

Some yeasts produce mannose-containing polysaccharides which contain such additional sugar residues as fucose, glucose, rhamnose, and pentoses. Most of these polysaccharides form too viscous a water solution for the pmr spectra to be determined. However, some of these yeasts form heteropolymers whose spectra can be determined. *Candida buffonii* (Fig. 10, Group 21a) forms one containing galactose, mannose, fucose, and rhamnose. Its spectrum has high-field signals at $\tau 4.62$ and $\tau 4.72$, like the spectrum of *Rhodotorula glutinis* mannan, but has an additional signal at $\tau 4.40$. There are also signals at $\tau 8.12$ and $\tau 8.18$ representing the C-methyl protons of fucose. The spectrum of *Candida diffluens* heteropolymer has points of similarity to that of *C. buffonii* polysaccharide, but has signals at $\tau 4.44$ and $\tau 4.50$.

Ustilago maydis, a smut which has a yeastlike phase under some conditions, forms a glucomannan whose spectrum has only high-field signals, as far as can be determined. Two of these, also, at $\tau 4.66$ and $\tau 4.76$, may correspond to the signals of the spectrum of *R. glutinis* mannan. The other signals are at $\tau 4.94$ and $\tau 5.00$.

Group 22. *Ceratocystis* and *Graphium* Species Producing Rhamnomannans and Glucomannans

Subgroups 22a to 22h (Fig. 10) include species which produce rhamnomannans whose spectra have a major H-1 signal at approximately $\tau 4.50$ and C–CH$_3$ signals of $\tau 8.12$ and $\tau 8.18$. The spectra have also up to five minor signals which are characteristic of the different subgroups. *Ceratocystis ulmi*, for instance, the pathogen of Dutch Elm disease, forms a rhamnomannan whose spectrum has signals at $\tau 4.28$, $\tau 4.65$, $\tau 4.73$, and $\tau 4.81$, besides the major signal at $\tau 4.50$, and the spectrum of *Ceratocystis pluriannulata* has an additional signal at $\tau 4.12$. The *C. ulmi* rhamnomannan contains a $(1 \rightarrow 6)$-linked α-D-mannopyranosyl main-chain substituted on the 3-positions mainly by single-unit α-L-rhamnopyranosyl side-chains (Gorin and Spencer, 1970b).

A much smaller group of *Ceratocystis* species forms glucomannans with a major signal at approximately $\tau 4.33$, $\tau 4.37$, and $\tau 4.41$, and a minor signal at $\tau 4.52$ (Fig. 10, Group 22j). *Ceratocystis brunnea* of this group forms a glucomannan with $(1 \rightarrow 6)$-linked α-D-mannopyranosyl main-chains substituted in the 2-positions mainly by α-D-glucopyranosyl side-chains (Gorin and Spencer, 1970b). *Ceratocystis coerulescens* and *Ceratocystis polonica* form glucomannans whose spectra are not so well defined, but which appear to belong in this group.

Group 23. *Ceratocystis* Species which Form Mannose-Containing Polysaccharides Having Unusual Spectra

Of this group, *Ceratocystis fimbriata* forms a galactomannan, and

Ceratocystis galeiformis forms a heteropolymer containing galactose, glucose, rhamnose, and mannose. *Ceratocystis moniliformis* forms a mannan whose spectrum is like that of *Saccharomyces fragilis* and *Endomycopsis capsularis* mannans. *Ceratocystis nigra* and *Ceratocystis obscura* form rhamnomannans whose spectra bear little resemblances to those of the rhamnomannans formed by other species of *Ceratocystis*.

The species of *Graphium*, the imperfect forms of *Ceratocystis*, are placed in the different groups of the latter, according to the spectra of their mannose-containing polysaccharides. Eventually it may be possible to place them in definite species of *Ceratocystis* on this basis.

While the major groups of *Ceratocystis* species form polysaccharides which contain mainly rhamnose and mannose, or glucose and mannose, many of the polysaccharides contain small amounts of galactose. Considerable proportions of glucose or galactose are present in some of the polysaccharides containing rhamnose and mannose. The monosaccharide composition of the polysaccharides, as well as the pmr spectra, may be used as a taxonomic aid.

Group 24. Miscellaneous Filamentous Fungi

Few of the species of filamentous fungi so far tested form mannans or mannose-containing polysaccharides. However, some species of *Alternaria*, *Microsporum*, *Mortierella* and *Nectria*, do.

Four species of *Alternaria* formed recoverable polysaccharides containing varying amounts of galactose, glucose, and mannose. All of the spectra had a major signal at approximately $\tau 4.52$. The spectrum of the polysaccharide from *Alternaria raplani*, which contained mannose, glucose, and a trace of galactose, had an additional major signal at $\tau 4.37$, with a shoulder at $\tau 4.34$. The other mannose-containing polysaccharides of *Alternaria* species had spectra with minor signals of various chemical shifts.

Microsporum quinckeanum forms two types of galactomannan having easily distinguishable spectra (Fig. 10, Groups 24e and 24f). Other dermatophytes of the genus *Trichophyton* have been investigated by Bishop and associates (1962, 1965, 1966; Blank and Perry, 1964) and found to each contain two galactomannans. These have not yet been examined using pmr spectroscopy.

Mortierella renispora formed a polysaccharide containing glucose, mannose, and a trace of galactose, whose spectrum had a major signal at $\tau 4.28$ and a smaller one at $\tau 4.50$. Further studies will be required before significant statements can be made concerning the taxonomic value of the mannose-containing polysaccharides formed by these and related fungal species.

IV. Summary and Conclusions

Groups 1, 2, and 3, composed mostly of *Saccharomyces* species, include at least two sets of interbreeding species. One of these includes those species recently placed in the genus *Kluyveromyces*, by van der Walt (1965b) or *Fabospora* (Phaff et al., 1966), and the other, consists of strains and species of *Saccharomyces cerevisiae*, *Saccharomyces carlsbergensis*, *Saccharomyces chevalieri*, *Saccharomyces italicus* and others. A number of species (Group 2a) have been placed in synonymy with *S. cerevisiae* and others may yet be reclassified in the same way. In addition, some of the species placed in these three groups include strains of different serotypes. The spectra of the mannans appear in Groups 2 and 3 *(S. chevalieri)* and Groups 1 and 3 *(S. exiguus)*. Other serotypic strains of *S. exiguus* produce mannans whose spectra are placed in Group 5. The pmr spectra of these mannans can probably be correlated with the serological reactions of the yeasts forming them. The genus *Saccharomyces* is a heterogeneous one, as was emphasized by the many types of pmr spectra obtained from the mannans formed by the different species (Spencer and Gorin, 1969c). Further study is required to show whether the chemical structures of the mannans are genetically connected with the fermentative and assimilative properties according to which these species are now classified.

The genera *Hansenula* and *Pichia* form a natural overlapping group, which differ only in the ability to assimilate nitrate (Wickerham, 1970). The various types of mannans they form, as exemplified by the pmr spectra, are shared by members of both genera (Spencer and Gorin, 1969b). It may eventually be shown that the utilization or nonutilization of nitrate is insufficient as a criterion for maintaining them as separate genera. *Pichia pinus*, for instance is identical with *Hansenula glucozyma* except for this characteristic (Spencer et al., 1970). The genus *Endomycopsis*, likewise, includes species which are similar to members of the genera *Hansenula* (Kreger-van Rij, 1964). The spectra of the mannans of these species [*Endomycopsis bispora* and *Endomycopsis platypodis*, classified by Wickerham (1951, 1968) as *Hansenula* species, for instance] are the same as those of a number of *Hansenula* and *Pichia* species (Fig. 10, Groups 9b, 9d, 11g), and support the probability of a close relationship of some species of *Endomycopsis* to *Hansenula* and *Pichia*.

The apiculate yeasts include the genera *Saccharomycodes*, *Nadsonia*, and *Hanseniaspora*, together with the asporogenous genus *Kloeckera*, which includes imperfect forms of *Hanseniaspora* species.

These three genera differ widely (Miller and Phaff, 1958) which is confirmed by the considerable differences in the pmr spectra of the mannose-containing polysaccharides (Spencer and Gorin, 1968). The species of *Hanseniaspora* and *Kloeckera* themselves, form mannans whose pmr spectra differ considerably. They resemble the spectra of the mannans formed by a number of species of *Saccharomyces* (Fig. 10, Parts 1,5g,6g). The *Brettanomyces* species also form mannans whose spectra are like those of the mannans of *Hanseniaspora osmophila*, *Kloeckera africana*, and *Kloeckera magna*. Van der Walt and van Kerken (1960) observed ascospore formation by *Brettanomyces bruxellensis*, *Brettanomyces intermedius*, and *Brettanomyces schanderlii*, and suggested that these and other *Brettanomyces* species would have to be transferred to an ascosporogenous genus. The spore shape coupled with the spectra of the mannans suggest that they may prove to be species of *Hanseniaspora*.

The genera *Metschnikowia* and *Debaryomyces* can be considered together on the basis of the similarities in the pmr spectra of their mannans (Spencer and Gorin, 1970d). While one genus consists of species which produce needle-shaped spores, and the other, round, warty spores, the spectra of the mannans of *Metschnikowia kamienskii*, *Metschnikowia krissii*, and *Metschnikowia zobellii* are the same as those of the mannans of *Debaryomyces hansenii* and other *Debaryomyces* species tested. *Metschnikowia pulcherrima* and *Metschnikowia reukaufii* form mannans with a slightly different spectrum, which is almost identical with the spectra of the mannans of *Pichia vini* and *Pichia haplophila*. *Pichia vini* was originally classified in the genus *Debaryomyces* but was removed from it by Kreger-van Rij (1964). The closeness of the relationship among these genera is not known. However, on the basis of the spectra of the mannans, the genus *Pichia* is heterogeneous, while the other two are nearly homogeneous. In spite of the difference in spore shape, the possibility of a relationship between the genera *Debaryomyces* and *Metschnikowia* should be investigated. In view of the heterogeneity of the genus *Pichia*, some of its species are probably quite closely related to *Metschnikowia* and *Debaryomyces* species as well.

Pigment production by *Rhodotorula* species was not regarded by Phaff and Spencer (1966) as a valid taxonomic criterion, which is in accordance with the data obtained from the study of the mannans. The group of species which form mannans having $\beta\text{-}(1 \to 3)$ and $\beta\text{-}(1 \to 4)$ linkages includes "normal" species of pigmented *Rhodotorula*, a nonpigmented asporogenous species *(Torulopsis ingeniosa)* and pigmented and nonpigmented species which discharge ballistospores

(*Sporobolomyces roseus* and *Bullera tsugae*). Also included in the group, on the basis of the mannans formed, are true Basidiomycetes, the parasites *Puccinia graminis tritici* (Klöker et al., 1965) and *Gymnosperma virginiae* (Gorin and Tulloch, unpublished results). Thus it is necessary either to accept a multiphyletic relationship for the yeasts or transfer this entire group to the Basidiomycetes, as was done with *Sporobolomyces* by Alexopoulos (1962).

The form genera *Candida* and *Torulopsis*, as previously stated, consist of species for which no reason can be found for transferring them to ascosporogenous genera. On the basis of the spectra of their mannose-containing polysaccharides, they can be grouped as (1) forms which produce mannose-containing polysaccharides whose spectra are like the spectra of those of ascosporogenous species, (2) forms producing mannose-containing polysaccharides with spectra like those of the polysaccharides of the species of other asporogenous genera, and (3) forms producing mannose-containing polysaccharides whose spectra do not resemble those of the polysaccharides of any other yeast studied so far.

Candida and *Torulopsis* species in the first group fall mostly into groups containing species of *Saccharomyces, Hansenula, Pichia*, and *Debaryomyces*, and will probably eventually prove to be imperfect forms of species in these genera. In the second group are a number of species which form galactomannans whose spectra are identical with that of *Trichosporon fermentans* galactomannan. *Torulopsis vanderwaltii*, in particular, resembles *T. fermentans* in the spectrum of its galactomannan and in its carbon and nitrogen assimilation patterns.

Pmr spectra of mannose-containing polysaccharides may prove useful as an aid in classification of filamentous fungi as well as yeasts, for those species which produce such polysaccharides. They are formed by *Ceratocystis, Graphium, Alternaria, Nectria, Trichophyton*, and *Mortierella* species, and differences in the pmr spectra of the polysaccharides formed by these species have been observed.

Pmr spectroscopy of invariant biopolymers may eventually be extended to compounds other than mannose-containing polysaccharides. Cohen (1969) has determined 100-MHz pmr spectra of human and hen egg-white lysozyme denatured with excess 2-mercaptoethanol in 8 M urea, which show differences in the aromatic region. This finding suggests that the method could be extended to make use of pmr spectra of microbial proteins as a taxonomic aid. Similar refinements of technique may eventually permit the determination of well-resolved pmr spectra of nucleic acids as well. The taxonomic position of a microbial species might then be determined primarily on the basis of chemical structure of its macromolecular constituents, as indicated by their pmr spectral "fingerprints."

REFERENCES

Abercrombie, M. J., Jones, J. K. N., Lock, M. V., Perry, M. B., and Stoodley, R. J. (1960). *Can. J. Chem.* **38**, 1617.
Alexopoulos, C. J. (1962). "Introductory Mycology," 2nd ed. Wiley, New York.
Aschner, M., Mager, J., and Leibowitz, J. (1945). *Nature* **56**, 295.
Bak, A. L., and Stenderup, A. (1969). *Abstr. 3rd Int. Symp. Yeasts, Antonie van Leeuwenhoek Suppl.* **35**, A11.
Banno, I. (1967). *J. Gen. Appl. Microbiol.* **13**, 167.
Barnett, J. A. (1968). *In* "The Fungi" (G. C. Ainsworth and A. S. Sussmann, eds.), Vol. III. Academic Press, New York.
Belozersky, A. N., and Spirin, A. S. (1960). *In* "The Nucleic Acids" (E. Chargaff, and J. N. Davidson, eds.), Vol. 3, p. 147. Academic Press, New York.
Benham, R. W. (1931). *J. Infec. Dis.* **49**, 185.
Bhattacharjee, S. S., and Gorin, P. A. J. (1969). *Can. J. Chem.* **47**, 1207.
Bishop, C. T., Blank, F., and Hranisovljevic-Jakovljevic M. (1962). *Can. J. Chem.* **40**, 1816.
Bishop, C. T., Perry, M. B., Blank, F., and Cooper, F. P. (1965). *Can. J. Chem.* **43**, 30.
Bishop, C. T., Perry, M. B., and Blank, F. (1966). *Can. J. Chem.* **44**, 2291.
Blank, F., and Perry, M. B. (1964). *Can. J. Chem.* **42**, 2862.
Bradbury, E. M., and Crane-Robinson, C. (1968). *Nature* **220**, 1079.
Claisse, M. L. (1969). *Abstr. 3rd Int. Symp. on Yeasts, Antonie van Leeuwenhoek Suppl.* **35**, I 21.
Clare, B. G. (1963). *Nature* **200**, 803.
Cohen, J. S. (1969). *Nature* **223**, 43.
Einbinder, J. M., Benham, R. W., and Nelson, C. T. (1954). *J. Invest. Dermatol.* **22**, 279.
Elinov, N. P., and Vitovskaya, G. A. (1965). *Biokhimiya* **30**, 933.
Estabrook, R. W. (1966). *In* "Hemes and Hemoproteins" (B. Chance, R. W. Estabrook, and T. Yonetani, eds.), p. 405. Academic Press, New York.
Falcone, G., and Nickerson, W. J. (1956). *Science* **124**, 272.
Falkow, S., Ryman, I. R., and Washington, O. (1962). *J. Bacteriol.* **83**, 1318.
Foda, M. S., and Phaff, H. J. (1969). *Abstr. 3rd Int. Symp. Yeasts, Antonie van Leeuwenhoek Suppl.* **35**, H9.
Frechet, J., and Schuerch, C. (1969). *J. Amer. Chem. Soc.* **91**, 1161.
Fukazawa, Y., and Tsuchiya, T. (1969). *Abstr. 3rd Int. Symp. Yeasts, Antonie van Leeuwenhoek Suppl.* **35**, E7.
Garber, E. D., and Rippon, J. W. (1968). *Advan. Appl. Microbiol.* **10**, 137.
Gillespie, D., and Spiegelman, S. (1965). *J. Mol. Biol.* **12**, 829.
Goldstein, I. J., Hay, G. W., Lewis, B. A., and Smith, F. (1959). *Abstr. Pap. Amer. Chem. Soc. Meet.* **135**, 3D.
Gorin, P. A. J., and Perlin, A. S. (1956). *Can. J. Chem.* **34**, 1796.
Gorin, P. A. J., and Spencer, J. F. T. (1967). *Can. J. Chem.* **45**, 1543.
Gorin, P. A. J., and Spencer, J. F. T. (1968a). *Can. J. Chem.* **46**, 2299.

Gorin, P. A. J., and Spencer, J. F. T. (1968b). *Can. J. Chem.* **46**, 3407.
Gorin, P. A. J., and Spencer, J. F. T. (1970a). *Can. J. Chem.* **48**, 198.
Gorin, P. A. J., and Spencer, J. F. T. (1970b). *Carbohyd. Res.* **13**, 339.
Gorin, P. A. J., Horitsu, K., and Spencer J. F. T. (1965). *Can. J. Chem.* **43**, 950.
Gorin, P. A. J., Spencer, J. F. T., and MacKenzie, S. L. (1966). *Can. J. Chem.* **44**, 2087.
Gorin, P. A. J., Mazurek, M., and Spencer, J. F. T. (1968). *Can. J. Chem.* **46**, 2305.
Gorin, P. A. J., Spencer, J. F. T., and Bhattacharjee, S. S. (1969a). *Can. J. Chem.* **47**, 1499.
Gorin, P. A. J., Spencer, J. F. T., and Magus, R. J. (1969b). *Can. J. Chem.* **47**, 3569.
Gorin, P. A. J., Spencer, J. F. T., and Eveleigh, D. E. (1969c). *Carbohyd. Res.* **11**, 387.
Haworth, W. N. (1915). *J. Chem. Soc.* p. 8.
Haworth, W. N., Heath, R. L., and Peat, S. (1941). *J. Chem. Soc.* p. 833.
Hudson, C. S. (1938). *J. Amer. Chem. Soc.* **60**, 1537.
Ingle, T. R., and Whistler, R. L. (1965). *Methods Carbohyd. Chem.* **5**, 411.
Jeanes, A. R., and Watson, P. R. (1962). *Can. J. Chem.* **40**, 1318.
Jeanes, A. R., Pittsley, J. E., Watson, P. R., and Dimler, R. J. (1961). *Arch. Biochem. Biophys.* **92**, 343.
Jeanes, A. R., Pittsley, J. E., Watson, P. R., and Sloneker, J. H. (1962). *Can. J. Chem.* **40**, 2256.
Jones, G. H., and Ballou, C. E. (1968a). *J. Biol. Chem.* **244**, 1052.
Jones, G. H., and Ballou, C. E. (1968b). *J. Biol. Chem.* **244**, 1043.
Klöker, W., Ledingham, G. A., and Perlin, A. S. (1965). *Can. J. Biochem.* **43**, 1387.
Kreger-van Rij, N. J. W. (1964). A Taxonomic Study of the Yeast Genera *Endomycopsis*, *Pichia* and *Debaryomyces*. Thesis, Leiden.
Kuhn, R., Trischmann, H., and Löw, I. (1955). *Angew. Chem.* **67**, 32.
Lee, K. Y., Wahl, R., and Barbu, E. (1956). *Ann. Inst. Pasteur Paris* **91**, 212.
Lee, Y-C., and Ballou, C. E. (1965). *Biochemistry* **4**, 257.
Lodder, J., and Kreger-van Rij, N. J. W. (1952). "The Yeasts, A Taxonomic Study." Wiley (Interscience), New York.
Lodder, J., Slooff, W. Ch., and Kreger-van Rij, N. J. W. (1958). In "The Chemistry and Biology of Yeasts" (A. H. Cook, ed.), pp. 1–62. Academic Press, New York.
Margoliash, E., and Schejter, A. (1966). *Advan. Protein Chem.* **21**, 113.
Marmur, J., and Doty, P. (1962). *J. Mol. Biol.* **5**, 109.
Marmur, J., Falkow, S., and Mandel, M. (1963). *Ann. Rev. Microbiol.* **17**, 329.
Masler, L., Sikl, D., and Bauer, S. (1968). *Collect. Czech. Chem. Commun.* **33**, 942.
Mill, P. J. (1966). *J. Gen. Microbiol.* **44**, 329.
Miller, M. W., and Phaff, H. J. (1958). *Mycopathol. Mycol. Appl.* **10**, 113.
Miyazaki, T. (1961). *Chem. Pharm. Bull.* **9**, 826.
Nakase, T., and Komagata, K. (1968). *J. Gen. Appl. Microbiol.* **14**, 345.
Nakase, T., and Komagata, K. (1969). *J. Gen. Appl. Microbiol.* **15**, 85.
Northcote, D. H., and Horne, R. W. (1952). *Biochem. J.* **51**, 232.
Peat, S., Whelan, W. J., and Edwards, T. E. (1961). *J. Chem. Soc.* p. 29.
Phaff, H. J., and do Carmo-Sousa, L. (1962). *Antonie van Leeuwenhoek J. Microbiol. Serol.* **28**, 193.
Phaff, H. J., and Spencer, J. F. T. (1966). *Proc. 2nd Symp. Yeasts,* p. 59. Publishing House of the Slovak Academy of Sciences, Bratislava.
Phaff, H. J., Miller, M. W., and Mrak, E. M. (1966). "The Life of Yeasts." Harvard Univ. Press, Cambridge, Massachusetts.
Rolfe, R., and Meselson, M. (1959). *Proc. Nat. Acad. Sci. U.S.* **45**, 1039.
Scheda, R., and Yarrow, D. (1966). *Arch. Mikrobiol.* **55**, 209.

Schildkraut, C. L., Marmur, J., and Doty, P. (1961). *J. Mol. Biol.* **3**, 595.
Sentandreu, R., and Northcote, D. H. (1968). *Biochem. J.* **109**, 419.
Sentandreu, R., and Northcote, D. H. (1969). *Carbohyd. Res.* **10**, 584.
Slodki, M. E. (1962). *Biochim. Biophys. Acta* **57**, 525.
Slodki, M. E. (1963). *Biochim. Biophys. Acta* **69**, 96.
Slodki, M. E., and Wickerham, L. J. (1966). *J. Gen. Microbiol.* **42**, 381.
Slodki, M. E., Wickerham, L. J., and Cadmus, M. C. (1961). *J. Bacteriol.* **82**, 269.
Spencer, J. F. T., and Gorin, P. A. J. (1968). *J. Bacteriol.* **96**, 180.
Spencer, J. F. T., and Gorin, P. A. J. (1969a). *Antonie van Leeuwenhoek J. Microbiol. Serol.* **35**, 33.
Spencer, J. F. T., and Gorin, P. A. J. (1969b). *Can. J. Microbiol.* **15**, 375.
Spencer, J. F. T., and Gorin, P. A. J. (1969c). *Antonie van Leeuwenhoek J. Microbiol. Serol.* **35**, 361.
Spencer, J. F. T., and Gorin, P. A. J. (1970a). *Antonie van Leeuwenhoek J. Microbiol. Serol.*, in press.
Spencer, J. F. T., and Gorin, P. A. J. (1970b). *Antonie van Leeuwenhoek J. Microbiol. Serol.*, in press.
Spencer, J. F. T., and Gorin, P. A. J. (1970c). *Mycologia.*, in press.
Spencer, J. F. T., and Gorin, P. A. J. (1970d). *Antonie van Leeuwenhoek J. Microbiol. Serol.* **36**, 135.
Spencer, J. F. T., Gorin, P. A. J., and Wickerham, L. J. (1970). *Can. J. Microbiol.* **16**, 445.
Stenderup, A., and Bak, A. L. (1968). *J. Gen. Microbiol.* **52**, 231.
Storck, R. (1966). *J. Bacteriol.* **91**, 227.
Sueoka, N., Marmur, J., and Doty, P. (1959). *Nature* **183**, 1427.
Tsuchiya, T., Fukazawa, Y., and Kawagita, S. (1965). *Mycopathol. Mycol. Appl.* **26**, 1.
van der Walt, J. P. (1965a). *Antonie van Leeuwenhoek J. Microbiol. Serol.* **31**, 189.
van der Walt, J. P. (1965b). *Antonie van Leeuwenhoek J. Microbiol. Serol.* **31**, 341.
van der Walt, J. P., and van Kerken, A. E. (1960). *Antonie van Leeuwenhoek J. Microbiol. Serol.* **26**, 292.
van Uden, N., and Buckley, H. (1970). "The Yeasts, A Taxonomic Study," 2nd ed., in press. Wiley (Interscience), New York.
van Uden, N., and Vidal-Lieria, M. (1970). "The Yeasts, A Taxonomic Study," 2nd ed., in press. Wiley (Interscience), New York.
Walker, P. M. B. (1969). *Prog. Nucleic Acid Res. Mol. Biol.* **9**, 301.
Wickerham, L. J. (1951). *U.S. Dep. Agr. Washington, D.C., Tech. Bull. No.* **1029**.
Wickerham, L. J. (1968). *Mycopathol. Mycol. Appl.* **32**, 15.
Wickerham, L. J. (1970). "The Yeasts, A Taxonomic Study," 2nd ed., in press. Wiley (Interscience), New York.
Wickerham, L. J., and Burton, K. A. (1948). *J. Bacteriol.* **56**, 363.
Wickerham, L. J., and Burton, K. A. (1961). *J. Bacteriol.* **82**, 265.
Wolf, F. A., and Wolf, F. T. (1947). "The Fungi," Vols. I and II. Wiley, New York.
Yarrow, D. (1969). *Antonie van Leeuwenhoek J. Microbiol. Serol.* **35**, 24.

Large-Scale Cultivation of Mammalian Cells

R. C. Telling and P. J. Radlett

*Animal Virus Research Institute,
Pirbright, Surrey, England*

I.	Introduction	91
II.	Selection, Storage, and Maintenance of Cells	92
	A. Cell Types	92
	B. Spontaneous Cell Transformation	93
	C. Maintenance and Storage	94
III.	Production of Culture Media	99
	A. Composition	99
	B. Preparation	101
	C. Sterilization	101
	D. Sterility Testing	103
IV.	Mass Cultivation of Cells	104
	A. Monolayer Cultures	104
	B. Micro-Carrier Culture	105
	C. Submerged Culture	107
V.	Conclusions	115
	References	116

I. Introduction

Since the time Earle and his associates pioneered the mass *in vitro* cultivation of mammalian cells, many types of cells of both normal and tumor origin have been grown in a variety of culture systems. The early demands for mass culture arose from the expanding fields of virus and cancer research. These required large quantities of cells for the isolation and identification of viruses, the production of viral vaccines, and the screening of antiviral and anticancer agents. Major emphasis has been on the use of tissue culture as a tool for virus production, which has provided a prime impetus for the developments in large-scale operations. Perhaps of no less importance is the potential use of mass culture for the production of chemicals unique to animal cells such as hormones, interferons, and antibodies. Many cells will grow only as monolayers attached to a surface and, until recently, their mass cultivation was confined to large individually sealed bottles. However, many of the problems associated with large-scale monolayer culture have been overcome by developing equipment containing multi-surfaces within a single vessel (Molin and Hedén, 1969; Weiss and Schleicher, 1968).

Because the growth of mammalian cells and microorganisms is similar, there was early interest in growing cells in an agitated suspension culture. Not all cells will grow in this manner but, for those that will, the expertise gained from the submerged culture of bacteria and fungi has obvious application and has been used successfully by many workers. Currently an important practical limitation of submerged culture is that it cannot be used for the production of virus for human vaccines. The human diploid cells considered suitable for this purpose will not grow in suspension culture. Unless there is a fundamental difference in the way normal cells divide, it is reasonable to speculate that this is a transient difficulty which may possibly be resolved by better definition of nutritional and environmental requirements. Meanwhile the work of van Wezel (1967) has demonstrated the feasibility of growing cells on micro-carriers of Sephadex beads kept in suspension in a stirred vessel. In this way the improved environmental control obtainable in stirred culture vessels can be utilized for the cultivation of those cells that will not grow in suspension.

In general, the problems of growing mammalian cells are similar to those encountered in the cultivation of microorganisms. These are essentially: the selection, maintenance, and storage of cells; the selection of the nutrient medium and the control of cultural conditions; and the design of equipment and process operation to reduce the risks of microbial contamination.

II. Selection, Storage, and Maintenance of Cells

Since the development of cell culture techniques early in the present century, cell populations capable of multiplying indefinitely *in vitro* have been derived from many different mammalian tissues, and the number of more finite cultures investigated is countless. The use of cell cultures in large-scale operations has emphasized the need for careful control over the conditions under which cells are stored and maintained; this section outlines some of the associated problems.

A. Cell Types

1. Primary Cultures

Primary cultures are those employing cells obtained directly from embryonic, adult, or tumor tissues. Free cells are normally obtained from primary tissue by enzymatic disaggregation, resuspended in nutrient medium, and inoculated into vessels to form monolayers on suitable surfaces. Primary cells kept in free suspension may remain viable but do not generally multiply. The techniques used for the cultivation of primary cells have been described in detail by Merchant *et al.* (1960), and Paul (1965) and discussed by Perlman and Guiffre (1963) and Perlman (1967).

2. *Cell Strains*

Hayflick and Moorhead (1961) have defined a cell strain as a population of normal cells subcultivated *in vitro* more than once but lacking the property of indefinite serial passage, while retaining the karyotype of the original tissues. These workers have concluded that the history of such a strain may be divided into an early growth phase, followed by a period of rapid cell multiplication, leading to a phase of declining mitotic activity. To explain this they advance a concept of senescence at the cellular level and postulate the existence of a factor necessary for cell survival but with a rate of duplication less than that of the cell. The fact that cell strains have a diploid karyotype, are inexpensive to procure, and can be tested for the presence of extraneous viruses and tumorigenic activity has led several workers to suggest their use for the production of human virus vaccines in place of primary cell cultures (Hayflick, 1963; Majer, 1967). So far, attempts to grow normal human cells with diploid karyotypes in suspension culture have failed and mass cultivation is confined to monolayer systems.

3. *Cell Lines*

A cell line has been defined as a population of cells which may be grown *in vitro* for an indefinite period of time. Cell lines are rarely diploid and it has been suggested that the heteroploid condition is a necessary characteristic. However, Macpherson and Stoker (1962) have described a line of hamster fibroblasts (BHK 21) that are diploid but capable of growth for an apparently indefinite period of time. Likewise, Moore and Ulrich (1965) have grown a diploid human tumor cell line with a normal karyotype in various culture vessels for many weeks.

B. Spontaneous Cell Transformation

1. *Diploid Cell Strain to Cell Line*

When a primary cell is passaged in culture for some weeks a phase of declining mitotic activity usually develops. In most cases the cells die out, but sometimes after a variable period new cells emerge which have a more uniform morphology and altered growth properties. Pulvertaft and his colleagues (1959) reported morphological transformation of their human thyroid cultures in 3-6 weeks and considered that this was a true transformation and not just the selective survival of a few cells. The mouse embryo cells studied by Todaro and Green (1963) developed into cell lines within 3 months of primary culture and, although the cells associated with the increase in growth rate were diploid, chromosome abnormalities soon developed.

Stoker and Macpherson (1964) have pointed out that the frequency with which cell lines emerge from normal tissue seems to be a function of their species of origin. Although normal human fibroblasts survive for many months, cell lines are not formed (Hayflick and Moorhead, 1961). On the other hand, mouse cells frequently give rise to aneuploid variants capable of indefinite growth (Levan and Biesele, 1958; Todaro and Green, 1963). Cell lines derived from a third group—including cells from the Chinese hamster (Yergaman and Leonard, 1961), Syrian hamster (Macpherson and Stoker, 1962), and pigs (Ruddle, 1962)—are not markedly aneuploid.

2. *Adaptation to Suspension Culture*

Both primary cells and diploid cell strains lack the ability to grow in suspension culture and it has been suggested that only cells with carcinogenic potential will propagate in this way. It was noticed at an early stage that some cells will readily adapt to suspension culture (Cherry and Hull, 1960; Cooper *et al.*, 1959; Earle *et al.*, 1954; Graff and McCarty, 1957; Ziegler *et al.*, 1958); others prove more difficult and it may be necessary to resuspend cells in fresh medium many times before growth occurs.

The adaptation of hamster kidney cells (BHK 21) to suspension culture has been described by Capstick *et al.* (1962), who reported irregular growth with considerable clumping and growth on the vessel wall during the first 21 days in culture. In a later report, Capstick *et al.* (1966) described the adaptation of a diploid and an aneuploid substrain of BHK 21 cells to suspension culture. The aneuploid substrain adapted readily with little evidence of clumping and a regular growth rate was established in 22 days. In contrast, the diploid substrain grew irregularly and 40 days elapsed before consistent growth was achieved. At this time the strain remained predominantly diploid but there followed a rapid increase in aneuploidy until at 75 days most cells were aneuploid.

C. MAINTENANCE AND STORAGE

The many difficulties associated with the continuous cultivation of mammalian cells can be broadly divided into two groups: (1) the prevention of contamination, and (2) the retention of desired characteristics. These problems can be minimized by establishing a convenient handling scheme for cell production and instituting suitable quality control procedures.

1. Microbial Contamination of Cultures

Despite the protection afforded by modern antibiotics, microbial contamination is still a constant problem and of particular importance in large-scale operations, where failures can be costly. Although the use of antibiotics has greatly simplified the handling of tissue cultures, it should be emphasized that their use is no substitute for aseptic technique.

a. Bacteria, Fungi, and Yeasts

Many laboratories routinely add penicillin and streptomycin to culture media to suppress bacterial growth. The use of neomycin has been suggested to suppress those bacteria resistant to penicillin and streptomycin (Capstick et al., 1962; Cooper et al., 1959; Cruikshank and Lawburg, 1952), while nystatin (Mycostatin) and amphotericin B have been used to control fungal and yeast contamination (Capstick et al., 1962; Cooper et al., 1959; Perlman et al., 1961). Since many cell lines are themselves sensitive to these antibiotics, care must be taken to use only the minimum effective concentration.

b. Mycoplasma

The first reported isolation of mycoplasmas from mammalian cell lines was by Robinson and his associates (1956). They have been shown to reduce cell growth rate, produce cytopathogenic effects, alter cell morphology, and produce chromosome changes in tissue cultures. The source of primary contamination in many cases is probably the upper respiratory tract of man and the presence of contamination may be undetected for several months.

Mycoplasma may be detected by changes in the appearance of cells, by arginine depletion of the medium (Pollock et al., 1963), by other biochemical reactions, or by isolation on artificial medium. These methods have been reviewed by Macpherson (1966). It will suffice here to say that no one method will detect all mycoplasma types and attempts should be made to isolate contaminants by a number of methods before declaring a culture free from contamination.

Many methods have been suggested for the elimination of mycoplasmas from contaminanted tissue cultures. These methods include heat treatment (Hayflick, 1960) and the use of antibiotics such as kanamycin (Fogh and Hacker, 1960; Pollock *et al.*, 1960), tetracycline (Carski and Shepard, 1961), and novobiocin (Johnson and Orlando, 1967). One of the most useful guides to the selection of suitable antibiotics is provided by the work of Perlman and his associates (1967), who investigated the antibiotic sensitivity of eight mycoplasma strains against numerous antibiotics and studied the toxicity of these antibiotics for cells in cultures.

c. Viruses

The use of tissue culture for the production of human virus vaccines has accentuated the need to screen tissue cultures for latent viruses. No less than 20 serologically distinct viruses have been recovered from apparently normal primary monkey kidney tissue culture (Hull *et al.*, 1958; Sweet and Hilleman, 1960). The occurrence of contamination by latent virus is, however, not restricted to primary cell cultures and it has been shown that certain viruses, notably SV 40 and polyoma virus, can transform diploid cell strains to heteroploid cell lines (Hull *et al.*, 1958; Macpherson, 1963). The detection of latent virus remains a problem, although the technique of co-cultivation with other, virus-susceptible cells has been successfully used by a number of workers (Black, 1966; Gerber, 1966; Watkins and Dulbecco, 1967).

d. Other Cells

Contamination of one cell culture by another may go undetected for months. There have been reports of "primate" cell lines which turned out to be mixtures of primate and murine cells (Kunin *et al.*, 1960) and the ERK/KD cells said to have been isolated from embryonic rabbit kidney are now believed to be HeLa cells (Sargeant, 1968). Identification of cell lines is difficult but karyological and immunological studies are probably the most useful techniques presently available. Of the more interesting tests which have been used, Coombs and his associates (1961) performed mixed agglutination reactions to identify cell strains derived from pig and rabbit tissue, while Stulberg and workers (1961) used immunofluorescence to determine the species of origin for their cell line.

2. Serial Subculture

Cell cultures in routine use are generally maintained by serial subculture. The frequency of subculture depends on the growth rate under the particular conditions used but frequently either a change of medium or subculture is required twice weekly. Cells may be stored for short periods of time at +4°C, when metabolism is greatly reduced and the cells require less frequent medium changes to remain viable. A useful practice is to store any surplus production of cells at +4°C ready for immediate emergency use, should the main line become contaminated.

3. Low-Temperature Storage

Dipold cell strains have a finite lifespan, and to maintain stocks of any particular cell, some form of long-term storage is needed. Similarly, when growing cell lines, it is necessary to maintain a uniform cell stock of unchanging characteristics should the line be lost by contamination or its desirable characteristics be lost on continuous subcultivation. To surmount these problems some form of low-temperature storage is generally employed.

At a temperature of +4°C metabolic processes are slowed but not stopped and purely physical processes such as diffusion still proceed. To obviate the disadvantages of a finite lifespan *in vitro*, it is necessary to reduce the temperature to below −40°C; convenient temperatures for storage are −70° and −196°C (liquid nitrogen). The processes occurring during freezing have been reviewed by Robinson (1966) and Farrant (1966) and only a brief outline will be given here.

When an aqueous salt solution is frozen, only a proportion of the water is converted to ice and the concentration of salt dissolved in the remaining water increases. It is this increase in electrolyte concentration that seems to cause most cell damage during freezing, however, the formation of large intracellular ice crystals may also be responsible. The addition of glycerol or dimethyl sulfoxide (DMSO) to the culture medium helps to protect cells against these damaging changes by lowering the percentage of water frozen at any given temperature. Since both glycerol and DMSO may damage cells, it is desirable to add them to the medium immediately prior to freezing and dilute them out as quickly as possible on thawing.

Robinson (1966) has pointed out that growth of ice crystals still occurs at −70°C; HeLa cells, for example, rapidly lose viability when stored at this temperature. Since ice crystal growth ceases at −130°C he suggests that storage below this temperature is desirable. The rate of cooling is also considered critical and, although the optimum cooling rate probably varies with the medium and cell concerned, it is commonly held that a cooling rate of 1°C per minute is satisfactory.

A number of rules for the frozen storage of living cells were proposed by Muggleton (1963) and are listed below.
1. Freezing down to −20°C should be slow (1°C per minute is optimal).
2. Thawing should be as rapid as possible.
3. Cultures should be kept to small volumes to permit rapid heat transfer.
4. The addition of a protective substance (5–20% glycerol or 10% DMSO) to the suspending medium aids survival.
5. Reliable refrigeration equipment should be used.

It is interesting to note that, despite the obvious difficulties, BHK 21 cells stored for up to 3 years at −70°C in our laboratory have grown well on first revival and shown no change in growth characteristics or susceptibility to foot-and-mouth disease virus. The cells are stored in the presence of either 10% v/v glycerol or DMSO, the ampuls having been placed on the floor of the −70°C unit (thus aiding heat transfer) until frozen, after which they may be transferred to any part of the cabinet.

4. Handling Schemes

We have already suggested that some of the difficulties associated with the continuous cultivation of mammalian cells can be minimized by establishing a suitable handling scheme. Shedden and Wildy (1966) and Walker et al. (1968) have described schemes used in their own laboratories that have a number of points in common. In both cases a selected strain of cells is grown for use as a stock culture and stored at a low temperature in several aliquots. The cells forming this master bank are screened for contaminants and tested for their ability to produce the desired product. From the master bank a number of substrains are prepared, each of which is tested for purity and stored to form a secondary bank. An ampul from the secondary bank is then revived at regular intervals and grown up to produce a sufficient quantity of cells for use. It seems generally agreed that the addition of antibiotics to early cultures is inadvisable because their use increases the risk of masking contamination.

In our laboratory a similar scheme with BHK 21 cells has operated satisfactorily for many years. A master bank is stored in liquid nitrogen and the secondary bank is stored in 20-ml aliquots, each containing about 2×10^9 cells, in a number of mechanical refrigerators operating at −70°C. Cells from the secondary bank are revived at about monthly intervals and examined for microbial contamination, general characteristics, growth, and viral susceptibility. For use, the cells are subcultured by serial transfer until the next revival; in this way cell age during each successive month remains approximately constant. A method for the buildup of cells from cold storage to provide inoculum for large-scale suspension culture has been described by Telling and Elsworth (1965) and Telling (1969).

III. Production of Culture Media

A. COMPOSITION

The early systematic studies of the nutritional requirements for the cultivation of mammalian cells *in vitro* were begun by Fischer and continued during the 1950's by Parker, Eagle, Rappaport, Waymouth, and others. These workers developed chemically defined media of varying complexity, each consisting essentially of a balanced salt solution with additions of amino acids, vitamins, glucose, and sometimes galactose. Perlman and Guiffre (1963) have summarized the development of these chemically defined media which, however, still require the addition of serum for the growth of most cells.

The simplest of these early media, that of Eagle (1955, 1959), is still commonly used, although often modified for particular applications by the addition of further components such as peptones, lactalbumin hydrolyzate, yeast extract, tryptose phosphate broth, insulin, or increased amounts of amino acids and vitamins (Birch and Pirt, 1969; Griffiths and Pirt, 1967; Kilburn and Webb, 1968; Macpherson and Stoker, 1962; Pumper *et al.*, 1965; Telling and Radlett, 1969).

Most media have been developed for the growth of cells and contain nutrients essential for cell multiplication. Provided that other factors such as pH and oxygen are not inhibitory, the cell yields obtained will depend upon the amount of nutrients available and will be restricted as soon as any single nutrient becomes a limiting factor. The depletion of a nutrient can be overcome by changing the medium frequently, a device exemplified by the report of Bryant (1966), who obtained a fifty fold increase of mouse liver cells to a population density of 29×10^6 cells/ml by a daily change of medium. However, such a procedure is inconvenient in large-scale operations, particularly for submerged cultures, and it is obviously desirable to design the culture medium so that high cell densities can be achieved without further replenishment of nutrients. Griffiths and Pirt (1967) have drawn attention to the need for determining growth yields, i.e., the yield of cells per unit weight of nutrient utilized. These workers determined by analysis the minimum amounts of essential amino acids required to produce 10^6 mouse LS cells and found that leucine, isoleucine, and particularly glutamine could become growth-limiting in several standard tissue culture media.

Restriction of growth due to glutamine deficiency poses a difficulty in the development of improved media. Tritsch and Moore (1962) reported the instability of glutamine in tissue culture media. Incubating their medium at 37°C, they determined the rate of spontaneous decomposition of glutamine to pyrrolidone-carboxylic acid and ammonia to be 10% per day. Similar decomposition was reported by Griffiths and Pirt (1967), who also found that glutamine in excess of 10 m moles per liter inhibited the growth of their mouse LS cells even though the tonicity of the medium was kept constant.

In a later report on the nutritional requirements of mouse LS cells, Birch and Pirt (1969) stated that the maximum population of the cells was linearly proportional to the choline concentrations, with a growth yield of 3.2×10^5 cells/μg of choline chloride. Serum was the major source of choline in their original medium but, when the amount of choline chloride was increased, bovine serum albumen fraction V could be substituted for serum without reduction in growth. Limitation of growth due to choline deficiency has also been reported by Nagle (1969). By increasing the choline chloride in his chemically defined medium from 1 to 50 mg/liter, he found that the peak density of mouse L cells was more than doubled and the growth of HeLa cells improved.

In our laboratory growth-limiting factors in suspension culture of BHK 21 cells were found to be glutamine, glucose, and vitamins, and when these nutrients were approximately doubled there was more than a twofold increase in peak cell density to a maximum of 6×10^6 cells/ml (Telling and Radlett, 1969).

In addition to being a source of nutrition, serum protects cells from mechanical stress. Several workers have observed the occurrence of cell damage as a consequence of prolonged shaking, stirring, or sparging of gases in cultures containing little or no serum. Damage has been manifest in various ways, including complete disintegration of cells, increased permeability to trypan blue dye without immediate lysis, loss or reduction in metabolic activity, and greater fragility at low temperatures. As alternatives to serum, methylcellulose and Pluronic Polyol F68 have been used to protect cells against mechanical damage (Bryant, 1966; Bryant et al., 1960; Holmström, 1968; Kilburn and Webb, 1968; Runyon and Geyer, 1963; Telling and Radlett, 1969).

B. Preparation

Hayflick et al. (1964) have drawn attention to an important aspect of media preparation — that of its consistent manufacture to give reproducible cell growth within and between laboratories. They reported that the difficulty was often traceable to variations in reagent purity and in compounding a published formula. In their view a more uniform product is obtained by using powdered media prepared in large quantities. They advocate mixing the dry ingredients in the correct proportions and grinding to a free-flowing powder which, after desiccation, can be packaged until required for reconstitution. They found their method successful for several defined media and balanced salt solutions.

The production and storage of high purity water is an important and costly feature of large-scale media preparation. At the Roswell Park cell culture plant, double-distilled water is stored in a tin-lined tank fitted with an ultraviolet sterilizing system (Vosseller and Moore, 1969). Our practice is to use distilled water produced over stainless steel. Water with a resistivity of 4–5×10^5 ohms/cm has been found suitable for the range of primary cells and cell lines cultivated in our laboratory. The water is stored in closed stainless steel reservoirs which are vented through bacterial filters and fitted with an Ultraviolet source in the head space. Distribution is through stainless steel pipes with sterilizing grade filters at the outlets; both reservoirs and lines can be steam sterilized.

C. Sterilization

1. Filtration

Most tissue culture media contain heat-labile components and are therefore sterilized by filtration. Although the method is often restricted to the thermolabile constituents, filtration of complete medium is more convenient and economical for large-scale operations.

Portner et al. (1967) examined the probability of sterilizing by filtration and concluded that, if a high degree of certainty is required, the liquid should be passed through two or more filters as a safeguard against a defective filter sheet. Double filtration of antibiotic-free medium through Seitz-type sheets is used for the large-scale continuous cell culture plant at the Roswell Park Memorial Institute (Vosseller and Moore, 1969). At the Pirbright laboratories reliable sterilization of complete growth medium by single filtration of volumes up to 200 liters has been demonstrated by Telling and Elsworth (1965) and Telling et al. (1966). These reports emphasize the need for proper attention to design and operation of equipment and restriction of the degree of microbial contamination in the unfiltered medium, particularly when using asbestos-cellulose depth filters. Features of the system include permanently piped filter presses sterilized by steam injection, with the added precaution of maintaining an internal positive pressure throughout the cycle of sterilization, cooling, and filtration. In our opinion, these features are essential if risks of contamination are to be negligible.

The sterilizing grade filter sheets commonly used are either the Seitz-type neutral asbestos–cellulose or membranes produced from cellulose esters. Both types have been shown to release substances inhibitory to cell growth and virus replication (Brown et al., 1965; House, 1964). Cahn (1967) reported that many membranes contain 2–3% of their dry weight as detergent. He found that the concentration of detergent in eluates of these filters is high enough to cause damage to cells cultured in filtered medium. Hot water effectively removes most of the detergent and our observations lead us to conclude that this is achieved in our systems by the hot condensate continually produced and drained during sterilization by steam injection.

The development of the cartridge filter, which can be sterilized *in situ* by steam injection, provides an alternative to the filter press for large-scale sterile filtration. The disposable filter cartridge is mounted in a stainless steel housing which can be fitted into pipe lines with screw or compression fittings. One such filter made of asbestos fibers bonded to a cellulose sheet (Ultipor .12, Pall Filter Corporation, Glen Cove, Long Island, New York) is now used to sterilize tissue culture media in our laboratory.

2. Chemical

Some observations on the chemical sterilization of tissue culture medium using β-propiolactone (BPL) have been reported by Toplin and Gaden (1961). They found that serum treated with BPL at a concentration of 0.15% v/v could be incorporated in Eagle's medium at the 10% level without apparent deleterious effects on HeLa or Hep #2 cells over 5–7 days. These workers emphasized the importance of ensuring complete hydrolysis of BPL by demonstrating that a concentration of about 0.001% unhydrolyzed BPL was toxic.

Sterilization of media by BPL has been examined at Pirbright (unpublished work). After allowing up to 96 hours for hydrolysis, growth of BHK 21 cells was reduced by half in medium treated with 0.02% BPL and completely inhibited in medium treated with 0.06%. Cell growth was also inhibited by the addition of hydrolyzed BPL to culture medium at a concentration of 0.1% v/v, the lowest concentration tested.

3. Heat

The convenience and reliability of autoclaving has encouraged several workers to develop serum-free and heat-stable media for the growth of several cell types (Nagle, 1968; Nagle et al., 1963; Orlando et al., 1968; Pumper et al., 1965). Richards (1968) draws attention to the effect of heat on the nutritional value of culture media. While heat sterilization of short duration may be beneficial in that it causes changes that would otherwise be brought about by metabolism, prolonged heat treatment can be seriously detrimental and result in decrease of cell yield. Thus, when sterilizing larger volumes of these heat-stable media it may be necessary to make some allowance for the increased heating and cooling times that occur.

D. Sterility Testing

Failures in large-scale tissue culture are costly, therefore, closer attention must be paid to sterility testing of nutrient medium than is normally necessary at the laboratory bench. Since no form of sterility testing can demonstrate the absence of all living organisms, only an operational definition of sterility can have meaning and, in practice, testing is usually restricted to the detection of bacteria and fungi likely to proliferate under working conditions.

As pointed out by Tritsch and Moore (1962), it is unwise to check sterility of tissue culture medium by incubating the whole batch at 37°C because of the spontaneous decomposition of glutamine that occurs at that temperature. Hayflick et al. (1964) proposed a sterility test procedure of holding medium (without glutamine and serum) at room temperature for 3–4 weeks. This method is clearly impractical when dealing with large volumes of medium, particularly if using factory-scale culture vessels, where of necessity sterility control must be based on sampling and the results interpreted in terms of probability.

A sampling procedure will detect gross contamination, but statistical considerations show that a low degree of contamination may not be revealed. This has been discussed by Elsworth et al. (1955), Herbert et al. (1956), and considered in detail by Peto and Maidment (1969). The latter briefly describe the theory of assigning an upper limit to the estimate of the degree of contamination and give examples of the conclusions that can be drawn from sampling. The argument is based on the assumption that contaminants are randomly distributed throughout the medium and that the possible upper limit of contamination is defined by the probability level (or confidence limit) of a mischance of sampling fixed by the investigator. They have compiled useful tables which show, for example, that if a sample volume of 1% is sterile there could be in the remaining medium a possible upper limit of 299 contaminants at the 5% confidence limit and up to 459 contaminants at the 1% confidence limit. Increasing the sample volume to 10% reduces the possible upper limits to 29 and 44 contaminants at the 5 and 1% confidence limits, respectively. In large-scale operations, sample volumes will rarely exceed 1% of the culture volume and, while the medium may well be sterile, sampling cannot prove it.

IV. Mass Cultivation of Cells

A. Monolayer Cultures

1. Static and Roller Bottles

The methodology of monolayer culture is well established and, despite the risks of contamination arising from multiple handling, large-scale production has been accomplished using both static and roller bottles. Examples of such systems have been described in detail by Bachrach and Polatnick (1968) and Ubertini et al. (1960, 1963).

A major disadvantage of monolayer culture in bottles is that cell growth is frequently limited by the development of an unfavorable pH value or the depletion of oxygen or nutrient supply. To surmount this difficulty, medium is often changed or the cultures gassed with a mixture of air and carbon dioxide, but in large-scale operations such procedures are laborious and introduce additional risks of contamination. Kruse and Miedema (1965) have shown the importance of a controlled environment to increase cell densities of monolayer cultures in bottles. Using a perfusion system (Kruse et al., 1963), they grew a variety of cells, layer upon layer, reporting, in one instance, a cell membrane thickness equivalent to 17 layers of cells.

2. Multi-Surface Propagators

These propagators provide within a single vessel a large surface area for the attachment and growth of cells as monolayers. For the mass cultivation of those cells that will not grow in suspension, this type of apparatus offers several advantages over the conventional static and roller bottle systems.

The apparatus described by Molin and Hedén (1969) is a liter jar containing 35 titanium disks, each of 9-cm diameter. Their method is to add the cell inoculum of 5×10^5 cells per disk in sufficient medium to fill the vessel, which is then incubated in the vertical position for 1 hour. This allows about 90% of the cells to settle and attach to the disks. Approximately half of the medium is then removed and the vessel placed horizontally so that each disk is vertical and about half-immersed. The disks are then rotated at one revolution every 2 minutes. In this system Molin and Hedén obtained a tenfold increase of a human diploid cell in 5 days.

In the propagators of Weiss and Schleicher (1968) a tier of glass disks provide the surface for cell attachment. The vessel is inoculated with cells in sufficient medium to immerse the disks and the culture is aerated for 15 minutes to distribute the cells and equilibrate the medium. The culture is aerated by directing sparged gases into an airlift pump which discharges just above the level of the culture. This arrangement forces medium through the pump, thus providing both aeration and circulation of the culture. After the initial mixing, aeration is halted for 2 hours to allow the cells to settle and attach to the glass disks. Control of pH, dissolved oxygen, and carbon dioxide during the culture period is by manual adjustment of the composition and flow rate of the air and carbon dioxide gas mixture.

Once confluent monolayers are formed, the growth medium is removed and the cells either harvested by trypsinization or left undisturbed for virus production. Schleicher and Weiss (1968) have used their apparatus for the growth of both primary cells and cell lines and for the production of viruses, interferon, and mycoplasma. Their propagators range in capacity from 1 to 200 liters and apparently can be readily adapted for continuous or intermittent flow of nutrients and fitted with electrodes to measure pH and dissolved oxygen. As an indirect measurement of cell growth, Schleicher and Weiss (1968) followed the utilization of glucose, which was reduced from 1.5 to 0.5 gm/liter to attain confluent monolayers on all surfaces.

These reports indicate that multi-surface units offer both a convenient method of mass cell production with a minimum of labor and a facility for improving the environmental conditions of monolayer cultures.

B. Micro-Carrier Culture

The use of micro-carriers was first reported by van Wezel (1967) and van Hemert *et al.* (1968) have reviewed some of the problems and possibilities of the technique. The method consists of growing cells as monolayers on the surface of positively charged Sephadex granules suspended in culture medium in a stirred vessel. In this manner different cell lines, human diploid cells, and primary rabbit kidney cells were successfully cultivated.

The stirred vessel used was the Bilthoven microbial culture unit (van Hemert, 1964) operating at a stirrer speed of 100 rpm. Culture volume was a maximum of 3 liters, using Eagle's minimum essential medium supplemented with calf serum and "Methocel." The micro-carriers were sterilized granules of DEAE-Sephadex about 100 μ in diameter. Maximum cell densities depend upon the available area of Sephadex but if the quantity of granules exceeded 2 mg/ml of culture a toxicity phenomenon was encountered. This difficulty was initially overcome by treating the Sephadex with serum, but later work suggested that treatment with collodion may be a satisfactory alternative.

The cell growth rates obtained in micro-carrier cultures were about the same as in monolayer culture but higher densities are possible by changing or perfusing the medium. To change the medium, the Sephadex is allowed to settle and the spent medium is siphoned off. For perfusion a volume of medium about 5 times the culture volume is continuously recycled, using a screen on the outlet pipe to prevent loss of Sephadex beads. This method yielded the best results, with maximum cell densities of 3×10^6/ml in a culture containing 4 mg/ml of Sephadex.

It is evident that the micro-carrier system is able to provide the advantages of controlled environment for those cells that will not grow in conventional suspension culture. The method has the added advantage that restricted growth due to nutrient deficiency may be overcome by changing or perfusing the medium. In the view of van Hemert *et al.* (1968), the technique provides an excellent solution for diploid cells but the preparation of primary tissue to give an inoculum of single viable cells, substantially free from debris, is a major difficulty. Treating the trypsinized cell suspension in a Vibromixer has improved the viable cell yields, but cell debris is not completely removed and the problem apparently needs further study.

The production of polio virus in micro-carrier culture was examined by van Wezel (1967) and he found that virus multiplication was essentially similar to that in monolayer culture. van Hemert *et al.* (1968) have discussed the potential use of the system for virus production and have outlined the different methods which can be used for the cultivation of cytopathogenic and noncytopathogenic viruses.

C. SUBMERGED CULTURE

1. Development and Uses

Since the first report on the growth of mammalian cells in "tumbling tube" suspension culture by Owens *et al.* (1954), many workers have described successful cell growth in submerged culture, but mostly in small volumes and often using equipment which cannot conveniently be scaled up. Much of the published work has been reviewed by Perlman and Guiffre (1963) and Perlman (1967), who summarized the equipment and cell strains used.

For small-scale laboratory studies most workers have found it convenient to use shake flasks or to agitate cultures with a simple rotating magnet (Cherry and Hull, 1960; McLimans *et al.*, 1957; Moore *et al.*, 1968; Smith and Burrows, 1963, 1966; Weirether *et al.*, 1968). When culturing larger volumes, the stirred vessels developed for the deep culture of bacteria are, with minor modifications, finding increasing application because they provide a facility for environmental control and a basis for scale-up (Holmström, 1968; Kilburn and Webb, 1968; McLimans *et al.*, 1957; Moore *et al.*, 1968; Telling and Stone, 1964; Telling and Elsworth, 1965; Ziegler *et al.*, 1958).

On the industrial scale, BHK 21 cells are now grown in submerged culture at volumes of up to 2000 liters and are used for the production of virus vaccines against foot-and-mouth disease and Newcastle disease. A large-scale culture plant for the growth of various cells of human origin has been described by Vosseller and Moore (1969). This plant may be operated on a semicontinuous or batch process and is said to be capable of producing 1–3 kg (presumably wet weight) of cells per day. The cell production section consists of four culture vessels ranging from 20 to 1250 liters in capacity and, as the associated piping and valving is complex, a valve programmer has been incorporated to minimize operating errors. Temperature and pH are controlled, while cell count, viability, glucose, lactic acid, gas phase CO_2, and dissolved oxygen tension are all measured.

Much of the published work on submerged cultivation of mammalian cells has concerned either a single batch culture or a succession of batch cultures in which a residuum of the old culture serves as inoculum for the next—a type of operation which Pirt and Callow (1964) have termed "solera culture." True continuous flow culture has been studied by several workers (Cohen and Eagle, 1961; Cooper *et al.*, 1969; Merchant *et al.*, 1960; Pirt and Callow, 1964). These authors reported irregular and unpredictable growth even under apparently constant environmental conditions. More recently, Griffiths and Pirt (1967) have reported predictable population densities for mouse LS cells. This enabled them to determine the effect of different cell growth rates on the quantitative amino acid requirements for growth. Their work illustrates the usefulness of the continuous flow culture technique for the correlation of cell growth with environmental conditions.

Apart from the obvious immediate uses of submerged culture for the production of cells to be used in virus vaccine manufacture, an important potential use lies in the biochemical conversion and production of hormones by mammalian cells. It has been established that cells from certain hormone-producing tissues will continue to produce hormones when grown in suspension culture. Examples are the production of gonadotrophin by human pituitary cells (Thompson et al., 1959) and the conversion of cortisol and progesterone to a wide range of steriod products by uterine fibroblastic cells (Sweat et al., 1958). More recently, Tashjian (1969) has described the production of a number of hormones from monolayer cell cultures and suggested that the scale of hormone production could be expanded by suspension culture techniques.

2. Environmental Conditions

a. Temperature

Cell growth rate and yield depend on culture temperature and for maximal growth this must be closely controlled, usually between 35° and 37°C. Capstick (1963) found the optimal incubation temperature of BHK 21 cells was between 33° and 37°C and that yields were depressed at 31°C, with little growth occurring at 39°C.

A common method for controlling the temperature of large culture vessels is by circulating water at incubation temperature from a constant temperature tank through the jacket of the culture vessel. Moore et al. (1968) have described a large, closed, water-circulating system piped to all vessels in their plant, but some workers consider the water circulation method too complicated (Elsworth and Stockwell, 1968; Telling and Elsworth, 1965). Elsworth and Stockwell (1968) describe a system of two-step control action with overlap, using electric heaters and a water film cooler on vessels up to 150 liters and, at the 400-liter scale, supplying heat to the vessel wall by steam. At Pirbright we have used electric heaters for cultures of BHK 21 cells up to the 100-liter scale and found the method satisfactory, indicating that any medium degradation or cell death caused by local overheating is not of practical significance.

b. pH value

It has been demonstrated many times that for maximum growth rate and cell production pH must be maintained within close limits. Optimal conditions are often about pH 7.2–7.4 and the development of electrode assemblies which can be steam sterilized repeatedly *in situ* have made pH control a routine matter.

It is common practice in tissue culture to incorporate a bicarbonate buffer into the medium and regulate pH by passing an air/CO_2 mixture over the culture during growth. A major difficulty associated with this technique is the varying concentration of CO_2 required to control pH as the growth cycle proceeds. McLimans et al. (1957) found that incorporation of 0.01 M phosphate in place of bicarbonate in the medium eliminated the need for a CO_2 overlay but it proved necessary to add alkali occasionally to keep the pH above 7.0.

The control system devised by Telling and Stone (1964) overcomes some of the problems of the bicarbonate–CO_2 system. Their method is to inject CO_2 into the culture automatically when pH rises too high and to inject air when pH falls too low. These workers used this method to control the pH of cultures of BHK 21 cells and obtained yields which were maximal and constant between pH 7.2 and 7.6. Telling and Radlett (1969) found that a practical limitation of the method was loss of pH control in cultures of BHK 21 cells when concentrations reached about 3×10^6/ml. They overcame this difficulty by adding more sodium bicarbonate during the growth cycle. In this way reasonable pH control was provided, exponential growth was maintained, and densities of 6×10^6 cells/ml were obtained. Disadvantages of the method are that it precludes the control of dissolved oxygen and the measurement of the carbon dioxide produced by metabolism.

The control of culture pH by automatic addition of acid or alkali has been applied by microbiologists for many years. More recently, Pirt and Callow (1964) and Kilburn and Webb (1968) have reported satisfactory control in cell cultures by this method for cell concentrations of 1.5 and 0.9×10^6 cells/ml, respectively. A number of workers have reported the importance of CO_2 in cell metabolism (Geyer and Chang, 1958; Swim and Parker, 1958; Chang et al., 1961), but the minimal essential concentration was not determined. This is a matter of some importance when replacing a bicarbonate–CO_2 with an acid–

alkali pH control system. Kilburn and Webb (1968), however, found for their mouse LS cells that growth was independent of pCO_2 in the range 1-70 mm Hg and that the normal pCO_2 in their fermentor after inoculation was 4 mm Hg. There was thus no need for pCO_2 control in the absence of the bicarbonate-CO_2 buffer system.

c. Oxygen

Oxygen is frequently supplied by passing a stream of air or an air/CO_2 mixture over the culture, but rates of oxygen solution from surface aeration are generally low. Hence, oxygen supply may limit both growth rate and peak cell densities obtainable, particularly since with increasing scale there is usually a smaller surface area relative to culture volume for gas diffusion.

Many studies on the influence of oxygen in animal cell cultures have been based on analysis of gas in the head space, but such estimates are not always valid because there may not be equilibrium between the gas and liquid phases. Cooper et al. (1958) recognized this difficulty when they reported oxygen concentrations of 25-30% to be rapidly cytocidal and concluded that the liquid phase oxygen level for fastest growth of their ERK cell line is much less than the air equilibrium value.

Definite confirmation of the importance of oxygen in the growth and metabolism of mammalian cells requires the measurement and control of dissolved oxygen tension (pO_2) during growth. Such an investigation was undertaken by Kilburn and his associates (Kilburn and Webb, 1968; Kilburn et al., 1969; Self et al., 1968), using a diffusion current electrode of the type described by Mackareth (1964). These workers found pO_2 markedly influenced the cell growth and glucose metabolism of their mouse LS cells. Maximum cell densities were obtained when pO_2 was controlled at values between 40 and 100 mm Hg and cell growth was seriously inhibited at 320 mm Hg. In cultures with oxygen supplied from a head space of air initially at 810 mm Hg absolute pressure, cell growth was limited by the availability of oxygen, pO_2 falling to less than 2 mm Hg before maximum cell density was reached.

At Pirbright, Telling and Radlett (1969) have examined the effect of different aeration systems on the growth of BHK 21 cells at the 4-liter scale and determined the cell yields per unit of glucose consumed (Table I). With surface air only, pO_2 fell to zero in 24 hours, both growth and growth rate were limited, and all the glucose was used. The continuous sparging of air at the rate of about 6 ml/liter of culture

TABLE I
EFFECT OF DIFFERENT AERATION SYSTEMS ON THE GROWTH OF
BHK 21 CELLS AT THE 4-LITER SCALE[a]

Aeration	Growth yields for glucose		
	Cells produced ($\times 10^{-6}$/ml)	Glucose used[b] (mg/ml)	10^6 Cells produced per mg glucose used
Surface air (500 ml/minute)	1.9	3.6	0.53
Continuous air sparging (25 ml/minute)	1.3	2.6	0.50
Air/CO_2 pH control	2.8	3.0	0.93
Controlled pO_2			
$\geq$ 8 mm Hg	3.2	3.5	0.91
8 mm Hg	3.1	3.4	0.91
$\geq$ 40 mm Hg	3.1	2.3	1.35
40 mm Hg	3.3	2.4	1.38
80 mm Hg	3.1	2.3	1.35

[a]Telling and Radlett, 1969.
[b]Starting glucose 3.6 mg/ml.

per minute was found to be detrimental to growth, growth rate, and cell viability. In cultures with pH controlled by the automatic injection of air and CO_2 gas, the pO_2 oscillated irregularly throughout the period of culture, but cell yield per unit of glucose consumed was similar to that obtained when pO_2 was controlled at 8 mm Hg. The most efficient utilization of glucose occurred when pO_2 was controlled at 40 and 80 mm Hg and was independent of whether the oxygen was supplied by the controlled sparging of air or by increasing the oxygen tension in the head space above the culture.

It is clear that cell growth in industrial-scale cultures may become limited by oxygen availability and improvements in culture media leading to higher cell densities will enhance this difficulty. Surface aeration is unlikely to be adequate and sparged air must be used with care. Difficulties with sterilization and stability of oxygen electrodes, however, preclude their routine use at the present time. Until these problems are resolved, pH control by intermittently injecting air into a bicarbonate buffered system when pH falls below the desired value is one method of increasing the oxygen availability in large-scale cultures.

d. Agitation

Mammalian cells are often irreparably damaged if agitated too vigorously. This fact, therefore, has prime consideration when selecting the agitator system. However, different cell types vary in their susceptibility to mechanical damage, in their tendency to aggregate, and in their rate of growth and, hence, rate of oxygen demand. It is thus clearly necessary to determine empirically the most suitable agitator system for a particular cell type.

Several cell types have been grown satisfactorily in culture vessels stirred by a single turbine impeller at low speeds, usually about 200–400 rpm (Kilburn and Webb, 1968; McLimans et al., 1957; Telling and Stone, 1964; Telling and Elsworth, 1965; Ziegler et al., 1958). However, other devices have been used successfully. Examples of these are the vibromixers found to give satisfactory agitation without foaming in 200-liter hemispherical vessels by Moore et al. (1968), and the helical screw rotating at 150 rpm in a 150-liter culture vessel used by Ubertini et al. (1967). Holmström (1968) agitated a 10-liter culture in a round-bottomed flask with a two-bladed propeller at 250 rpm and found that this method provided a gentle lifting action sufficient to keep the culture well mixed and prevent aggregation.

Agitation may be increased by "baffling" the system. This is generally achieved by fitting vertical baffle plates radially to the vessel wall and in this way stirrer speeds may be reduced without decreasing mixing efficiency. At the Pirbright laboratories, the fitting of four baffle plates to both 30- and 100-liter culture vessels improved cell dispersion and permitted a reduction in stirrer speed. However, a similar baffling arrangement was unsatisfactory at the 4-liter scale because cells adhered to the baffles.

3. Design and Operation of Culture Equipment

Vessel design and operation for the submerged culture of bacteria is well established and several workers have shown that this expertise can be successfully applied to the growth of mammalian cells. Generally, the essential modification has been to minimize turbulence by reducing the stirrer speed or changing the type of agitator (Holmström, 1968; Kilburn and Webb, 1968; McLimans et al., 1957; Moore et al., 1968; Telling and Stone, 1964; Telling and Elsworth, 1965; Ubertini et al., 1967; Ziegler et al., 1958).

These reports suggest that the requirements of equipment design and operation for large-scale tissue cultures are no more stringent than those found satisfactory for deep bacterial cultures. Adventitious microbial contamination, although always a risk, is more likely to grow to detectable levels in tissue cultures owing to the relatively slow growth rate of mammalian cells. Design factors affecting secondary infection have been discussed in detail by Elsworth (1960).

A key problem in the maintenance of sterility is the reliability of the agitator shaft seal. The double radial mechanical seal described by Elsworth and his associates (Elsworth, 1960; Elsworth and Stockwell, 1968; Elsworth et al., 1958; Telling and Elsworth, 1965) has proved satisfactory and is widely used. Moore and his co-workers have elected to use vibromixers on vessels up to 1250 liters on the grounds that these give adequate agitation in their vessels at low power inputs and that the shaft seal is simpler (Moore et al., 1968; Ulrich and Moore, 1965; Vosseller and Moore, 1969). Another alternative is to use powerful ceramic magnets mounted externally which provide agitation by indirect transmission (Cameron and Godfrey, 1968; Holmström, 1968). Cameron and Godfrey found this method successful in cultures up to 300 liters. Vessels, media filter, and ancillary pipe work are conveniently sterilized by steam injection. The risk of toxic residues from steam will depend largely upon the quality of the boiler-feed water, particularly on the scale-preventing additives used, and prior testing is obviously necessary. Some workers consider it desirable to use only steam generated from distilled water but steam produced from mains water softened by the soda-lime process, has been found satisfactory for all our applications. The only precaution we take is to filter the steam through porous stainless steel (average pore diameter 20 μ) and, thereafter, to distribute in stainless steel pipe. The principles of laying out process piping to ensure proper sterilization have been discussed by Walker and Holdsworth (1958) and Elsworth (1960).

Sterilization of gases is usually by filtration. Gas flow rates are much lower than for bacterial cultures (our maximum gas flow rate for a 100-liter culture of BHK cells is 6 liters per minute) and relatively small commercially available filters can be used. Our preference is to reduce aseptic connections to a minimum and so use filters which can be permanently connected to the vessel and sterilized *in situ* by steam.

A steam-sealed draw-off valve similar to that used on bacterial culture vessels is satisfactory for sampling. Such a system consists of a valve having on the outlet a jet fitted with a shroud, the whole of which can be steam-sterilized. Convenient sampling procedures have been described in detail by Elsworth (1960) and Elsworth and Stockwell (1968). Small volumes (e.g., cell seed, antifoam) can be easily inoculated into a vessel through self-sealing rubber diaphragms, and a suitable housing, which may be modified for independent steam sterilization, has been described by Elsworth (1960). A pressure of sterile air is a convenient way to transfer large volumes of culture from one vessel to another. This method eliminates the need for pumps, which are difficult to sterilize and may subject the cells to damaging shear stresses.

Valves similar to those used on microbiological plant are suitable for use on mammalian cell culture equipment. Jackson (1958) considers diaphragm valves fitted with heat-resistant diaphragms to be the valves of choice, but Elsworth (1960) preferred sleeve-packed cocks, which required little maintenance over prolonged periods of use. We find diaphragm valves satisfactory, provided they are regularly maintained, particularly in the smaller sizes where suitable alternatives are often not available. Sleeve-packed cocks are used extensively on steam lines but where there is contact with culture medium we favor stainless steel ball plug valves. These have the advantage that both the plug and operating handle rotate through 90°. Our preference is for the type in which leakage via the spindle is prevented by a self-adjusting polytetrafluoroethylene diaphragm seal (Sabal, Type M, Saunders Valve Co. Ltd.). In our hands these valves have required no maintenance in nearly 2 years of routine operation, including regular and frequent steam sterilization.

4. Scale-up

A number of relationships have been suggested for the scale-up of fermentations. These have been described and discussed by Aiba et al. (1965) and Bartholomew (1960), who point out some of the problems associated with their use. At the present time a useful concept for scale-up is on the basis of oxygen solution rates. Using this approach we have obtained satisfactory results in batch cultures of BHK 21 cells. In this work (unpublished) an oxygen electrode was used to measure the oxygen solution rates for various flow rates of sparged air and to estimate the oxygen demand of growing BHK cells. From these data, estimates could be made of the air rates required to satisfy the oxygen demand at any stage of the culture. In practice the air rate was adjusted so that the oxygen solution rate was just in excess of the calculated requirement and was periodically readjusted to meet the increased demand as the culture progressed. This method avoided unnecessarily high rates of sparging and maintained the pO_2 between 40 and 100 mm Hg. Under these conditions cell growth,

growth rate, and glucose utilization were the same in culture volumes of 4, 30, and 100 liters. About five adjustments of the air flow rate were made for a twelvefold increase in cell density to a maximum of approximately 6×10^6 cells/ml in a culture period of 60 hours.

Johnson (1964) described a method for calculating the oxygen requirement for bacterial cell growth and Elsworth et al. (1968) reported its use for calculating the oxygen requirement for a continuous flow culture of E. coli. If there are common factors influencing assimilative production of both bacterial and mammalian cells, it may be possible to calculate the oxygen requirement of the latter by Johnson's method.

V. Conclusions

In summary, the cultivation of mammalian cells has been accomplished successfully on the large scale by several different methods. Factors found to restrict cell growth have been depletion of nutrients, including oxygen, and unfavorable environment such as pH, temperature, dissolved oxygen tension, and (in suspension cultures) mechanical stress. It is necessary to emphasize, however, that the cultural conditions for optimum growth are not necessarily the same as those required for other physiological activities.

The continuing need for serum for the growth and survival of most cells has several disadvantages. These include its high cost, the risk of introducing adventitious viruses, the possible presence of specific and nonspecific inhibitors, and its surfactant properties, which encourage foaming in stirred, aerated cultures.

Contamination of cultures by bacteria and fungi is a constant hazard and the risks arising from a small number of contaminants are increased when mass cultivation is to be in a single large vessel. Provided that proper attention is paid to the design, sterilization, and operation of large-scale equipment, then in our experience with a cell line the greatest risk of contamination occurs in the laboratory during the initial preparation and buildup of the cell inoculum. If large-scale operations are to be successful, it is essential that these early manipulations are performed in a sterile area, using strict aseptic precautions and with adequate screening procedures for the detection of contamination.

To conclude, large-scale cultivation of mammalian cells is now an established operation. Possible future application of the technology is in the production of cells for chemical, physical, and genetic studies, for cell antigens and other components, and for the biosynthesis of pharmaceuticals.

References

Aiba, S., Humphrey, A. E., and Millis, N. F. (1965). "Biochemical Engineering." Academic Press, New York.
Bachrach, H. L., and Polatnick, J. (1968). *Biotechnol. Bioeng.* **10**, 589–599.
Bartholomew, W. H. (1960). *Advan. Appl. Microbiol.* **2**, 289–300.
Birch, J. R., and Pirt, S. J. (1969). *J. Cell Sci.* **5**, 135–142.
Black, P. H. (1966). *J. Nat. Cancer Inst.* **37**, 487–493.
Brown, F., Cartwright, B., and Newman, J. F. E. (1965). *Nature (London)* **205**, 310–311.
Bryant, J. C. (1966). *Ann. N. Y. Acad. Sci.* **139**, 143–161.
Bryant, J. C., Schilling, E. L., Earle, W. R., and Evans, V. J. (1960). *J. Nat. Cancer Inst.* **24**, 859–871.
Cahn, R. D. (1967). *Science* **155**, 195–196.
Cameron, J., and Godfrey, E. I. (1968). *J. Appl. Bacteriol.* **31**, 405–410.
Capstick, P. B. (1963). *Proc. Roy. Soc. Med.* **56**, 1062–1064.
Capstick, P. B., Telling, R. C., Chapman, W. G., and Stewart, D. L. (1962). *Nature (London)* **195**, 1163–1164.
Capstick, P. B., Garland, A. J., Masters, R. C., and Chapman, W. G. (1966). *Exp. Cell Res.* **44**, 119–128.
Carski, R. R., and Shepard, C. C. (1961). *J. Bacteriol.* **81**, 626–635.
Chang, R. S., Liepins, H., and Margolish, M. (1961). *Proc. Soc. Exp. Biol. Med.* **106**, 149–152.
Cherry, W. R., and Hull, R. N. (1960). *J. Biochem. Microbiol. Technol. Eng.* **2**, 267–285.
Cohen, E. P., and Eagle, H. (1961). *J. Exp. Med.* **113**, 467–474.
Coombs, R. R. A., Gurner, B. W., Beale, A. J., Christofinis, G., and Page, Z. (1961). *Exp. Cell Res.* **24**, 604–605.
Cooper, P. D., Wilson, J. N., and Burt, A. M. (1959). *J. Gen. Microbiol.* **21**, 702–720.
Cruikshank, C. N. D., and Lawburg, E. J. L. (1952). *Brit. Med. J.* **2**, 1070.
Eagle, H. (1955). *Science* **122**, 501–504.
Eagle, H. (1959). *Science* **130**, 432–437.
Earle, W. R., Bryant, J. C., Schilling, E. L., and Evans, V. J. (1954). *J. Nat. Cancer Inst.* **14**, 1159–1171.
Elsworth, R. (1960). *In* "Progress in Industrial Microbiology" (D. J. D. Hockenhull, ed.), Vol. 2, pp. 103–130. Heywood, London.
Elsworth, R., and Stockwell, F. T. E. (1968). *Process Biochem.* **3** [No. 3], 15–18.
Elsworth, R., Telling, R. C., and Ford, J. W. S. (1955). *J. Hyg.* **55**, 445–457.
Elsworth, R., Capell, G. H., and Telling, R. C. (1958). *J. Appl. Bacteriol.* **21**, 80–85.
Elsworth, R., Miller, G. A., Whitaker, A. R., Kitching, D., and Sayer, P. D. (1968). *J. Appl. Chem.* **17**, 157–166.
Farrant, J. (1966). *Lab. Pract.* **15**, 402–404.
Fogh, J., and Hacker, C. (1960). *Exp. Cell Res.* **21**, 242–244.
Gerber, P. (1966). *Virology* **28**, 501–509.
Geyer, R. P., and Chang, R. S. (1958). *Arch. Biochem. Biophys.* **73**, 500–506.
Graff, S., and McCarty, K. S. (1957). *Exp. Cell Res.* **13**, 348–357.
Griffiths, J. B., and Pirt, S. J. (1967). *Proc. Roy. Soc.* **B168**, 421–438.
Hayflick, L. (1960). *Nature (London)* **185**, 783–784.
Hayflick, L. (1963). *Proc. Symp. Characterization and Uses of Human Diploid Cell Strains. Permanent Section of Microbiological Standardization. Opatija, Yugoslavia, 1963*, pp. 37–54.
Hayflick, L., and Moorhead, P. S. (1961). *Exp. Cell Res.* **25**, 585–621.

Hayflick, L., Jacobs, P., and Perkins, F. (1964). *Nature (London)* **204,** 146–147.
Herbert, D., Elsworth, R., and Telling, R. C. (1956). *J. Gen. Microbiol.* **14,** 601–622.
Holmström, B. (1968). *Biotechnol. Bioeng.* **10,** 373–384.
House, W. (1964). *Nature (London)* **201,** 1242.
Hull, R. N., Minner, J. R., and Mascoli, C. C. (1958). *Amer. J. Hyg.* **68,** 31–44.
Jackson, I. (1958). *In* "Biochemical Engineering" (R. Steel, ed.), p. 210. Heywood, London.
Johnson, M. J. (1964). *Chem. Ind. (London)* pp. 1532–1537.
Johnson, R. W., and Orlando, M. D. (1967). *Appl. Microbiol.* **15,** 209–210.
Kilburn, D. G., and Webb, F. C. (1968). *Biotechnol. Bioeng.* **10,** 801–814.
Kilburn, D. G., Lilly, M. D., Self, D. A., and Webb, F. C. (1969). *J. Cell Sci.* **4,** 25–37.
Kruse, P. F., Jr., and Miedema, E. (1965). *J. Cell Biol.* **27,** 273–279.
Kruse, P. F., Jr., Myhr, B. C., Johnson, J. E., and White, P. B. (1963). *J. Nat. Cancer Inst.* **31,** 109–120.
Kunin, C. M., Emmons, L. R., and Jordan, W. S. (1960). *J. Immunol.* **85,** 203–219.
Levan, A., and Biesele, J. J. (1958). *Ann. N. Y. Acad. Sci.* **71,** 1022.
Mackareth, F. J. H. (1964). *J. Sci. Instrum.* **41,** 38–41.
McLimans, W. F., Giardinello, F. E., Davis, F. E., Kuceram, C. J. and Rake, G. W. (1957). *J. Bacteriol.* **74,** 768–774.
Macpherson, I. A. (1963). *J. Nat. Cancer Inst.* **30,** 795–815.
Macpherson, I. A. (1966). *J. Cell Sci.* **1,** 145–148.
Macpherson, I. A., and Stoker, M. G. P. (1962). *Virology* **16,** 147–151.
Majer, M. (1967). *Process Biochem.* **2** [No. 12], 25–31.
Merchant, D. F., Kahn, R. H., and Murphy, W. H. (1960). "Handbook of Cell and Organ Culture." Burgess, Minneapolis, Minnesota.
Merchant, D. J., Kuchler, R. J., and Munyon, W. H. (1960). *J. Biochem. Microbiol. Technol. Eng.* **2,** 253–265.
Molin, O., and Hedén, C. G. (1969). *In* "Progress in Immunobiological Standardization," Vol. 3, pp. 106–110. Karger, Basel.
Moore, G. E., and Ulrich, K. (1965). *J. Surg. Res.* **5,** 270–282.
Moore, G. E., Hasenpusch, P., Gerner, R. E., and Burns, A. A. (1968). *Biotechnol. Bioeng.* **10,** 625–640.
Muggleton, P. W. (1963). *In* "Progress in Industrial Microbiology" (D. J. D. Hockenhull, ed.), Vol. 4, pp. 189–214. Heywood, London.
Nagle, S. C., Jr. (1968). *Appl. Microbiol.* **16,** 53–55.
Nagle, S. C., Jr. (1969). *Appl. Microbiol.* **17,** 318–319.
Nagle, S. C., Jr., Tribble, H. R., Jr., Anderson, R. E., and Gary, N. D. (1963). *Proc. Soc. Exp. Biol. Med.* **112,** 340–344.
Orlando, M. D., Bowersox, O. C., and Riley, J. M. (1968). *Biotechnol. Bioeng.* **10,** 61–67.
Owens, O. V. H., Gey, M. K., and Gey, G. O. (1954). *Ann. N. Y. Acad. Sci.* **58,** 1039–1055.
Paul, J. (1965). "Cell and Tissue Culture," 3rd ed. Livingstone, London.
Perlman, D. (1967). *Process Biochem.* **2** [No. 4], 42–46.
Perlman, D., and Guiffre, N. A. (1963). *In* "Progress in Industrial Microbiology" (D. J. D. Hockenhull, ed.), Vol. 4, pp. 61–94. Heywood, London.
Perlman, D., Guiffre, N. A., and Brindle, S. A. (1961). *Proc. Soc. Exp. Biol. Med.* **106,** 880–883.
Perlman, D., Rahman, S. B., and Sernar, J. B. (1967). *Appl. Microbiol.* **15,** 82–85.
Peto, S., and Maidment, B. J. (1969). *J. Hyg.* **67,** 533–538.
Pirt, S. J., and Callow, D. S. (1964). *Exp. Cell Res.* **33,** 413–421.

Pollock, M. E., Kenny, G. F., and Syverton, J. T. (1960). *Proc. Soc. Exp. Biol. Med.* **105**, 10–15.
Pollock, M. E., Treadwell, P. E., and Kenny, G. E. (1963). *Exp. Cell Res.* **31**, 321–328.
Portner, D. M., Phillips, C. R., and Hoffman, R. K. (1967). *Appl. Microbiol.* **15**, 800–807.
Pulvertaft, R. J. V., Davies, J. R., Weiss, L., and Wilkinson, J. A. (1959). *J. Pathol. Bacteriol.* **77**, 19–32.
Pumper, R. W., Yamashiroya, H. M., and Molander, L. T. (1965). *Nature (London)* **207**, 662–663.
Richards, J. W. (1968). "Introduction to Industrial Sterilization," pp. 76–92. Academic Press, New York.
Robinson, D. M. (1966). *Lab. Pract.* **15**, 410–412.
Robinson, L. B., Wichelhausen, R. H., and Roizman, B. (1956). *Science* **124**, 1147.
Ruddle, C. H. (1962). *Proc. Soc. Exp. Biol. Med.* **109**, 116–119.
Runyon, W. S., and Geyer, R. P. (1963). *Proc. Soc. Exp. Biol. Med.* **112**, 1027–1030.
Sargeant, K. (1968). *Process Biochem.* **3** [No. 4], 51–53.
Schleicher, J. B., and Weiss, R. E. (1968). *Biotechnol. Bioeng.* **10**, 617–624.
Self, D. A., Kilburn, D. G., and Lilly, M. D. (1968). *Biotechnol. Bioeng.* **10**, 815–828.
Shedden, W. I. H., and Wildy, P. (1966). *Nature (London)* **212**, 1068–1069.
Smith, H. M., and Burrows, T. M. (1963). *Lab. Pract.* **12**, 451–453.
Smith, H. M., and Burrows, T. M. (1966). *Lab. Pract.* **15**, 864–866.
Stoker, M. G. P., and Macpherson, I. A. (1964). *Nature (London)* **203**, 1355–1357.
Stulberg, C. S., Simpson, W. F., and Berman, L. (1961). *Proc. Soc. Exp. Biol. Med.* **108**, 434–439.
Sweat, M. L., Grosser, B. I., Berliner, D. L., Swim, H. E., Nabors, C. J., and Dougherty, T. F. (1958). *Biochem. Biophys. Acta* **28**, 591–596.
Sweet, B. H., and Hilleman, M. R. (1960). *Proc. Soc. Exp. Biol. Med.* **105**, 420–427.
Swim, H. E., and Parker, R. F. (1958). *J. Biophys. Biochem. Cytol.* **4**, 525–528.
Tashjian, A. H. (1969). *Biotechnol. Bioeng.* **11**, 109–126.
Telling, R. C. (1969). *Process Biochem.* **4** [No. 6], 49–52.
Telling, R. C., and Elsworth, R. (1965). *Biotechnol. Bioeng.* **7**, 417–434.
Telling, R. C., and Radlett, P. J. (1969). *Paper presented at the Meeting of the Standing Tech. Comm., European Comm. for the Control of Foot-and-Mouth Disease, Brescia, September, 1969.*
Telling, R. C., and Stone, C. J. (1964). *Biotechnol. Bioeng.* **6**, 147–158.
Telling, R. C., Stone, C. J., and Maskell, M. A. (1966). *Biotechnol. Bioeng.* **8**, 153–165.
Thompson, K. W., Vincent, M. M., Jensen, F. C., Price, R. T., and Shapiro, E. (1959). *Proc. Soc. Exp. Biol. Med.* **102**, 403–413.
Todaro, G., and Green, H. (1963). *J. Cell Biol.* **17**, 299–313.
Toplin, I., and Gaden, E. L. (1961). *J. Biochem. Microbiol. Technol. Eng.* **3**, 311–323.
Tritsch, G. L., and Moore, G. E. (1962). *Exp. Cell Res.* **28**, 360–364.
Ubertini, B., Nardelli, L., Santero, G., and Panina, G. (1960). *J. Biochem. Microbiol. Technol. Eng.* **2**, 327–338.
Ubertini, B., Nardelli, L., Dal Prato, A., Panina, G., and Santero, G. (1963). *Zentrabl. Veterinaermed.* **10**, 93–101.
Ubertini, B., Nardelli, L., Dal Prato, A., Panina, G., and Barei, S. (1967). *Zentrabl. Veterinaermed.* **14**, 432–441.
Ulrich, K., and Moore, G. E. (1965). *Biotechnol. Bioeng.* **7**, 507–515.
van Hemert, P. A. (1964). *Biotechnol. Bioeng.* **6**, 381–401.
van Hemert, P. A., Kilburn, D. G., and van Wezel, A. L. (1968). *Paper presented at the 3rd Int. Fermentation Symp., New Brunswick, New Jersey.*

van Wezel, A. L. (1967). *Nature (London)* **216**, 64–65.
Vosseller, G. V., and Moore, G. E. (1969). *Res./Develop.* **20** [No. 5], 20–24.
Walker, J. A. H., and Holdsworth, H. (1958). *In* "Biochemical Engineering" (R. Steel, ed.), pp. 249–251. Heywood, London.
Walker, J. S., Lincoln, R. E., and Weirether, F. J. (1968). *Biotechnol. Bioeng.* **10**, 557–566.
Watkins, J. F., and Dulbecco, R. (1967). *Proc. Nat. Acad. Sci. U. S. A.* **58**, 1396–1403.
Weirether, F. J., Walker, J. S., and Lincoln, R. E. (1968). *Appl. Microbiol.* **16**, 841–844.
Weiss, R. E., and Schleicher, J. B. (1968). *Biotechnol. Bioeng.* **10**, 601–615.
Yergaman, G., and Leonard, M. J. (1961). *Science* **133**, 1600–1601.
Ziegler, D. W., Davis, E. V., Thomas, W. J., and McLimans, W. F. (1958). *Appl. Microbiol.* **6**, 305–310.

Large-Scale Bacteriophage Production

K. SARGEANT

*Microbiological Research Establishment,
Porton, Salisbury, Wiltshire, England*

I.	Introduction	121
II.	General Considerations	122
	A. Growth Methods	122
	B. Counting of Bacteriophage Particles	123
	C. Calculation of Bacteriophage Yield	123
	D. Special Precautions to Prevent the Unwanted Bacteriophage Infection of Cultures	124
	E. Limitation of Foaming	124
III.	DNA Bacteriophages	125
	A. General	125
	B. Bacteriophage T7	125
	C. Bacteriophage ϕX174	128
IV.	RNA Bacteriophages	128
	A. General	128
	B. Bacteriophage MS2	129
	C. Bacteriophage μ2	130
	D. Bacteriophage R17	133
V.	Bacteriophage-Infected Cells	133
	References	136

I. Introduction

Bacteriophages were discovered by Twort and independently by d'Herelle. It is now well recognized that they are bacterial viruses. For detailed information on their discovery and on their properties the reader is referred to the works of Adams (1959) and Stent (1963).

Probably because there is no large industrial interest in their manufacture little is known about their large-scale production. At one time there were hopes that phage therapy, in which doses of a bacteriophage were administered to a patient suffering from a disease caused by a phage-sensitive organism, would prove a useful method of disease control. However, early experiments in this direction were unsuccessful and the method fell into disrepute. It is possible that the methods of phage production used to support this early work resulted in inadequate products, and perhaps further attempts to control some intractable infections either by phage therapy alone or in combination with antibiotic treatment, are now timely in view of the better bacteriophage preparations which could be made available.

The main stimulus to work on the large-scale production of bacteriophage in recent years has come from research workers interested in detailed studies of bacteriophage components. A considerable interest has also been shown in the products induced in bacterial cells as a result of bacteriophage infection, and this has led to the development of methods for the production of bacteria, infected with bacteriophage, but harvested before cell lysis has occurred.

Little effort has been devoted to devising large-scale methods for the isolation of bacteriophage and bacteriophage components. In most cases laborious repetitive small-scale procedures have been employed.

II. General Considerations

A. Growth Methods

Since bacteriophages are bacterial viruses and their production depends on using bacterial cells as host organisms it is necessary to provide a good supply of living bacteria in order to obtain a large quantity of bacteriophage.

Relatively small quantities of many bacteriophages can readily be made by using bacteria grown on an agar surface (Hershey et al., 1943; Swanstrom and Adams, 1951). The method is very convenient for obtaining a small amount of high-titer bacteriophage, since a small volume of liquid can be used for washing off the product. It is also a useful way of obtaining an adequate seed stock for larger scale work.

Shake flasks or aerated bottles have often been used in the preparation of bacteriophages, and the need to use rapidly growing bacteria has long been recognized. Thus Thomas and Abelson (1966) in listing growth conditions for the production of the bacteriophages, T2, T4, T6, T*2, T5st, T7, λ, and P22, recommend that the bacterial cultures should be "growing logarithmically at the time of infection." Hence the recommended bacterial concentrations at infection are all in the range 1–5×10^8/ml. In all cases it is recommended that cultural conditions be maintained for 1–3 hours so that bacterial lysis will occur, with the release of progeny bacteriophage. It is also recommended that whenever lysis does not occur spontaneously, chloroform (1/50 volume) should be employed to promote it.

Stirred deep-culture vessels have been used for the production of a number of bacteriophages, but in the opinion of this author their full potential has not often been realized. It has now been shown that under the conditions of high aeration attainable in these vessels bacteriophages MS2, μ2, R17, and T7 can be produced at very high concentrations by infecting growing bacterial cultures at cell densities in the range 4–30 × 10⁹/ml (Rushizky et al., 1965; Sargeant and Yeo, 1966, 1969; Sargeant et al., 1968).

When a bacteriophage is produced in a deep-culture vessel it is important to estimate the viable bacterial cell density in the culture just prior to infection. This can only be determined directly after a considerable time lapse and so indirect methods have to be employed. The most favored are either to measure the optical density of the culture or the carbon dioxide evolution rate. In either case the result is compared with that obtained on a previous similar culture, for which plate counts were carried out, and the cell density estimated. Examples of the use of both methods will be mentioned later.

B. Counting of Bacteriophage Particles

The most commonly used method of bacteriophage assay is the agar layer method which is described in detail by Adams (1959). A few milliters of melted 0.6% agar is cooled to 45°C and inoculated with a concentrated suspension of the host bacteria. A suitable volume of the diluted bacteriophage suspension is added and the whole mixture is spread over the surface of a nutrient agar plate. When the upper agar layer has set the plate is incubated. The bacteria grow in the upper agar, but the dense background is broken by clear plaques where bacteriophage growth has caused cell lysis. The clear plaques are counted, and the result is expressed in "plaque-forming units" (PFU).

C. Calculation of Bacteriophage Yield

The plaque assay method described above gives the number of bacteriophage particles capable of forming plaques. This is generally less than the total number of particles present, and the ratio, p, of plaque-forming particles to total particles is called the absolute efficiency of plating of the plaque assay method. The p is critically dependent on the conditions employed in the assay, and its value may be determined by the method of Luria et al. (1951).

From a knowledge of p, and the particle mass M (in daltons) of the bacteriophage, the mass, m (in grams), in a given sample may be calculated from the plaque assay by using Eq. (1)

$$m = Mx/Ap \tag{1}$$

where x is the number of PFU in the sample, and A is Avogadro's number (6.02×10^{23}).

If the value of p for the plaque assay method is unknown the minimum mass of bacteriophage in a sample may be calculated from Eq. (1) by assuming that $p = 1$, i.e., that all the virus particles are plaque formers.

D. Special Precautions to Prevent the Unwanted Bacteriophage Infection of Cultures

It is essential to ensure that unwanted bacteriophage infections do not occur and this is relatively easy using standard bacteriological techniques. The following four points are worthy of special attention.

1. *The Preparation of Bacterial Seed Cultures.* This is best done in a laboratory remote from the one in which the bacteriophage is to be produced. A sensible further precaution is to lay down a stock of freeze-dried ampuls of the host organism before the bacteriophage is brought into use. By doing this it is possible to ensure a continuing supply of virus-free host cells.

2. *The Sterilization of Apparatus After Use.* Bacteriophages are sensitive to heat. Exposure to steam at 15 psi for 30 minutes is sufficient to destroy residual bacteriophage in culture vessels or in other apparatus.

3. *The Containment of Culture Vessels.* It is not necessary to use culture vessels which have been specially isolated, providing that adequate steps are taken to filter the air entering and leaving them. For air-filtration requirements see Elsworth (1969).

4. *Sterilization of the KCl Bridge to the Calomel Electrode.* In pH-controlled cultures a problem which needs special attention is the destruction of bacteriophage particles which have penetrated the glass sinter at the culture vessel end of the KCl bridge to the calomel electrode. Such particles are not all destroyed when the culture vessel is sterilized by normal steam-sterilization techniques, and survivors reenter the culture vessel later, thus causing an unwanted contamination of the next culture. This problem is unique to bacteriophage and other virus cultures because the glass sinters are impervious to bacteria. The solution is to dismantle and separately sterilize the KCl bridge system and calomel electrode after each experiment (Sargeant and Yeo, 1966).

E. Limitation of Foaming

There is a strong tendency for cultures to produce foam, especially during lysis, and this needs to be controlled. Rushizky *et al.* (1965) found that Dow-Corning antifoam AF gave satisfactory foam control in the production of MS2 and Sargeant and Yeo (1966) successfully used MS antifoam emulsion RD in the production of $\mu 2$.

III. DNA Bacteriophages

A. General

This group contains the well-studied "T even" viruses, which are among the easiest to produce in large quantities. These large bacteriophages (particle mass, 200×10^6 daltons) were grown by Wyatt and Cohen (1953) in 350-ml culture volumes in vigorously shaken 2-liter flasks. By infecting cultures which contained 3×10^9 host organisms/ml final bacteriophage titers of up to 10^{12}/ml were obtained, equivalent to not less than 330 mg of bacteriophage per liter even if the absolute efficiency of plating were unity. There is little doubt that similar, or even better yields could be obtained by carrying out the preparation in stirred deep-culture vessels.

The only detailed investigation of growth conditions for the production of a DNA bacteriophage in culture vessels is that of Sargeant et al. (1968) who grew bacteriophage T7. This is described below together with some information on the production of the important bacteriophage, ϕX174, which contains single-stranded DNA.

B. Bacteriophage T7

Bacteriophage T7 has a particle mass of 38×10^6 daltons (Davison and Freifelder, 1962). It is virulent for *Escherichia coli* B and was grown by Lunan and Sinsheimer (1956) in 1-liter batches in the glycerol-casamino acids medium of Frazer and Jerrel (1953). Growing cultures of the host organism, at a density of 10^9/ml, were infected with T7 at an input ratio of 1 PFU per 10 bacteria, and aerated for an additional 2–3 hours. The average yield of T7 in 14 such batches was 2.43×10^{11} PFU/ml, equivalent to 243 PFU for each bacterial cell present when the culture was infected. Later Davison and Freifelder (1962) used the same cultural conditions, with the exception that in place of air, pure oxygen was bubbled through the culture. They obtained twice the yield of T7 reported above.

Sargeant et al. (1968), who also used a medium based on glycerol and casamino acids, reported on the preparation of T7 in 3- and 20-liter stirred deep-culture vessels. Later Sargeant and Yeo (1969) extended this work to the 150-liter scale.

A series of experimental cultures was grown in the 3-liter vessel, which had the capacity under the conditions used to dissolve 220 mmoles of oxygen per liter per hour (Elsworth *et al.*, 1957). The results (which are summarized in Table I) show that cultures infected at cell densities in the range 2×10^9/ml to 3×10^{10}/ml continued to grow and eventually lysed giving rise to satisfactory yields of progeny T7. The highest yields (an average of 1.43×10^{12}/ml from three experiments) were obtained from cultures infected at the highest cell densities, and under these conditions the pH value was controlled at 7.0 by the automatic addition of 2 N NaOH. (The pH value of cultures infected at the lower cell densities did not change greatly from the initial value of 7.0.) The yields of progeny T7 per cell were in the approximate range 50–100, and the input ratio of T7 to bacteria was normally 0.7–0.8. It was found that by increasing this input ratio to an average value of 10.5 in three experiments there was no increase in yield of progeny T7 per cell.

Thus the higher culture lysate titers obtained in the 3-liter vessel cultures as compared with the 1-liter cultures of Lunan and Sinsheimer (1956) were the result of infecting cultures at a very high cell density and accepting a somewhat lower T7 yield per cell from many more cells. Part of the reason for this reduced yield per cell is undoubtedly that in the 1-liter cultures an input ratio of 1 phage to 10 bacteria was employed. Thus at least 90% of the organisms present when the cultures were infected were free to multiply, but would, together with any issue, become infected by progeny phage later. Thus a greater number of bacteria than were present originally were actually involved in phage production. A clue as to a further reason for this lower yield per cell is given by the observation of Davison and Freifelder (1962, *vide infra*) that the use of oxygen instead of air doubles the yield of phage. Possibly, despite the high aeration capacity of the 3-liter culture vessel, there was less oxygen available per cell during the later stages of the culture, and this relative shortage was responsible for the reduced yield.

Twenty-liter cultures were grown in a culture vessel which supported a sulfite oxidation rate slightly higher than that of the 3-liter vessel. Thus no pH control was necessary and a series of four cultures, infected at an average cell density of 2.85×10^{10}/ml gave an average yield of 9.5×10^{11} PFU/ml of T7 in the culture lysate, or 33 PFU/cell.

One hundred and fifty-liter cultures were grown in a culture vessel having a sulfite oxidation capacity of about 80 mmoles of oxygen per liter per hour. No pH control was used, and cultures were infected when the pH value had fallen to 6.5. Seven cultures were grown, infected at an average cell density of 1.67×10^{10}/ml, and gave an average yield of 7.1×10^{11} PFU/ml, or 43 PFU/cell.

TABLE I
PRODUCTION OF BACTERIOPHAGE T7

Cultural equipment	No. of cultures grown	Average values			
		Viable bacteria at infection per milliliter ($\times 10^9$)	No. of infective particles per bacterium at infection	No. of infective particles in final culture supernatant fluid per milliliter ($\times 10^{11}$)	No. of infective particles produced per bacterium
3-Liter stirred vessel[a]	3	2.9	—	3.0	103
	3	11.0	10.5	6.3	57
	3[b]	17.9	0.8	12.6	70
20-Liter stirred vessel[a]	4	29.1	0.7	14.3	49
150-Liter stirred vessel[c]	4	28.5	0.7	9.5	33
	7	16.7	0.7	7.1[d]	43
1 Liter[e]	14	1	0.1	2.43	243

[a]Sargeant et al. (1968).
[b]pH control was used in these three cultures.
[c]Unpublished data, Sargeant and Yeo (1969).
[d]Bacteriophage assays were carried out on whole culture samples.
[e]Lunan and Sinsheimer (1956).

The bacteriophage was purified by a simple method, based on centrifugation techniques, which was applied to 2.4-liter lots of culture lysate. The average yield of pure T7 was 143 mg/liter from 3-liter cultures and 131 mg/liter from the 20-liter cultures. The efficiency of plating varied from 18–42%.

C. Bacteriophage ϕX174

Bacteriophage ϕX174 has a particle mass of 6.2×10^6 daltons (Sinsheimer, 1959a) and contains a single-stranded DNA (Sinsheimer, 1959b). Its preparation is mentioned here in spite of the fact that no large-scale growth method for its production has been described, because considerable research interest is still being shown in it.

A conventional method for preparing ϕX174 was described by Sinsheimer (1966). The host organism, *E. coli* C406, was grown to a cell density of $2-3 \times 10^8$/ml, in 2 liters of an amino acids, pyruvate, salts medium contained in a vigorously aerated 10-liter flask. The culture was inoculated with wild-type ϕX174 at an input ratio of 3 PFU/cell and cultural conditions were maintained. Thirty minutes later some disodium versenate was added. Lysis was complete at 5–7 hours when the lysate titer was $4-12 \times 10^{11}$/ml. From this, 5–10 mg of purified bacteriophage were isolated by a method which is described in detail.

Sinsheimer (1966) goes on to describe the use of lysis-defective mutants in the preparation of larger quantities of ϕX174. These mutants are ϕX174p^- which lyses poorly in concentrated cell cultures, but which can be assayed on the usual host, *E. coli* C on which it forms small, irregular plaques; and ϕX174am3, which will not lyse *E. coli* C, but which can be assayed on strain CR34/C416 on which it forms normal plaques. For the preparation, *E. coli* C was grown to a cell density of $2-3 \times 10^9$/ml, in 2 liters of a glycerol, casamino acids, salts medium and infected with either mutant at an input ratio of 3 PFU/cell. After 3 hours the cells were recovered and resuspended in 100 ml of tris buffer. Lysis was effected by the use of lysozyme and disodium versenate in a freeze-thawing process. This gave $2-5 \times 10^{13}$ PFU/ml, equivalent to $1-2.5 \times 10^{12}$ PFU/ml of culture, or approximately a 2.5-fold increase on the yield from the wild-type bacteriophage.

IV. RNA Bacteriophages

A. General

The discovery of an RNA-containing bacteriophage, f2, was first reported by Loeb and Zinder (1961). This announcement started a search for similar viruses and now a range of related bacteriophages is known. All have particle masses in the range 3.6–4.3×10^6 daltons, and they are very closely related to each other (Isenberg, 1963; Kaesberg, 1968).

The first detailed account of the large-scale production of an RNA bacteriophage was given by Rushizky et al. (1965), who produced MS2 on a 300-liter scale. This was followed by a report from Sargeant and Yeo (1966) on the production of $\mu 2$ in 3-, 20-, and 150-liter lots. It is reasonable to expect that methods similar to those reported for MS2 and $\mu 2$ would be satisfactory for other RNA bacteriophages and this has been found to be the case for R17 (Sargeant and Yeo, 1969).

B. BACTERIOPHAGE MS2

The brief report of Rushizky et al. (1965) is still the only paper on bacteriophage preparation which includes details of a method for the large-scale recovery of pure virus.

The host organism, E. coli C3000, was grown in 300 liters of a medium containing yeast extract, glucose, and salts, in a 380-liter stirred aerated culture vessel. The bacteriophage was added to the growing culture at an input ratio of 10 PFU/cell when optical measurements indicated that the cell density was 4×10^9/ml. The pH values of the growing culture fell both before and after the addition of bacteriophage.

This fall was moderated by two additions of Na_2HPO_4, and by using increased aeration and stirring after the bacteriophage was added. Spontaneous cell lysis, which was accompanied by increased foaming and a decrease in the optical density of the culture, was complete 4 hours after bacteriophage addition, when the titer was 10^{13} PFU/ml, equivalent to 2500 PFU per cell.

Chloroform (2 liters) and EDTA (500 gm free acid) were added to the culture to complete the cell lysis and to stabilize the product. A simple purification process which depended on ammonium sulfate precipitation, centrifugation, dialysis, and ultracentrifugation was used to give 19.1 gm of MS2, or about 65 mg/liter of culture. The recovery of phage was equivalent to 0.7×10^{13} PFU/ml of culture or 71% of that present prior to the addition of chloroform. The absolute efficiency of plating [calculated from eq. (1)] for the product was 66%.

This large-scale method for the preparation of MS2 must be regarded as a highly successful development of the earlier small-scale work of Strauss and Sinsheimer (1963), who obtained pure phage equivalent to 40 mg/liter and having an absolute efficiency of plating of 15–20% from 7-liter cultures. It is probable that the method would

C. BACTERIOPHAGE $\mu2$

1. The Importance of Aeration

The most extensive investigation of the growth conditions necessary for the production of a bacteriophage was that reported by Sargeant and Yeo (1966) for $\mu2$. They used a medium which contained glycerol, casamino acids, and salts to grow the host organism, *E. coli* K12 58-161 $F^+Fim\sigma^+$.

A series of experiments was carried out in which a 3-liter culture vessel was deliberately operated under conditions of relatively low aeration. It was established that the pH value of an uncontrolled culture fell rapidly from when the cell density reached about 5×10^8/ml. At this stage the carbon dioxide concentration in the culture effluent gas (ca. 0.4%) ceased to rise, indicating that growth was limited by the oxygen supply rate. After about 20 hours, growth had stopped, the cell density was about 3×10^9/ml, and the pH value of the culture had fallen to 5.0. A similar culture was infected with $\mu2$ at an input ratio of 3.4 PFU/cell when the pH value had reached 6.0 and the cell density was 2.9×10^9/ml. Cultural conditions were maintained for a further 18 hours, but the final bacteriophage concentration was only 4.2×10^{10}/ml, or 14.5 PFU/cell.

Later cultures in the series were operated under pH control in the range 7.0–7.1, afforded by the addition of sodium hydroxide and phosphoric acid, as required. This treatment prolonged bacterial growth to a final cell density of 10^{10}/ml at 36 hours. One of these cultures was infected with $\mu2$ at an input ratio of 3.2 PFU/cell when the cell density was 2.9×10^9/ml, and after cultural conditions had been maintained for only a further 12 hours the bacteriophage concentration was 2.4×10^{12}/ml, or 820 PFU/cell.

These and similar experiments showed that the dominating factor restricting the bacteriophage yield under conditions of limiting aeration was the drop in pH. They demonstrated the value of pH control in enhancing bacteriophage yield. It is probable that the generally accepted necessity to infect shake flask or aerated bottle cultures when the cell density is in the low range $1-5 \times 10^8$/ml *(vide infra)* in order to obtain a high bacteriophage yield is a result of the adversely acidic environment which develops as the cell density increases.

2. The Production of $\mu 2$

To study the conditions for optimum production of $\mu 2$ the host organism was grown in a 3-liter culture vessel under conditions of high aeration (220 mmoles oxygen dissolved/liter/hour). A preliminary culture of the host organism was grown in order to establish the relationship between carbon dioxide evolution rate and bacterial cell density. The carbon dioxide concentration of growing cultures was subsequently used as a guide to their cell density.

A series of eight cultures was grown and infected with $\mu 2$ at successively later stages of growth. Cultural conditions were maintained for a further 3 hours, chloroform (50 ml) was added to assist lysis, and stirring without aeration was continued for another hour. The $\mu 2$ content of whole culture samples was then determined by the plaque assay technique. The results are given in Table II (Cultures 1–8). These show that for cultures infected when the cell density was less than about 7×10^9/ml the yield of progeny bacteriophage increased roughly in proportion to the cell density. For cultures infected at higher cell densities than this there was a sharp fall in the yield of bacteriophage per cell. These results are quite different from those found for T7 (vide infra) and they show that there is a well-defined optimum cell density at which cultures should be infected if a maximum bacteriophage yield is to be obtained.

To confirm this, and to investigate the influence of the bacteriophage/cell input ratio on $\mu 2$ yield, a series of six cultures was grown under what were judged from Cultures 1–8 to be optimal conditions. The results, given in Table II (Cultures 9–14), showed a high bacteriophage yield (average value 2.07×10^{13} PFU/ml) which was not influenced by the input ratio over a very wide range.

This work was extended to 20- and 150-liter cultures, which were grown under similar conditions in vessels which had successively lower oxygen-solution rates (170 and 80 mmoles oxygen dissolved per liter per hour respectively). Because of this lower aeration it was thought necessary to infect these larger cultures at successively lower cell densities to maintain a very high yield per cell. The cell densities at which infection was carried out were chosen after observing the carbon dioxide evolution pattern of a culture of the host organism, grown to completion in the absence of $\mu 2$. The results are summarized in Table II and they show that a series of four cultures grown in the 20-liter vessel gave an average final $\mu 2$ yield of 1.68×10^{13} PFU/ml. A similar series grown in the 150-liter vessel gave 1.26×10^{13} PFU/ml.

TABLE II
PRODUCTION OF BACTERIOPHAGE $\mu 2$

Cultural equipment	Culture no.	Viable bacteria at infection per milliliter	No. of infective particles per bacterium at infection	No. of infective particles in final whole culture per milliliter	No. of infective particles produced per bacterium	Minimum mass of bacteriophage per liter of culture lysate (mg)
3-Liter stirred vessel	1	1.3×10^8	33	3.8×10^{11}	2900	
	2	5.3×10^8	3.7	3.7×10^{11}	700	
	3	1.9×10^9	1.4	1.0×10^{12}	530	
	4	3.0×10^9	4.9	7.9×10^{12}	2600	
	5	5.0×10^9	7.7	7.1×10^{12}	1400	
	6	6.9×10^9	6.4	10.2×10^{12}	1500	
	7	9.9×10^9	2.4	2.9×10^{11}	30	
	8	16.7×10^9	0.96	2.0×10^{10}	1.2	
3-Liter stirred vessel	9	6.9×10^9	1.8	3.2×10^{13}	4600	
	10	7.4×10^9	5.6	1.8×10^{13}	2400	
	11	6.4×10^9	19	2.1×10^{13}	3300	
	12	6.9×10^9	34	1.6×10^{13}	2300	
	13	9.6×10^9	46	2.0×10^{13}	2100	
	14	5.4×10^9	410	1.7×10^{13}	3200	
	Average 9–14	7.1×10^9	—	2.07×10^{13}	2900	124
20-Liter stirred vessel	Average of 4 cultures	6.3×10^9	27	1.68×10^{13}	2700	100
150-Liter stirred vessel	Average of 4 cultures	4.3×10^9	3.8	1.26×10^{13}	2900	75

In both cases the average yield per cell was almost identical with that obtained in the series of six 3-liter cultures.

Assuming a plating efficiency of 100% and a particle mass for $\mu 2$ of 3.6×10^6 daltons, the average minimum masses of bacteriophage present in the 3-, 20-, and 150-liter cultures, calculated from Eq. (1) (*vide infra*) were 124, 100, and 75 mg/liter, respectively.

3. The Purification of $\mu 2$

The whole culture lysate was separated from the bulk of insoluble debris by a slow-speed centrifugation and crude $\mu 2$ was precipitated by adding $(NH_4)_2SO_4$ to a concentration of 2 M. The precipitated bacteriophage was removed by centrifugation and purified by CsCl equilibrium density gradient ultracentrifugation on a small scale (Sargeant, 1969). From one 15-liter batch of culture lysate which was purified in this way, 15 gm of pure $\mu 2$, with a plating efficiency of 19%, was obtained.

D. BACTERIOPHAGE R17

This bacteriophage, which was discovered by Paranchych and Graham (1962), was grown in a 20-liter culture on *E. coli*, strain 526 by Sargeant and Yeo (1969). Growth conditions were similar to those used for $\mu 2$ (*vide infra*), and the growing culture was infected at an input ratio of 28 PFU/cell when the cell density was 4.8×10^9/ml. Cultural conditions were maintained for 3 hours and the R17 content of a whole culture sample was found to be 1.7×10^{13} PFU/ml, equivalent to 3540 PFU/cell. From Eq. (1) it can be calculated (assuming that the particle mass of R17 is 3.6×10^6 daltons, and that the absolute efficiency of plating is 1) that not less than 102 mg of R17 were produced per liter.

V. Bacteriophage-Infected Cells

There is an increasing interest in the production of bacteria that have been infected with bacteriophage and allowed to continue growing to a stage prior to lysis. Such cells are used as a source of bacteriophage-induced enzymes and other products involved in the bacteriophage replication process.

Materials for which detailed large-scale isolation procedures have been described include: DNA polymerase from T2-infected *E. coli* B or *E. coli* B/1,5 (Aposhian, 1966), a range of DNA-glucosylating enzymes from T2r$^+$-, T4r$^+$-, and T6r$^+$-infected *E. coli* B or *E. coli* B/1,5 (Zimmerman, 1966), RNA synthetase (Weissman *et al.*, 1966) and double-stranded RNA (Billeter and Weissman, 1966) from MS2-infected *E coli.* Hfr3000.

The T even bacteriophage-infected cells used in the above procedures were grown on a 25- or 50-liter scale with maximal aeration, in a half-filled 50- or 100-liter "Biogen" (American Sterilizer Company) in a glucose, ammonia, and salts medium. They were infected with bacteriopage at an input ratio of 3–5 PFU/cell when the cell density had reached the "late logarithmic phase, 2–4×10^9 cells/ml depending on conditions of aeration." After bacteriophage addition aeration was continued for 15 minutes, the culture was rapidly cooled to 0°–5°C, and harvested by centrifugation. The yield of cells was 2–2.5 gm wet weight per liter from cultures infected at 2×10^9 cells/ml (Zimmerman, 1966, see Table III).

The bacteria infected with MS2 were grown on a 150-liter scale with aeration at 2 ft³/minute before being infected with bacteriophage, and at 4 ft³/minute thereafter, in a tryptone, yeast extract, glucose, and salts medium in a 200-liter fermenter. They were infected with bacteriophage at an input ratio of 33:1 when the cell density was 2×10^8 cells/ml. After a further 30 minutes at 37°C, the whole culture was cooled to 20°C (time taken = 15 minutes) and the cells were recovered by centrifugation. The yield was 350–400 gm wet weight, equivalent to 2.3–2.7 gm/liter (Billeter and Weissman, 1966, see Table III).

In the opinion of this author much higher yields of satisfactory bacteriophage-infected cells could have been obtained in both the cases quoted by employing culture vessels having a very high capacity to dissolve oxygen and by carrying out the bacteriophage infection when the bacterial cell density was much higher.

In support of this suggestion the reader is referred to the work reported earlier in this article, in which cultures were infected with T7, MS2, μ2, or R17 when the cell densities were in some cases very much higher than those used to make the bacteriophage-infected cells. Since these denser cultures lysed satisfactorily and gave rise to very high yields of progeny bacteriophage, it is reasonable to assume that prior to lysis they contained a full complement of bacteriophage-induced products at satisfactory concentrations.

Some work carried out recently on the production of a viral RNA-dependent RNA polymerase, from *E. coli* Q.13 cells infected with the RNA bacteriophage Qβ, demonstrated that over 400 gm of wet cells could be recovered from 20 liters of culture grown in the 20-liter culture vessel described by Sargeant *et al.* (1968). This had a sulfite oxidation capacity of over 220 mmoles oxygen per liter per hour. Cultures were grown at 37°C in a glycerol, casamino acids, and salts medium

TABLE III
BACTERIOPHAGE-INFECTED CELLS

Bacteriophage	Host organism	Scale of operation	Viable bacteria at infection per milliliter	No. of infective particles per bacterium at infection	Recovery of wet bacteria per liter of culture	Reference
T2r$^+$, T4r$^+$ or T6r$^+$	E. coli B or E. coli B/1,5	25 and 50 liters	2×10^9	3–5	2.0–2.5 gm	Zimmerman, 1966
MS 2	E. coli Hfr 3000	150 Liters	2×10^8	33	2.3–2.7 gm	Billeter and Weissman, 1966
QB	E. coli Q13	20 Liters	10^{10}	7	20.3 gm	Sargeant et al (1970)[a]

[a] Average values for 5 experiments are given.

to a cell density of about 10^{10}/ml and infected with bacteriophage at an average input ratio of 7:1. Cultural conditions were maintained for 35 minutes, the cultures were cooled to 5°C by passing them through a heat exchanger during 8–10 minutes, and the cells were recovered by centrifugation. These cells were a richer source of the RNA polymerase than similar cells produced at much lower cell densities in smaller scale cultures (Sargeant *et al.*, 1970).

The results obtained in the preparation of the three different types of bacteriophage-infected cells are summarized in Table III.

REFERENCES

Adams, M. H. (1959). "Bacteriophages." Wiley (Interscience), New York.
Aposhian, H. V. (1966). *In* "Procedures in Nucleic Acid Research" (G. L. Cantoni, and David R. Davies, eds.), pp. 277–283. Harper and Row, New York.
Billeter, M. A., and Weissman, C. (1966). *In* "Procedures in Nucleic Acid Research" (G. L. Cantoni, and David R. Davies, eds.), pp. 498–512. Harper and Row, New York.
Davison, P. F., and Freifelder, D. (1962). *J. Mol. Biol.* **5**, 635–642.
Elsworth, R. (1969). *In* "Methods in Microbiology" (J. R. Norris, and D. W. Ribbons, eds.), pp. 123–136. Academic Press, New York.
Elsworth, R., Williams, V., and Harris-Smith, R. (1957). *J. Appl. Chem.* **7**, 261–268.
Frazer, D., and Jerrel, E. A. (1953). *J. Biol. Chem.* **205**, 291–295.
Hershey, A. D., Kalmanson, G., and Bronfenbrenner, J. (1943). *J. Immunol.* **46**, 267–279.
Isenberg, H. (1963). M.Sc. Thesis, University of Birmingham, England.
Kaesberg, P. (1968). *In* "The Molecular Biology of Viruses" (J. S. Colter and W. Paranchych, eds.), pp. 241–250. Academic Press, New York.
Loeb, T., and Zinder, N. D. (1961). *Proc. Nat. Acad. Sci. U.S.* **47**, 282–289.
Lunan, K. D., and Sinsheimer, R. L. (1956). *Virology* **2**, 455–462.
Luria, S. E., Williams, R. C., and Backus, R. C. (1951). *J. Bacteriol.* **61**, 179–188.
Paranchych, W., and Graham, A. F. (1962). *J. Cell Comp. Physiol.* **60**, 199–208.
Rushizky, G. W., Greco, A. E., and Rogerson, D. L., Jr. (1965). *Biochim. Biophys. Acta* **108**, 142–3.
Sargeant, K. (1969). *In* "Methods in Microbiology" (J. R. Norris and D. W. Ribbons, eds.), pp. 505–520. Academic Press, New York.
Sargeant, K., and Yeo, R. G. (1966). *Biotechnol. Bioeng.* **8**, 195–215.
Sargeant, K., and Yeo, R. G. (1969). Unpublished Data.
Sargeant, K., Yeo, R. G., Lethbridge, J. H., and Shooter, K. V. (1968). *Appl. Microbiol.* **16**, 1483–1488.
Sargeant, K., Yeo, R. G., and Vickers, T. G. (1970). *Biochim. Biophys. Acta* In Preparation.
Sinsheimer, R. L. (1959a). *J. Mol. Biol.* **1**, 37–42.
Sinsheimer, R. L. (1959b). *J. Mol. Biol.* **1**, 43–53.
Sinsheimer, R. L. (1966). *In* "Procedures in Nucleic Acid Research" (G. L. Cantoni and David R. Davies, eds.), pp. 569–576. Harper and Row, New York.
Stent, G. S. (1963). "Molecular Biology of Bacterial Viruses." Freeman, London.
Strauss, J. H., Jr., and Sinsheimer, R. L. (1963). *J. Mol. Biol.* **7**, 43–54.
Swanstrom, M., and Adams, M. H. (1951). *Proc. Soc. Exp. Biol. Med.* **78**, 372–5.

Thomas, C. A., Jr., and Abelson, J. (1966). *In* "Procedures in Nucleic Acid Research" (G. L. Cantoni and David R. Davies, eds.), pp. 553–561. Harper and Row, New York.

Weissman, C., Borst, P., and Ochoa, S. (1966). *In* "Procedures in Nucleic Acid Research" (G. L. Cantoni and David R. Davies, eds.), pp. 340–346. Harper and Row, New York.

Wyatt, G. R., and Cohen, S. S. (1953). *Biochem. J.* **55**, 774–782.

Zimmerman, S. B. (1966). *In* "Procedures in Nucleic Acid Research" (G. L. Cantoni and David R. Davies, eds.), pp. 307–322. Harper and Row, New York.

Microorganisms as Potential Sources of Food[1]

JNANENDRA K. BHATTACHARJEE

Department of Microbiology,
Miami University, Oxford, Ohio

I.	Introduction	139
II.	Inadequacy of Cereal and Vegetable Protein	141
III.	Desirable Characteristics of Microorganisms	142
IV.	Algae	142
	A. Growth Requirements	143
	B. Mass Cultivation	144
	C. Cost of Production	146
	D. Food Value	146
	E. Total Yield	147
	F. Use as Human Food	147
	G. Animal Food	149
	H. Algae in Relation to Space Travel	149
V.	Fungi	149
VI.	Yeast	152
	A. Fermentation of Industrial Wastes and Hydrocarbons	153
	B. Food Value	154
	C. Use as a Protein Supplement	155
VII.	Lichens	155
VIII.	Bacteria	156
IX.	Chemosynthesis	158
X.	Palatability	158
XI.	Conclusion	158
	References	159

I. Introduction

The gap between food supply and human population has widened steadily in recent years and the balance is greatly in favor of population growth. Absence of equilibration is attributed mainly to the ever increasing rate of growth of world population, low death rate (especially among children), and no significant increase in the production of food stuffs, compounded by inadequate distribution. The population growth in the twentieth century is a soaring parameter, whereas the source and the production of conventional food supply is very much at a steady state. The shortage of food supply is more dramatic in some countries than in others. Evidence for this is too numerous and well known to mention here (Ehrlich, 1969; Scrimshaw, 1968). To

---

[1]Supported by Grant No. GB-12130 from the National Science Foundation and the Faculty Research Committee, Miami University, Oxford, Ohio.

more than half of the population of the world, not only the quantity of food is lacking, but the quality of food available is far below the minimum requirement for health and survival. It is estimated that the world's food supply must be doubled by 1980 and tripled by the turn of the century to provide adquate food and nutrition to the increasing world population. Multiple approaches are needed both to overcome this enormous shortage of food and to limit the rate of growth of population. This "concern" has fostered research in recent years in limiting the rate of growth of population. Implementation of some of these efforts has been disappointing. Even if this endeavor is successful, the results will not be achieved for some decades when it would be too late to make up for the food shortage. On the other hand, all available pieces of land hitherto not cultivated, arid, semiarid, otherwise waste, and even polar lands, for example, may be brought under cultivation. Knowledge and resources of modern agriculture and technology can be applied to obtain a maximum utilization of land and highest yield of foodstuff. Yet, all experts agree that the gap between men and the traditional food supply is too large to bridge. Means of improving food supplies are outlined in Fig. 1. In a document entitled "Increasing the Production and Use of Edible Protein" (E/4343, May 25, 1967), The United Nations Economic and Social Council has taken the view that the protein shortage of the world is so serious that every possible means of helping to eliminate this shortage must be investigated—and the sooner the better.

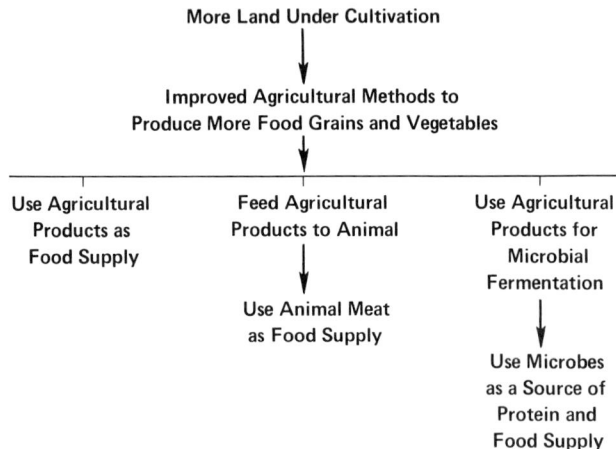

FIG. 1. Means to produce more food.

It is true that the supply of food is critical in this regard, but no less critical is the nutritional value of the food itself. Today a large proportion of the world population is undernourished, sometimes in regions where raw material is available for microbial fermentation. In this article I shall emphasize the potential of microorganisms as a new horizon for human food supply. The importance of microorganisms as potential sources of food stems from the finding that microbial cell matter is especially rich in most B-group vitamins and in protein that contains essential amino acids. They, therefore, constitute potential enrichment for deficient diets. Proteins of microbial origin compare well with conventional protein sources such as egg, milk, meat, and fish in terms of overall pattern of nutrients (Miller, 1968).

The literature on this subject contains so many reports that it is not possible to review all of the literature in this article. The reader may find a broad review of the subject in "Single-Cell Protein" (R. I. Mateles and S. R. Tannenbaum, ed.), M. I. T. Press, Cambridge, Massachusetts, 1968. The term single-cell protein (protein from microorganisms) has been wisely used in the above title and elsewhere in the literature. Perhaps in this way the reluctance of any culture to accept food from microbial origin can be overcome.

II. Inadequacy of Cereal and Vegetable Protein

For a long time it was believed that all proteins had the same nutritional value. It is now well known that proteins vary in their amino acid composition so the emphasis has shifted from the whole protein molecule to the nutritional value of individual amino acids. A list of essential amino acids in the diet varies with the species of animal. For human beings, eight of the amino acids—namely, isoleucine, leucine, lysine, methionine, phenylalanine, threonine, tryptophan, and valine—are essential (Rose et al., 1955). Lack of essential amino acids in the diet causes malnutrition and disease. One such fatal disease is "kwashiorkor" caused by a deficiency of lysine in the diet. Cereal and vegetable diets are invariably lacking in one or more of the essential amino acids, various vegetables being deficient in different amino acids. According to Flodin (1953), the amino acids most needed for improving these deficient diets are lysine, methionine, threonine, and tryptophan. With wheat as a major source of protein, addition of lysine is needed to maintain the balance. Similarly for maize, lysine and tryptophan are required; for rice, lysine and threonine are required; for roots and leguminous vegetables, methionine, lysine, and tryptophan need to be added. The development of an improved variety of corn with high lysine content at Purdue University, Indiana, is an important step toward improving the food value of grain and vegetable protein.

The world's total annual protein supply for human consumption is approximately 82 million tons, of which cereals contribute about 40 million tons. Although cereals are the major source of protein, the protein from cereals is not as high in quality for human food as is meat protein (Altschul, 1968).

III. Desirable Characteristics of Microorganisms

Certain characteristics of a microorganism should be carefully evaluated before it can be considered for the purpose of food. These requirements are as follows[2]:

   a. Rapid growth in a medium of inexpensive, indigenous materials;
   b. Simple nutritional requirements;
   c. Suspended and continuous culture;
   d. Simple separation and harvest;
   e. Resistance to contamination and stable fermentation;
   f. Known genetic and physiological properties and ability to improve genetically;
   g. Efficient utilization of energy source;
   h. Disposable effluence with little or no waste;
   i. Nontoxic and nonallergenic properties;
   j. Palatability and economy;
   k. Protein, fat, and carbohydrate content of high quality;
   l. Simple storage and packaging requirements.

IV. Algae

A food derived from autotrophic microorganisms would be most desirable since their requirement for growth is atmospheric carbon dioxide instead of carbohydrate. The possibility of using algae for food has been studied closely. The greatest interest in algae arises from the fact that they (1) synthesize considerable protein of high quality; (2) can be cultivated continuously; (3) are a rich source of vitamin C and vitamins of the B complex; (4) utilize solar energy more efficiently than ordinary land plants; (5) accumulate appreciable amounts of carbohydrates and lipids; (6) can utilize carbon dioxide as a source of carbon; (7) do not require excessive amounts of water;

---

[2]Adapted from Bunker (1968).

(8) and can be grown in arid and waste land. Some of the blue-green algae can also fix atmospheric nitrogen. Verduin and Schmid (1966) estimated that under optimal conditions algae will yield 7000 gm of protein/meter^2/year as compared to 150 gm for a good crop of wheat or maize. Krauss and Thomas (1954) and Krauss (1955) predicted on the basis of laboratory results a yield of 0.1 ton/acre/day for *Scenedesmus*. In the view of this research, depletion of micronutrients accentuated by the regimen of continuous harvest are the major limiting factors in the growth of *Scenedesmus*.

A. Growth Requirements

The requirements for growth of algae are minimal: light, water, carbon dioxide, and inorganic salts. As for habitat, algae exist mainly as freshwater or marine species. Although proper control of temperature, sunlight, agitation, and harvest are important considerations for the satisfactory yield and continuous cultivation of algae, most of these factors can be resolved by processing and harvesting algae from its natural habitat. This has been practiced for many years in the Orient and in the South Pacific. This technique, however, severely limits its use on a worldwide scale. A more universal approach would be to grow algae under controlled laboratory conditions. Excessive use of DDT and other pollutants is already a serious threat to the natural habitat of algae and other resources in water. This loss of water vegetation may be reflected by the possible imbalance of the earth's total oxygen supply.

Algae convert energy from sunlight very efficiently into useful food material with little or no inedible waste, an important economic consideration. Algae can be cultivated even in agriculturally useless areas and can be easily adapted to an industrial process for mass cultivation, a fact not applicable to all microorganisms.

Depletion of nutrients, slow gas exchange, possible accumulation of toxic products, accessibility to a good light source, harvesting, and drying are some of the difficulties to be considered in the controlled cultivation of algae. Continuous-culture devices and constant agitation, either mechanical or forced air, are recommended for mass cultivation. Agitation serves to disperse the cells for more uniform growth and helps each cell to receive adequate sunlight. These techniques reduce the chance for depletion of nutrient and accumulation of any toxic product, although precaution should be taken against contamination by other algae, bacteria, fungi, or protozoa. Carbon dioxide added to the medium improves gas exchange and helps to agitate the culture.

A constant supply of light is important for photosynthetic organisms and can be supplied externally by special fluorescent lamps either surrounding the culture or with the culture tube surrounding the lamp.

Sorokin and Myers (1953) developed a strain of *Chlorella*, 7-11-05, suitable for growth in a simple medium. The optimal pH is about 6.0 to 6.5 and the optimal temperature for this strain is about 37°C. During active growth of algae, the entire synthetic activity is directed toward the formation of fresh cell material. Analysis of this *Chlorella* strain reveals that as much as 60% of the organic matter of the cell is protein; carbohydrate and fat represent 35% and 5%, respectively (McDowell and Leveille, 1963). *Chlorella* strain 7-11-05 was found to contain more protein than an equivalent amount of dried beef, soy bean meal, or even yeast on a dry weight basis.

B. Mass Cultivation

Cultivation of algae on a large scale would require a very large surface area. According to Oswald and Golueke (1968), present volume of algal culture in the world is about 10^8 liters. They estimated that the entire protein requirement of the United States could be satisfied with 10^{13} liters and all of the world's protein requirement could be met by 10^{15} liters of algal culture (Fig. 2).

Some large-scale production of algae is carried out presently in Czechoslovakia, Japan, and in the United States. At Trebon in Czechoslovakia, the largest unit has a surface area of 900 meters^2. The organism favored is *Scenedesmus quadricauda*. This species has a larger (30 μ) and heavier cell and is more easily recovered than *Chlorella*. In Japan, units in excess of 100,000 liters are in operation (Tamiya, 1955, 1959). Algae are also cultivated in Leningrad (U.S.S.R.), in a 10,000-liter unit. A one-million-liter pilot algae production plant is in operation in the University of California Engineering Field Station, Richmond, California (Oswald and Golueke, 1968).

Controlled, large-scale culture of algae began in the late 1940's and in the early 1950's with the pioneering work of Jorgensen and Convit (1953), Ketchum et al. (1949), and Spoehr and Milner (1949). In early research, the culture varied from 1 to 1000 liters in volume. Arthur D. Little Company in Cambridge (England) in 1951 experimented with *Chlorella* cultures varying from 2000 to 5000 liters, and in 1953 the University of California operated cultures of 10,000 liters (Gotaas et al., 1954). Small industries have been developed on the west coast of Ireland and in Scotland for processing seaweed, chiefly *Laminaria*,

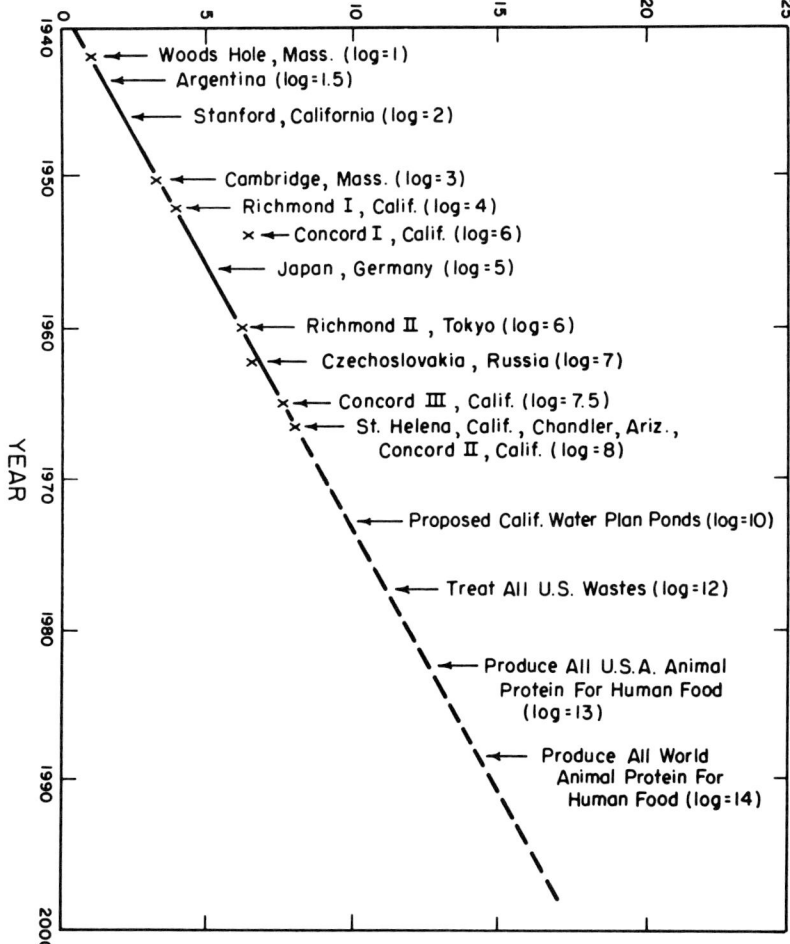

FIG. 2. Past, present, and future large-scale algal cultures. (Oswald and Golueke, 1968). Copyright © 1968, The M.I.T. Press; reprinted by permission.

Fucus, and *Ascophylum,* into a feeding meal. Milner (1955) estimated the possible yield of algae under ordinary conditions of light and heat in different parts of the world, for example, 120 tons dry wt/acre/year in California and 80 tons in Holland.

C. Cost of Production

On the basis of earlier research, Fisher (1955) concluded that the cost of production of *Chlorella* would be high; the price was estimated at 20–50 cents per pound. The high price was attributed to the costly temperature control systems, expensive pond linings and mixing equipment, elaborate nutrient-compounding systems, and highly specialized separation and drying systems. However, Oswald *et al.* (1963) have subsequently studied the algal growth in waste products and proposed that they could be produced at a much lower cost than had been previously estimated. According to these researchers, algae can serve as a low-cost source of protein and the large-scale cultivation of algae can contribute significantly and economically to the future protein resources of the world.

Production of algae may also play a significant role in sewage disposal. The organic matters in sewage are utilized in shallow ponds for the production of algae (Myers, 1951). Lodge and Isaac (1958) estimated that a ton or more of algae could be produced from a million gallons of sewage, and a yearly yield of 30 tons dry wt of algae per acre is possible at a cost of 3–6 cents a pound.

D. Food Value

Fink *et al.* (1954; Fink and Herold, 1955) have demonstrated the high nutritive value of algal protein. Algal food is nutritionally satisfactory and its chief use would be its contribution to dietary protein for humans. The protein content of algae such as *Chlorella* is considerably higher than that of terrestrial plants including legumes. The protein content of alfalfa hay is about 15% on a dry weight basis while that of *Chlorella* cells is 40–60%. On the basis of protein content algae have at least three- to four-fold advantage over leguminous crops. McDowell and Leveille (1963) reported that *Chlorella pyrenoidosa* has a protein content of 40–60%, depending on growth and processing conditions. All essential amino acids are present in algal protein; however, the level of methionine is low, lysine and threonine are high, and levels of the rest of the essential amino acids are adequate. There are no exceptional features about the algal amino acids. Vitamins B and C are present in adequate amounts, and β-carotene and vitamin K are plentiful. It is estimated that 600 gm of algae per day would be nutritionally adequate with regard to proteins and vitamins for an individual. By selection and breeding process along with genetic analysis an improved variety of a species in terms of nutritive value and protein content can be obtained. Research in this area of algal study is very much lacking. Palatability of algae is not considered here as a serious problem; so far, the poor odor and taste of algae have been found to be associated with bacterial contamination. A proper method of cultivation and improved processing may contribute positively to the aesthetic value of algal food.

E. TOTAL YIELD

Verduin and Schmid (1966) projected a rough estimate of the earth's daily yield, assuming one-half of its land surface to be covered with panels of algal culture. One-half of the earth's land surface is 2.5×10^{14} meter^2. Each square meter would yield 50 gm dry wt of algae per day of which 25 gm are protein; the caloric value is approximately 190 K calories. This amounts to a daily global yield of 125×10^{14} gm of protein per day and 470×10^{14} K calories of energy fixed per day. If the dry matter is used directly for human consumption, it would support a population of 10^{13}, or 3000 times the present world population. If it becomes necessary, the panels of algal culture can be floated on the ocean, thus making an additional 73% of the earth's surface available for the cultivation of algae for human consumption. Oswald and Golueke (1968) estimated that with proper cultivation, at least 20 tons (dry wt) of algae having a protein concentration of 50%, will be produced per acre of pond per year. This yield is 10 to 15 times greater than that of an acre of land planted with soy beans and 25 to 50 times that of planted with corn (Table I).

F. USE AS HUMAN FOOD

Algal food is nutritionally satisfactory and its chief use would be its contribution to dietary protein. Algal protein is rich in lysine and threonine and poor in methionine. The level of methionine can be increased by either selecting a suitable constitutive mutant or by effecting amino acid substitution due to point mutation (DeZeeuw, 1968). It has been shown for tryptophan synthetase (Yanofsky, 1963) and in the case of cytochrome *c* (Margoliash and Schejter, 1966) that some amino acid substitutions can be obtained without impairing the biological activity of the protein. Also, product fortification with methionine-rich protein from a different microorganism such as *Fusarium* or *Rhizopus* can overcome this deficiency.

A blue-green alga, *Spirulina maxima,* is known to have been consumed in the Island of Chad since ancient times (Clement, 1968). Algal soups fed to patients in a lepers' hospital in Cabo Blanca, Venezuela, were reported palatable, nutritious, and beneficial

TABLE I
USE OF LAND FOR FOOD AND FEED PRODUCTION WITH CONVENTIONAL CROPS AND WITH ALGAE[a]

Crop	Land use total area (millions of acres)	Area of microalgae culture required for equal production (millions of acres)	
		Total free-energy basis	Protein basis
Corn	82	4.1	1.67
Hay	69	3.2	1.00
Wheat	53	1.1	0.55
Oats	27	0.8	0.25
Soybeans	23	0.4	1.10
Sorghum	15	0.4	0.20
Barley	14	0.3	0.14
Totals	283	10.3	4.91

[a] Oswald and Golueke, 1968. Copyright ©1968, The M.I.T. Press; reprinted by permission.

(Jorgensen and Convit, 1961). Experiments on large-scale algal culture for human feeding are being conducted at the U. S. Army Natick Laboratories (Mathern, 1965). A systematic feeding of 50–100 gm of dry biomass of algae in human diets is known to have no adverse effect. It was concluded in one report of the U. S. Department of Commerce in 1967 that up to 100 gm of unicellular algae could be included in human diets with no significant disadvantage.

Only in the Far East have algae been regularly used as human food. In the Pacific Islands, the raw algae are chopped and added to other dishes. *Laminaria*, primarily used in Japan, is popularly known as "Kombu." Young stipes of *Laminaria* have also been eaten without much further preparation in Europe. The most prolific users of seaweeds are the coastal populations of China and Japan. In Japan, *Chlorella* is used as an additive to yoghurt, ice cream, green bread, and related products (Mitsuda *et al.*, 1967). The alga *Porphyra* is made into a pulp and used in bread, with oatmeal, and in other dishes. *Porphyra* is also used as an edible wrapper in cooking fish and is popularly known as "Sushi." It is also used in preparation of Japanese macaroni, soups, and sauces. *Caulerpa racimosa* and *Porphyra* are cultivated for food in the Philippines. "Dulse," prepared in several countries from red algae, is eaten like candy and used as a relish in potatoes and in soups. In spite of these broad uses of algae as food in different parts of the world, much research needs to be done in the area of the use of algae as a principal item in the diet. The cell wall of *Chlorella* is quite resistant to breakdown and may not be utilized efficiently by man. Both cooking and enzymatic digestion have been used to effect the breakdown. Use of a second organism to feed on algae is yet another possibility to effect the digestion of algal wall.

G. Animal Food

Early research in animal feeding mainly involved work with rats and chicks. In the early 1960's, the grain processing corporation of Muscatine, Iowa, produced about two million pounds of the green algae *Spongeococcum* on corn-steep liquor for the purpose of improving yellow pigmentation in poultry and poultry products. The process has been discontinued, at least temporarily, because a more conventional and less costly source of the pigments has been found. Pilot-scale chicken feeding experiments with waste-grown algae were also carried out by the North American Rockwell Corporation. Large-scale feeding experiments to higher animals with waste-grown algae are now being carried out at the University of California (Oswald and Golueke, 1968). The addition of small quantities of vitamin B_{12} to algal–barley mixtures containing 10% algae have been found to be adequate for swine. Algae up to 10% level of food also have been found satisfactory for sheep. The ruminants, with the help of bacteria, are known to digest algae more readily than nonruminants. Feeding algae to produce animal protein is an inefficient process. However, an exploitation of this possibility will release additional land for crop production.

H. Algae in Relation to Space Travel

The use of algae in bioregeneration and as a food during long or distant space travel may offer some advantages over other microorganisms (Lachance, 1968). However, the feasibility of such an ecological system for a shorter space flight is not apparent. A good discussion of the problems related to food and nutrition during space travel appears in a panel discussion organized by the Federation of American Societies for Experimental Biology [*Federation Proc.* **22**, 1451 (1963)].

V. Fungi

Various fungi have been used as food for many years. Edible mushrooms belong to mankind's earliest foods. *Agaricus campestris,* a common mushroom, along with other edible fungi, was familiar to the Romans and Greeks. Even today in some parts of the world (U.S.S.R., Continental Europe, Scandinavia, Canada, and parts of the United States) mushrooms constitute an important item of food. Chinese for several centuries cultivated annually millions of pounds of mushrooms, *Cortinellus berkelyanus*, on logs. The "Padi-straw" mushroom, *Volvaria volvacea*, also has been cultivated in large quantities in South China, Malaya, and the Philippines. In Germany during World War II a fungus, *Geotrichum candidum*, was used to supplement human foods because of a shortage of vitamins and proteins. *Geotrichum lactis*, regarded as a useful source of food, is able to utilize such substrates as whey and sulfite waste liquor (Bunker, 1968). Fungi are used today to provide supplements such as methionine, glutamic acid, riboflavin, and vitamin B_{12}.

The possibility of using members of the class Fungi Imperfecti as sources of protein food was investigated by Gray et al. (1964). This particular class of fungi has a rapid growth rate and also can utilize different crude, raw products, such as sweet potatoes, corn, rice, cassava root, beet molasses, paper pulp, and waste products of food processing industries (Gray and Abou-El-Seoud, 1966). Thus, various species of this group are able to convert these cheap carbohydrates and inorganic salts of nitrogen into protein-rich fungal mycelia which, in turn, can be used as food. Large amounts of water are generally required for the growth of fungi imperfecti. Gray and his co-workers substituted seawater for freshwater in culture medium and reported a higher yield of protein compared to controls in distilled water in 18 of 20 species tested. They attributed higher production of protein to the magnesium ion content of seawater (Gray et al., 1963). These researchers also noted that the amino acid composition of fungal mycelia was as satisfactory nutritionally as casein as a source of essential amino acids. The mycelial preparations in most cases were odorless and tasteless with a high caloric value. Bradley (1962) has described "hybrid vigor" in fungi.

Fusarium, Candida, and *Rhizopus* were incorporated into human diets during the last world war and were found satisfactory. *Fusarium* and *Rhizopus* contain high concentration of cystine and methionine. Thus, protein from these organisms can be used to enrich algal and bacterial proteins which generally have a low methionine content. Vinson et al. (1945) showed that species of *Fusarium* were better than brewers' yeast as a source of protein. It may be noted that for the mass cultivation of fungi, no elaborate equipment need be designed since the equipment and techniques already used in the antibiotic and distilling industries can be applied directly. Application of submerged culture procedure for mass cultivation of fungi has been discussed by Litchfield (1968). Gray (1962) made an interesting comparison of yields of protein per acre of corn with yields per acre when the grain carbohydrate is used as feed for livestock or used as a source of carbon and energy in the production of fungal protein (Fig. 3). The efficiency

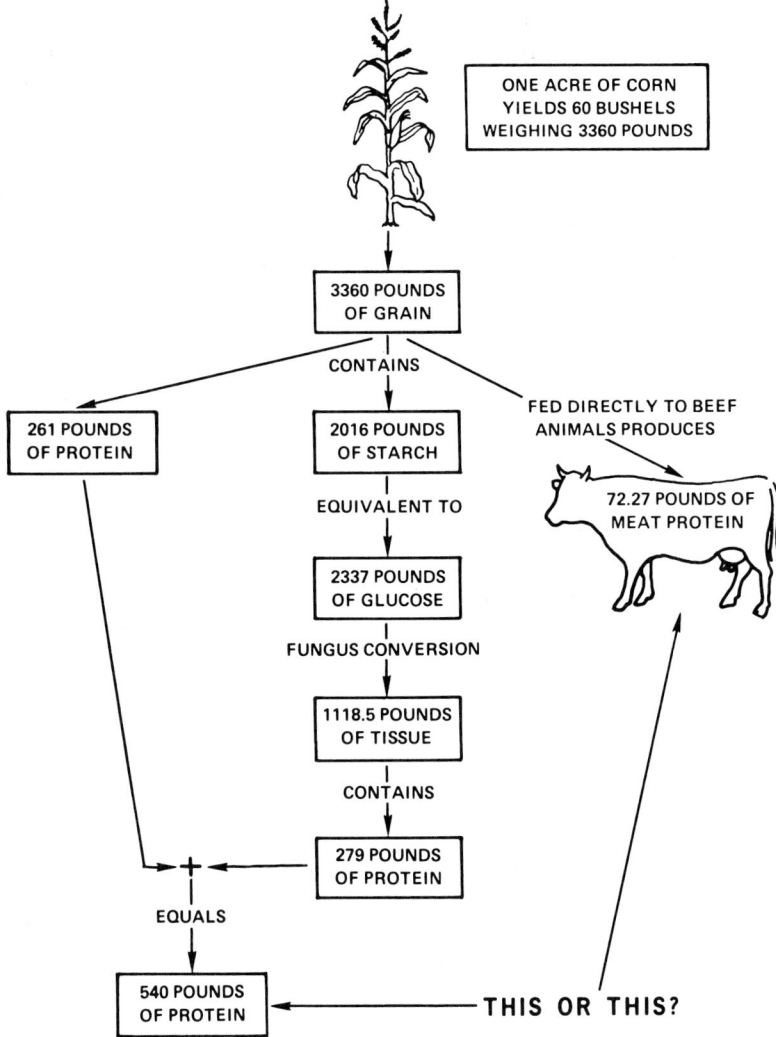

FIG. 3. Efficiency in the production of protein from an acre of corn when fed directly to beef animals and when corn carbohydrate is used in the synthesis of fungal protein (Gray, 1962). Reprinted from *Develop. Ind. Microbiol.* **3,** 63 with permission.

of microbial conversion appears to be much higher than animal conversion. Minimum daily protein requirement per person is approximately 65 gm. This daily protein requirement can be met by 150 gm of algae or fungi, 8 lb of potatoes, or 32 lb of fresh sugar cane. If man depended on potatoes alone as a protein source, a crop of 15 billion acres (75% of all available land) would be needed to meet this demand. Average annual meat consumption per person in the United States is an estimated 178 lb, which is far less than adequate in terms of protein requirement. A maintenance of the present level of consumption would require an increase of beef cattle by one million head annually.

VI. Yeast

Of all the microorganisms studied, yeasts have probably the most favorable characteristics for use as a major source of food. However, the use of this group as a means of meeting food and protein needs in particular has lagged far behind existing knowledge of its nutritive value. Yeasts have been used since ancient times in baking and the large-scale cultivation of yeast for nutritional use was carried out in Germany during both world wars. Germans incorporated about 16,000 tons of *Candida utilis* per year into human food during World War II. It is estimated that in Britain alone, over 30,000 tons of surplus brewers' yeasts are used annually for animal food, yeast extract, and in dietary supplements. The present annual global production of food and feed yeast from carbohydrate sources such as sulfite waste liquor, molasses, and agricultural wastes is about 250,000 tons (Bunker, 1964; Humphrey, 1968). A 100,000-ton mash capacity fermentation plant is under construction in Czechoslovakia to produce protein from certain unwanted components of crude oil. Large-scale yeast fermentation plants are already in operation in the Soviet Union, France, China, Japan, and the United States.

Yeasts contain large proportions of high quality protein, carbohydrate, and lipid. Yeast also is one of the richest sources of the vitamin B complex group. The yeast most commonly used as a food today is *Candida utilis* because of its ability to utilize sulfite waste liquor and pentoses. Another point in its favor is that its growth requirements are less exacting, being able to utilize nitrates and urea. The variability of the protein-to-vitamin ratio from different species under varied growth conditions is small. The aerobic yeast *Rhodotorula gracilis* is an active producer of lipid. Protein content of *Saccharomyces cerevisiae* is higher than *Saccharomyces carlsbergensis*.

A. FERMENTATION OF INDUSTRIAL WASTES AND HYDROCARBONS

Whey, molasses, and spent sulfite liquor are fermented by *Saccharomces fragilis*, *Candida utilis*, and *Saccharomyces cerevisiae*, respectively. The wastes of the paper industry consist of pentoses and hexoses. The presence of sulfurous acid and lignin creates problems in the disposal of the wastes. Industrial wastes can be utilized in appropriate fermentations by yeast. In the United States, about one quarter of all sulfite waste liquor solids are fermented by *Candida utilis*, representing a production of about 50,000 tons of yeast per year (Mead, 1958). After the fermentation, cells are removed and vacuum-packed into drums. The waste water may be used for irrigation of soil. Spraying of the unfermented pulp water has an unfavorable effect on the soil; however, the fermented waste water results in high agricultural yields.

Whey, a byproduct of the cheese industry, has also been used as a substrate for the production of food yeasts (Wasserman, 1960). Such yeast contains thiamine, riboflavin, and ascorbic acid while the protein is nutritionally similar to high grade plant protein. The discarded whey in the United States represents a reservoir of approximately 204,000 tons of sugar and 36,000 tons of protein. Conversion of this reserve into utilizable products may also solve the problem of disposal.

Some yeasts are also capable of fermenting hydrocarbons. Many reports have been published on hydrocarbon-assimilating yeasts (Yamada *et al.*, 1968). The tremendous value of hydrocarbons as substrates lies in their abundance and insignificant cost. It is estimated that at the present time yeast suitable for animal consumption can be produced at 5–10 cents per pound. *Candida tropicalis* has been successfully cultivated in a hydrocarbon medium at a very high temperature (Raymond, 1961). Production of yeast cells on straight- or branched-chained hydrocarbons does not require any different approach compared to growth on nonhydrocarbon substrates. The world's annual yeast production is estimated at 150,150 tons of bakers' yeast and 187,700 tons of dried yeast. A conversion of all available molasses, sulfite liquor, and whey would result in 915,000 tons molasses yeast, 510,000 tons sulfite yeast, and 400,000 tons whey yeast (Peppler, 1968). Cellulose, numerous agricultural products, and waste materials of certain industries can be added to the vast quantities of spent sulfite liquor and whey for fermentation by yeast. Vast quantities of cheap fermentable substrates, which can be easily converted into yeast proteins and vitamins, are available in countries with an acute food shortage.

B. Food Value

All of the essential amino acids except methionine and cystine are present in adequate quantities in yeast protein (Carter and Phillips, 1944). *Candida* serves as a good source of riboflavin and pantothenic acid. High production of lipids is possible with such species as *Trichosporon pullulans, Candida utilis,* and *Saccharomyces cerevisiae. Rhodotorula gracilis* is, however, the most promising microbial lipid producer.

Nutritional studies on yeast reveal that crude protein accounts for 45–55% of dry weight. Feeding experiments with rats showed yeast being readily digested and absorbed and that these nutrients could provide up to 94% caloric value in the diets. Feeding experiments with dogs and men showed a respective utilization of 80 and 90%. Yeast must be killed and dried before consumption to avoid any undesirable effect. Live yeast depletes the B vitamins in the intestines of mammals causing avitaminosis.

Yeast has been used as fodder successfully for certain animals such as horses and cows and as feed for poultry. According to Carter, when fed to cows, yeast will increase the production of high quality milk. Pigs also have been noted to utilize yeast satisfactorily. In humans, the high purine content of yeast causes production and excretion of large amounts of uric acid. This results in an increased level of uric acid in the blood. Also, high blood pressure has been attributed to food yeast (Carter and Phillips, 1944). However, an increased purine intake may not necessarily result in a corresponding increase in blood uric acid level. Undoubtedly, further research is needed before marketing yeast or yeast products for human consumption. Utilization of yeasts as fodder can certainly play an important role in relieving the world's food shortage. The Medical Research Council of Great Britain has been conducting useful research in relation to utilization of yeast by human beings (Bunker, 1968).

The potential of yeast as a food supplement lies not only in its ability to utilize cheap raw material, but also in its rapid growth rate, palatability, and lack of pathogenicity. Both genetic and biochemical principles of yeasts have been extensively studied. Genetically stable strains of specific nutritional quality can be selected easily (Braun, 1964).

A young pig or chicken may double its weight in a month, but a yeast cell does this in less than 2 hours. A yeast factory with ten large-sized fermenting tanks can furnish 10 tons of yeast per day on a continuous basis. The protein equivalent of these yeasts would require killing 80 pigs a day or 30,000 pigs in a year. Microorganisms like yeasts can circumvent the limitations of time. The large quantity of cheap raw materials for the growth of microbes can be realized, among other things, in wastes from factories and sewage plants.

According to Willcox (1959), the best sugar cane varieties can produce up to 44,662 lb of dry matter/acre/year. This dry matter is 80% fermentable and can yield approximately 20,000 lb of dry yeast containing 10,000 lb of protein. This protein can be combined with edible nonprotein materials such as cassava or any other high-starch foods grown in another acre of land. This process would result in enough protein food to feed approximately 60,000 people per square mile.

C. Use as a Protein Supplement

Cereal grains contain low amounts of protein, deficient mainly in lysine. Among cereal grains, wheat flour supplemented with various strains of yeast has been studied extensively. Several researchers reported an increased biological value of the wheat flour enriched with proteins from dried food yeast. This enrichment was attributed to the essential amino acid lysine. Similar enrichment of corn meal with dried food yeast was attributed to lysine and tryptophan of yeast protein (Sure, 1948; Kon and Markuze, 1931). Cremer *et al.* (1951) used human subjects to demonstrate the increase in biological value of white bread by supplementing DL-lysine or dried yeast. According to Sure (1948) and Tsien *et al.* (1957), enrichment of wheat flour supplemented with 3 to 5% dried yeast was greater than could be obtained with a quantity of lysine equal to the dried yeast. It was concluded that in addition to lysine, yeast provides essential nutrients in terms of total protein and other nutrients. Yeast has also been incorporated directly into human food (Bressani, 1968). Lysine-rich yeast can be produced in the molasses media supplemented by 2-ketoadipic acid (Jensen and Shu, 1961).

VII. Lichens

Lichens are symbionts composed of algae and fungi and an extreme case of mixed cultures. Lichens, like mushrooms, have been used as food for many years, dating back to the ancient Egyptians who added it to bread. Lichens are widely distributed. Perez-Llano (1944) described a species of lichens, *Centraria islandica,* which he claimed as suitable for human consumption. However, lichens, in general, have a bitter taste and they also cause irritation to the digestive tract. Perez-Llano has reported various carbohydrates from lichens. Polysaccharides are the most common. Lichens are not particularly rich in vitamins, but they may be of value for their inorganic salts and carbohydrates. Studies on the short and long growth forms of lichen reveal that man can tolerate short forms better than the long ones. Short forms also appear to be more palatable.

VIII. Bacteria

Some of the important considerations that make bacteria prospective candidates as a potential source of food are rapid growth rate, high protein content, and ability to grow on hydrocarbons. The specific growth rate of various microorganisms has been compared by Bunker (1968). *Escherichia coli,* for example, multiplies about 4 times as fast as yeast and about 40 times as fast as algae.

Protein content of bacteria is relatively high compared to yeast or algae. Several nonpathogenic species of bacteria are known to have a higher than 80% protein content: *Lactobacillus fermentans,* 87%; *Alcaligenes viscosus,* 84%; *Escherichia coli,* 82%. The greater ability of bacteria to produce protein in comparison to other sources of food is given in Table II. Bacterial protein includes all the essential amino

TABLE II
COMPARISON OF PROTEIN PRODUCTION[a]

Origin	Protein produced per day (lb)
1000 lb Steer	1
1000 lb Soybean	100
1000 lb Yeast	100,000
1000 lb Bacteria	100,000,000,000,000

[a]Ogur, 1966. Reprinted from *Develop. Ind. Microbiol.* 7, 216 with permission.

acids. In nutritional studies with rats, Kaufman *et al.* (1957) have shown that drying the cells would significantly increase the digestibility of bacterial protein. These authors also compared the digestibility of two different bacterial species in rats and demonstrated that *Lactobacillus arabinosus* was more easily digested than *Escherichia coli.* Roberts (1950, 1953) reported that *E. coli* grown in a simple medium under aerated conditions yields a good protein supplement for rats and chicks. He found that dried *E. coli* cells were not toxic, but the culture fluid was toxic. Roberts concluded that *E. coli* protein was as nutritive as animal protein. Reusser *et al.* (1957) compared the protein content of nineteen species of bacteria, actinomyces, and yeast. They showed that all of the essential amino acids were present in bacterial protein, but in varying amounts. The amount of lysine was high, those of tryptophan and threonine were low, and the methionine content was poor. The level of methionine can be improved either by strain selection or by varying conditions during growth or by mixed cultures. Bacteria also produce substantial amounts of vitamins. Further knowledge regarding the toxicity and digestibility of bacteria should be gained before using bacteria as a source of food. Suitable methods for the commercial cultivation of bacteria are well within the knowhow of modern technology and research. A great deal of knowledge has been gained in recent years regarding the genetics and physiology of bacteria. Information gained from these studies is sometimes useful in nutritional and related studies with bacteria.

Ability of bacteria to grow on hydrocarbons is well known and the literature in this area has been reviewed periodically (Zobell, 1946; Treccani, 1963; Yamada *et al.*, 1968). Microbiology of coal and petroleum has been reviewed by Davis (1956). Takahashi *et al.* (1963) reported production of bacterial cells in hydrocarbon, particularly kerosene. One particular strain of *Pseudomonas aeruginosa*, isolated from soil, was reported to assimilate *n*-docosane and 1-octadecene. This strain could utilize 89% of the hydrocarbon added in the media for the production of cell mass. The amount of protein produced by this strain was 58% on the basis of dry weight of cells. Utilization of 1-octadecene for the growth of *Pseudomonas aeruginosa* was also reported by Ertola *et al.* (1965). The metabolic pathways employed by microorganism have been discussed by Treccani (1963).

Usefulness of bacteria to eliminate petroleum wastes and to be subsequently used as a foodstuff merit discussion. Most petroleum or petroleum waste products undergo some microbial decomposition. Aerobic microorganisms are capable of growing at the sole expense of paraffins. Utilization of inexpensive industrial wastes as media for the production of possible food and protein demands further research and development. This may also partially solve the problem of waste disposal and pollution. Ability of bacteria to grow on sewage and in other wastes can be beneficial in terms of waste disposal and food supply. Bacteria grown on sewage wastes and used as food for livestock are known to have no harmful effect.

IX. Chemosynthesis

At least a brief discussion of the possibility of chemical synthesis of food seems appropriate. According to experts, the synthesis of bulk food constituents is unlikely to offer a lasting solution to the problem of food shortage (Borgstrom, 1964). There is not enough cheap raw material available. Coal and oil may be mentioned as most promising raw materials for food, but these raw materials are already in great demand as fuel and precursors for various other products, including fibers. The present annual rate of growth of the world population is above 70 million. This annual increase alone would require an estimated 15 million tons of food each year based on approximately 220 kg of food calculated as dry matter per year per person. This quantity is above the present annual production capacity of the synthetic organic chemical industry of the United States. The investment required to synthesize this additional food material would be $16 billion, which is more than double the value of the synthetic organic chemical industry and more than eight times the annual investment in the entire organic chemical industry of the United States. Nevertheless, chemical synthesis may and should play an important role in supplementing food. The United States is producing synthetically no less than 5.5 million kg of vitamins. Ironically, very little of this reaches the poor and hungry where it is needed most. More can and should be done in the field of chemical synthesis with respect to amino acid, carbohydrates, lipids, proteins, and vitamins (Mrak, 1964).

X. Palatability

What can be said about palatability? The major problem so far with microbial foods is neither the quality nor the quantity, but the acceptibility. It is obvious that the taste, flavor, and texture of foodstuffs, no matter from what source they are derived, must be acceptable to the consumers. It is important from aesthetic and emotional aspects, as well as from the nutritional point of view. People must eat a given food to benefit from its nutritional value. The present state of research and development in the artificial flavoring and coloring of food should be able to satisfy this need. Once a particular food is aesthetically acceptable, eating customs can be changed with very little effort.

XI. Conclusion

Microorganisms provide an excellent supplement for conventional foodstuffs. Microorganisms can be utilized to furnish an unlimited supply of nutrient without requiring the cultivation of more and more land. Food supplements from microorganisms can be cheap, palatable, wholesome, and without any toxic effect. The need for further research and development in the field of microbial food must be met with the same urgency and intensity as that devoted to the space program. Food industries and the governments of the developing countries should be encouraged to remedy deficient diets with enriching supplements of microbial origin. The food supply of the world and the potential of microbial foods need be constantly evaluated and discussed on an international basis within and beyond the scientific community.

Acknowledgement

The author is grateful to the M. I. T. Press, Cambridge, Massachusetts and the Society for Industrial Microbiology for their permission to reproduce Fig. 2, Table I, and Fig. 3, Table II, respectively. The author is also indebted to Dr. Robert J. Brady for helpful criticism of the manuscript.

References

Altschul, A. M. (1968). *In* "Single-Cell Protein (R. I. Mateles and S. R. Tannenbaum, ed.), p. 48. M. I. T. Press, Cambridge, Massachusetts.
Borgstrom, G. (1964). *In* "Global Impacts of Applied Microbiology" (M. P. Starr, ed.), p. 130. Wiley, New York.
Bradley, S. G. (1962). *Ann. Rev. Microbiol.* **16**, 35.
Braun, W. (1964). *In* "Global Impacts of Applied Microbiology" (M. P. Starr, ed.), p. 75. Wiley, New York.
Bressani, R. (1968). *In* "Single-Cell Protein" (R. I. Mateles and S. R. Tannenbaum, ed.), p. 90. M. I. T. Press, Cambridge, Massachusetts.
Bunker, H. J. (1964). *In* "Global Impacts of Applied Microbiology" (M. P. Starr, ed.), p. 234. Wiley, New York.
Bunker, H. J. (1968). *In* "Single-Cell Protein" (R. I. Mateles and S. R. Tannenbaum, ed.), p. 67. M. I. T. Press, Cambridge, Massachusetts.
Carter, H. E., and Phillips, G. E. (1944). *Fed. Proc.* **3**, 123.
Clement, G. (1968). *In* "Single-Cell Protein" (R. I. Mateles and S. R. Tannenbaum, ed.), p. 306. M. I. T. Press, Cambridge, Massachusetts.
Cremer, H. D., Lang, K., Hubbe, I., and Kulik, U. (1951). *Biochem. Z.* **322**, 58.
Davis, J. B. (1956). *Bacteriol. Rev.* **20**, 261.
DeZeeuw, J. R. (1968). *In* "Single-Cell Protein" (R. I. Mateles and S. R. Tannenbaum, ed.), p. 181. M. I. T. Press, Cambridge, Massachusetts.
Ehrlich, P. R. (1969). *The Biologists* **51**, 8.
Ertola, R. J., Lilly, M. D., and Webb, F. C. (1965). *Biotech. Bioeng.* **7**, 309.
Fink, H., and Herold, E. (1955). *Naturwissenschaften* **42**, 516.
Fink, H., Schlie, I., and Herold, E. (1954). *Naturwissenschaften* **41**, 169.
Fisher, A. W., Jr. (1955). *Proc. World Symp. Appl. Solar Energy, Stanford Res. Inst., Menlo Park, California*, p. 245.

Flodin, N. W. (1953). *J. Agr. Food Chem.* **1**, 222.
Gotaas, H. B., Oswald, W. J., and Golueke, C. G. (1954). *Sanit. Eng. Res. Lab., Univ. Calif., Berkeley, Calif., Ser.* **44**, No. 5.
Gray, W. D. (1962). *Develop. Ind. Microbiol.* **3**, 63.
Gray, W. D., and Abou-El-Seoud, M. (1966). *Develop. Ind. Microbiol.* **7**, 221.
Gray, W. D., Pinto, P. V. C., and Pathak, S. G. (1963). *Appl. Microbiol.* **11**, 501.
Gray, W. D., Oach, F. F., and Abou-El-Seoud, M. (1964). *Develop. Ind. Microbiol.* **5**, 384.
Humphrey, A. E. (1968). *In* "Single-Cell Protein" (R. I. Mateles and S. R. Tannenbaum, ed.), p. 330. M. I. T. Press, Cambridge, Massachusetts.
Jensen, A. L., and Shu, P. (1961). *Appl. Microbiol.* **9**, 12.
Jorgensen, J., and Convit, J. (1961). *In* "Algal Culture from Laboratory to Pilot Plant" J. S. Burlew, ed.), p. 190. Carnegie Inst., Washington, D.C.
Kaufman, B., Nelson, W. O., Brown, R. E., and Forbes, R. M. (1957). *J. Dairy Sci.* **40**, 847.
Ketchum, B. H., Lillick, L., and Redfield, A. C. (1949). *J. Cell. Comp. Physiol.* **33**, 267.
Kon, S. K., and Markuze, Z. (1931). *Biochem. J.* **25**, 1476.
Krauss, R. W. (1955). *Sci. Mon.* **80**, 21.
Krauss, R. W., and Thomas, W. H. (1954). *Plant Physiol.* **29**, 205.
Lachance, P. A. (1968). *In* "Single-Cell Protein" (R. I. Mateles and S. R. Tannenbaum, ed), p. 122. M. I. T. Press, Cambridge, Massachusetts.
Litchfield, J. H. (1968). *In* "Single-Cell Protein" (R. I. Mateles and S. R. Tannenbaum, ed.), p. 309. M. I. T. Press, Cambridge, Massachusetts.
Lodge, M., and Isaac, P. C. G. (1958). *Verh. Int. Ver. Theoret. Angew Limnol.* **13**, 828.
McDowell, M. E., and Leveille, G. A. (1963). *Fed. Proc.* **22**, 1431.
Margoliash, E., and Schejter, A. (1966). *Advan. Protein Chem.* **21**, 113.
Mathern, R. O. (1965). "The Potential of Algae for Food," U. S. Army Laboratories, Natick, Massachusetts, Vol. 73, p. 1.
Mead, S. W. (1958). *Rep. Sulphite Pulp Mfg. Res. League.*
Miller, S. A. (1968). *In* "Single-Cell Protein" (R. I. Mateles and S. R. Tannenbaum, ed.), p. 79. M. I. T. Press, Cambridge, Massachusetts.
Milner, H. W. (1955). *Sci. Mon.* **80**, 15.
Mitsuda, H., Yasumota, K., Nakamura, H. (1967). Paper presented at the *Eng. Res. Found. Conf., Santa Barbara, California.*
Mrak, E. M. (1964). *In* "Global Impacts of Applied Microbiology" (M. P. Starr, ed.), p. 115. Wiley, New York.
Myers, J. (1951). *Ann. Rev. Microbiol.* **5**, 157.
Ogur, M. (1966). *Develop. Ind. Microbiol.* **7**, 216.
Oswald, W. J., and Golueke, C. G. (1968). *In* "Single-Cell Protein" (R. I. Mateles and S. R. Tannenbaum, ed.), p. 271. M. I. T. Press, Cambridge, Massachusetts.
Oswald, W. J., Golueke, C. G., Cooper, R. C., Gee, H. K., and Bronson, J. C. (1963). *Int. J. Air Water Pollut.* **7**, 6.
Peppler, H. J. (1968). *In* "Single-Cell Protein" (R. I. Mateles and S. R. Tannenbaum, ed.), p. 229. M. I. T. Press, Cambridge, Massachusetts.
Perez-Llano, G. A. (1944). *Bot. Rev.* **10**, 1.
Raymond, R. L. (1961). *Develop. Ind. Microbiol.* **2**, 23.
Reusser, F., Spencer, J., F. T., and Sallans, H. R. (1957). *Canad. J. Microbiol.* **3**, 721.
Roberts, L. P. (1950). *Nature* **165**, 494.
Roberts, L. P. (1953). *Nature* **172**, 351.

Rose, W. C., Wixom, R. L., Lockhart, H. B., and Lambert, G. F. (1955). *J. Biol. Chem.* **217,** 987.
Scrimshaw, N. S. (1958). *In* "Single-Cell Protein" (R. I. Mateles and S. R. Tannenbaum, ed.), p. 3. M. I. T. Press, Cambridge, Massachusetts.
Sorokin, C., and Myers, J. (1953). *Science* **117,** 330.
Spoehr, H. A., and Milner, H. W. (1949). *Plant Physiol.* **24,** 120.
Sure, B. (1948). *J. Nutr.* **36,** 59.
Takahashi, J., Kobayashi, K., Kawabata, Y., and Yamada, K. (1963). *Agr. Biol. Chem.* **27,** 836.
Tamiya, H. (1955). *Proc. World Symp. Appl. Solar Energy, Stanford, Res. Inst., Menlo Park, California,* p. 231.
Tamiya, H. (1959). *Rep. Japan Micro-algae Res. Inst. Jap. Nut. Ass. Tokyo* **I,** 1.
Treccani, V. (1963). *Prog. Ind. Microbiol.* **4,** 1.
Tsien, W. S., Johnson, E. L., and Liener, I. E. (1957). *Arch. Biochem. Biophys.* **71,** 414.
Verduin, J., and Schmid, W. E. (1966). *Develop. Ind. Microbiol.* **7,** 205.
Vinson, L. J., Cerecedo, L. R., Mull, R. P., and Nord, F. F. (1945). *Science* **101,** 388.
Wasserman, E. A. (1960). *Appl. Microbiol.* **8,** 291.
Willcox, O. W. (1959). *J. Agr. Food Chem.* **7,** 813.
Yamada, K., Takahashi, J., Kawabata, Y., Okada, T., and Onihara, T. (1968). *In* "Single-Cell Protein" (R. I. Mateles and S. R. Tannenbaum, ed.), p. 192. M. I. T. Press, Cambridge, Massachusetts.
Yanofsky, C. (1963). *Cold Spring Harbor Symp. Quant. Biol.* **28,** 581.
Zobell, C. E. (1946). *Bacteriol. Rev.* **10,** 1.

Structure–Activity Relationships Among Semisynthetic Cephalosporins

M. L. Sassiver and Arthur Lewis

Lederle Laboratories, American Cyanamid Company,
Pearl River, New York

I.	Introduction	163
	A. Cephalosporin C	163
	B. 7-Aminocephalosporanic Acid	164
II.	Semisynthetic Cephalosporins	165
	A. 7-Acylaminocephalosporanic Acids	167
	B. Modification of the 3-Side Chain	181
	C. Miscellaneous Modifications	205
III.	Resistance to Cephalosporinases	219
	A. Occurrence	219
	B. Chemistry	220
	C. Competitive Inhibition	222
	D. Structure Activity Studies; Miscellaneous	224
IV.	Pharmacology, Metabolism, and Mode of Action	225
	A. Metabolism and Absorption of Cephalosporins	225
	B. *In Vivo* Activity of Cephalosporins	228
	C. Mode of Action of Cephalosporins	230
	References	230

I. Introduction

The purpose of this review is to present *in vitro* data on the antibacterial activity of a large number of semisynthetic cephalosporins organized in such a way that where structure–activity relationships exist they may be readily discerned. Other reviews of the cephalosporins have dealt in detail with their chemistry, their biosynthesis and, in varying but lesser degrees, their biological activity. An authoritative review by Abraham (1967) is concerned mainly with the chemistry of the cephalosporin C group. A comprehensive review of the cephalosporins which is particularly well referenced is that of Van Heyningen (1967).

A. Cephalosporin C

Cephalosporin C (Fig. 1a) is produced by a species of *Cephalosporium*, originally identified as *Cephalosporium acremonium* (Brotzu, 1948), but later determined to be a new species (Crawford *et al.*, 1952). So far, cephalosporin C has been isolated in significant yield only from mutants derived from this one species.

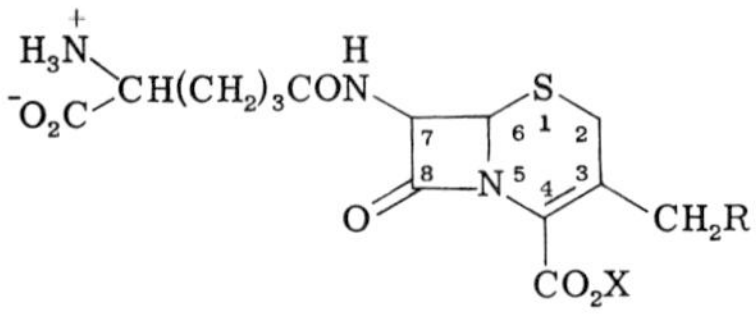

FIG. 1. (a) Cephalosporin C: R = OCOCH$_3$, X = H; (b) desacetylcephalosporin C: R = OH, X = H; (c) desacetoxycephalosporin C: R = H, X = H; (d) cephalosporin C$_A$ (pyridine): R = [pyridinium structure]; X = -; (3) cephalosporin C Bunte salt: R = S$_2$O$_3$Na, X = H.

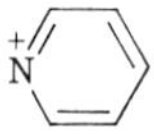

The *Cephalosporium* sp. was found to produce a number of antibiotics, five of which were active only against gram-positive organisms and were named cephalosporin P$_1$ to P$_5$ (Burton and Abraham, 1951). A hydrophilic antibiotic with activity against gram-negative bacteria was isolated and named cephalosporin N (Abraham *et al.*, 1954). It was later characterized as D-(4-amino-4-carboxybutyl)penicillin (Fig. 2) and renamed penicillin N (Newton and Abraham, 1954). Chrom-

FIG. 2. Penicillin N.

atography on an anion-exchange resin of the crude penicillic acid, formed from penicillin N in acidic solution, revealed for the first time the presence of cephalosporin C (Newton and Abraham, 1955). The new antibiotic was isolated as its crystalline sodium salt and structure elucidation studies were begun (Newton and Abraham, 1956). These proved to be unexpectedly difficult due to the presence of the novel fused β-lactam-dihydrothiazine ring system in the molecule. Eventually, the structure (Fig. 1a) was suggested based on the chemical and spectral properties of cephalosporin C (Abraham and Newton, 1961) and was confirmed by X-ray analysis (Hodgkin and Maslen, 1961).

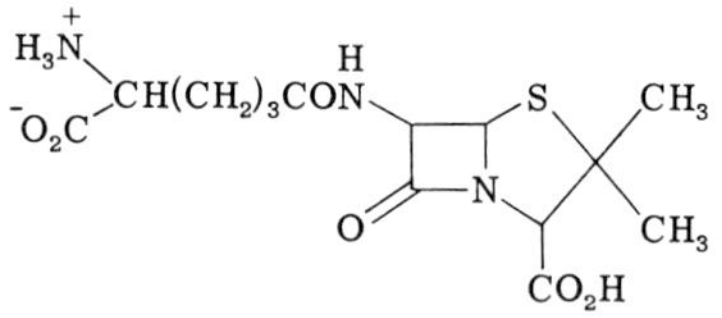

B. 7-AMINOCEPHALOSPORANIC ACID

Although cephalosporin C had a broad antibacterial spectrum, the level of activity was only moderate, particularly against gram-positive bacteria (see Table I). The structure of cephalosporin C, however, presented several possible sites for chemical modification. A key transformation of cephalosporin C was the removal of the aminoadipoyl side chain to give 7-aminocephalosporanic acid (7-ACA, or ACA as it will be referred to in this chapter) (Fig. 3a).

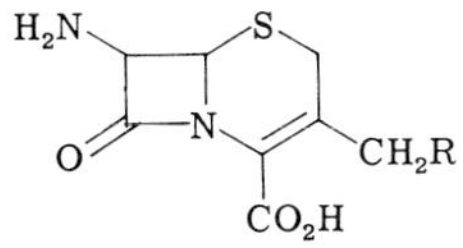

FIG. 3. (a) 7-Aminocephalosporanic acid: R = OCOCH$_3$; (b) 7-aminodesacetoxycephalosporanic acid: R = H.

The antibacterial activity of ACA is negligible (see Table I); however, its availability made possible the preparation of a vast number of 7-acylaminocephalosporins, many of which possessed enhanced activity.

ACA was first obtained in very low yield by mild acid hydrolysis of cephalosporin C (Loder et al., 1961). Under these conditions, several other reactions occurred including hydrolysis of the O-acetyl group, lactonization, and fragmentation of the cephalosporin ring system.

The first practical method for the preparation of ACA was reported by Morin and co-workers at the Lilly Research Laboratories (Morin et al., 1962, 1969a). It depended on the formation of an easily hydrolyzed iminolactone by an intramolecular cyclization of cephalosporin C. Other methods for the preparation of ACA depend on intramolecular aminolysis of cephalosporin C or formation of a readily hydrolyzed iminoester from an intermediate imide-chloride of carboxy-protected cephalosporin C derivatives (Fechtig et al., 1968).

Extensive searches have failed to locate an enzyme capable of removing the aminoadipoyl side chain of cephalosporin C to produce ACA (Claridge et al., 1963; Demain et al., 1963b). The failure is clearly associated with the structure of the side chain and not the cephalosporin nucleus, as other N-acyl cephalosporins have been successfully cleaved enzymatically. For example, acylases of *Nocardia* and *Proteus* cleaved 7-phenoxyacetyl cepahlosporin to ACA in appreciable yield (Huang et al., 1963; see also Sjöberg et al., 1967), discussed later. The aminoadipoyl side chain of penicillin N is similarly inert to removal by enzymes (Claridge et al., 1963).

II. Semisynthetic Cephalosporins

In the following discussions, broad structure activity areas will be categorized and illustrated with specific and general structural types.

Correlations will be made which are entirely empirical; the molecular basis for differences in biological activity is not known.

It is impractical to evaluate the initial antimicrobial properties of derivatives with more than a representative selection of ten to twenty gram-positive and gram-negative organisms. A somewhat different *in vitro* screen has been used by each laboratory; in some of the literature, data is supplied for only one or two organisms. Nevertheless, general patterns of susceptibility to semisynthetic cephalosporins have been established and are discernible in the publications of different laboratories. Throughout the discussions, comparisons will be made with the four semisynthetic cephalosporins that have achieved clinical significance: cephalothin (Fig. 4a), cephaloridine (Fig. 4b), cephaloglycin (Fig. 5a), and cephalexin (Fig. 5b).

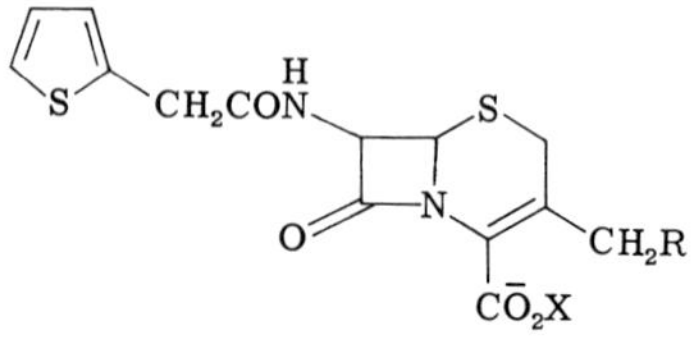

FIG. 4. (a) Cephalothin: $R = OCOCH_3$, $X = Na^+$; (b) cephaloridine: $R = $ [pyridinium], with no X.

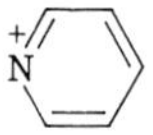

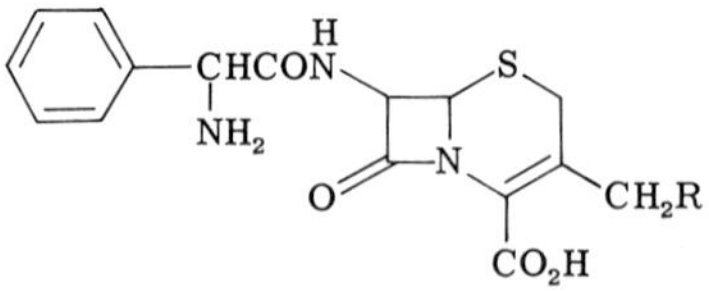

FIG. 5. (a) Cephaloglycin: $R = OCOCH_3$; (b) cephalexin: $R = H$.

In the following collection of structure-activity tables, a minimal inhibitory concentration (MIC) range will be given (if available) for the susceptible gram-positive and gram-negative spectrum, to facilitate comparisons of different derivatives and groups. Susceptible gram-positive organisms include, for example, *Staphylococcus aureus* (including penicillinase-producers) and *Streptococcus* species such as *S. pyogenes*; representative gram-negative organisms are chosen from *Klebsiella*, *Salmonella*, *Shigella*, and some *Escherichia* and *Proteus* species. A typical description of the individual organisms and patterns of susceptibility in a laboratory screen is given in the recent paper of Sassiver *et al.* (1969).

TABLE I
RING-SUBSTITUTED PHENYLACETYL-ACA

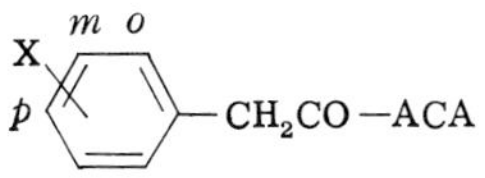

X-(m,o,p positions)-C₆H₃-CH$_2$CO—ACA

X	MIC (μg/ml) Gram-pos. range	MIC (μg/ml) Gram-neg. range	Reference[a]
H	0.4–3	3–25	(1)
p-HO$_2$CCH$_2$; p-H$_2$NCOCH$_2$	0.8–3	0.4–3	(2)
o-HO$_2$CCH$_2$	0.8–3	1.6–12	(2)
m-HO$_2$CCH$_2$	1.6–6	25–50	(2)
o-NO$_2$, m-F, Br, p-Cl, CH$_3$O	0.05–3	25	(1)
o-F, Cl, I, CH$_3$O, CH$_3$, m-I, CH$_3$O, CH$_3$, p-F, Br, I, CF$_3$S, NO$_2$	0.05–3	25–100	(1)
Polysubstituted, as			
o-NO$_2$ + p-Cl	0.1–0.8	100	(1)
Pentafluoro	0.2–1.6	100	(1)
Cephalosporin C	25–100	12–25	(3)
		38–79	(4)
7-ACA	>100	>50	(4)

[a]References: (1) Sassiver et al. (1969); (2) Lewis et al. (1969); (3) Lederle, unpublished; (4) Flynn (1967b).

A. 7-ACYLAMINOCEPHALOSPORANIC ACIDS

Thousands of semisynthetic cephalosporins have been synthesized by acylation of ACA with carboxylic acids, in an effort to find antibiotics with superior antimicrobial and pharmacological properties. It has been possible to correlate the structure of the 7-acyl side chain with *in vitro* antimicrobial activity, with respect to both potency and spectrum of activity.

Many ring-substituted phenylacetylcephalosporanic acids have been synthesized, due both to the great variety of readily available phenylacetic acids and to the obvious relationship to benzylpenicillin. With few exceptions, these derivatives (Table I) are highly active only against gram-positive organisms. Substituent effects are unpredictable. Polysubstituted derivatives are notably weak in gram-negative activity. Among *p*-carboxymethyl substituents, the corre-

TABLE II
Ring-Substituted Phenoxyacetyl-ACA

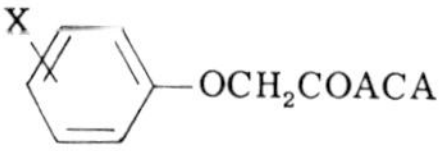

X—⟨benzene⟩—OCH$_2$COACA

X	MIC (μg/ml)		Reference[a]
	Gram-pos. range	Gram-neg. range	
H	0.1–0.4	25–100	(1)
p-HO$_2$CCH$_2$	0.4–3	1.6–3	(2)
p-HO$_2$CCH$_2$O	12–100	50–>100	(2)
p-Br	0.05–0.2	≥100	(2)
o-Cl, NO$_2$, HC(=O)[b] m-F p-Cl, CH$_3$O	0.05–0.8	50–>100	(1)
p-(C$_2$H$_5$)$_2$NSO$_2$	1.6–12	≥100	(1)

[a]References: (1) Sassiver et al. (1969); (2) Lewis et al. (1969).
[b]From Chauvette et al. (1963).

TABLE III
Ring-Substituted Phenylthioacetyl-ACA

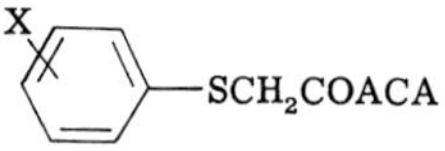

X—⟨benzene⟩—SCH$_2$COACA

X	MIC (μg/ml)		Reference[a]
	Gram-pos. range	Gram-neg. range	
H	0.1–3	12–25	(1)
p-HO$_2$CCH$_2$	0.4–1.6	0.8–3	(1)
m-NO$_2$	0.05–0.4	25–50	(2)
p-F	3–6	50–>100	(2)
Pentafluoro	0.2–1.6	>100	(2)

[a]References: (1) Lewis et al. (1969); (2) Lederle, unpublished.

sponding ester derivative (not shown) is weaker than the acid or amide; also interesting, but inexplicable, is the weak gram-negative activity of the m-carboxymethyl analog. Chauvette et al. (1963) have compared the activity in Oxford units of many monosubstituted derivatives. The activity of several amino-substituted phenylacetyl cephalosporins is discussed later (Table VI).

TABLE IV
EFFECT OF CHAIN LENGTH IN 7-ACYL-ACA

7-Acyl	n	MIC (μg/ml) Gram-pos. range	MIC (μg/ml) Gram-neg. range	Reference[a]
 C$_6$H$_5$-(CH$_2$)$_n$CO-	1	0.4–3	12–25	(1)
	2, 4	0.4–3	50	(1)
	5, 6, 10	0.2–6	≧ 100	(1)
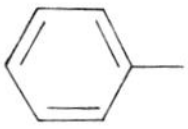 C$_6$H$_5$-O(CH$_2$)$_n$CO-	1, 2	0.1–0.8	50–100	(1)
	3, 5	0.4–6	≧ 100	(1)
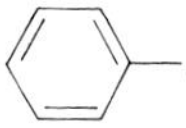 C$_6$H$_5$-S(CH$_2$)$_n$CO-	1	0.1–0.4	12–25	(1)
	2	0.3–0.6	–	(2)
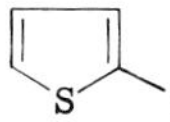 (thienyl)(CH$_2$)$_n$CO	1	0.2–3	3–6	(1)
	2	0.2–6	25–50	(1)
	3, 4	0.1–1.6	50–100	(1)
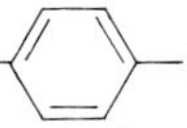 HO$_2$C-C$_6$H$_4$-(CH$_2$)$_n$CO	1	1.6–3	1.6–3	(3)
	2	3–12	12–25	(3)
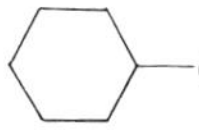 (cyclohexyl)(CH$_2$)$_n$CO	1	3–6	50–100	(1)
	2, 3, 4	0.1–1.6	50–100	(1)
CH$_3$(CH$_2$)$_n$CO	0	6–50	50–100	(4)
	5	0.1–1.8	–	(2)
	6	0.2–0.8	≧ 100	(4)
HO$_2$C(CH$_2$)$_n$CO	2	50–100	≧ 100	(3)
	5	6–12	6–12	(3)
	10	0.8–6	≧ 100	(3)

[a] References: (1) Sassiver et al. (1969); (2) Chauvette et al. (1963); (3) Lewis et al. (1969); (4) Lederle, unpublished.

Loss of potency against gram-negative organisms, but maintenance of high gram-positive activity, is the most notable difference between the phenoxyacetylcephalosporanic acids (Table II) and the phenylacetyl analogs just described. Certain carboxy-ring substituents such as p-carboxymethyl confer high gram-negative activity (Lewis et al., 1969); the capriciousness of this effect is seen in the data for the p-carboxymethoxy derivative. Polysubstituted analogs showed no difference from monosubstituted ones (Sassiver et al., 1969).

In comparing the gram-positive activity of phenylacetyl-, phenoxyacetyl-, and phenylthioacetylcephalosporanic acids, Chauvette et al. (1963) concluded that only in the phenylthio series was activity not

TABLE V
ALPHA-SUBSTITUTED PHENYLACETYL-ACA

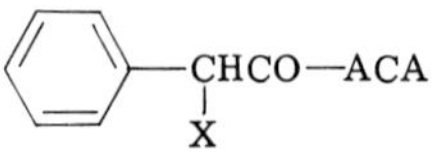

$$\text{C}_6\text{H}_5\text{—CHCO—ACA}$$
$$|$$
$$\text{X}$$

X	MIC (μg/ml) Gram-pos. range	Gram-neg. range	Reference[a]
H	0.4–3	12–25	(1)
CH$_3$	1–6	–	(1), (2)
Cl, Br	0.2–6	10–50	(1), (3)
HO	3–25	6–25	(1)
CH$_3$O	0.8–6	20–100	(1), (3)
(CH$_3$)$_3$CO	3–12	>100	(1), (3)
CH$_3$CO$_2$	0.8–12	20–>100	(1), (3)
C$_6$H$_5$CH$_2$O$_2$CO	10	20	(4)
CH$_3$S	1	20	(3)
NH$_2$CO	1	>40	(5)
HO$_2$C	0.4–1.6	6–12	(6)
(CH$_3$)$_2$NCH$_2$	1.6–6	50–100	(1)
C$_6$H$_5$CH$_2$O$_2$CNH	1	20	(3)
N$_3$	1	10	(3)

[a]References: (1) Sassiver et al. (1969); (2) Chauvette et al. (1963); (3) Takano et al. (1967); (4) Fujisawa (1967l); (5) Fujisawa (1966a); (6) Lewis et al. (1969).

enhanced over the parent compound by means of introduction of a suitable substituent. The phenylthio series (Table III) does exhibit a wide range of antimicrobial potency, as noted by both Chauvette et al. and Lewis et al. (1969, and unpublished data) similar to the oxygen analogs of Table II. The activity of the m-nitro compound was greater than that of the parent, falling outside the aforementioned generalization. Again, certain acidic substituents, as p-carboxymethyl, conferred high in vitro gram-negative activity.

Extensive data on the effect of chain length on antimicrobial activity is available (Table IV). When the first member of a series has high gram-negative activity, extension of the chain causes a dramatic decrease in this property. An exception is the series derived from aliphatic dicarboxylic acids; here activity peaks at intermediate chain length. With the aromatic derivatives, however, the increased aliphatic nature of the higher homologated members must adversely influence the gram-negative response.

The effect of introducing an α-substituent X in phenylacetylcephalosporanic acids is shown in Table V. This change creates a new asymmetric center in the side chain. However, the data has been reported for the racemic (DL) mixtures, and it is likely that in some cases higher activity will be present in one of the two possible isomers. The lowering of activity by introduction of an α-methyl group has been noted for phenoxy-, phenylthio-, and phenylacetyl-ACA's (Chauvette et al., 1963), but this effect is not generalizable to all substituents. Takano et al. (1967) found the following gram-positive activity relationship for substituents: phenoxy > methoxy > acetoxy > hydroxy (least active). The good gram-negative activity of the hydroxy and carboxy analogs should be noted. The α-amino substituent is described in Table VI. Introduction of substituents on the phenyl ring had little additional effect on activity in this series (Fujisawa, 1966e). Also, the thiophene analog of the α-hydroxy compound has been recently reported to have good activity (Crast, 1969).

Much effort has been spent in the preparation of cephalosporanic acids bearing a free amino group in the side chain, presumably stimulated by the success of ampicillin (D-α-aminobenzylpenicillin) as a "broad-spectrum" penicillin. Cephaloglycin, D-α-aminophenylacetyl-ACA (Table VI, top entry), is much more active against gram-negative organisms that its L-isomer, a characteristic which is repeated in related isomer pairs illustrated in Table VI. The location of the side chain amino function can be varied somewhat without loss of gram-negative activity. Substitution of the thiophene ring for benzene retains broad-spectrum activity; substitution of naphthalene does not. Introduction of substituents in the phenyl ring can be compatible with broad-spectrum activity. A number of acetone condensation products are as active as their precursors, as is true in the penicillins (Hardcastle et al., 1966; Bunn et al., 1966). Aliphatic amino derivatives are poorly active.

When the size of the aromatic ring in phenylacetyl-ACA is increased (Table VII), gram-positive activity rises and gram-negative activity falls. This pattern is repeated in the oxyacetyl and thioacetyl series and in derivatives derived from aromatic diacetic acids. In some cases (Sassiver et al., 1969) gram-positive potency is so high that *Streptococcus faecalis*, an organism generally resistant to cephalosporins, is inhibited at 12 µg/ml or less. Mere presence of an assemblage of aromatic rings, however, does not insure high gram-positive activity (Table VIII).

Some potential for activity against gram-negative organisms exists

TABLE VI
Aminoacyl-ACA

Aminoacyl	MIC (μg/ml) Gram-pos. range	MIC (μg/ml) Gram-neg. range	Reference[a]
D(−) 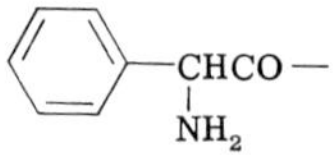 C₆H₅–CH(NH₂)CO–	1	2[b]	(1)
L-Isomer	0.7	>50	(1)
"CA" D-phenyl-imidazolidinone-cephalosporin (CH₂OCOCH₃, CO₂H) 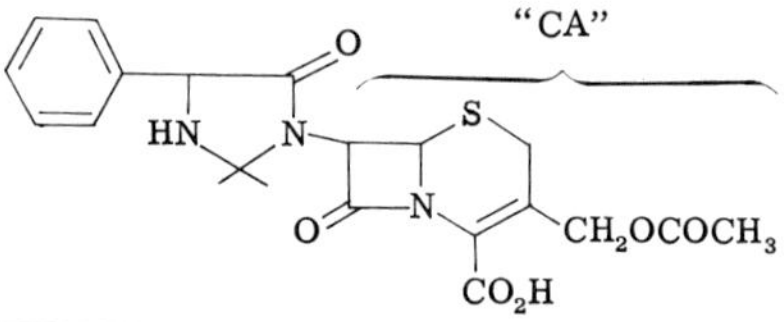	1–6	3–6	(2)
D 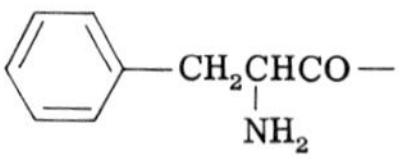 C₆H₅–CH₂CH(NH₂)CO–	1.1	>50	(1)
L-Isomer	12	>50	(1)
D 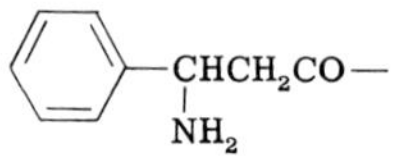 C₆H₅–CH(NH₂)CH₂CO–	1	16	(1)
L-Isomer	2	>50	(1)
D 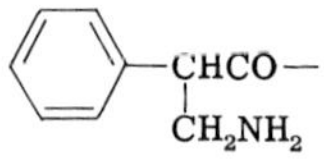 C₆H₅–CH(CH₂NH₂)CO–	>20	>50	(1)
L-Isomer	6	>50	(1)
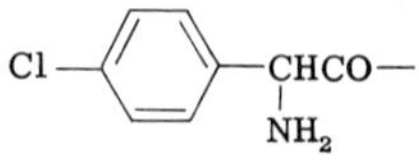 Cl–C₆H₄–CH(NH₂)CO–	5	20	(3)
D(−) 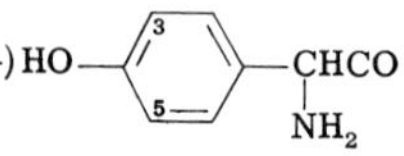 HO–C₆H₄–CH(NH₂)CHO (3,5)	0.08–5	2–6	(4)
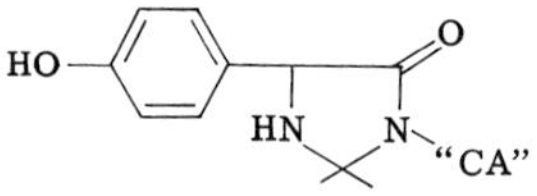 HO–C₆H₄– imidazolidinone "CA"	0.08–6	1–6	(4)
3-Cl and 3,5 diCl	0.08–3	1–6	(4)

TABLE VI (continued)

Aminoacyl	MIC (µg/ml) Gram-pos. range	MIC (µg/ml) Gram-neg. range	Reference[a]
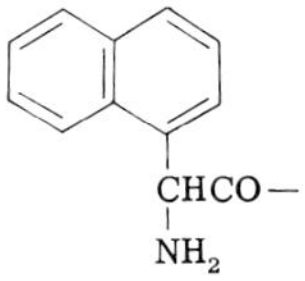 (1-naphthyl)-CHCO—, NH$_2$	5	40	(5)
D-(2-thienyl)-CHCO—, NH$_2$ 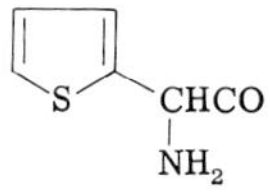	1–6	6–12	(6)
DL-Mixture	5	10	(5)
D-(2-thienyl) β-lactam "CA" 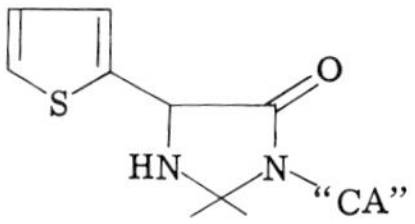	1.6–3	1.6–6	(6)
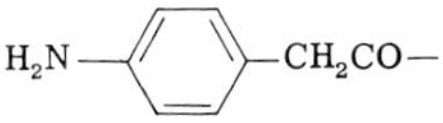 H$_2$N—C$_6$H$_4$—CH$_2$CO—	0.2–0.6	20	(7), (8)
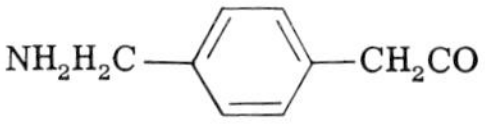 NH$_2$H$_2$C—C$_6$H$_4$—CH$_2$CO	0.4–1	5–36	(9)
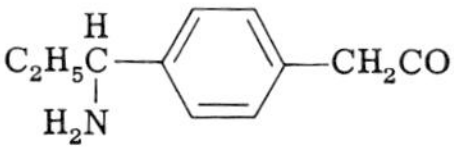 C$_2$H$_5$CH(NH$_2$)—C$_6$H$_4$—CH$_2$CO	0.2–0.4	10–24	(9)
L-Isomer	0.2–0.4	10–24	(9)
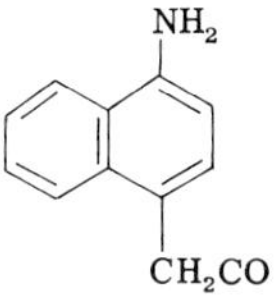 (4-amino-1-naphthyl)-CH$_2$CO	0.3–1	—	(7)
NH$_2$CH$_2$CO	25–100	>100	(10), (7)
NH$_2$CHCO, CH$_3$	250	—	(7)
NH$_2$CH$_2$CH$_2$CO	25–100	>100	(7)
CH$_3$(CH$_2$)$_5$CHCO, NH$_2$	10	>40	(5)

TABLE VI (continued)

Aminoacyl	MIC (μg/ml)		Reference[a]
	Gram-pos. range	Gram-neg. range	
$(CH_3)_2CHCHCO$ $\quad\quad\quad\;\;\;|$ $\quad\quad\quad\;\;NH_2$	40	>40	(5)
$NH_2CH(CH_2)_2CO$ $\quad\;\;|$ $\quad\;\;CO_2H$	6–12	–	(7)
$NH_2CH(CH_2)_3CO$ (Cephalosporin C) $\quad\;\;|$ $\quad\;\;CO_2H$	25–100	12–25	(10)
$NH_2CHCONHCH_2CO$ 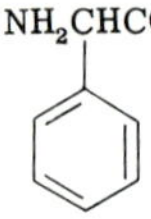 $\quad\;\;|$ $\quad\;\;C_6H_5$	10	40	(11)

[a]References: (1) Spencer et al. (1966); (2) Godfrey (1967); (3) Fujisawa (1966b); (4) Bristol-Myers Co. (1969); (5) Kurita et al. (1966); (6) Crast and Essery (1967); (7) Glaxo (1964b); (8) Fujisawa (1967a); (9) Flynn (1968); (10) Lederle, unpublished; (11) Fujisawa (1969d).

[b]For Spencer et al. (1966), gram-negative MIC's for many organisms were averaged.

in low molecular weight halogenated aliphatic acyl-ACA's (Table IX). A few members are as active as cephalothin in the gram-negative spectrum; further exploration of this group would appear attractive.

Another series of aliphatic acyls, substituted with alkylthio groups, exhibits good gram-negative activity in low molecular weight and especially carboxyl-substituted members (Table X). The combination of a small aliphatic acyl with an electronegative substituent such as halogen or alkylthio seems to promote, in certain instances, a high gram-negative activity. This correlation is further exemplified by a cyano-substituted derivative and some azi derivatives (Table XI).

Modification of cephalothin (top entry, Table XII) by substitution of simple heterocyclic rings for thiophene has been quite successful in retaining similar broad-spectrum activity (Table XII). Other sydnone-containing derivatives related to those in the table have been reported (Fujisawa, 1968b, 1969e). Introduction of ring substituents has often decreased activity; the carboxymethyl group is an exception (Lewis et al., 1969). Likewise, benzoheterocyclic analogs are less active (Chauvette et al., 1963). Some derivatives have been derived from cephalosporin C itself by modification of the aminoadipoyl side chain (Ciba, Ltd., 1963; 1964a,b; Patchett, 1966) and are of no great interest.

Except for perhaps a pyridine-3-thioacetyl compound, a series of

TABLE VII
Fused Ring Acyl-ACA

Ring $\begin{Bmatrix} SCH_2 \\ OCH_2 \\ CH_2 \end{Bmatrix}$ CO—ACA

Structure	MIC (μg/ml) Gram-pos. range	MIC (μg/ml) Gram-neg. range	Reference[a]
 phenyl—CH$_2$CO	0.4–3	12–25	(1)
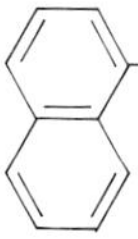 naphthyl—CH$_2$CO	0.05–2	50–100	(1)
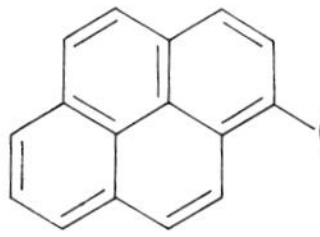 pyrenyl—(CH$_2$)$_3$CO	0.05–0.4	100	(1)
 phenyl—SCH$_2$CO	0.1–0.4	12–25	(1)
 naphthyl—SCH$_2$CO	0.05–0.2	100	(1)
 phenyl—OCH$_2$CO	0.1–0.4	25–100	(1)
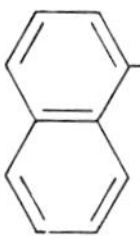 naphthyl—OCH$_2$CO	0.05–0.2	50–100	(1)
 HO$_2$CCH$_2$—phenyl—CH$_2$CO	0.8–3	0.4–3	(2)
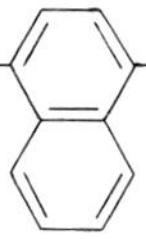 HO$_2$CCH$_2$—naphthyl—CH$_2$CO	0.2–3	3–12	(2)
 HO$_2$CCH$_2$—anthryl—CH$_2$CO	0.1–1.6	50–100	(2)

[a]References: (1) Sassiver et al. (1969); (2) Lewis et al. (1969).

TABLE VIII
Phenylated Acyl-ACA[a]

	RCO—ACA	
	MIC (μg/ml)	
R	Gram-pos. range	Gram-neg. range
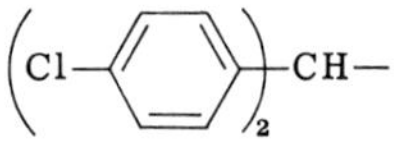 (Cl—C₆H₄—)₂CH—	0.2–1.6	100
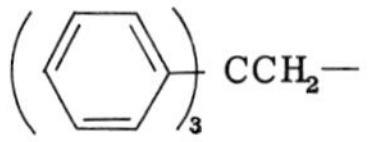 (C₆H₅—)₃CCH₂—	12–50	≧ 100
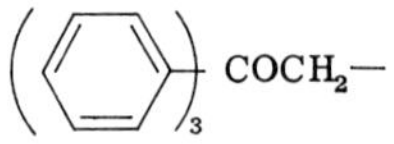 (C₆H₅—)₃COCH₂—	12–50	>100
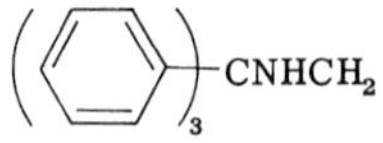 (C₆H₅—)₃CNHCH₂	25–50	>100

[a] From Lederle, unpublished.

TABLE IX
Haloacyl-ACA

	MIC (μg/ml)		
	Gram-pos. range	Gram-neg. range	Reference[a]
R of RCHCO (Br)			
CH_3	1.6–3	25	(1)
C_2H_5, n-C_3H_7	0.8–3	25–50	(1)
n-$C_{10}H_{21}$	0.1–1.6	100	(2)
$BrCH_2$	6–12	25–50	(2)
$HO_2CCHCH_2CH_2$ (Br)	6–25	50–100	(3)
R of RCHCO (Cl)			
H	—	8–50	(4)
CH_3	6–100	50–100	(4)
$ClCH_2$	1.6–3	4–15	(4)
$CH_3O_2C(CH_2)_3$	3–12	12–25	(1)
ICH_2CH_2CO	0.8–6	3–12	(1)

[a] References: (1) Sassiver et al. (1969); (2) Lederle, unpublished; (3) Lewis et al. (1969); (4) Chauvette et al. (1963).

TABLE X
Substituted Thioalkylacyl-ACA

| | | RS(CH$_2$)$_n$COACA | | |
| | | MIC (μg/ml) | | |
R	n	Gram-pos. range	Gram-neg. range	Reference[a]
C$_2$H$_5$	1	0.2–1.6	3–12	(1)
n-C$_4$H$_9$	1	0.2	1–70	(2)
t-C$_4$H$_9$	1	0.4–1.6	100	(1)
C$_2$H$_5$O(CH$_2$)$_2$	1	0.4–3	12–25	(1)
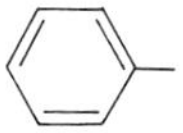 C$_6$H$_5$–	1	0.1–0.4	12–25	(1)
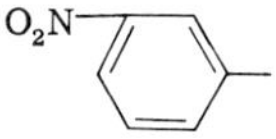 O$_2$N-C$_6$H$_4$–	1	0.05–1	25	(1)
CH$_3$C(=O)–	1	3–12	12–25	(1)
C$_2$H$_5$	3	1.6–6	25–100	(1)
HO$_2$CCH$_2$	1	0.8–25	25–50	(3)
HO$_2$CCH$_2$	2	6–12	3–12	(3)
HO$_2$CCH$_2$SCH$_2$	1	3–12	3–12	(3)
HO$_2$CCH$_2$S	1	6–12	1.6–6	(3)
HO$_2$C(CH$_2$)$_3$S	3	0.8–6	6–25	(3)
HO$_2$CCH$_2$SC(=O)–	1	3–12	1.6–6	(3)
(CH$_3$)$_2$NC(=S)–	1	0.25–0.8	12	(4)
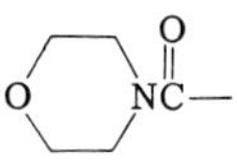 morpholino-NC(=O)–	1	0.5–1.6	12	(4)

[a]References: (1) Sassiver et al. (1969); (2) Chauvette et al. (1963); (3) Lewis et al. (1969); (4) Gottstein and Eachus (1968).

TABLE XI
AZI AND CYANO ALKYLACYL-ACA

Structure	MIC (μg/ml)		Reference[a]
	Gram-pos. range	Gram-neg. range	
$CH_3-\underset{N=N}{C}-(CH_2)_n CO$			
$n = 1$	0.8–6	6–25	(1)
$n = 2$	0.8–6	12–25	(1)
$n = 3$	0.2–3	25–50	(1)
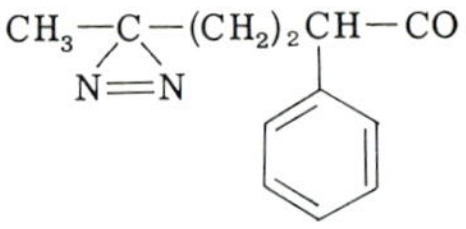 $CH_3-\underset{N=N}{C}-(CH_2)_2CH-CO$ (with phenyl)	0.8–25	100	(1)
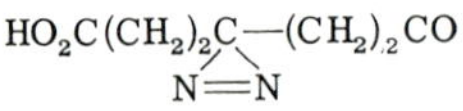 $HO_2C(CH_2)_2\underset{N=N}{C}-(CH_2)_2CO$	0.3–25	6–25	(2)
$N\equiv CCH_2CO$	2	4–15	(3)

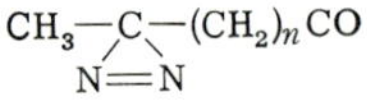

[a]References: (1) Sassiver et al. (1969); (2) Lewis et al. (1969); (3) Ciba (1965).

heterocyclic thioacetylaminocephalosporanic acids (Table XIII) is much less active than the cephalothinlike analogs of Table XII.

A relatively small number of semisynthetic cephalosporins with unsaturated acyl side chains (Table XIV) have been reported. When the unsaturated function is alpha to the side-chain carbonyl, low activity is the likely consequence. Movement of the unsaturated function out of conjugation with the carbonyl improves gram-positive activity.

Chauvette et al. (1963) stated that ACA derivatives of aromatic acids were much less active than their acetyl homologs. A possible use of some of these compounds as competitive inhibitors of β-lactamase enzymes is discussed later. Some unpublished data of the Lederle group is given in Table XV. In a number of instances of electronegatively substituted benzoyl derivatives, moderate gram-positive activity is seen. In over fifty randomly selected heterocarboxyl types, only 10% showed activity of 3 μg or less against gram-positive organisms (Lederle, unpublished data).

Saturated ring acyl-ACA's (Table XVI) generally reflect the relationship found in aromatic types; the ring must be separated from the side-chain carbonyl group by at least one CH_2 unit for good gram-positive activity to be achieved.

Carbonyl-containing acyl-ACA's (Table XVII) have poor gram-

TABLE XII
Heteroacetyl-ACA

| | | Heterocyclic ring-CH₂COACA | | |
| | | MIC (µg/ml) | | |
Ring		Gram-pos. range	Gram-neg. range	Reference[a]
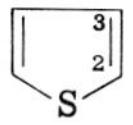 thiophene	2-	0.2–3	5–15	(1)
	3-	–	2–30	(1)
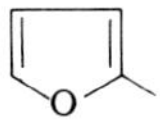 furan		–	7–17	(1)
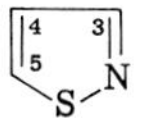 thiadiazole N-oxide	3, 4, or 5-	0.1–0.8	1.6–12	(2)
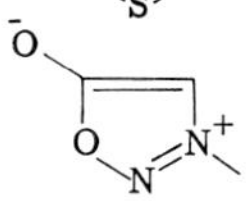 (Sydnone-1-)		0.4–1	3–6	(3)
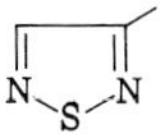 thiadiazole		0.1–0.8	1.6	(4)
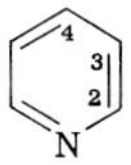 pyridine	2-	1–2.5	8–31	(5)
	3 or 4	0.2–1	2–8	(5)
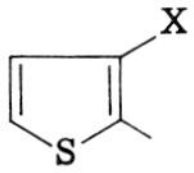 thiophene-X	X = CH₃	–	7–52	(6)
	(CH₃)C	–	17–106	(6)
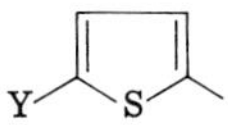 thiophene-Y	Y = Cl	–	0.5–31	(6)
	CH₃CO	–	14–34	(6)
	HO₂CCH₂	0.4–1.6	0.4–3	(7)
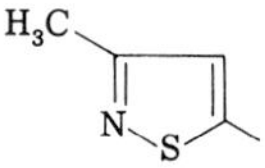 methylisothiazole		0.3–0.8	3–25	(2)
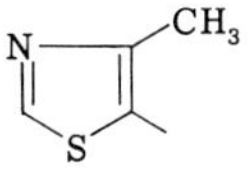 methylthiazole		0.1–1.6	12–25	(8)
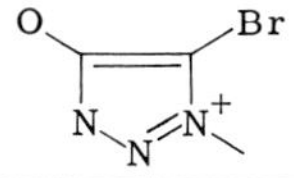 bromo-methyl-triazole oxide		0.4–1.6	12–25	(3)

TABLE XII (continued)

Ring		MIC (μg/ml) Gram-pos. range	MIC (μg/ml) Gram-neg. range	Reference[a]
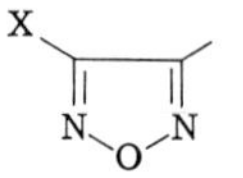 (oxadiazole)	X = CH$_3$	0.8	1.6–12	(9)
	CH$_3$O	0.3–0.8	1.6	(10)
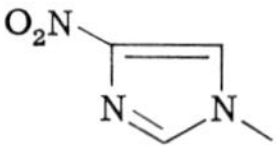 O$_2$N-imidazole		0.4–3	12–50	(8)
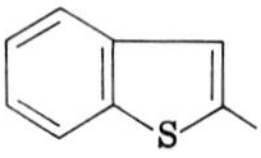 benzothiophene		—	3–67	(6)
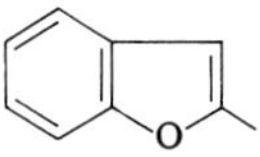 benzofuran		—	6–69	(6)
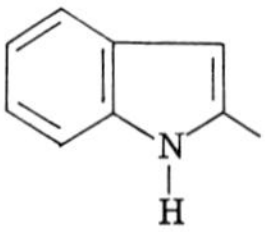 indole		6–50	>100	(8)
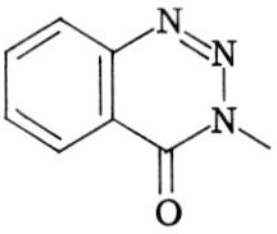 benzotriazinone-N-methyl		0.1–1.6	100	(11)
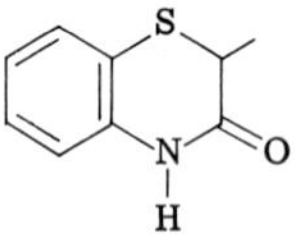 benzothiazinone		1	>40	(12)
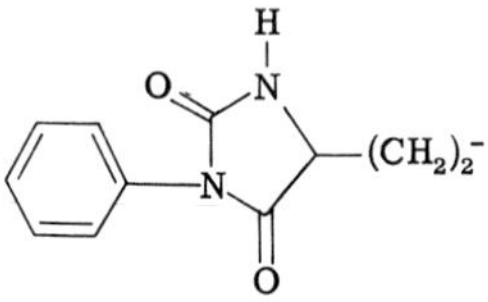 phenylhydantoin–(CH$_2$)$_2$–		3–12	—	(13)
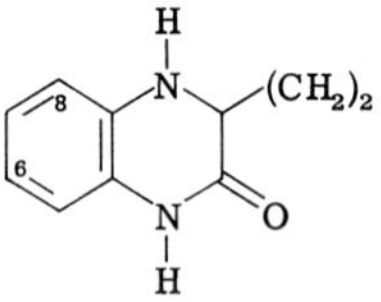 dihydroquinoxalinone–(CH$_2$)$_2$–	6-NH$_2$	2–125	—	(14)
	6,8-diNH$_2$	8–16	—	(14)
	6-CO$_2$CH$_3$	1.6–12	—	(14)
Ferrocene		3–12	>100	(8)

[a]References: (1) Chauvette et al. (1962); (2) Rapp and Micetich (1968); (3) Naito et al. (1968); (4) Crast (1967a); (5) Stedman et al. (1967); (6) Chauvette et al. (1963); (7) Lewis et al. (1969); (8) Sassiver et al. (1969); (9) Crast (1967b); (10) Crast (1967c); (11) Lederle, unpublished; (12) Fujisawa (1968a); (13) Ciba (1964a); (14) Ciba (1964b).

TABLE XIII
HETEROTHIOACETYL-ACA

Heterocyclic ring-SCH$_2$COACA

Ring		MIC (μg/ml) Gram-pos. range	Gram-neg. range	Reference[a]
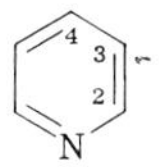 pyridyl (positions 2,3,4 numbered)	2-	0.06–0.8	12–100	(1)
	3-	0.01–0.2	3–12	(1)
	4-	0.06–0.25	12–50	(1)
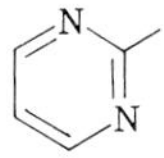 pyrimidinyl		6–50	12–50	(2)
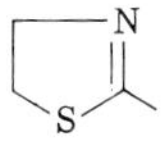 thiazolyl (N, S)		1.6–3	25–50	(2)
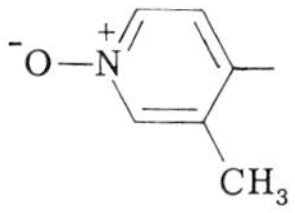 N-oxide pyridyl with CH$_3$ ($^-$O—N$^+$, CH$_3$)		25	>40	(3)
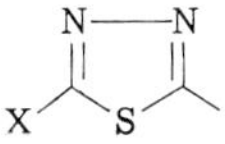 thiadiazolyl (N—N, S, X)		2–20	>40	(4)
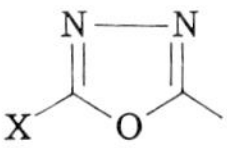 oxadiazolyl (N—N, O, X)		0.4–1	20–40	(4)

[a]References: (1) Bristol-Myers (1968); (2) Lederle, unpublished; (3) Fujisawa (1966c); (4) Fujisawa (1967b).

negative activity, but may have good gram-positive activity in certain cases.

Several derivatives containing the urea moiety are listed in Table XVIII.

Miscellaneous N-substituted acyl derivatives (Table XIX) exhibit a range of potency against gram-positive organisms. Interestingly, in two cases, chain extension from one to two methylenes improves activity significantly.

A final group of miscellaneous acyl analogs is given in Table XX. Reaction of ACA with benzenesulfonyl chloride to form a sulfonamide gives a particularly poor cephalosporin (Fujisawa, 1968d). A formyl derivative with good gram-negative activity has been reported (Chauvette et al., 1963).

B. MODIFICATION OF 3-SIDE CHAIN

In addition to the 7-acyl side chain, there are several other positions in the cephalosporin molecule where chemical modification presents

TABLE XIV
UNSATURATED ACYLAMINOCEPHALOSPORANIC ACIDS

Acyl	MIC (μg/ml)		Reference[a]
	Gram-pos. range	Gram-neg. range	
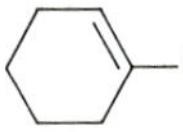 cyclohexenyl–CH$_2$CO	0.4–1.6	50–100	(1)
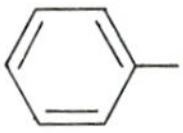 C$_6$H$_5$–CH=CH–CO	0.8	>40	(2)
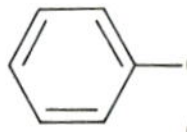 C$_6$H$_5$–C(=CH$_2$)–CO	5	>40	(3)
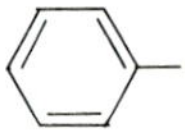 C$_6$H$_5$–CH=CHCH$_2$CO	0.1–0.4	50–100	(1), (4)
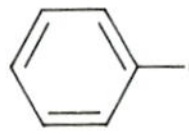 C$_6$H$_5$–C≡C–CO	3–50	100	(1)
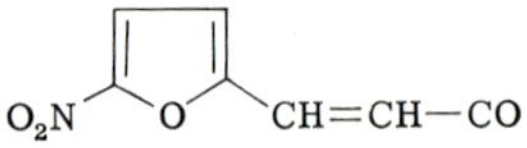 O$_2$N-furyl-CH=CH–CO	6–12	12–50	(4)
CH$_2$=CH–CO	1.6–6	25–100	(4)
CH≡C–CO	12–50	50–100	(4)
C$_2$H$_5$CH=CH–CH$_2$CO	0.8–3	25–50	(4)
CH$_3$CH=CH–CH=CH–CO	10	>40	(5)
CH$_2$=CH–(CH$_2$)$_8$CO	0.1–1.6	>100	(1)
HO$_2$CCH$_2$–CH=CH–CH$_2$CO	6–12	25–50	(6)
HO$_2$CCH$_2$–C≡C–CH$_2$CO	3–12	25–50	(6)

[a]References: (1) Lederle, unpublished; (2) Fujisawa (1968c); (3) Fujisawa (1967c); (4) Fujisawa (1967d); (5) Fujisawa (1967e); (6) Lewis *et al.* (1969).

TABLE XV
AROMATIC AND HETEROAROMATIC CARBOXY-ACA

Ring	Ring CO—ACA		
	MIC (μg/ml)		
	Gram-pos. range	Gram-neg. range	Reference[a]
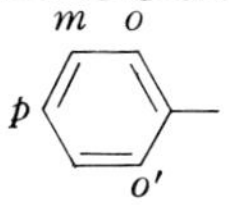 *m o* *p*–⟨ ⟩– *o'*			
Unsubstituted	12–16	—	(1)
p-Phenyl	1–25	100	(2)
o, *m*, or *p* (F, Cl, Br, CF$_3$)	1.6–3	—	(2)
m- or *p*-HO$_2$C—	50–100	50–100	(3)
m-CH$_3$O$_2$C	6–25	100	(2)
p-Phenyl-SO$_2$NH	25–50	100	(2)
p-CH$_3$O, CF$_3$S	3–6	100	(2)
Pentafluoro	6–50	100	(2)
o,*o'*-(CH$_3$O)$_2$	12–100	100	(2)
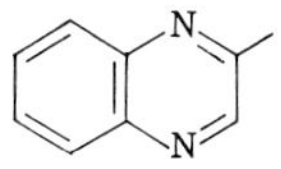 quinoxaline	12–100	100	(2)
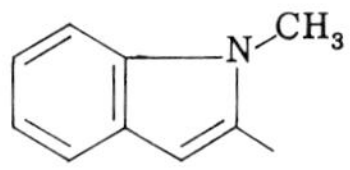 N-methylindole	0.8–6	100	(2)
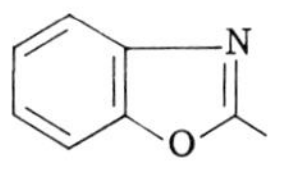 benzoxazole	3–25	100	(2)
Ferrocene	50–>100	100	(2)

[a] References: (1) Chauvette *et al.* (1963); (2) Lederle, unpublished; (3) Lewis *et al.* (1969).

TABLE XVI
Cycloaliphatic Acyl-ACA

	RCO—ACA		
	MIC (μg/ml)		
R	Gram-pos. range	Gram-neg. range	Reference[a]
(CH$_2$)$_{3-7}$—CHCH$_2$	3–12	50–100	(1)
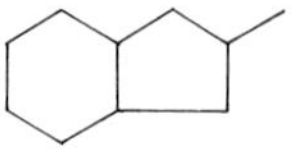 [decahydronaphthalenyl-CH]	0.4–1.6	50–100	(1)
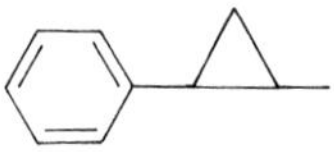 [phenyl-cyclopropyl-CH]	0.4–0.8	>100	(1)
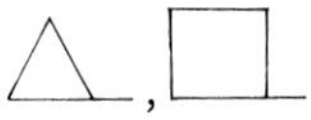 [cyclopropyl, cyclobutyl]	6–25	≧100	(1)
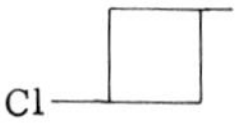 [Cl-cyclobutyl]	0.8–6	25–50	(1)
 [cyclohexyl]	3	–	(1)
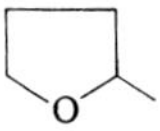 [2-methyltetrahydrofuranyl]	6–50	50–>100	(1)
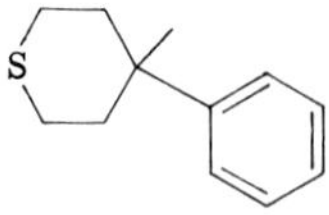 [4-phenyl-thiopyranyl]	12–50	≧100	(1)
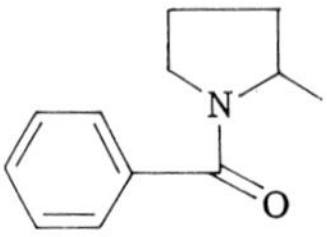 [phenyl-N-methylpyrrolidinyl-CO]	25–50	≧100	(1)
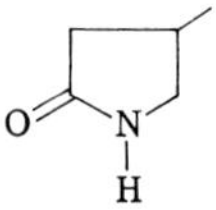 [pyrrolidinone-NH]	40	40	(2)
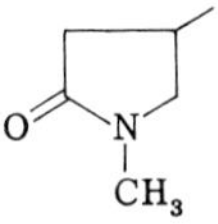 [pyrrolidinone-N-CH$_3$]	1	>40	(2)
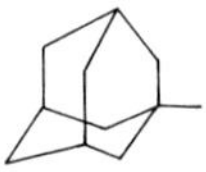 [adamantyl]	1	≧100	(1)

[a]References: (1) Lederle, unpublished; (2) Fujisawa (1967m).

TABLE XVII
Miscellaneous Carbonylacyl-ACA

R	RCO—ACA MIC (μg/ml) Gram-pos. range	Gram-neg. range	Reference[a]
X—C₆H₄—CO, X = H	0.8–3	50–100	(1)
X = p-Br, m-Br, m-I, CH_3, CH_3O	0.02–0.5	–	(2)
X = diCH_3O	2–16	–	(2)
$(CH_3)_2CH-CO-C(CH_3)_2CH_2$	10	>40	(3)
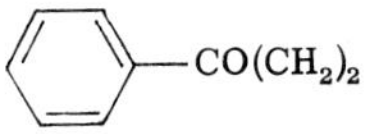 C₆H₅—CO(CH₂)₂	6–25	≧ 100	(1)

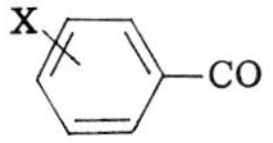

[a]References: (1) Lederle, unpublished; (2) Glaxo (1968a); (3) Fujisawa (1967g).

TABLE XVIII
Substituted-Urea Acylamino-ACA

R	RCOACA MIC (μg/ml) Gram-pos. range	Gram-neg. range	Reference[a]
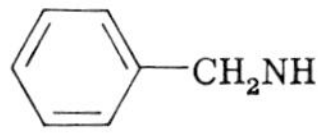 C₆H₅—CH₂NH	1	40	(1)
$C_2H_5OCOCH_2NH$	5	40	(1)
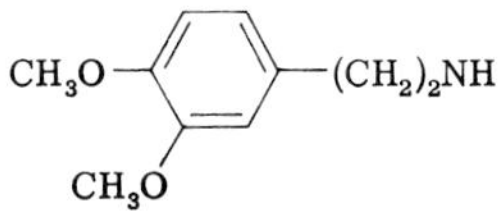 CH_3O—C₆H₃(CH_3O)—$(CH_2)_2NH$	5	–	(2)
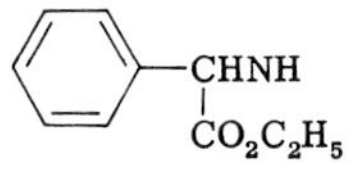 C₆H₅—CHNH—$CO_2C_2H_5$	2	>40	(3)
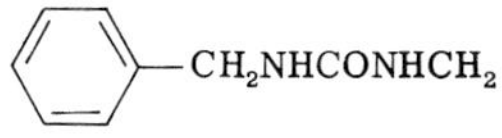 C₆H₅—$CH_2NHCONHCH_2$	10	>40	(4)

[a]References: (1) Fujisawa (1967h); (2) Fujisawa (1967i); (3) Fujisawa (1967j); (4) Fujisawa (1966d).

TABLE XIX
N-SUBSTITUTED AMINOACYL-ACA

RCOACA

R	MIC (μg/ml)		Reference[a]
	Gram-pos. range	Gram-neg. range	
$(CH_3)_2NCH_2$	40	>40	(1)
$(C_2H_5)_2NCH_2$	12–50	25–100	(2)
$(C_2H_5)_2N(CH_2)_2$	3–6	50–100	(2)
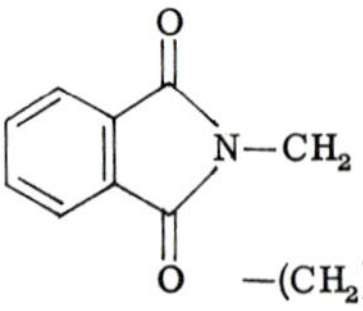 phthalimido-N—CH$_2$	25–100	>100	(2)
phthalimido—(CH$_2$)$_2$	0.4–1.6	12–50	(3)
C$_6$H$_5$—SO$_2$NHCH$_2$	25–100	≧100	(2)
C$_6$H$_5$—SO$_2$—N(CH$_3$)—CH$_2$	1.6–3	25–100	(2)
$(CH_3)_3C$—O—(CO)NHCH$_2$	0.8–6	>100	(2)
thiazol-2-yl—NHCH$_2$	20	>40	(4)

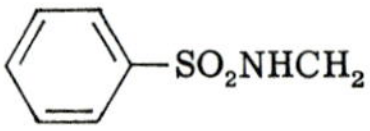

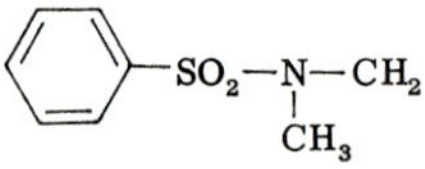

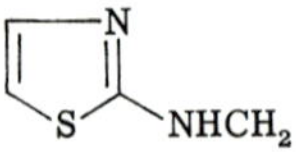

[a]References: (1) Fujisawa (1967k); (2) Lederle, unpublished; (3) Sassiver et al. (1969); (4) Fujisawa (1969b).

TABLE XX
MISCELLANEOUS DERIVATIVES

R	R—ACA MIC (µg/ml)		Reference[a]
	Gram-pos. range	Gram-neg. range	
$\overset{O}{\underset{}{\|}}$ HC—	—	1–15	(1)
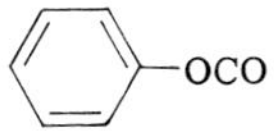 C₆H₅—OCO	10	40	(2)
C_2H_5O-CO	12–16	25–>100	(1)
CH_3OCH_2CO	3–12	50–100	(3)
$CH_3O_2CCH_2CO$	1–6	50	(3)
$CH_3(CH_2)_5\underset{N_3}{CH}-CO$	0.2	>40	(4)
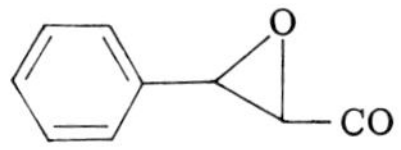 C₆H₅-epoxide-CO	5	>40	(5)
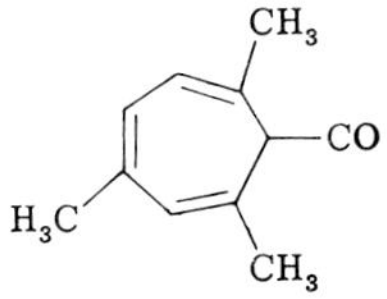 (cycloheptatriene-CH₃,CH₃,H₃C)-CO	0.8–12	—	(6)
C₆H₅—$\overset{O}{\underset{\|}{S}}$CH₂CO	1–6	50–100	(7)
$(CH_3)_3Si(CH_2)_2CO$	1.6–12	≧ 100	(7)
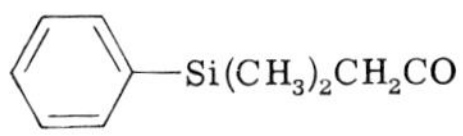 C₆H₅—$Si(CH_3)_2CH_2CO$	6–50	≧ 100	(7)
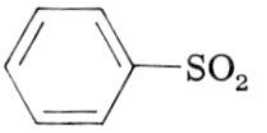 C₆H₅—SO₂	>40	>40	(8)

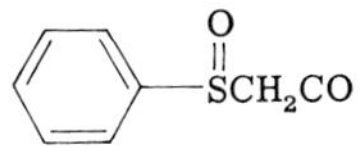

[a]References: (1) Chauvette et al. (1963); (2) Fujisawa (1967f); (3) Lederle, unpublished; (4) Fujisawa (1966f); (5) Fujisawa (1969c); (6) Parke, Davis (1969); (7) Sassiver et al. (1969); (8) Fujisawa (1968d).

the possibility of enhanced biological activity. The 3-acetoxymethyl side chain has proved to be a fertile site for such transformations.

1. Desacetylcephalosporins

Citrus acetylesterase catalyzed the hydrolysis of the O-acetyl group in cephalosporin C to give desacetylcephalosporin C (Fig. 1b) (Jeffery et al., 1961). The desacetyl derivative was readily converted into the lactone (Fig. 6) by treatment with dilute acid. This method of pre-

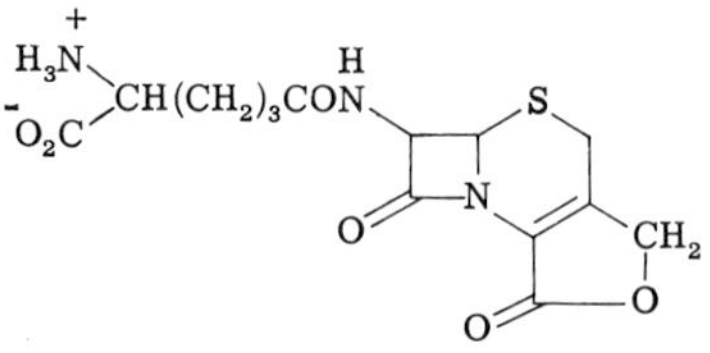

FIG. 6. Cephalosporin C lactone.

paring desacetylcephalosporins has been used with a variety of 7-acylcephalosporins. The corresponding lactones have been prepared by direct acid treatment of the cephalosporins (Chauvette and Flynn, 1966; Cocker et al., 1965) or by treatment of the desacetyl derivatives with dilute acid or acetic anhydride (Van Heyningen, 1965).

For a small series of cephalosporins, the desacetyl derivatives had about half the activity of the parent compounds against Staphylococci, while the lactones were comparable to the unmodified cephalosporins. When serum was added to the assay medium, there was a very marked decrease in the activity of the lactones (see Table XXI). From a comparison of cephalothin with its desacetyl derivative, it appeared that the acetyl function was essential for high activity against gram-negative bacteria (Chauvette et al., 1962). This was also true for desacetylcephaloglycin which was shown to be a metabolic product of cephaloglycin in blood and urine. It was about as active as the parent cephalosporin against Staphylococci but only about one-fifth as active against gram-negative organisms (Kukolja, 1968).

Acylation of the 3-hydroxymethyl group of desacetylcephalosporins is difficult due mainly to the ease of cyclization to the corresponding lactones. Some acyl derivatives have been prepared under selected mild conditions or with protection of the 4-carboxyl group. For a small number of aroyl derivatives (see Table XXII), the activity against gram-negative organisms was significantly less than that of the corresponding cephalosporins (Van Heyningen, 1965). For a more extensive

TABLE XXI
DESACETYLCEPHALOSPORINS AND THEIR LACTONES

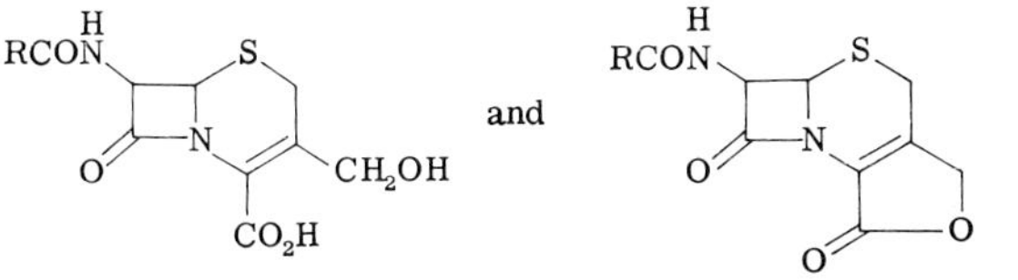

[Structures: RCONH-cephalosporin with CH₂OH and CO₂H substituents, and the corresponding lactone form]

R	MIC (μg/ml) Gram-pos. range	MIC (μg/ml) Gram-neg. range	Reference[a]
Desacetyls			
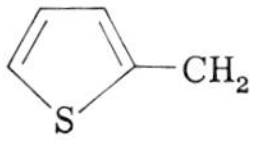 2-thienyl-CH$_2$	0.5–4		(1)
		3–134	(2)
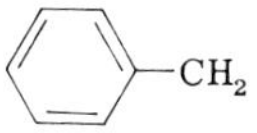 phenyl-CH$_2$	0.5–1.5		(1)
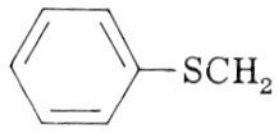 phenyl-SCH$_2$	0.1–0.3		(1)
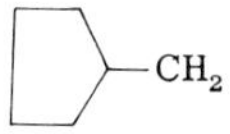 cyclopentyl-CH$_2$	2–3		(1)
CH$_3$(CH$_2$)$_3$SCH$_2$	0.1		(1)
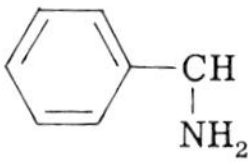 phenyl-CH(NH$_2$)	1.3–3.4	11.6–16.4	(3)
Lactones			
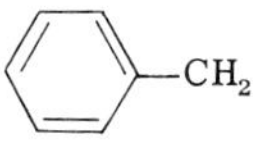 phenyl-CH$_2$	1–9 (60–>200)[b]		(1)
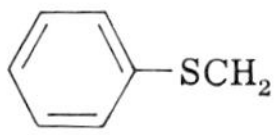 phenyl-SCH$_2$	0.1 (>100)[b]		(1)
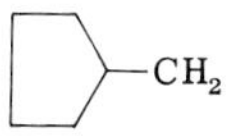 cyclopentyl-CH$_2$	0.6 (>100)[b]		(1)
CH$_3$(CH$_2$)$_3$SCH$_2$	0.2–2 (23–37)[b]		(1)

[a]References: (1) Chauvette et al. (1962); (2) Lilly (1969b); (3) Kukolja (1968).
[b]With serum.

series with a 7-phenylacetyl side chain, increasing the length of a 3-alkanoyloxymethyl side chain decreased activity across the spectrum, while for 3-benzoyloxymethyl and 3-arylacetoxymethyl compounds

TABLE XXII
ACYL DERIVATIVES OF DESACETYLCEPHALOSPORINS

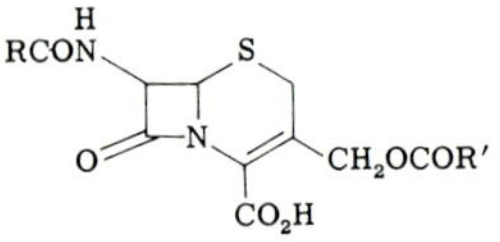

[Structure: RCONH-β-lactam-S-ring with CH$_2$OCOR' and CO$_2$H]

R	R'	MIC (µg/ml) Gram-pos. range	MIC (µg/ml) Gram-neg. range	Reference[a]
C$_6$H$_5$CH$_2$–	n-C$_3$H$_7$	0.16–1	16–125	(1)
	(CH$_2$)$_4$CH$_3$	0.62–8	31–250	(1)
	CH$_2$CH(CH$_3$)$_2$	0.31–4	31–>250	(1)
	CH(CH$_2$)$_2$CH$_3$ | C$_2$H$_5$	0.62–2.5	>250	(1)
	 C$_6$H$_5$– (phenyl)	0.31–2	125–>250	(1)
	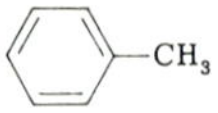 C$_6$H$_5$–CH$_3$ (tolyl, via CH$_3$)	0.08–>0.5	16–250	(1)
	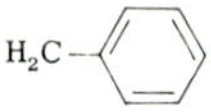 H$_3$C–C$_6$H$_5$–	0.16–2	4–125	(1)
	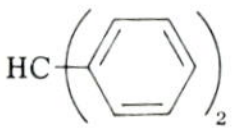 HC(C$_6$H$_5$)$_2$	1.25–2.5	125–>250	(1)
	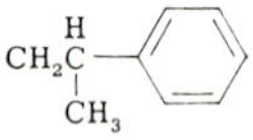 CH$_2$CH(CH$_3$)–C$_6$H$_5$	0.3–125	>250	(1)
	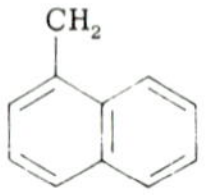 1-methylnaphthyl (CH$_2$–naphthyl)	0.16–>0.5	16–125	(1)
	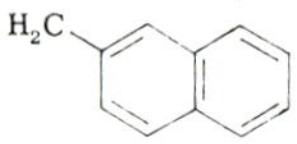 2-methylnaphthyl (H$_2$C–naphthyl)	0.31–4	≥250	(1)
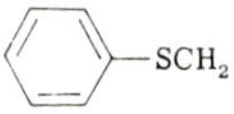 C$_6$H$_5$–SCH$_2$–	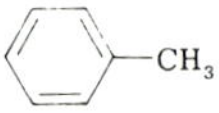 C$_6$H$_5$–CH$_3$	0.006–>0.1	20–54	(2)
	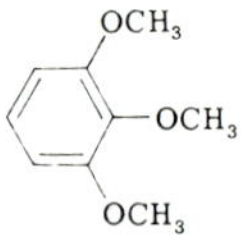 2,3-dimethoxyphenyl (OCH$_3$, OCH$_3$, OCH$_3$)	0.1–0.3	110–192	(2)
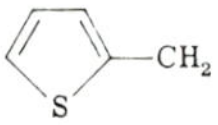 2-thienyl–CH$_2$–	 C$_6$H$_5$–	0.025–0.1	18–37	(2)
	 2-thienyl	0.2–6	50–63	(2)

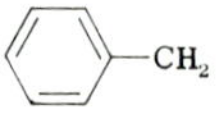

[a] References: (1) Glaxo (1966a); (2) Van Heyningen (1965).

TABLE XXIII
O-CARBAMYL DERIVATIVES OF DESACETYLCEPHALOSPORINS[a]

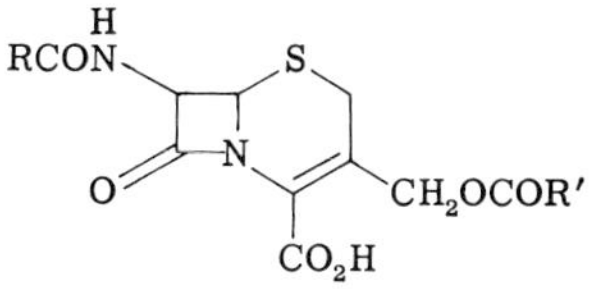

RCONH—[β-lactam-cephem]—CH$_2$OCOR'
CO$_2$H

R	R'	MIC (μg/ml) Gram-pos. range	MIC (μg/ml) Gram-neg. range	Reference[a]
Cl(CH$_2$)$_2$NH	CH$_3$	6–8	30–125	(1)
	NHCH$_3$	30	15–125	(1)
	NHCH$_2$CH$_3$	30	15–60	(1)
	NHCH$_2$CH$_2$Cl	4–8	2–250	(1)
ClCH$_2$C(CH$_3$)$_2$NH	NCH$_3$	60	4–30	(1)
	NCH$_2$CH$_3$	125	15–60	(1)
	NHCH$_2$Cl	125	15–125	(1)
	NHCH$_2$CH$_2$Cl	60	4–60	(1)

[a]From Ciba (1965).

the activity against gram-positive bacteria was often comparable to or occasionally even better than that of 7-phenylacetyl-ACA though the activity against gram-negative organisms were generally much poorer (Glaxo, 1966a). In series of O-carbamyl derivatives of 7-[N'-(β-chloroethyl)ureido]cephalosporins the reverse appeared to occur (see Table XXIII). Activity against gram-negative organisms was increased, while activity against Staphylococci was generally appreciably less (Ciba, 1965).

2. Desacetoxycephalosporins

Hydrogenation of cephalosporin C in the presence of palladium gave the desacetoxy derivative (Fig. 1c) which on hydrolysis was converted to 7-aminodesacetoxycephalosporanic acid (7-ADCA or ADCA) (Fig. 3b). A limited series of 7-acyl derivatives of ADCA (Table XXIV) exhibited appreciably lower activity than the parent cephalosporins (Stedman et al., 1964), however, a series of substituted 7-phenylacetyl derivatives appeared to have comparable activity to the corresponding compounds with a 3-acetoxymethyl side chain (Ryan et al., 1969). The high potency and broad spectrum of activity of cephaloglycin (Fig. 5a) prompted the preparation of series of substituted phenylglycine

TABLE XXIV
7-ACYLAMINODESACETOXYCEPHALOSPORINS

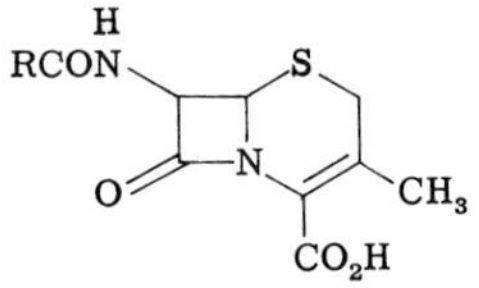

$$\text{RCONH} - \text{[β-lactam-cephem]} - CH_3, CO_2H$$

R	MIC (µg/ml)		Reference [a]
	Gram-pos. range	Gram-neg. range	
NH_2	125–250	500–>1000	(1)
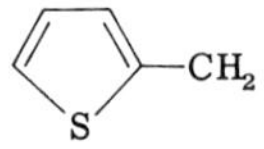 (2-thienyl)–CH_2	5–20	31–62	(1)
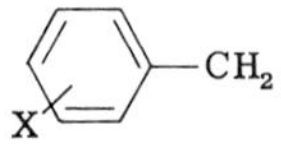 (X-phenyl)–CH_2			
X = H	5–20	250	(1)
3-Br	0.5	42.7–>200	(2)
3-NO_2	1	>50	(2)
3-CF_3	0.7	36–>50	(2)
4-Cl	0.8	>200	(2)
4-NO_2	1.4	>50	(2)
4-CN	8.1	>50	(2)
4-SCH_3	12.1	>50	(2)

[a] References: (1) Stedman et al. (1964); (2) Ryan et al. (1969).

derivatives of ADCA (Ryan et al., 1969). The D-phenylglycyl derivative of ADCA (cephalexin, Fig. 5b) retained about half the activity of cephaloglycin against gram-positive organisms and about one-fifth to one-half of its activity against gram-negative organisms (see Table XXV). A 3-hydroxy or 3-methoxy substituent in the phenyl ring of cephalexin did not appreciably alter the antibacterial activity, though other substituents produced a loss of potency particularly against gram-negative organisms.

An important property of cephalexin is its efficient absorption from the gastrointestinal tract (ED_{50} = 1.7 mg/kg mice versus *Streptococcus pyogenes*). Of the other analogs, only the 4-nitrophenylglycyl compound showed appreciable oral absorption (Ryan et al., 1969; Lilly, 1969c).

TABLE XXV
D-PHENYLGLYCYL DERIVATIVES OF ADCA[a]

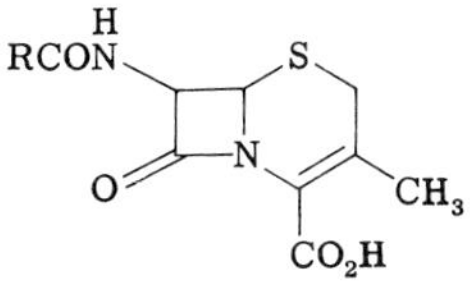

RCONH—[β-lactam-cephem]—CH$_3$, CO$_2$H

R	MIC (µg/ml)	
	Gram-pos. range	Gram-neg. range
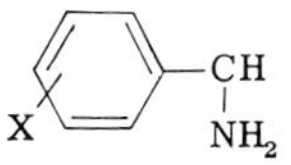 X—C$_6$H$_4$—CH(NH$_2$)—		
X = H	3.5	9.3–19
3-F	7.2	15.4–>50
3-Cl	1.9	17–>50
3-Br	1.3	14.5–>50
3-OH	1.7	8.8–20.9
3-OCH$_3$	4.1	12–26
4-Cl	4.4	19.8–>50
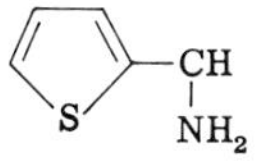 2-thienyl—CH(NH$_2$)—	8.1	12.2–8.9

[a] From Ryan et al. (1969).

3. Derivatives Produced by Displacement of the Acetoxy Group by Nucleophiles

Cephalosporin C was found to react with pyridine in neutral aqueous solution to give a derivative with enhanced antibacterial activity. It was named cephalosporin C_A (pyridine) (Hale et al., 1961). The new derivative (Fig. 1d) was formed by nucleophilic displacement of the acetoxy group. Subsequently, a large number of other nucleophilic reagents have been used for the displacement reaction in a variety of 7-acylaminocephalosporanic acids.

The displacement reaction is a simple S_N1 type where the rate is dependent solely on the concentration of the cephalosporin. Only protic solvents of high dielectric constant are suitable for the reaction. In practice, solvents are limited to mixed aqueous solutions or formamide. Yields of pyridinium compounds obtained by reaction of cephalosporins with aqueous pyridine under a variety of conditions

TABLE XXVI
Alkyl-Substituted Cephaloridines

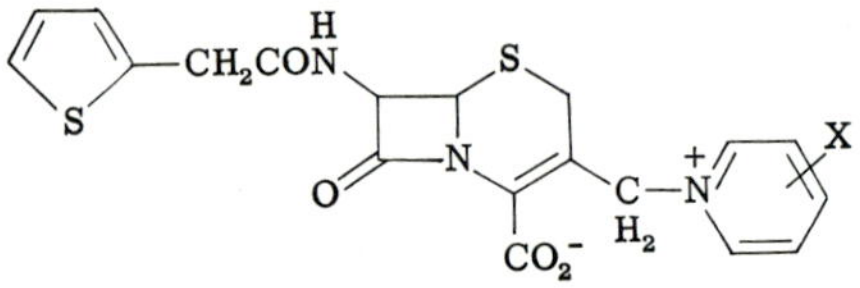

	MIC (μg/ml)		
X	Gram-pos. range	Gram-neg. range	Reference[a]
H	0.3	3.8	(1)
	0.02−6.2	4−8	(2)
	0.4	2.4−5.6	(3)
2-CH_3	0.7	8.8	(1)
	0.04−4	16	(2)
3-CH_3	1.6	4.8	(1)
	0.6−4.8	3.8−9	(3)
4-CH_3	2.3	2.2	(1)
	1.2−4.6	1−13.1	(3)
2-C_2H_5	0.3	11.1	(3)
3-C_2H_5	2.9	6.2	(3)
4-C_2H_5	6.4	5.5	(3)
4-C_3H_7-n	7.3	3.4	(3)
2,4-di-CH_3	0.08−2.5	16−62	(2)
3,5-di-CH_3	4.2	9.3	(1)

[a]References: (1) Spencer et al. (1967a); (2) Glaxo (1964d); (3) Lilly (1969a).

were normally less than 50%. The "theoretical" maximum yield was calculated to be 54% (Taylor, 1965). When the acetoxy group was displaced initially by a thio compound, and the thio derivative subsequently reacted with a tertiary amine in the presence of silver or mercury salts, higher yields were obtained (Glaxo, 1965a). Similarly, when thiocyanate or iodide salts were used to saturate the solvent mixture, yields of displacement products exceeded the "theoretical maximum" (Spencer et al., 1967b). The role of thiocyanate or iodide in increasing the yield is not defined.

The product of displacement of the acetoxy group in cephalothin by

TABLE XXVII
HALOGEN-SUBSTITUTED CEPHALORIDINES

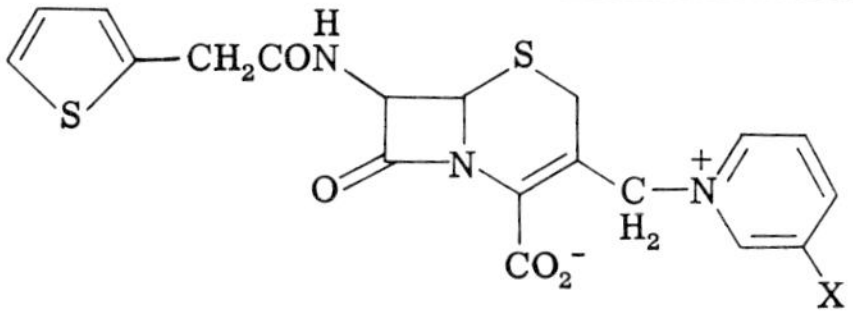

[Structure: 2-thienyl—$CH_2CON(H)$—β-lactam-cephem with CO_2^- and CH_2—pyridinium bearing substituent X]

X	MIC (μg/ml) Gram-pos. range	Gram-neg. range	Reference[a]
F	0.2	9.1	(1)
	0.1–0.2	7.1–34.2	(2)
Cl	0.2	8.3	(1)
	0.1–0.3	6.2–13.6	(2)
Br	<0.1	6.0	(1)
	<0.1–0.2	4.4–14.5	(2)
I	0.3	6.1	(1)
	–	3.8–11.8	(2)

[a]References: (1) Spencer et al. (1967a); (2) Lilly (1969a).

pyridine was named cephaloridine (Fig. 4b). This derivative was about two to four times as active as the parent compound against gram-positive bacteria and about equal against gram-negative bacteria. The spectrum of activity of the two compounds is essentially the same though minor variations do occur (Barber and Waterworth, 1964).

Among cephaloridines, alkyl substitution of the pyridine ring causes small but definite changes in biological activity (Spencer et al., 1967a). Comparing 2-, 3-, and 4-substituted derivatives there is a trend to decreased activity against gram-positive organisms and increased activity against gram-negative organisms (Table XXVI). Introducing a halogen into the 3-position of the pyridine ring may produce increased activity against gram-positive bacteria accompanied by a slight reduction in activity against gram-negative bacteria (see Table XXVII). Of other subsituted pyridinium derivatives of cephalothin, those with 2- and 3-substituents had generally lower activity than cephaloridine, while many 4-substituted derivatives such as 4-hydroxymethyl, 4-carboxamido, and 4-monoalkylcarboxamido pyridinium derivatives had activity similar to that of cephaloridine (Tables XXVIII–XXXI). The 4-carboxamide derivative had the greatest

TABLE XXVIII
AMIDE-SUBSTITUTED CEPHALORIDINES

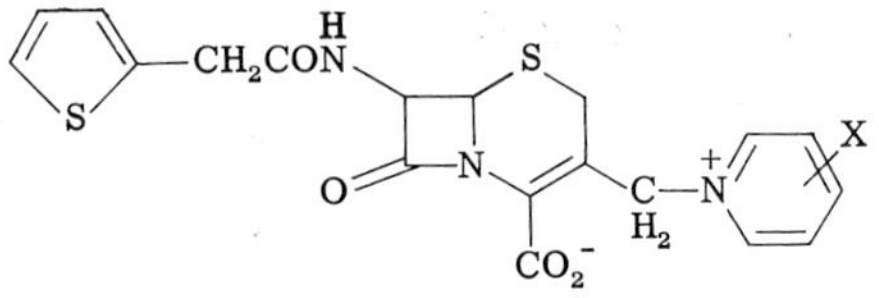

	MIC (μg/ml)		
X	Gram-pos. range	Gram-neg. range	Reference[a]
3-$CONH_2$	0.5	4.8	(1)
	0.2	4.2–9.2	(2)
	0.04–0.32	8–16	(3)
4-$CONH_2$	0.4	2	(1)
	0.4–0.9	3.6–5	(2)
3-$CONHCH_3$	0.4	6	(1)
	0.4	6.6–17.8	(2)
4-$CONHCH_3$	0.4	4	(1)
	0.3–0.4	5–11.8	(2)
3-$CONHC_2H_5$	0.04–1	31–125	(3)
4-$CONHC_2H_5$	0.3	8.0	(1)
3-$CON(CH_3)_2$	0.31–2	16–250	(3)
4-$CON(CH_3)_2$	0.5	22	(1)
3-$CON(C_2H_5)_2$	0.5	21	(1)
	0.16–4	31–>250	(3)
4-$CONH(CH_2)_2CH_3$	0.4	6	(1)
4-$CONHCH(CH_3)_2$	0.6	9	(1)
4-CONHC(H)(CH_2-CH_2) (cyclopropyl)	0.3	9	(1)
3-$CONHCH_2OH$	0.3–1	<4–8	(3)
4-$CONHCH_2OH$	0.3	4	(1)
	0.08–<0.5	<4–8	(3)
4-CH_2CONH_2	0.08–2.5	4–16	(3)

[a]References: (1) Spencer et al. (1967a); (2) Lilly (1969a); (3) Glaxo (1964d).

activity against gram-negative organisms of any pyridinium derivative. This was true in the 7-phenylacetyl series (Table XXIX) also (Spencer et al., 1967a).

The effect of various 7-acyl side chains on pyridinium-substituted cephalosporins is shown in Tables XXXII and XXXIII. Derivatives

TABLE XXIX
CARBOXAMIDE-SUBSTITUTED 7-PHENYL-ACA PYRIDINIUM DERIVATIVES

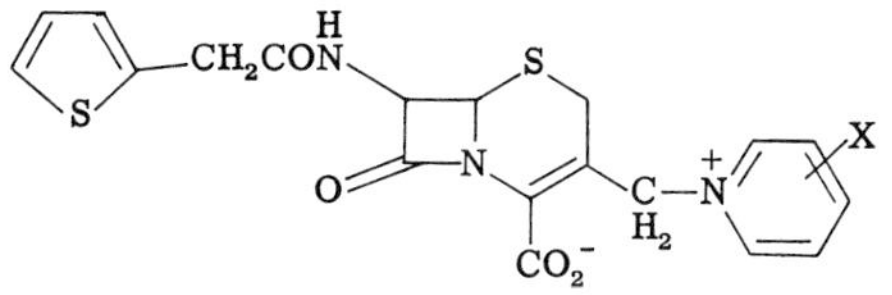

X	MIC (µg/ml)		Reference[a]
	Gram-pos. range	Gram-neg. range	
H	1.8	9	(1)
3-CONH$_2$	0.8	10	(1)
		8–25	(2)
4-CONH$_2$	0.5	5	(1)
		3.6–36	(2)
4-CONHCH$_3$	0.3	6	(1)
4-CONHCH$_2$OH	0.3	6	(1)

[a]References: (1) Spencer et al. (1967a); (2) Lilly (1965b).

TABLE XXX
CARBOXYLIC ACID-, ESTER-, OR NITRILE-SUBSTITUTED CEPHALORIDINES

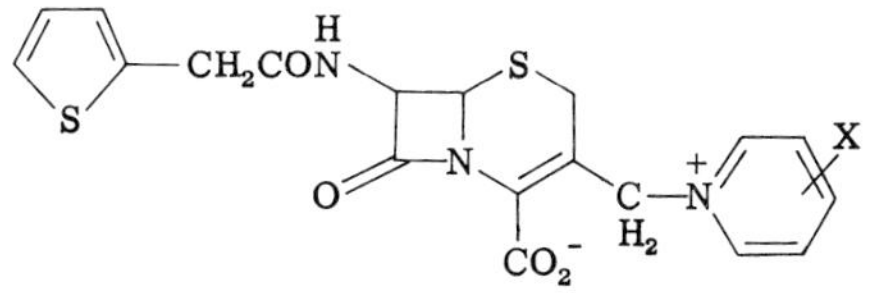

X	MIC (µg/ml)		Reference[a]
	Gram-pos. range	Gram-neg. range	
3-CO$_2$H	2.1	7	(1)
4-CO$_2$H	2.7	16	(1)
3-CO$_2$CH$_3$	0.16–4	<4–8	(2)
4-CO$_2$CH$_3$	0.8	8	(1)
	0.25–1	2–4	(2)
	0.6–1.3	2–43	(3)
3-CO$_2$C$_2$H$_5$	0.08–1	<4–31	(2)
3-CN	0.4	15	(1)
	0.3–0.5	11–44	(3)
4-CN	0.5	7	(1)
	0.4–0.5	5.6–12.2	(3)
4-CH$_2$CO$_2$H	5.7	8	(1)

[a]References: (1) Spencer et al. (1967a); (2) Glaxo (1964d); (3) Lilly (1969a).

TABLE XXXI
MISCELLANEOUS SUBSTITUTED CEPHALORIDINES

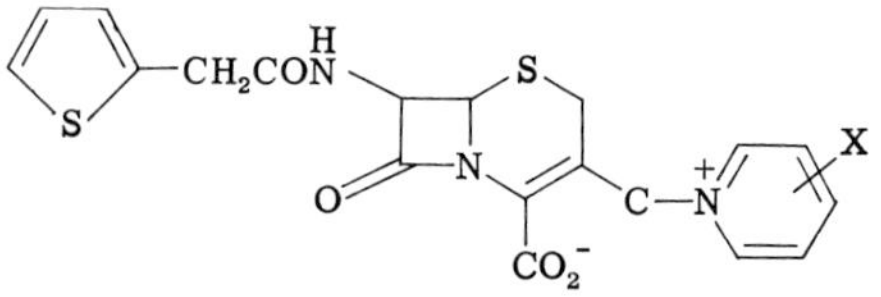

[Structure: thiophene-CH$_2$CONH-cephem with C—N$^+$—X pyridinium at 3-position, CO$_2^-$]

X	MIC (μg/ml) Gram-pos. range	Gram-neg. range	Reference[a]
3-OH	1.9	16	(1)
	0.9–3.4	4.8–57.5	(2)
2-CH$_2$OH	0.04–<0.5	<4–31	(3)
3-CH$_2$OH	1.5	5.3	(1)
	0.4–1.9	4.2–9.2	(2)
4-CH$_2$OH	0.9	3	(1)
	1.8–3.1	4.1–16.8	(2)
3-COCH$_3$	0.1–0.4	7–46.3	(2)
4-COCH$_3$	<0.1–0.2	3.3–8.1	(2)
4-CF$_3$	0.2	4	(1)
3-SO$_3$H	0.7	19.7	(1)

[a]References: (1) Spencer et al. (1967a); (2) Lilly (1969a); (3) Glaxo (1964d).

with the thiophene-2-acetyl side chain were clearly the most potent against gram-negative bacteria and, in most cases, against the gram-positive bacteria too.

Many other nucleophiles besides pyridines have been used to displace the acetoxy group of cephalosporins. Cephalosporin C reacted with sodium thiosulfate to form a Bunte salte (Fig. 1e) of higher activity than the parent antibiotic (Demain et al., 1963a). This explained an earlier observation that sodium thiosulfate appeared to stimulate the biosynthesis of cephalosporin C (Demain and Newkirk, 1962).

Subsequently, a large number of derivatives have been formed by the use of sulfur nucleophiles to displace the acetoxy group. A group of xanthate derivatives of cephalothin were prepared by Van Heyningen and Brown (1965). They were comparable to cephalothin against gram-positive organisms though their gram-negative potency was relatively poor (Table XXXIV). The same was true, in general, of

TABLE XXXII
Various 7-Acyl-ACA Pyridinium Derivatives

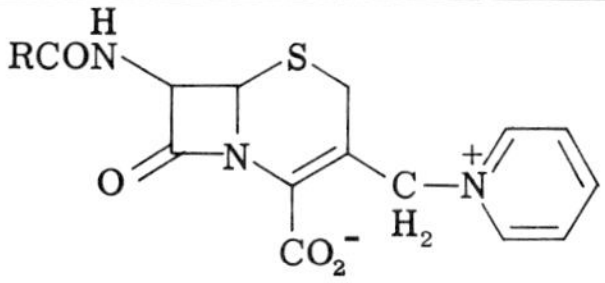

RCONH—[β-lactam-cephem]—CH$_2$—N$^+$(pyridinium), CO$_2^-$

R	MIC (μg/ml) Gram-pos. range	MIC (μg/ml) Gram-neg. range	Reference[a]
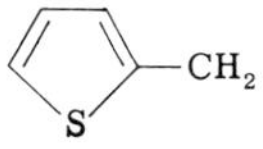 2-thienyl–CH$_2$	0.3	3.8	(1)
	0.02–6.2	4–8	(2)
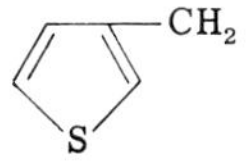 3-thienyl–CH$_2$	0.02–2.5	<4–16	(2)
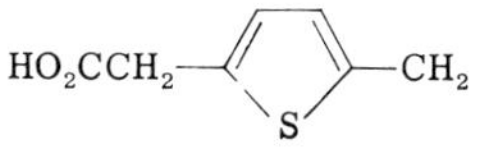 HO$_2$CCH$_2$–(thienyl)–CH$_2$	1.56–6.25	3.12–6.25	(3)
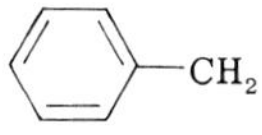 phenyl–CH$_2$	1.8	9	(1)
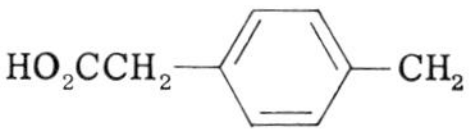 HO$_2$CCH$_2$–(phenyl)–CH$_2$	3.12–12.5	6.25–12.5	(3)
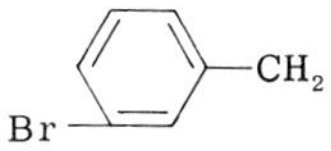 Br–(phenyl)–CH$_2$	7.9	>50	(1)
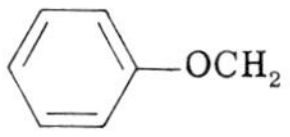 phenyl–OCH$_2$	0.2	38	(1)
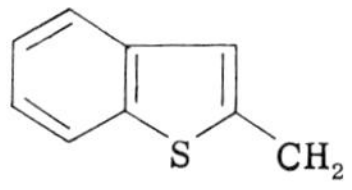 benzothienyl–CH$_2$	5.5	52	(1)
H	0.4	>50	(1)

[a]References: (1) Spencer et al. (1967a); (2) Glaxo (1964d); (3) Lewis et al. (1969a).

simple dithiocarbamates of cephalothin (Van Heyningen and Brown, 1965) (see Table XXXV), and of N,N-dimethyldithiocarbamates of other 7-acylcephalosporins (Glaxo, 1964c, Table XXXVI). Dithiocarbamate derivatives containing a piperazine ring had better activity, the most active being the 4-methyl derivative (compound 1, Table XXXVII). The zwitterionic derivatives of dithiocarbamates were, as a group, about half as active as the piperazino compounds (see Table

TABLE XXXIII
VARIOUS 7-ACYL-ACA 4'-CARBOXAMIDO PYRIDINIUM DERIVATIVES

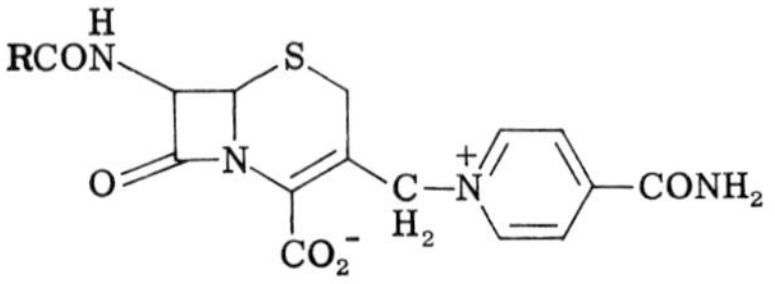

$$\text{RCONH—[β-lactam-cephem]—CH}_2\text{—N}^+\text{C}_5\text{H}_4\text{—CONH}_2$$

R	MIC (μg/ml)		Reference[a]
	Gram-pos. range	Gram-neg. range	
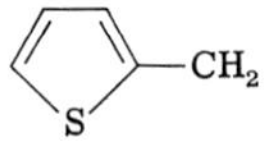 2-thienyl-CH$_2$	0.4	2	(1)
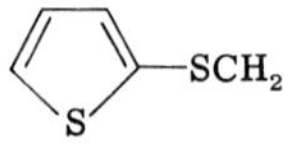 2-thienyl-SCH$_2$	0.8	9.7	(1)
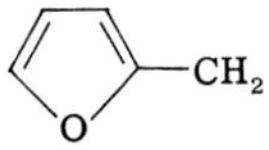 2-furyl-CH$_2$	0.9	8.6	(1)
	0.7–1.2	8.1–14	(2)
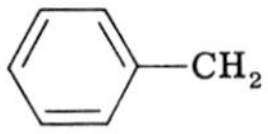 phenyl-CH$_2$	0.5	5	(1)
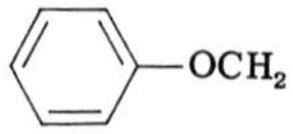 phenyl-OCH$_2$	2.1	23	(1)
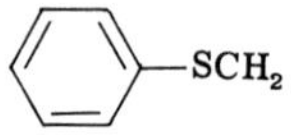 phenyl-SCH$_2$	1.5	17	(1)
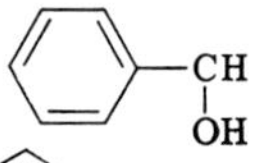 phenyl-CH(OH)-	4.3	5.9	(1)
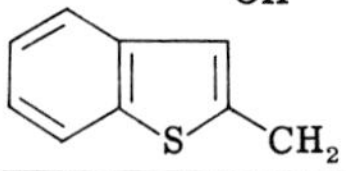 benzothien-2-yl-CH$_2$	0.5	>50	(1)

[b]References: (1) Spencer et al. (1967a); (2) Lilly (1969a).

TABLE XXXIV
XANTHATE DERIVATIVES OF CEPHALOTHIN[a]

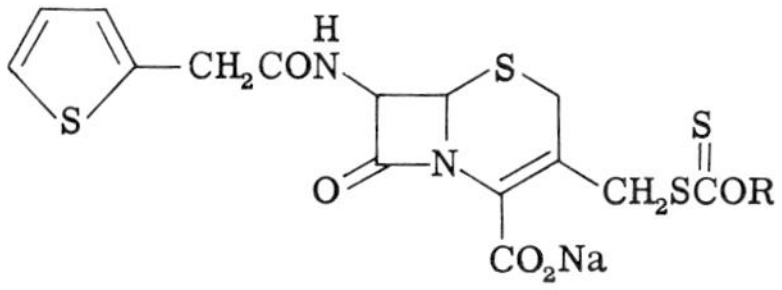

[Structure: 2-thienyl-$CH_2CON(H)$-β-lactam-cephem with CO_2Na and CH_2SCOR (with C=S) at position 3]

R	MIC (μg/ml)	
	Gram-pos. range	Gram-neg. range
$-C_2H_5$	0.05-0.7	6-83
$-(CH_2)_2CH_3$	0.025-1.3	5-84
$-CH(CH_3)_2$	0.05-0.7	9->200
$-(CH_2)_3CH_3$	0.05-0.3	8-108
$-(CH_2)_5CH_3$	0.2-0.4	14->200
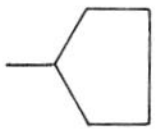 —cyclopentyl	0.05-0.4	6-107
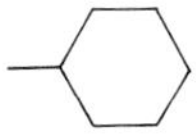 —cyclohexyl	0.025-0.7	6-116

[a] From Van Heyningen and Brown (1965).

XXXVIII). Quaternized piperazine derivatives (see Table XXXIX) were less potent than their parent compounds (Van Heyningen and Brown 1965).

Displacement of the acetoxy group by thiourea, or cyclic analogs of thiourea, gave cephalosporin derivatives (see Tables XL and XLI) with generally good gram-positive activity (Glaxo, 1963). High gram-positive and gram-negative potencies were claimed for a series of heterocyclic thiol derivatives (see Table XLII) of sydnone cephalosporin (Fujisawa, 1969f).

Thio acids gave cephalosporins with good gram-positive and moderate gram-negative potencies (Glaxo, 1965b). Where the comparison could be made, they were similar in activity to their oxygen analogs (see Table XLIII).

Azide displacement of the acetoxy group has given a series of 3-azidomethylcephalosporins (Glaxo, 1966b). Although they had good gram-positive potency, they were generally poorly active against gram-negative bacteria (see Table XLIV). Exceptions to this were the 7-methylthioacetyl, 7-ethylthioacetyl, 7-(p-amino) phenylacetyl, and 7-thienylacetyl analogs, which had good gram-negative potency. An

TABLE XXXV
DITHIOCARBAMATES OF CEPHALOTHIN[a]

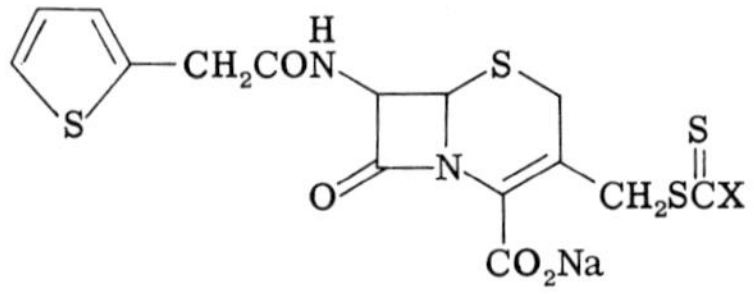

[structure: cephalothin core with —CH$_2$SCX where C=S, and CO$_2$Na, with 2-thienyl-CH$_2$CONH— side chain]

X	MIC (μg/ml)	
	Gram-pos. range	Gram-neg. range
—NHCH$_3$	0.4–1.5	9–>200
—NH(CH$_2$)$_2$CH$_3$	1.2	11–>200
—N(C$_2$H$_5$)$_2$	0.1–0.6	13–134
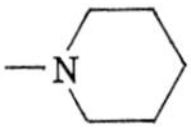 —N(piperidinyl)	0.1–0.8	4–44
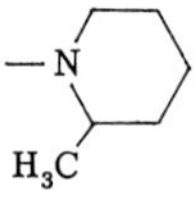 —N(2-methylpiperidinyl)	0.2–0.8	16–115
—N(CH$_3$)CH$_2$CH$_2$OH	0.1–0.4	12–42
—N(CH$_2$CH$_2$OH)$_2$	0.2–1	18–124
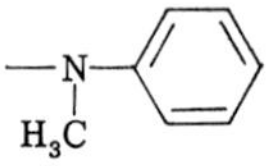 —N(CH$_3$)(C$_6$H$_5$)	0.05–0.2	13–>200
—N(CH$_3$)CH$_2$(CHOH)$_4$CH$_2$OH	0.4–2.4	72–141

[a] From Van Heyningen and Brown (1965).

TABLE XXXVI
DIMETHYLDITHIOCARBAMATES OF 7-ACYL-ACA'S[a]

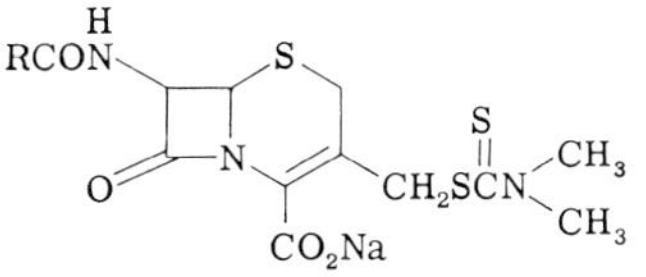

	MIC (μg/ml)	
R	Gram-pos. range	Gram-neg. range
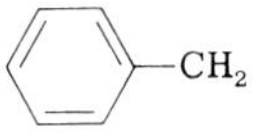 C$_6$H$_5$—CH$_2$	0.04–2.5	62–>25
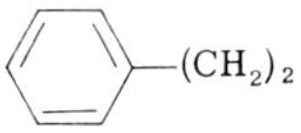 C$_6$H$_5$—(CH$_2$)$_2$	0.08–0.31	>250
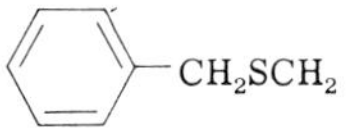 C$_6$H$_5$—CH$_2$SCH$_2$	0.08–>2.5	>250
CH$_3$(CH$_2$)$_3$	0.16–1.25	62–>250
CH$_3$(CH$_2$)$_3$SCH$_2$	0.02–0.3	125–>250
H$_2$C=CHCH$_2$SCH$_2$	0.16–0.62	125–250
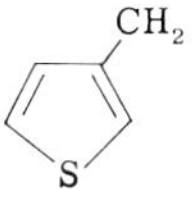 3-thienyl-CH$_2$	0.08–0.62	16–250
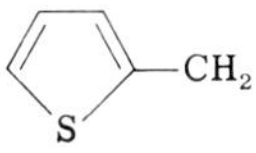 2-thienyl-CH$_2$	0.04–0.6	16–62.5

[a] From Glaxo (1964c).

TABLE XXXVII
PIPERAZINO DITHIOCARBAMATES OF CEPHALOTHIN[a]

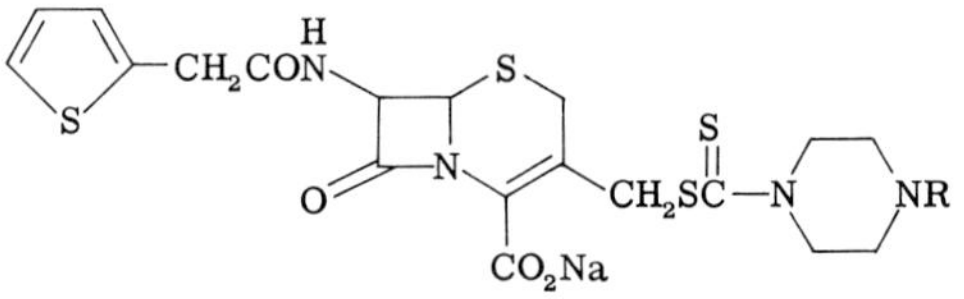

	MIC (μg/ml)	
R	Gram-pos. range	Gram-neg. range
CH_3	0.025–0.4	2–4
C_2H_5	0.5	2.8–12
$(CH_2)_2CH_3$	0.05–0.4	5–26
$CH(CH_3)_2$	0.5	5–46
$(CH_2)_3CH_3$	0.05–1	4.5–28
$(CH_2)_4CH_3$	0.05–0.6	5–72
$(CH_2)_7CH_3$	0.1–0.3	8–>100
CH_2CH_2OH	0.1–0.4	6–14
	<0.025–0.4	33–>100
1-Adamantane	<0.025–0.3	14–>100

[a] From Van Heyningen and Brown (1965).

TABLE XXXVIII
ZWITTERIONIC DITHIOCARBAMATES OF CEPHALOTHIN[a]

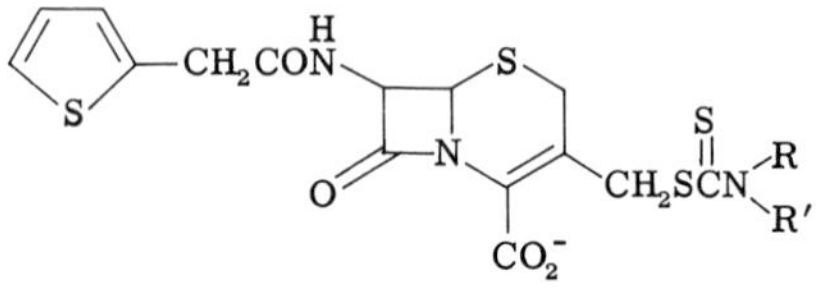

		MIC (μg/ml)	
R	R'	Gram-pos. range	Gram-neg. range
H	$CH_2CH_2N(C_2H_5)_2 \cdot H^+$	3.13–25	>250
H	$CH_2CH_2N(CH_3)C_6H_5 \cdot H^+$	0.78–6.7	>200
CH_3	$CH_2CH_2N(CH_3)_2 \cdot H^+$	0.4–2.4	18–110
CH_3	$CH_2CH_2N(C_2H_5)_2 \cdot H^+$	0.05–0.3	5–60
CH_3	$CH_2CH_2N(n\text{-}C_3H_7)_2 \cdot H^+$	0.8	11–118
CH_3	$CH_2CH_2CH_2N(C_2H_5)_2 \cdot H^+$	0.048–0.8	10–88
C_2H_5	$CH_2CH_2N(C_2H_5)_2 \cdot H^+$	0.04–0.7	16–100

[a] From Van Heyningen and Brown (1965).

TABLE XXXIX
Quaternized Piperazino Dithiocarbamates[a]

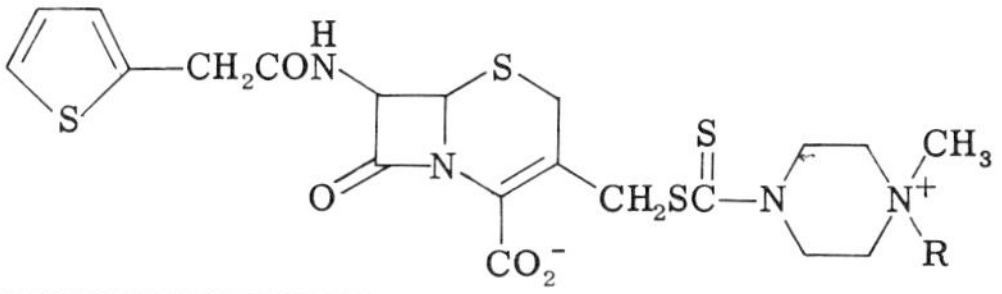

[Structure: thiophene-CH$_2$CONH-cephalosporin with 3-CH$_2$SC(=S)-N(piperazino)-N$^+$(CH$_3$)(R), CO$_2^-$]

R	MIC (μg/ml)	
	Gram-pos. range	Gram-neg. range
CH$_3$	0.05–0.4	7–37
—(CH$_2$)$_2$CH$_3$	0.05–0.5	7–13
—CH$_2$CH=CH$_2$	0.1–0.8	6–88
—(CH$_2$)$_3$CH$_2$	0.05–0.4	7–106

[a] From Van Heyningen and Brown (1965).

extensive series of 3-azidomethyl derivatives has been prepared where the 7-acyl side chain was derived from glyoxylic acids (Glaxo, 1968b). No information on the gram-negative potencies of these compounds was reported. The gram-positive activities were generally good particularly for p-substituted phenylglyoxalyl acid derivatives (see Table XLV).

Reduction of 3-azidomethylcephalosporins gave the corresponding 3-aminomethyl analogs (Glaxo, 1964a). These had very poor gram-negative activity but generally better gram-positive activity than the 3-azidomethylcephalosporins from which they were derived (Table XLVI). The 3-aminomethyl derivative of 7-phenylacetyl-ACA was acylated with a variety of acyl groups (Glaxo, 1964a). The series had only moderate gram-positive and poor gram-negative activities. The best compound in the series, the N-acetyl derivative (compound 2, Table XLVII) did not compare favorably with 7-phenyl-ACA.

A series of 3-alkoxymethyl derivatives of cephalothin had poor gram-negative activity (see Table XLVIII). The same was generally true of a series of 3-methoxymethylcephalosporins (Table XLIX), though those with 7-phenylglycyl, 7-(α-formyloxy)phenylacetyl, and 7-(p-fluoro)phenylacetyl side chains were exceptions (Glaxo, 1969).

C. Miscellaneous Modifications

1. Double-Bond Isomerization

Under certain conditions, the double bond of the thiazine ring of cephalosporins can shift from the 3,4- to the 2,3-position. The iso-

TABLE XL
Thiouronium Derivatives of 7-Acyl-ACA's[a]

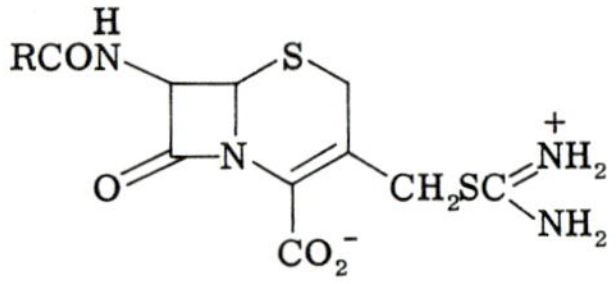

$$\text{RCON}\overset{H}{\underset{O}{\diagdown}}\text{—}\underset{\underset{CO_2^-}{N}}{\overset{S}{\diagup}}\text{—}CH_2SC\overset{+NH_2}{\underset{NH_2}{\diagdown}}$$

R	MIC (μg/ml) Gram-pos. range
$CH_3(CH_2)_2$	0.08–0.31
$CH_3(CH_2)_3$	0.16–0.62
$CH_3(CH_2)_4$	0.04–0.16
$CH_3(CH_2)_3SCH_2$	0.02–0.16
$H_2C=CHCH_2SCH_2$	0.04–0.15
$HC\equiv CCH_2SCH_2$	0.04–0.08
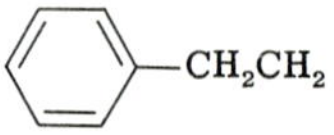 C₆H₅—CH₂CH₂	0.08–0.16
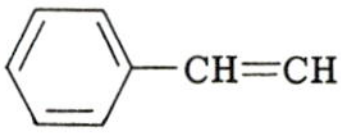 C₆H₅—CH=CH	0.5–2.0
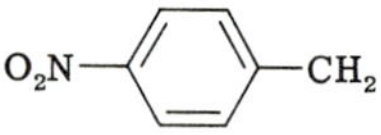 O_2N—C₆H₄—CH_2	0.01–0.15
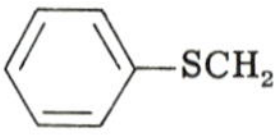 C₆H₅—SCH_2	0.01–0.08
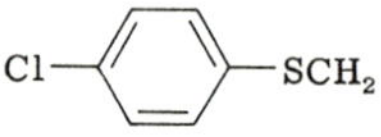 Cl—C₆H₄—SCH_2	0.02–0.08
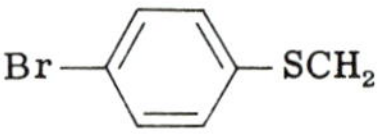 Br—C₆H₄—SCH_2	<0.01–0.31
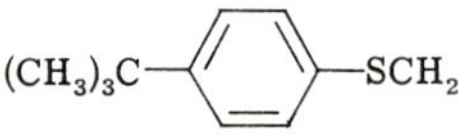 $(CH_3)_3C$—C₆H₄—SCH_2	0.05–0.62
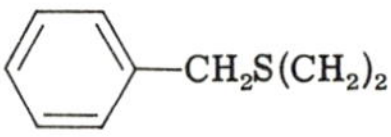 C₆H₅—$CH_2S(CH_2)_2$	0.01–0.16
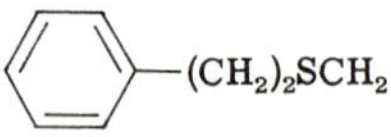 C₆H₅—$(CH_2)_2SCH_2$	0.08–0.62

TABLE XLI
CYCLIC ANALOGS OF THIOURONIUM DERIVATIVES OF 7-ACYL-ACA'S

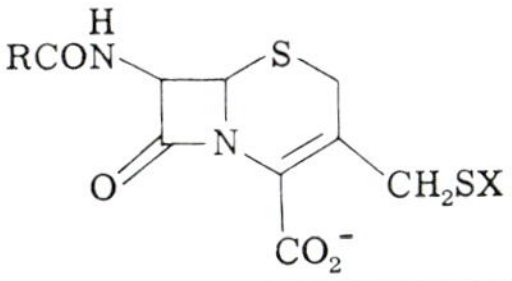

[Structure: cephalosporin core with RCONH- at 7-position and -CH₂SX at 3-position, with CO_2^-]

R	X	MIC (μg/ml) Gram-pos. range	Reference[a]
$HO_2CCH(CH_2)_4$ | NH_2	imidazolinium (HN⁺=C−NH, saturated)	20	(1)
	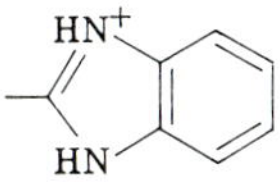 benzimidazolinium (HN⁺=C−NH, benzo-fused)	62	(1)
	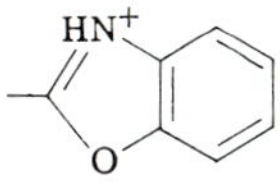 benzoxazolinium (HN⁺=C−O, benzo-fused)	31	(1)
	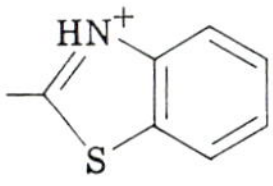 benzothiazolinium (HN⁺=C−S, benzo-fused)	0.5	(1)
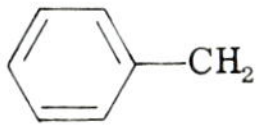 phenyl−CH_2	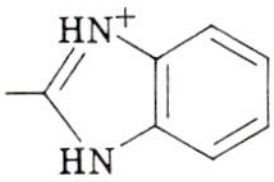 benzimidazolinium	0.04−0.08	(1)
	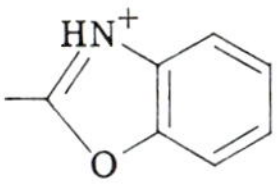 benzoxazolinium	0.04	(1)
	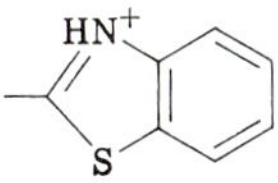 benzothiazolinium	0.01	(1)
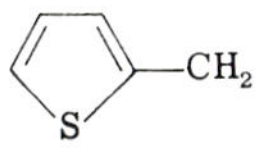 2-thienyl−CH_2	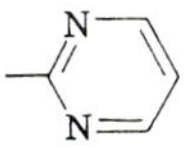 pyrimidinyl	0.4−0.5	(2)

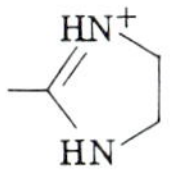

[a]References: (1) Glaxo (1963); (2) Van Heyningen (1967).

TABLE XLII
Heterocyclic Thiol Derivatives of Sydnone-ACA's[a]

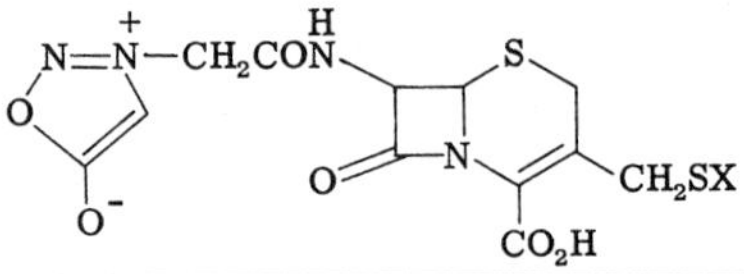

Structure:
$$\overset{+}{N}=N-CH_2CO\overset{H}{N}-[\beta\text{-lactam-cephem}]-CH_2SX$$
with CO_2H at position 4 and sydnone ring (O^-, O) on left.

X	MIC (μg/ml)	
	Gram-pos. range	Gram-neg. range
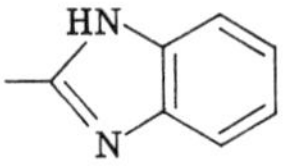 HN-benzimidazol-2-yl	0.5	2
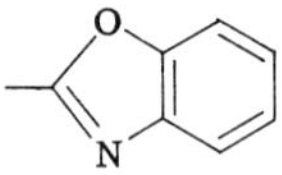 benzoxazol-2-yl	0.5	5
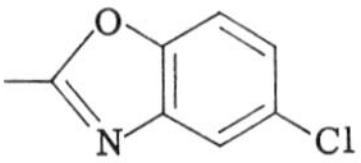 5-chloro-benzoxazol-2-yl	0.25	2.5
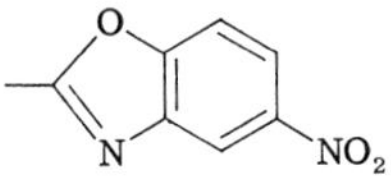 5-nitro-benzoxazol-2-yl	0.25	2.5
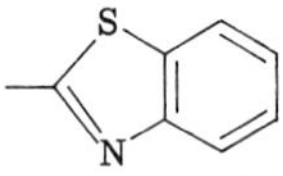 benzothiazol-2-yl	0.25	2.5

[a] From Fujisawa (1969f).

merization is base-catalyzed, and at equilibrium the ratio of isomers is usually about $\Delta^2:\Delta^3 = 7:3$ (Cocker et al., 1966). The Δ^2-isomers are essentially devoid of antibacterial activity and their β-lactam ring is much more stable to basic hydrolysis than the β-lactam ring of normal (Δ^3) cephalosporins.

It has been postulated that β-lactam antibiotics act by combining irreversibly with an enzyme essential for cell wall synthesis by β-lactam acylation of the active site (Cooper, 1956; Collins and Richmond, 1962). The essential lack of activity of Δ^2-cephalosporins, even though the corresponding Δ^3-cephalosporins are highly active, has been attributed to greater stability of the β-lactam ring in the Δ^2-cephalosporins. The possibility that the lack of activity is due to "incorrect" configuration of the carboxyl group was discounted when it was established that the 4-carboxyl group in Δ^2-cephalosporins has the

TABLE XLIII
THIO ACID DERIVATIVES OF CEPHALOTHIN[a]

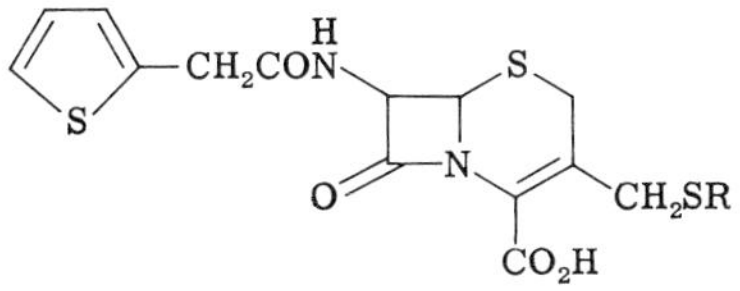

[Structure: thiophene-CH₂CONH-[β-lactam-dihydrothiazine]-CH₂SR, with CO₂H]

R	MIC (μg/ml)	
	Gram-pos. range	Gram-neg. range
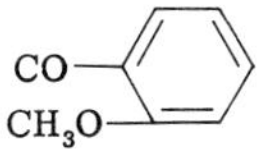 CO—C₆H₄(o-OCH₃)	0.16–4	14–250
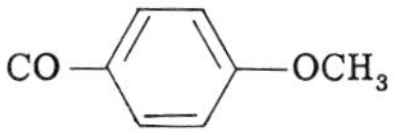 CO—C₆H₄(p-OCH₃)	0.02–4	4–125
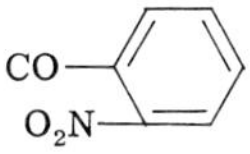 CO—C₆H₄(o-NO₂)	0.04–<0.5	<4–250
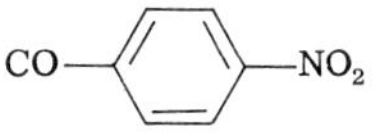 CO—C₆H₄(p-NO₂)	0.04–4	31–125
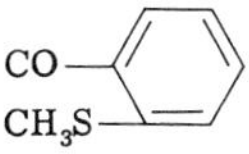 CO—C₆H₄(o-SCH₃)	0.08–8	16–>250
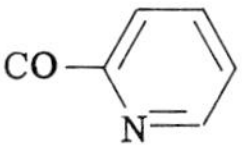 CO—(2-pyridyl)	0.08–9	<4–250
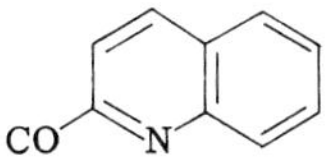 CO—(2-quinolyl)	0.04–8	125–250
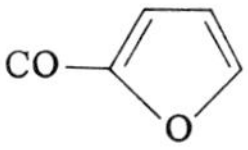 CO—(2-furyl)	0.04–2	<2–32
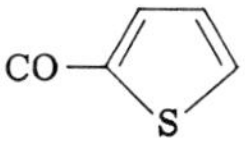 CO—(2-thienyl)	0.02–1	<2–>250
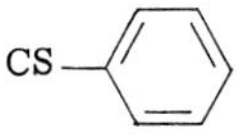 CS—C₆H₅	0.04–2	8–>250
 O₂S—C₆H₄—CH₃	0.31–4	125–250

[a] From Glaxo (1965b).

TABLE XLIV
3-AZIDOMETHYL CEPHALOSPORINS[a]

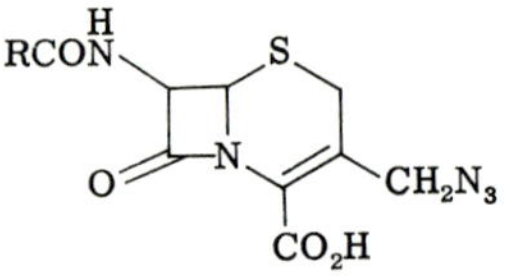

[Structure: cephalosporin core with RCONH-, S, CO₂H, and CH₂N₃ substituents]

R	MIC (μg/ml) Gram-pos. range	MIC (μg/ml) Gram-neg. range
$CH_2{=}CHCH_2$	0.04–<0.5	$\geq$250
$CH_3(CH_2)_2CH(SCH_3)$	0.62–4	250
CH_3SCH_2	0.16–0.6	4–31
$C_2H_5SCH_2$	0.02–0.08	16–31
$(CH_3)_2CHSCH_2$	0.6–1.25	31–250
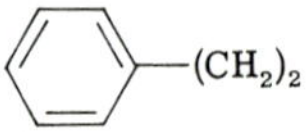 C₆H₅–(CH₂)₂	0.08–<0.5	125–250
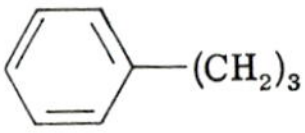 C₆H₅–(CH₂)₃	0.04–<0.5	$\geq$250
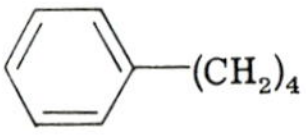 C₆H₅–(CH₂)₄	0.005–<0.5	>250
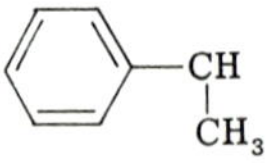 C₆H₅–CH(CH₃)	0.31–4	$\geq$250
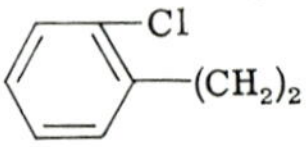 2-Cl-C₆H₄–(CH₂)₂	0.04–<0.5	2–250
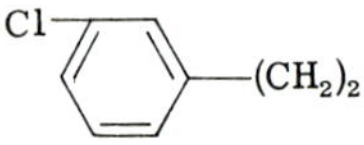 3-Cl-C₆H₄–(CH₂)₂	0.04–<0.5	62–25
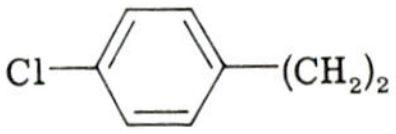 4-Cl-C₆H₄–(CH₂)₂	0.04–<0.5	$\geq$250
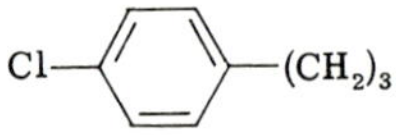 4-Cl-C₆H₄–(CH₂)₃	0.04–<0.5	>250
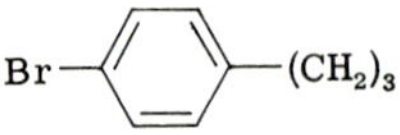 4-Br-C₆H₄–(CH₂)₃	0.04–<0.5	$\geq$250

TABLE XLIV (continued)

R	MIC (µg/ml)	
	Gram-pos. range	Gram-neg. range
 H$_2$N–C$_6$H$_4$–CH$_2$	0.62	4–62
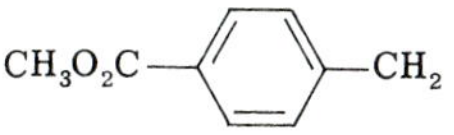 CH$_3$O$_2$C–C$_6$H$_4$–CH$_2$	<0.5–0.62	62–250
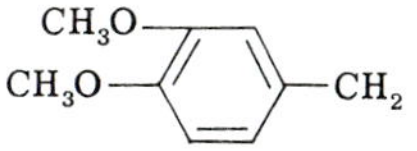 (CH$_3$O)$_2$–C$_6$H$_3$–CH$_2$	0.16–1	125–>250
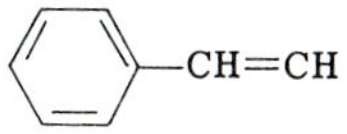 C$_6$H$_5$–CH=CH	0.01–0.6	≥ 250
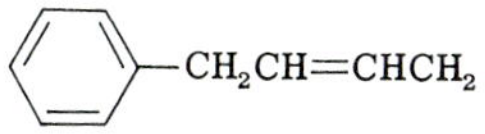 C$_6$H$_5$–CH$_2$CH=CHCH$_2$	0.08–<0.5	≥ 250
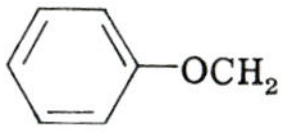 C$_6$H$_5$–OCH$_2$	0.04–<0.5	31–>250
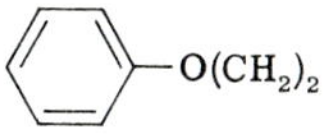 C$_6$H$_5$–O(CH$_2$)$_2$	0.04–<1	250
 H$_3$C–C$_6$H$_4$–OCH$_2$	0.04–0.16	>250
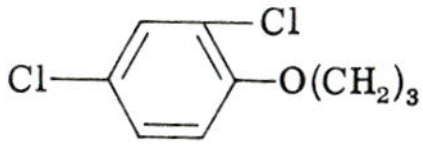 Cl,Cl–C$_6$H$_3$–O(CH$_2$)$_3$	0.08–<2	>250
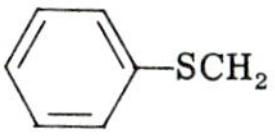 C$_6$H$_5$–SCH$_2$	0.08–0.16	125–250
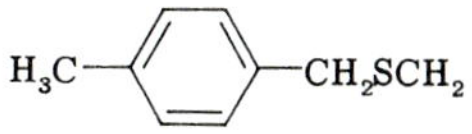 H$_3$C–C$_6$H$_4$–CH$_2$SCH$_2$	0.04–<0.5	>250
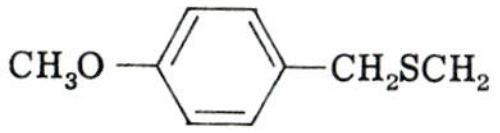 CH$_3$O–C$_6$H$_4$–CH$_2$SCH$_2$	0.04–0.16	>250
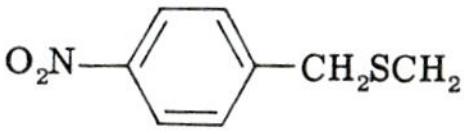 O$_2$N–C$_6$H$_4$–CH$_2$SCH$_2$	0.04–<0.5	>250
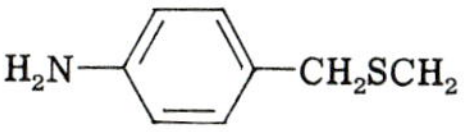 H$_2$N–C$_6$H$_4$–CH$_2$SCH$_2$	0.08–<0.5	≥ 250

TABLE XLIV (continued)

R	MIC (µg/ml)	
	Gram-pos. range	Gram-neg. range
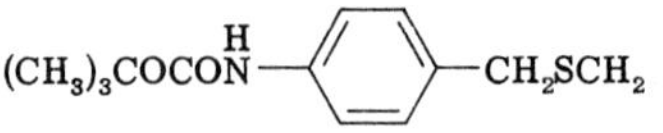 $(CH_3)_3COCON(H)$—C₆H₄—CH_2SCH_2	0.04–<0.5	>250
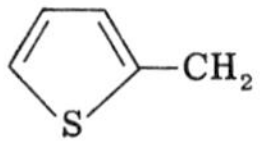 2-thienyl—CH_2	0.16–0.62	<8–62
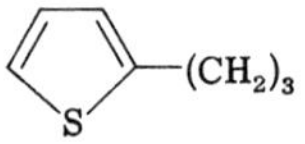 2-thienyl—$(CH_2)_3$	0.04–0.5	≧250
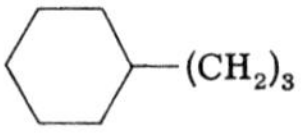 cyclohexyl—$(CH_2)_3$	0.02–<0.5	125–250
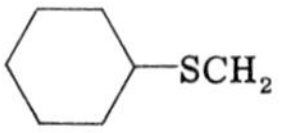 cyclohexyl—SCH_2	0.02–<0.5	62–125
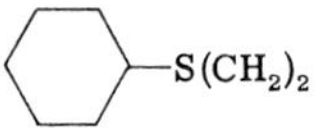 cyclohexyl—$S(CH_2)_2$	0.04–<0.5	250
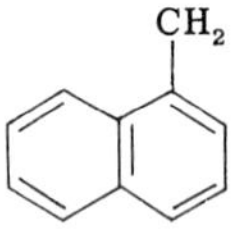 1-naphthyl—CH_2	0.08–0.31	125–>250
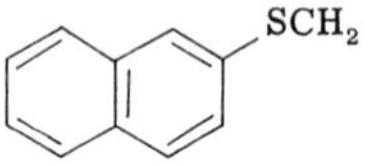 2-naphthyl—SCH_2	<0.05–31	≧250
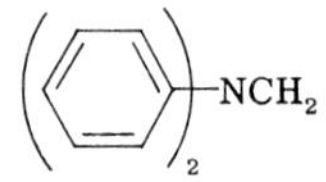 (C₆H₅)₂—NCH_2	0.04–0.5	≧250

[b] From Glaxo (1966b).

TABLE XLV
3-AZIDOMETHYL-7-GLYOXALYL CEPHALOSPORINS[a]

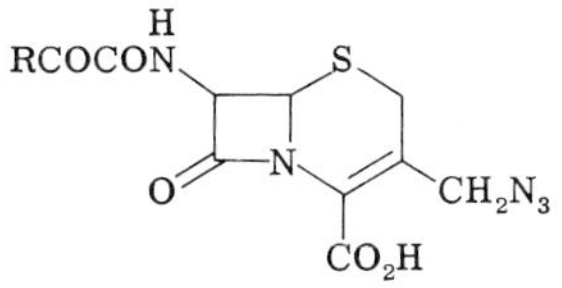

RCOCONH—[β-lactam structure with S, N, CH$_2$N$_3$, CO$_2$H]

R	MIC (μg/ml) Gram-pos. range	R	MIC (μg/ml) Gram-pos. range
X—phenyl		X,Y—phenyl	
X = H	0.31–1.25	X = 2-OCH$_3$, Y = 5-CH$_3$	2.5–16
4-CH$_3$	0.03–2	2-OCH$_3$, 4-OCH$_3$	>2.5–8
2-F	0.02–4	2-OCH$_3$, 5-OCH$_3$	2.5–16
4-F	0.04–1	2-OCH$_3$, 5-Cl	1.25–8
2-Cl	0.6–2	2-Cl, 4-Cl	0.31–4
3-Cl	>0.08–4	2-Cl, 5-Cl	0.31–4
4-Cl	0.08–<0.5	3-NO$_2$, 4-Cl	0.62–4
2-Br	1.25–8	3-HCONH, 4-Cl	0.62–4
3-Br	0.04–<0.5		
4-Br	0.08–4		
3-I	0.04–0.6	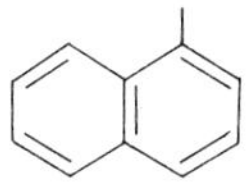 naphthyl	0.04–<0.5
2-OCH$_3$	1.25–4		
3-OCH$_3$	0.02–<0.05	anthracenyl	2.5–1.25
4-OCH$_3$	0.08–0.61 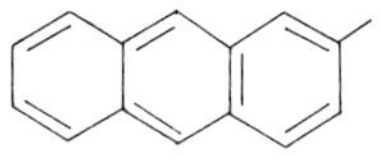		
4-NO$_2$	0.6–2		
4-NH$_2$	0.31–<0.5	thienyl	0.16–1.25
4-(CH$_3$)$_2$N	0.31–1.25		
4-HCONH	0.62–4		
4-CH$_3$CONH	>2.5–16		

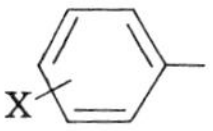

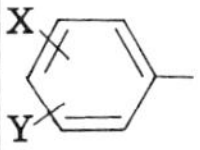

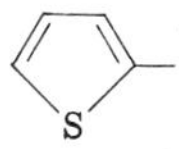

[a] From Glaxo (1968b).

TABLE XLVI
3-AMINOMETHYL CEPHALOSPORINS[a]

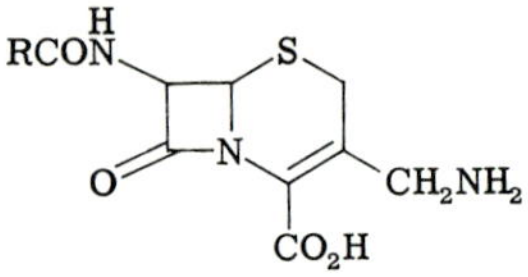

[Structure: RCONH-[β-lactam-thiazine]-CH$_2$NH$_2$ with CO$_2$H]

R	MIC (μg/ml)	
	Gram-pos. range	Gram-neg. range
$CH_3(CH_2)_3$	0.6->25	>250
CH_3SCH_2	0.04-0.08	>250
$(CH_3)_2CHSCH_2$	0.32-2.5	>250
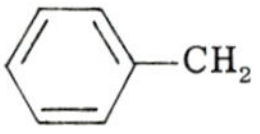 C$_6$H$_5$-CH$_2$	0.31-1.25	>250
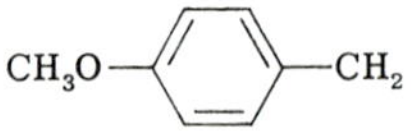 CH$_3$O-C$_6$H$_4$-CH$_2$	0.3->2.5	$\geq$250
C$_6$H$_5$-CH$_2$SCH$_2$	0.16->2.5	>250
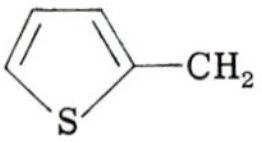 2-thienyl-CH$_2$	0.08-2.5	>250

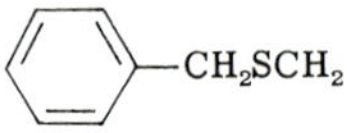

[a] From Glaxo (1964a).

same absolute configuration as the 3-carboxyl group in penicillins (Van Heyningen and Ahern, 1968).

2. Modification of the 4-Carboxyl Group

Attempts to modify the 4-carboxyl group in cephalosporins were complicated by isomerization to Δ^2-cephalosporins; however, some derivatives have been prepared without isomerization. For example, treatment of cephalosporanic acids with diazoalkanes gave pure Δ^3-esters, whereas cephalosporanate salts with active halo compounds gave isomeric esters mixtures (Cocker et al., 1966). The mixed anhydride formed by reaction of 7-phenyl-ACA and ethyl chloroformate gave Δ^3-carboxamides on treatment with amines. In contrast, cephalothin under the same conditions gave mixtures of Δ^2- and Δ^3-isomers due to rapid isomerization of the mixed anhydride. Simple esters and amides of cephalosporins are much less potent than the corresponding acids (Chauvette et al., 1963).

TABLE XLVII
3-ACYLAMINOMETHYL-7-PHENYLACETYL-ACA'S[a]

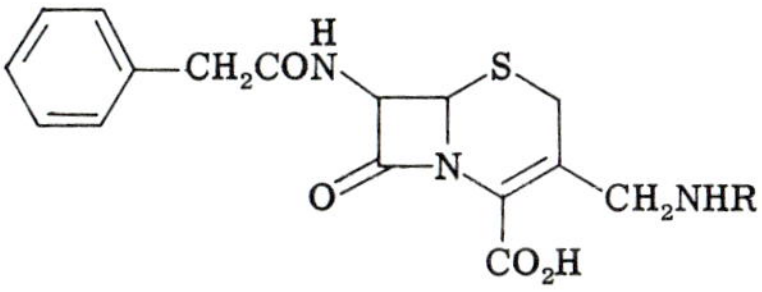

[Structure: phenyl-CH_2CONH- attached to cephalosporin nucleus with CH_2NHR at 3-position and CO_2H]

R	MIC (μg/ml)	
	Gram-pos. range	Gram-neg. range
COH	0.63–1.25	250
$COCH_3$	0.62–2.5	62–125
COC_2H_5	0.31–2.5	250
$CO_2C_2H_5$	0.32–1.25	250
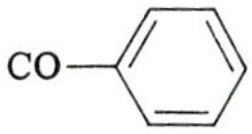 CO—C₆H₅	0.16–2.5	125–250
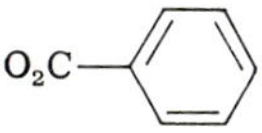 O_2C—C₆H₅	0.16–2.5	250
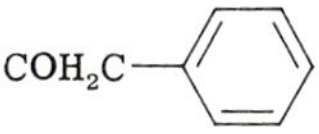 COH_2C—C₆H₅	0.4–1.6	250
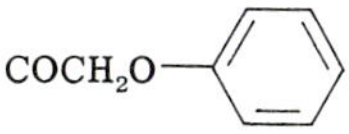 $COCH_2O$—C₆H₅	0.16–2.5	250
 CO—C₆H₄—NO_2	0.01–0.31	250
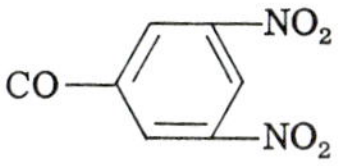 CO—C₆H₃(NO_2)$_2$	0.16–0.62	250
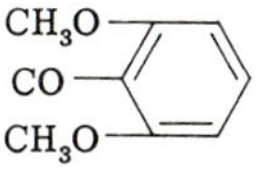 CO—C₆H₃(CH_3O)$_2$	1.25–2.5	250
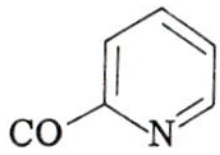 CO—(2-pyridyl)	0.3–1.25	250
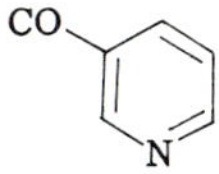 CO—(3-pyridyl)	0.63–2.5	250

TABLE XLVII (continued)

R	MIC (μg/ml) Gram-pos. range	MIC (μg/ml) Gram-neg. range
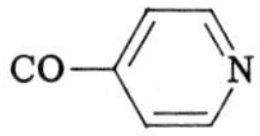 CO—⟨pyridyl-N⟩	0.32–2.5	250
O_2S—⟨phenyl-SO_2CH_3⟩	1.25	125–250
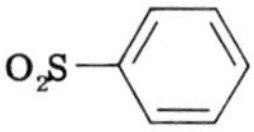 O_2S—⟨phenyl⟩	0.02–0.08	250
 O_2S—⟨phenyl⟩—CH_3	0.6–2.5	62–125
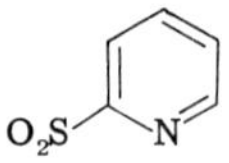 O_2S—⟨2-pyridyl⟩	1.25	125–250

[a] From Glaxo (1964a).

TABLE XLVIII
3-Alkoxymethyl Derivatives of Cephalothin[a]

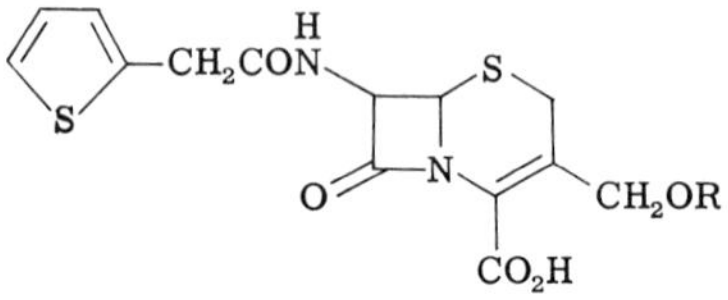

⟨thienyl⟩—CH_2CONH—[β-lactam]—CH_2OR, CO_2H

R	MIC (μg/ml) Gram-pos. range	MIC (μg/ml) Gram-neg. range
CH_3	0.04–4	125–250
C_2H_5	0.08–0.62	≧ 250
$(CH_2)_2CH_3$	0.16–16	125
$CH(CH_3)_2$	0.02–8	>250

[a] From Glaxo (1969).

TABLE XLIX
3-METHOXYMETHYL CEPHALOSPORINS[a]

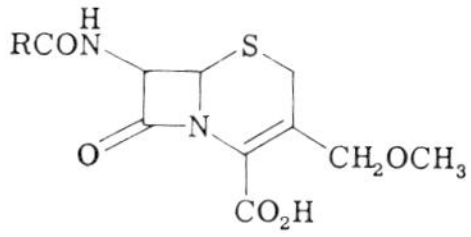

$$\text{RCONH—[β-lactam-cephem]—CH}_2\text{OCH}_3,\ \text{CO}_2\text{H}$$

R	MIC (µg/ml) Gram-pos. range[b]	Gram-neg. range
NCCH$_2$	0.16–0.62	≧250
CH$_3$SCH$_2$	0.31–16	125–250
BrCH$_2$	0.3–0.6	31–125
Cl$_3$C	1.25–4	250
Cl$_3$CCH$_2$O	2.5–4	≧250
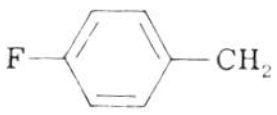 F–C$_6$H$_4$–CH$_2$	0.04–4	250
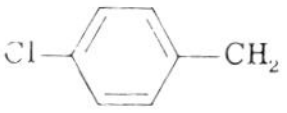 Cl–C$_6$H$_4$–CH$_2$	0.6–31	250
 CH$_3$O$_2$C–C$_6$H$_4$–CH$_2$	0.04–31	31–250
 H$_2$N–C$_6$H$_4$–CH$_2$	0.08–1.25	62–>250
 HCONH–C$_6$H$_4$–CH$_2$	0.16–2	62–125
 2,6-Cl$_2$–C$_6$H$_3$–CH$_2$	0.04–0.5	>250
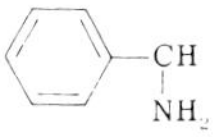 C$_6$H$_5$–CH(NH$_2$)–	0.31–4	8–16 (10 hours)
	0.62–8	16–62 (21 hours)
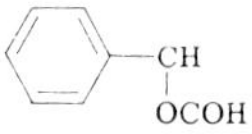 C$_6$H$_5$–CH(OCOH)–	0.62–4	16–31
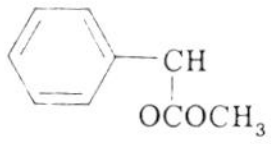 C$_6$H$_5$–CH(OCOCH$_3$)–	0.3–8	62–250
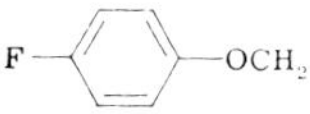 F–C$_6$H$_4$–OCH$_2$	0.03–0.6	31–125
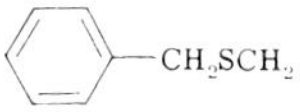 C$_6$H$_5$–CH$_2$SCH$_2$	0.04–<0.5	≧250

[a]From Glaxo (1969).

When the carboxyl group of cephalothin was activated directly by N,N'-dicyclohexylcarbodiimide, and then reacted with t-butylalaninate, no isomerization occurred. The Δ^3-alanine ester was hydrolyzed to give the Δ^3-cephalosporanoyl alanine (Fig. 7) which had poor antibacterial activity (Chauvette and Flynn, 1966). So far, no satisfactory general method for the preparation of cephalosporin esters or amides *via* an activated carboxyl derivative has been reported.

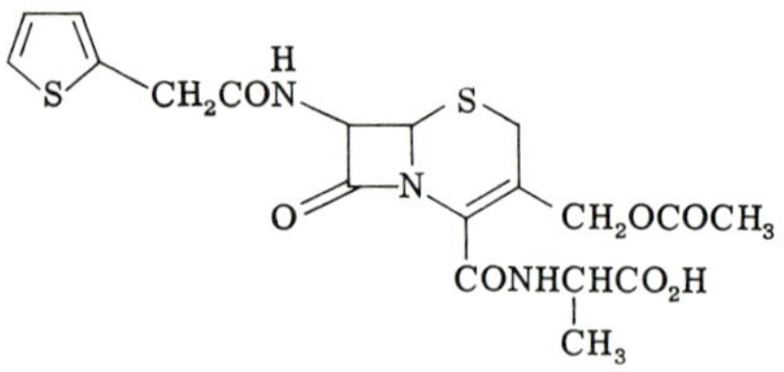

FIG. 7. Alanine amide of cephalothin.

3. Other Modifications

Cephalosporins may be oxidized by periodate to either the corresponding sulfoxides or sulfones. Both these modifications cause a drastic reduction in antibacterial activity (Cocker *et al.*, 1966).

Oxidation of the 4-methyl ester of desacetylcephalothin with manganese dioxide gave the 3-aldehyde which had weak antibacterial activity. Attempts to hydrolyze the aldehyde methyl ester to the free acid were unsuccessful (Chamberlain and Campbell, 1967).

Phenoxymethylpenicillin sulfoxide rearranges under acid catalysis to give a decarboxylated desacetoxycephalosporin (Morin *et al.*, 1963). Rearrangement of phenoxymethylpenicillin sulfoxide methyl ester gave, among other products, two cepalosporin derivatives (Fig. 8) (Morin *et al.*, 1969b). The ester (Fig. 8a) underwent limited hydrolysis in pH 7 buffer to the corresponding acid, however, mild alkaline hydrolysis of the ester (Fig. 8b) gave only Δ^2-acid. An attempt was made to correlate the acylating power of some of these cephalosporins

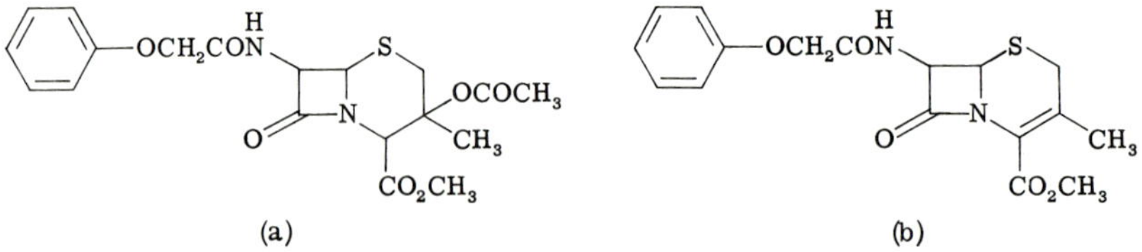

FIG. 8. Two products of phenoxymethyl penicillin sulfoxide rearrangement.

with their biological activity (see Table L) using the infrared frequency of the β-lactam as an indicator of the acylation ability (the more highly strained the β-lactam ring, the higher the infrared frequency, and also the more reactive the β-lactam is as an acylating agent).

III. Resistance to Cephalosporinases

A. OCCURRENCE

In 1963, Fleming et al. (also Goldner et al., 1968) reported the isolation of a filterable enzyme from a strain of *Enterobacter cloacae*. This

TABLE L
CORRELATION OF ACTIVITY WITH INFRARED FREQUENCY OF β-LACTAM[a]

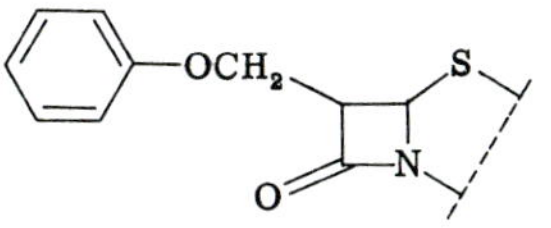

Structure	Bioassay (Oxford units)[b]	β-Lactam frequency,[c] cm^{-1}
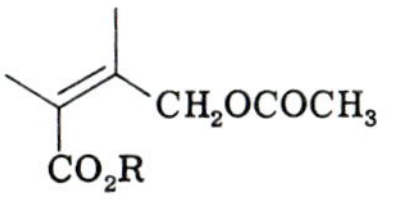 CH_2OCOCH_3, CO_2R	300	1792
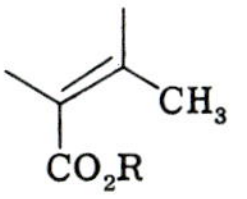 CH_3, CO_2R	25	1785
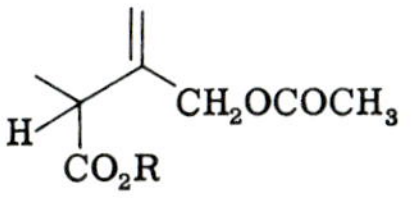 H, CH_2OCOCH_3, CO_2R	6	1784
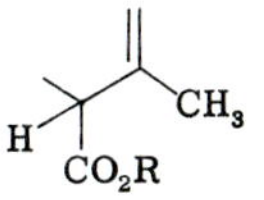 H, CH_3, CO_2R	15	1780
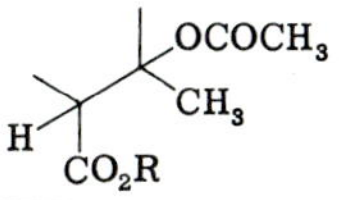 H, $OCOCH_3$, CH_3, CO_2R	low	1780

[a]From Morin et al. (1969b).
[b]Determined on the salts against a penicillin G-sensitive *Staphylococcus aureus* strain.
[c]Determined on the methyl esters in chloroform solution.

enzyme had β-lactamase activity because it caused loss of the typical 5.62 μ infrared band of the cephalosporin β-lactam; it was termed a cephalosporinase because it hydrolyzed cephalosporin C 200 times faster and phenylacetyl-ACA twelve times faster than benzylpenicillin. Study of 1000 strains of Enterobacteriacae showed that in many, but not all, instances bacterial resistance to cephalosporins was associated with a demonstrable production of cephalosporinase. Chang and Weinstein (1964) found cephalosporinase activity in a great number of gram-positive and gram-negative organisms; a strain of *Herella* was twelve times more active against cephalothin than against benzylpenicillin, but attempts to separate penicillinase and cephalosporinase activity failed. This observation was confirmed and quantitated (Bowman et al., 1965).

These and similar reports were included in Pollock's (1965) review and critique of the entire field of enzymatic degradation of penicillins and cephalosporins. Of the three possible sites of enzymatic cleavage of cephalosporins which Pollock discussed, the β-lactamase activity (site b, Fig. 9) has received far more attention than either the amidase (site a) or esterase activity (site c).

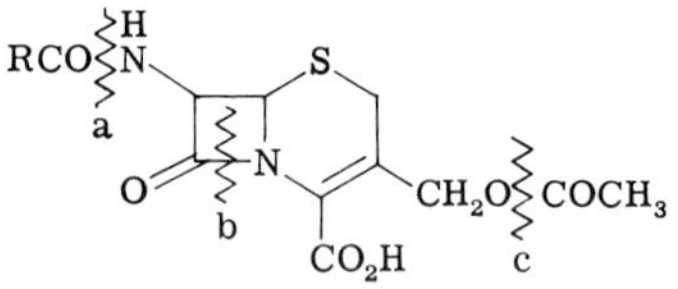

FIG. 9. Sites of enzymatic cleavage in cephalosporins.

B. CHEMISTRY

The chemistry of the β-lactam cleavage process has received some attention. Loss of the 5.62 μ infrared band was mentioned above. Sabath et al. (1965) found acidimetric evidence for the simultaneous expulsion of the 3-acetate in cephalosporin C and cephalothin, and of the 3-pyridine in cephaloridine. These results were similar to those of Eggers et al. (1965) who degraded phenylacetyl-ACA to a thiazine product (Fig. 10). The exact nature of the enzyme-degraded products was not elucidated, however. With the exception of cephalosporin C lactone (and presumably all cephalosporin lactones), which forms a spectrophotometrically stable species at 265 mμ on enzyme cleavage, 3-acetoxycephalosporins such as cephalosporin C and cephalothin give an unstable species, max 230 mμ. Recently, Newton et al. (1968) studied the enzyme reaction in greater detail and postulated the struc-

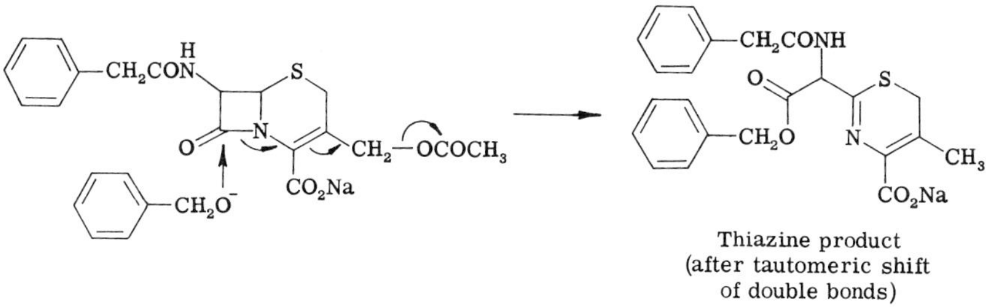

FIG. 10. Degradation of phenylacetyl-ACA with sodium benzyloxide.

ture (A) in Fig. 11 as the unstable species resulting from either hydrolysis of cephalosporins by *B. cereus* β-lactamase or, equivalently, by dilute aqueous ammonia solution. Structure (B) would result from the lactone. Structure (A) and the thiazine of Fig. 10 differ only by the addition of one mole of water and the position of the double bond in the heterocyclic ring. From the hydrolysis mechanism given, a rationalization of the slower rate of hydrolysis of desacetylcephalosporins (3-CH_2OH) can be made: the —OH group is a poorer leaving group than either acetate or pyridine, hence the simultaneous cleavage reaction is inhibited to some extent.

The question of whether or not discrete cephalosporinase and penicillinase enzymes coexist in the same organism has been considered. Many instances of a given enzyme preparation having a differential activity against penicillins and cephalosporins are re-

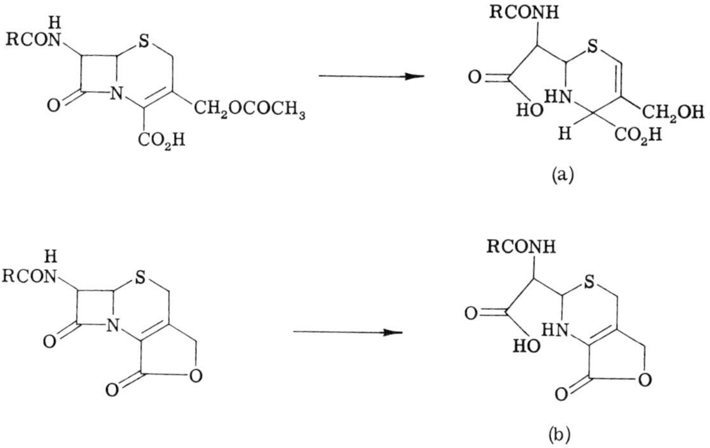

FIG. 11. Hydrolysis of cephalosporins with *Bacillus cereus* β-lactamase or with dilute NH_4OH.

corded. In addition to the earlier work already cited, Hamilton-Miller et al. (1965) showed that cephaloridine was inactivated 2 to 38 times faster than benzylpenicillin by strains of *E. coli* and *Proteus morganii*, 0.1 to 0.7 times slower by strains of *Klebsiella-Aerobacter* and *Aerobacter cloacae*, and concluded these activities result from a single enzyme. (Interestingly, cephaloridine was found to permeate every gram-negative organism tested, unlike ampicillin or benzylpenicillin). Multiple-resistant and methicillin-resistant Staphylococci were much less susceptible to cephaloridine that were merely benzylpenicillin-resistant organisms (Ridley and Phillips, 1965). Since each of these organisms were demonstrable penicillinase producers, the commonly accepted dictum that cephalosporins are resistant to penicillinase must be qualified. Cephalothin was found more inhibitory than cephaloridine against 100 strains of benzylpenicillin-resistant *Staphlococcus aureus* (Benner et al., 1965) using high inocula, and this was shown to be directly related to their relative rates of deactivation by penicillinase. A highly purified penicillinase, with only 0.01% as much cephalosporinase activity, has been prepared from *B. cereus* (Sabath and Abraham, 1966), but a correspondingly highly specific cephalosporinase from the same organism could not be made.

C. Competitive Inhibition

Cephalosporins have been found which competitively inhibit the effect of β-lactamase on other cephalosporins (O'Callaghan et al., 1967, 1968, 1969). These workers (1967) noted that β-lactamase-elaborating gram-negative organisms are fairly resistant to cephaloridine and cephalothin (MIC 250 μg/ml), whereas, non-β-lactamase-producing organisms are inhibited at 2–16 μg/ml. *Aerobacter aerogenes* and *Proteus morganii* β-lactamases are, in addition, much more active against cephalosporins than against penicillins. In screening some 90 cephalosporin analogs against these two cephalosporinases, three activity groups were discerned. A *very susceptible* class included arylacetyl-ACA's such as cephalothin, cephaloridine, and phenylacetyl-ACA. Of *intermediate resistance* were α-substituted arylacetyl-ACA's such as α-chlorophenylacetyl-ACA. A *highly resistant* group consisted of aroyl-ACA's as illustrated in Fig. 12, of which the 2,5-disubstituted benzoyl derivatives were described as *completely insusceptible* to enzyme degradation. These completely insusceptible derivatives protected cephalothin, cephaloridine, and cephaloglycin from enzymatic cleavage; combinations of an enzyme-

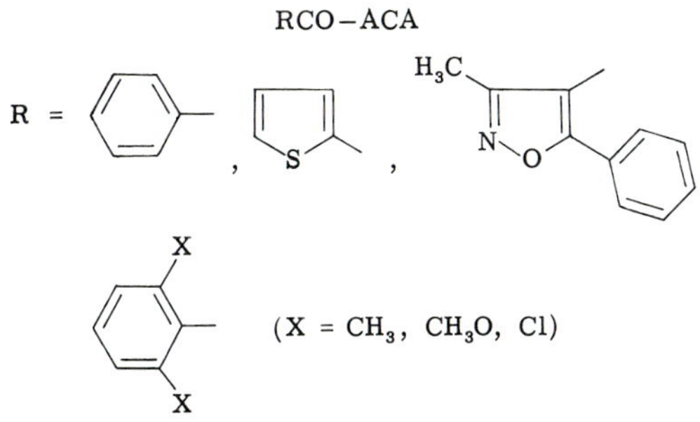

RCO—ACA

R = [phenyl], [thienyl-S], [3-methyl-5-phenyl-isoxazolyl]

[2,6-disubstituted phenyl] (X = CH$_3$, CH$_3$O, Cl)

FIG. 12. Aroyl-ACA's highly resistant to cephalosporinase.

resistant and an enzyme-sensitive cephalosporin were four- to sixtyfold more active *in vitro* than either analog alone. This synergism carried over to *Proteus* infections in mice. Further *in vivo* studies by O'Callaghan *et al.* (1969) were disappointing, however, when it was discovered that, in general, the aforementioned "insusceptible" cephalosporins were less active as competitive inhibitors *in vivo* than *in vitro*. Since incubation of these compounds with rat liver homogenate (a known means of deacetylating cephalosporins) brought about the same decrease in effect, it was concluded that 3-deacetylation was the cause. An order of decreasing activity as competitive inhibitors of various 3-substituted analogs of 2,6-dimethoxybenzoyl-ACA (Fig. 13) paralleled the order of decreasing ease of cleavage of the β-lactam function of each derivative.

The suggestion that the mechanism of competitive inhibition is a preferential acylation of the cephalosporinase by the inhibitor via its β-lactam function is supported by the failure of Δ^2-cephalosporins to act as competitive inhibitors, even though they are enzyme resistant. As pointed out by O'Callaghan *et al.* (1969), citing Van Heyningen

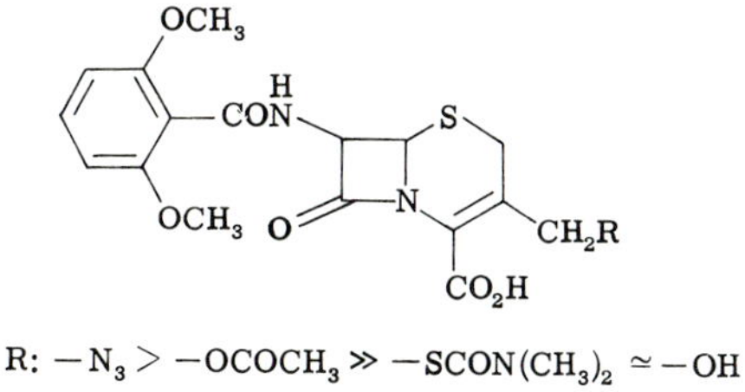

R: $-N_3 > -OCOCH_3 \gg -SCON(CH_3)_2 \simeq -OH$

FIG. 13. Decreasing order of effectiveness of certain competitive inhibitors of cephalosporinase.

and Ahern (1968), the β-lactam of Δ^2-cephalosporins is more stable than in normal Δ^3-cephalosporins, and therefore does not acylate the cephalosporinase. With these thoughts in mind it would appear attractive to prepare and evaluate the properties of the 3-methyl-cephalosporin analog of the structure shown in Fig. 13 (where R = H).

The penicillins nafcillin and oxacillin are also competitive inhibitors of the destruction of cephalothin by strains of *E. coli* and *Klebsiella-Aerobacter*, but only additive effects were observed with *Pseudomonas* strains (Farrar et al., 1967). Sabath (1968) has recently discussed some points of theoretical interest.

D. Structure Activity Studies; Miscellaneous

Sassiver et al. (1969) investigated the effect of *Aerobacter cloacae* β-lactamase on the *in vitro* antimicrobial activity of a large number of semisynthetic cephalosporins. Under conditions which destroyed >99% of cephalothin and cephaloridine, many derivatives were found to be highly resistant to this enzyme, as shown in Fig. 14. These results dovetail with those of O'Callaghan et al. (1967). Other correlations of structure with β-lactamase resistance are those of Hamilton-Miller (1967) and Sabath et al. (1965). No explanation was offered for the markedly different response of *m*-bromophenylacetyl- and α-phenylphenoxyacetyl-ACA to inactivation by various preparations from an *S. aureus* and an *E. coli* organism (Nishida et al., 1968). The former cephalosporin was more degraded by the enzyme from *E. coli*; opposite results were obtained with the latter compound.

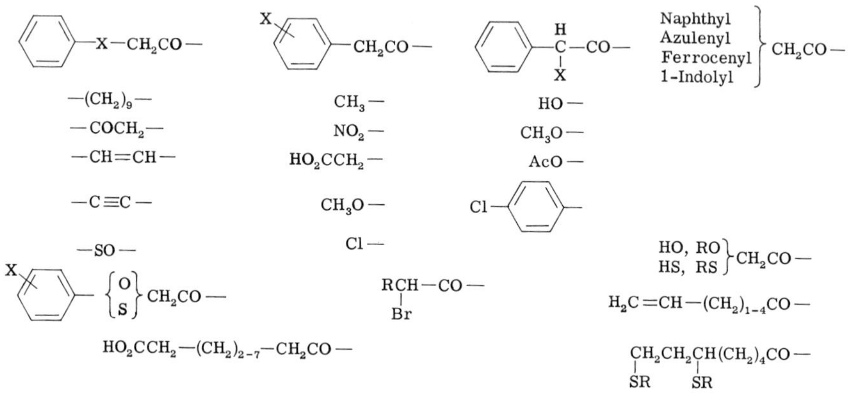

FIG. 14. 7-ACA derivatives highly resistant to cephalosporinase; loss in activity: 0–20% (limit of test accuracy). (>99% loss of activity: cephalothin, cepholoridine, cephalosporin C, phenylacetyl-7-ACA.)

Increased resistance to cephalothin in a clinical case has been associated with increased cephalosporinase production of the infecting *E. coli* organism (Kabins *et al.*, 1966).

Amidase ("acylase") activity (site a, Fig. 9) in a strain of *E. coli* was recently reported (Sjöberg *et al.*, 1967). This enzyme was specific for arylacetyl side chains, since benzylpenicillin and cephalothin were susceptible but cephalosporin C was not. The failure of cephalosporin C to undergo enzymatic cleavage of the 7-side chain was discussed in Section I.B.

IV. Pharmacology, Metabolism, and Mode of Action

A. METABOLISM AND ABSORPTION OF CEPHALOSPORINS

The metabolic fate of a number of semisynthetic cephalosporins has been studied, primarily by ^{14}C-labeling of the 7-carboxamide moiety, $R^{14}CONH$-ACA. Two pathways of metabolic degradation have been discovered. The 7-side chain can be cleaved to give various derived metabolites containing the ^{14}C label. Also, the 3-acetoxymethyl group, when present, is subject to hydrolysis to the hydroxymethylcephalosporin (desacetylcephalosporin).

In the earliest study of Culp *et al.* (1964), phenylacetyl(^{14}C)-ACA was administered orally to rats. Radioactivity slowly appeared in the urine, up to 50% of the administered dose after 24 hours. The metabolites and their relative amount as percentage of administered dose were phenylacetic and phenylaceturic acids (20%), unknown nonpolar metabolite (20%), unknown polar metabolite (10%), and desacetylcephalosporin (2%). It was suggested that the antibiotic is hydrolyzed by the multiflora of the gut; a by-pass experiment eliminated the stomach as the site of metabolism. On the other hand, intraperitoneally administered phenylacetyl-(^{14}C)-ACA gave mainly the desacetyl metabolite. Fifty percent of the radioactivity appeared in the urine after 2 hours and 86% after 24 hours consisting of desacetylcephalosporin (50% of administered dose) and 5–10% each of unchanged antibiotic, polar metabolite and, together, phenylacetic and phenylaceturic acids. Independent experiments with 3-desacetyl-7-phenylacetyl-(^{14}C)-ACA showed that it gave the same metabolic profile as the parent 3-acetoxymethyl compound on either oral or parenteral dosing, suggesting that for the parenteral route, deacetylation is rapid. The fate of a *m*-chlorophenylacetyl-ACA is similar to the unsubstituted analog (Okui *et al.*, 1967).

Intramuscular injection of unlabled cephalothin in humans gave a 67% average recovery of cephalothin and its desacetyl analog in a ratio of 1.5 : 1 within 6 hours (Lee et al., 1963). A much greater degree of deacetylation occurred in dogs. Wick (1966) has reasonably suggested that desacetylcephalothin contributes to the clinical effect of cephalothin. The metabolism of oral cephalothin-^{14}C in the rat (Sullivan and McMahon, 1967) resembled that of the phenyl analog discussed above. Radioactivity slowly appeared in the urine, up to 46% after 40 hours. A fleeting trace of desacetyl metabolite was observed, but the major metabolites as percentage of administered dose were unidentified component (23%), thienylacetylglycine (15%), and thienylacetamidoethanol (13%). Appearance of this latter compound suggested that the actual site of enzyme hydrolysis of cephalothin was at position 6 (site b, Fig. 15), not the side-chain carbonyl (site a), in

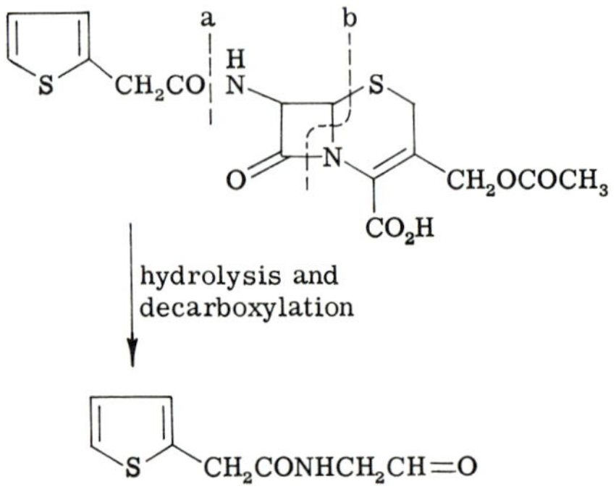

FIG. 15. Metabolic degradation of cephalothin.

which case the primary but unobserved metabolite would be thienylacetylaminoacetaldehyde, which could undergo both oxidation and reduction to the observed metabolites. In support of this, the aldehyde can be isolated from acid hydrolysis of cephalothin and it does metabolize to the observed cephalothin metabolites. In addition, Sullivan and McMahon (1967) could not demonstrate the presence of thienylacetic acid (the presumed precursor of thienylacetylglycine) in various in vitro enzyme hydrolyses of cephalothin. They did show, however, that this acid is converted in 86% yield to the glycine conjugate in the rat.

Oral cephaloridine-^{14}C undergoes the same metabolism as cephalothin (Sullivan and McMahon, 1967) except that some unchanged antibiotic (5% maximum) is recovered in the urine. A considerable difference from cephalothin is found on intramuscular injection of unlabeled cephaloridine to humans (Muggleton et al., 1964). Virtually all of the dose was recovered after 6–12 hours (microbiological assay); no active metabolites were noted and it would appear that the antibiotic is excreted unchanged.

When orally administered to rats, phenoxyacetyl-^{14}C-ACA is more readily absorbed than either cephalothin of cephaloridine (Sullivan and McMahon, 1967). The urinary metabolites, comprising a total of 62% of the dose at 24 hours, were desacetylcephalosporin (15%), phenoxyacetic acid (30%), phenoxyacetylaminoethanol (3%), and polar metabolite (9%). Whether or not deacetylation occurs before or after absorption is not established.

Cephaloglycin-^{14}C (Sullivan et al., 1969a), active orally in man, is not absorbed particularly well in the rat. After 24 hours, 70% of the radioactivity is in the feces with only a minor contribution from biliary excretion. The 24-hour urinary metabolites (20% of dose) are desacetylcephaloglycin (2%), 2-phenylglycine (8.5%), phenylglyoxylic acid (5%, C_6H_5-CO-CO_2H), mandelic acid (1.5%, C_6H_5CHOH-CO_2H), and polar metabolite (3%). The amount of desacetylcephaloglycin, cephaloglycin lactone, and cephaloglycin itself increase with increasing dose. The metabolic pattern of oral cephaloglycin is similar to that observed on administration of D(-)2-phenylglycine-^{14}C, of which 79% of the radioactive dose appears as the following urinary metabolites: 2-phenylglycine (50%), phenylglyoxylic acid (27%), and mandelic acid (2%). Hydrolysis of the phenylglycine side chain in cephaloglycin may occur both in the gut or in the absorbed phase. Intraperitoneal cephaloglycin gave a 71% yield of urinary metabolites at 24 hours: cephaloglycin (13.5%), desacetylcephaloglycin (28.6%), and phenylglycine plus its metabolites (25%). An interesting difference between D-cephaloglycin and its inactive L-isomer was noted. The L-2-phenlyglycine side chain of the L-isomer is much more readily hydrolyzed, regardless of mode of administration, than the D-isomer. Orally, 52% of radioactivity is present as the following urinary metabolites after 24 hours: phenylglyoxylic acid (40%), mandelic acid (4%), phenylglycine (4%), and unknown metabolite (4%). Parenteral results are similar. Both oral and parenteral L-2-phenylglycine is rapidly metabolized to phenylglyoxylic acid (84% in urine). The contribution of the greater rate of metabolism of L-cephaloglycin to its low intrinsic antimicrobial activity is not clear and requires investigation.

Cephalexin is more efficiently absorbed in all species than cephaloglycin (Sullivan et al., 1969b). After 24 hours, 84–89% of oral cephalexin appears in rat and mouse urine. The remainder of the oral dose appears in the feces and independent experiments showed this was due to absorption and biliary excretion. No metabolites of cephalexin were found.

A detailed discussion of other aspects of the pharmacology and clinical indications of various cephalosporins is beyond the scope of this review. Recent studies of cephalexin by Muggleton et al. (1969) have shown that this orally active cephalosporin is more slowly bactericidal than the other cephalosporins in current use, and its effect in rapidly progressing laboratory infections is poor unless the infection is moderated by reducing the size of the infecting dose. Man excretes cephalexin less rapidly than mice and thus achieves higher blood levels. The serum binding of cephalexin (20%), cephaloridine (15%), and cephalothin (65%) were measured by Kind et al. (1969). The general effect of serum on the antibacterial activities of cephalosporins has been mentioned occasionally in this review; MIC's of numerous acyl-ACA's with and without added serum have been given by Chauvette et al. (1963). Comparative studies of cephaloglycin and cephalexin have been reported by Griffith and Black, (1968) and Braun et al. (1968). Recent studies of cephaloglycin (Hogan et al., 1968; Pitt et al., 1968) only serve to show that cephalexin has a superior pharmacological profile.

B. In Vivo Activity of Cephalosporins

1. Correlation of In Vitro and In Vivo Data

Cephalosporins which looked promising *in vitro* would be further evaluated in experimental infections, usually in mice. A compound which is more active *in vitro* than, for instance, cephalothin or cephaloridine, may be less effective *in vivo*, due to any number of factors such as unfavorable metabolism, serum binding, or rapid excretion. Illustrative of this is the poor *in vivo* activity of a number of acyl-ACA's derived from dicarboxylic acids, which were highly potent in a broad spectrum *in vitro* screen (Lewis et al., 1969). This carry-over failure was particularly evident in gram-negative infections, and although as yet unexplained, was shown not to be due to either serum binding or drug toxicity.

In vivo data on cephalosporins, apart from the extensive investigations on the clinically used cephalosporins (Table LI), is often absent in the literature. Publications in which some reference to such data is made are those of Chamberlain and Campbell (1967); Lewis et al. (1969); Naito et al. (1968); Ryan et al. (1969); Sassiver et al. (1969); Spencer et al. (1967a); Van Heyningen (1965). Patents containing such data are Bristol-Myers (1968); Bristol-Myers (1969); Ciba (1965,

TABLE LI

ED_{50} (mg/kg) of Some Cephalosporins in Experimental Infections in Mice—Oral and Subcutaneous Administration

Cephalosporin	Gram-pos. infections		Gram-neg. infections		Reference[a]
	Oral	Subcutaneous	Oral	Subcutaneous	
Cephalothin	–	0.5–11	–	8–100, mainly	(1)
	–	–	–	15–39	(2)
	8	0.5	–	16–128, mainly	(3)
Cephaloridine	–	0.8–2.5	–	3–25	(4)
	1–>42	0.03–2.6	42–>166	5–37	(5)
	1–4	0.12–0.5	–	2–32, mainly	(6)
Cephaloglycin	2.6–29	–	8.6–42	–	(5)
	1.8–62	–	15–24	–	(7)
Cephalexin	1–58	–	5–22	–	(7)
	12–15	7–12	16–40	6–111	(8)

[a]References: (1) Boniece et al. (1962); (2) Wick (1966); (3) Sassiver et al. (1969); (4) Muggleton et al. (1964); (5) Wick and Boniece (1965); (6) Lederle, unpublished; (7) Wick (1967); (8) Muggleton et al. (1969).

1966); Crast (1967a,b,c, 1969); Flynn (1967a,b, 1968); Glaxo (1964a,c,d, 1965a, 1966a,b, 1968b, 1969); Godfrey (1967); Lilly (1965a,b, 1969a); and Rapp et al. (1966).

With reference to the activities of cephalothin, cephaloridine, cephaloglycin, and cephalexin in Table LI, the results of several investigators are given and naturally vary according to the actual organisms and methods used. The infecting organisms are frequently the same as those used for the *in vitro* screening.

2. Oral Activity in Cephalosporins

Structure–activity studies have not developed a *modus operandi* for devising orally active cephalosporins. Although cephalexin, and to a lesser extent cephaloglycin, are orally active in man, their investigators have not been able to offer any rationalization of this unique property among cephalosporins (Spencer et al., 1966; Ryan et al., 1969). The combination of the D-(2-phenylglycyl) side chain and various substituted phenyl analogs with a 3-methyl substituent gives a high degree of oral activity (Ryan et al., 1969), but when the 7-side chain is varied considerably, only one of ten 3-methyl cephalosporins made (7-*p*-nitrophenylacetyl) had high oral activity in the mouse. Thus the combination found in cephalexin has been termed "fortuitous."

C. Mode of Action of Cephalosporins

The exact molecular mechanism by which penicillins and cephalosporins inhibit cell wall synthesis in bacteria is not known. Cephalothin has been found to interfere with the dimerization of peptidoglycan fragments in *E. coli* cell walls. It may do this by irreversibly acylating a peptidoglycan transpeptidase (Strominger *et al.*, 1967) via the β-lactam function. A further suggestion that peptidoglycan transpeptidase falsely recognized penicillins and cephalosporins as the terminal D-alanyl-D-alanine fragment of the cell wall peptidoglycan has been advanced (Tipper and Strominger, 1965), but this is mere speculation and ignores stereochemical inconsistencies where convenient. In general, however, the dependence of biological activity on an intact β-lactam function is rationalized by the above arguments. Van Heyningen and Ahern (1968) and Morin *et al.* (1969b) have attempted to correlate activity with stability of the β-lactam of various cephalosporins. If the β-lactam is too stable, as in Δ^2-cephalosporins, the cephalosporin will presumably not acylate the transpeptidase and therefore will be biologically inactive. Such inactivity is indeed seen in the Δ^2-cephalosporins (Van Heyningen and Ahern, 1968).

References

Abraham, E. P. (1967). *Quart. Rev.* **21**, 231–248.
Abraham, E. P., and Newton, G. G. F. (1961). *Biochem. J.* **79**, 377–393.
Abraham, E. P., Newton, G. G. F., and Hale, C. W., (1954). *Biochem. J.* **58**, 94–102.
Barber, M., and Waterworth, P. M. (1964). *Brit. Med. J.* **2**, 344–349.
Benner, E. J., Bennet, J. V., Brodie, J. L., and Kirby, W. M. M. (1965). *J. Bacteriol.* **99**, 1599–1604.
Boniece, W. S., Wick, W. E., Holmes, D. H., and Redman, C. E. (1962). *J. Bacteriol.* **84**, 1292–1296.
Bowman, F. W., Knoll, E. W., and White, M. (1965). *In* "Antimicrobial Agents and Chemotherapy – 1964" (J. C. Sylvester, ed.), pp. 334–337. Williams & Wilkins, Baltimore, Maryland.
Braun, P., Tillotson, J. R., Wilcox, C., and Finland, M. (1968). *Appl. Microbiol.* **16**, 1684–1693.
Bristol-Banyu Res. Inst. (1969). *Farmdoc Complete Spec. Book* **887**, 187–192. (Belg. Pat. 724,072; Derwent Basic No. 38,472).
Bristol-Myers Co. (1968). *Farmdoc Complete Spec. Book* **761**, 101–109. (Netherlands Pat. 68,00176; Derwent Basic No. 32,864).
Bristol-Myers Co. (1969). *Farmdoc Complete Spec. Book* **845**, 413–428. (Netherlands Pat. 68,12382; Derwent Basic No. 36,496).

Brotzu, G. (1948). *Lav. Ist. Ig. Cagliari.*
Bunn, P. A., Milicich, S. and Lunn, J. S. (1966). *In* "Antimicrobial Agents and Chemotherapy–1965" (G. L. Hobby, ed.), pp. 947–950. Williams & Wilkins, Baltimore, Maryland.
Burton, H. S., and Abraham, E. P. (1951). *Biochem. J.* **50**, 168–174.
Chamberlain, J. W., and Campbell, J. B. (1967). *J. Med. Chem.* **10**, 966–968.
Chang, T-W., and Weinstein, L. (1964). *In* "Antimicrobial Agents and Chemotherapy–1963" (J. C. Sylvester, ed.), pp. 278–282. Williams & Wilkins, Baltimore, Maryland.
Chauvette, R. R., and Flynn, E. H. (1966). *J. Med. Chem.* **9**, 741–744.
Chauvette, R. R., Flynn, E. H., Jackson, B. G., Lavagnino, E. R., Morin, R. B., Mueller, R. A., Pioch, R. P., Roeske, R. W., Ryan, C. W., Spencer, J. L., and Van Heyningen, E. (1962). *J. Am. Chem. Soc.* **84**, 3401–3402.
Chauvette, R. R., Flynn, E. H., Jackson, B. G., Lavagnino, E. R., Morin, R. B., Mueller, R. A., Pioch, R. P., Roeske, R. W., Ryan, C. W., Spencer, J. L., and Van Heyningen, E. (1963). *In* "Antimicrobial Agents and Chemotherapy–1962" (J. C. Sylvester, ed.), pp. 687–694. Williams & Wilkins, Baltimore, Maryland.
Ciba, Ltd. (1963). Belg. Pat. 620,582. (*Chem. Abstr.* **59**, 10073 C.)
Ciba, S. A. (1964a). *Farmdoc Complete Spec. Book* **234**, 391–396. (Belg. Pat. 693,692. Derwent Basic No. 12,648).
Ciba, Ltd. (1964b). *Farmdoc Complete Spec. Book* **175**, 455–464. (Belg. Pat. 633,874; Derwent Basic No. 10,893).
Ciba, Ltd. (1965). *Farmdoc Complete Spec. Book* **366**, 353–367. (S. Afr. Pat. 64/4673; Derwent Basic No. 16,571).
Ciba, Ltd. (1966). *Farmdoc Complete Spec. Book* **517**, 111–131. (S. Afr. Pat. 65/6950; Derwent Basic No. 22,192).
Claridge, C. A., Luttinger, J. R., and Lein, J. (1963). *Proc. Soc. Exp. Biol. Med.* **113**, 1008–1012.
Cocker, J. D., Cowley, B. R., Cox, J. S. G., Eardley, S., Gregory, G. I., Lazenby, J. K., Long, A. G., Sly, J. C. P., and Somerfield, G. A. (1965). *J. Chem. Soc.* pp. 5015–5031.
Cocker, J. D., Eardley, S., Gregory, G. I., Hall, M. E., and Long, A. G. (1966). *J. Chem. Soc.*, pp. 1142–1151.
Collins, J. F., and Richmond, M. H. (1962). *Nature (London)* **195**, 142–143.
Cooper, P. D. (1956). *Bacteriol. Rev.* **20**, 28–48.
Crast, L. B., Jr., Assignor to Bristol-Myers Co. (1967a). *Farmdoc Complete Spec. Book* **624**, 167–171. (U.S. Pat. 3,322,749; Derwent Basic No. 26,882).
Crast, L. B., Jr., Assignor to Bristol-Myers Co. (1967b). *Farmdoc Complete Spec. Book* **624**, 173–178. (U.S. Pat. 3,322,750; Derwent Basic No. 26,883).
Crast, L. B., Jr., Assignor to Bristol-Myers Co. (1967c). *Farmdoc Complete Spec. Book* **624**, 179–183. (U.S. Pat. 3,322,751; Derwent Basic No. 26,884).
Crast, L. B., Jr., Assignor to Bristol-Myers Co. (1969). *Farmdoc Complete Spec. Book* **825**, 115–120. (U.S. Pat. 3,422,099; Derwent Basic No. 35,650).
Crast, L. B., Jr., and Essery, J. M., Assignors to Bristol-Myers Co. (1967). *Farmdoc Complete Spec. Book* **679**, 181–185. (U.S. Pat. 3,352,358; Derwent Basic No. 29,430).
Crawford, K., Heatly, N. G., Boyd, P. F., Hale, C. W., Kelly, B. K., Smith, G. A., and Smith, N. (1952). *J. Gen. Microbiol.* **6**, 47–59.
Culp, H. W., Marshall, F. J., and McMahon, R. E. (1964). *In* "Antimicrobial Agents and Chemotherapy–1963" (J. C. Sylvester, ed.), pp. 243–246. Williams & Wilkins, Baltimore, Maryland.
Demain, A. L., and Newkirk, J. F. (1962). *Appl. Microbiol.* **10**, 321–325.
Demain, A. L., Newkirk, J. F., Davis, G. E., and Harman, R. E. (1963a). *Appl. Microbiol.* **11**, 58–61.

Demain, A. L., Walton, R. B., Newkirk, J. F., and Miller, I. M. (1963b). *Nature (London)* **199,** 909–910.
Eggers, S. H., Kane, V. V., and Lowe, G. (1965). *J. Chem. Soc.*, pp. 1262–1270.
Farrar, W. E., Jr., O'Dell, N. M., and Krause, J. M. (1967). *Ann. Intern. Med.* **67,** 733–743.
Fechtig, B. W., Peter, H. H., Bickel, H., and Vischer, E. (1968). *Helv. Chim. Acta* **51,** 1109–1120.
Fleming, P. C., Goldner, M., and Glass, D. G. (1963). *Lancet* **i,** 1399–1401.
Flynn, E. H., Assignor to Eli Lilly and Co. (1967a). *Farmdoc Complete Spec. Book* **574,** 65–66. (U.S. Pat. 3,297,692; Derwent Basic No. 24,660).
Flynn, E. H. (1967b). *In* "Antimicrobial Agents and Chemotherapy – 1966" (G. L. Hobby, ed.), pp. 715–726. Williams & Wilkins, Baltimore, Maryland.
Flynn, E. H., Assignor to Eli Lilly and Co. (1968). *Farmdoc Complete Spec. Book* **738,** 265–269. (U.S. Pat. 3,382,241; Derwent Basic No. 31,947).
Fujisawa Pharm. Co., Ltd. (1966a). *Farmdoc Complete Spec. Book* **540,** 23–24. (Jap. Pat. 16951/66; Derwent Basic No. 23,237).
Fujisawa Pharm. Co., Ltd. (1966b). *Farmdoc Complete Spec. Book* **540,** 11–12. (Jap. Pat. 16871/66; Derwent Basic No. 23,231).
Fujisawa Pharm. Co., Ltd. (1966c). *Farmdoc Complete Spec. Book* **540,** 25–26. (Jap. Pat. 16952/66; Derwent Basic No. 23,238).
Fujisawa Pharm. Co., Ltd. (1966d). *Farmdoc Complete Spec. Book* **540,** 9–10. (Jap. Pat. 16870/66; Derwent Basic No. 23,230).
Fujisawa Pharm. Co., Ltd. (1966e). *Farmdoc Complete Spec. Book* **519,** 207–217. (Brit. Pat. 1,038,529; Derwent Basic No. 22,302).
Fujisawa Pharm. Co., Ltd. (1966f). *Farmdoc Complete Spec. Book* **540,** 17–18. (Jap. Pat. 16948/66; Derwent Basic No. 23,234).
Fujisawa Pharm. Co., Ltd. (1967a). *Farmdoc Complete Spec. Book* **593,** 39–40. (Jap. Pat. 2712/67; Derwent Basic No. 25,406).
Fujisawa Pharm. Co., Ltd. (1967b). *Farmdoc Complete Spec. Book* **581,** 481–489. (French Pat. 4,570M; Derwent Basic No. 24,930).
Fujisawa Pharm. Co., Ltd. (1967c). *Farmdoc Complete Spec. Book* **671,** 185–186. (Jap. Pat. 19466/67; Derwent Basic No. 29,022).
Fujisawa Pharm. Co., Ltd. (1967d). *Farmdoc Complete Spec. Book* **638,** 229–230. (Jap. Pat. 10996/67; Derwent Basic No. 27,458).
Fujisawa Pharm. Co., Ltd. (1967e). *Farmdoc Complete Spec. Book* **638,** 233–234. (Jap. Pat. 10998/67; Derwent Basic No. 27,460).
Fujisawa Pharm. Co., Ltd. (1967f). *Farmdoc Complete Spec. Book* **610,** 299–300. (Jap. Pat. 7475/67; Derwent Basic No. 26,212).
Fujisawa Pharm. Co., Ltd. (1967g). *Farmdoc Complete Spec. Book* **618,** 35–36. (Jap. Pat. 8877/67; Derwent Basic No. 26,559).
Fujisawa Pharm. Co., Ltd. (1967h). *Farmdoc Complete Spec. Book* **593,** 37–38. (Jap. Pat. 2711/67; Derwent Basic No. 25,405).
Fujisawa Pharm. Co., Ltd. (1967i). *Farmdoc Complete Spec. Book* **598,** 197. (Jap. Pat. 4155/67; Derwent Basic No. 25,633).
Fujisawa Pharm. Co., Ltd. (1967j). *Farmdoc Complete Spec. Book* **672,** 31–32. (Jap. Pat. 20064/67; Derwent Basic No. 29,068).
Fujisawa Pharm. Co., Ltd. (1967k). *Farmdoc Complete Spec. Book* **637,** 219–220. (Jap. Pat. 10,991/67; Derwent Basic No. 27,453).

Fujisawa Pharm. Co., Ltd. (1967l). *Farmdoc Complete Spec. Book* **638**, 227–228. (Jap. Pat. 10995/67; Derwent Basic No. 27,457).

Fujisawa Pharm. Co., Ltd. (1967m). *Farmdoc Complete Spec. Book* **591**, 481–488. (French Pat. 4,645M; Derwent Basic No. 25,348).

Fujisawa Pharm. Co., Ltd. (1968a). *Farmdoc Complete Spec. Book* **689**, 147. (Jap. Pat. 24429/67; Derwent Basic No. 29,841).

Fujisawa Pharm. Co., Ltd. (1968b). *Farmdoc Complete Spec. Book* **738**, 165–191. (Neth. Pat. 67,14888; Derwent Basic No. 31,936).

Fujisawa Pharm. Co., Ltd. (1968c). *Farmdoc Complete Spec. Book* **722**, 49–50. (Jap. Pat. 5888/68; Derwent Basic No. 31,262).

Fujisawa Pharm. Co., Ltd. (1968d). *Farmdoc Complete Spec. Book* **783**, 275–276. (Jap. Pat. 20,193/68; Derwent Basic No. 33,868).

Fujisawa Pharm. Co., Ltd. (1969a). *Farmdoc Complete Spec. Book* **876**, 11–12. (Jap. Pat. 10,556/69; Derwent Basic No. 37,866).

Fujisawa Pharm. Co., Ltd. (1969b). *Farmdoc Complete Spec. Book* **898**, 139–140. (Jap. Pat. 16,667/69; Derwent Basic No. 38,992).

Fujisawa Pharm. Co., Ltd. (1969c). *Farmdoc Complete Spec. Book* **898**, 137–138. (Jap. Pat. 16,666/69; Derwent Basic No. 38,991).

Fujisawa Pharm. Co., Ltd. (1969d). *Farmdoc Complete Spec. Book* **898**, 135–136. (Jap. Pat. 16665/69; Derwent Basic No. 38,990).

Fujisawa Pharm. Co., Ltd. (1969e). *Farmdoc Complete Spec. Book* **817**, 383–435. (Belg. Pat. 714,518; Derwent Basic No. 35,307).

Fujisawa Pharm. Co., Ltd. (1969f). *Farmdoc Complete Spec. Book* **895**, 553–566. (Can. Pat. 818,501; Derwent Basic No. 38,845).

Glaxo (1963). *Farmdoc Complete Spec. Book* **25**, 27–85 (Irish Pat. 415/62; Derwent Basic No. 7,044).

Glaxo (1964a). *Farmdoc Complete Spec. Book* **179**, 391–403 (Belg. Pat. 634,644; Derwent Basic No. 10,998).

Glaxo (1964b). *Farmdoc Complete Spec. Book* **183**, 409–420 (Belg. Pat. 635,137; Derwent Basic No. 11, 089).

Glaxo (1964c). *Farmdoc Complete Spec. Book* **208**, 473–487 (Belg. Pat. 637,547; Derwent Basic No. 11,848).

Glaxo (1964d). *Farmdoc Complete Spec. Book* **248**, 385–400 (Belg. Pat. 641,338; Derwent Basic No. 13,046).

Glaxo (1965a). *Farmdoc Complete Spec. Book* **331**, 189–234 (Belg. Pat. 650,445; Derwent Basic No. 15,534).

Glaxo (1965b). *Farmdoc Complete Spec. Book* **331**, 235–265 (Belg. Pat. 650,444; Derwent Basic No. 15,535).

Glaxo (1966a). *Farmdoc Complete Spec. Book* **512**, 119–135 (S. Afr. Pat. 65/6689; Derwent Basic No. 21,914).

Glaxo (1966b). *Farmdoc Complete Spec. Book* **567**, 453–464 (Neth. Pat. 6,606,820; Derwent Basic No. 23,984).

Glaxo (1968a). *Farmdoc Complete Spec. Book* **786**, 231–245. (Neth. Pat. 6,804,139; Derwent Basic No. 34,026).

Glaxo (1968b). *Farmdoc Complete Spec. Book* **786**, 247–256. (Neth. Pat. 6,804,140; Derwent Basic No. 34,027).

Glaxo (1969). *Farmdoc Complete Spec. Book* **841**, 359–407. (Belg. Pat. 719,710; Derwent Basic No. 36,338).

Godfrey, J. C., Assignor to Bristol-Myers Co. (1967). *Farmdoc Complete Spec. Book* **588**, 409–411. (U.S. Pat. 3,303,193; Derwent Basic No. 25,174).
Goldner, M., Glass, D. G., and Fleming, P. C. (1968). *Can. J. Microbiol.* **14**, 139–145.
Gottstein, W. J., and Eachus, A. H. (1968). *Farmdoc Complete Spec. Book* **758**, 291–302. (U.S. Pat. 3,391,141; Derwent Basic No. 32,770).
Griffith, R. S., and Black, H. R. (1968). *Clin. Med.* **75**(11), p. 14–22.
Hale, C. W., Newton, G. G. F., and Abraham, E. P. (1961). *Biochem. J.* **79**, 403–408.
Hamilton-Miller, J. M. T. (1967). *Nature (London)* **214**, 1333–1334.
Hamilton-Miller, J. M. T., Smith, J. T., and Knox, R. (1965). *Nature (London)* **208**, 235–237.
Hardcastle, G. A., Jr., Johnson, D. A., Panetta, C. A., Scott, A. I., and Sutherland, S. A. (1966). *J. Org. Chem.* **31**, 897–899.
Hodgkin, D. C., and Maslen, E. N. (1961). *Biochem. J.* **79**, 393–402.
Hogan, L. B., Jr., Holloway, W. J., and Jakubowitch, R. A. (1968). *In* "Antimicrobial Agents and Chemotherapy—1967" (G. L. Hobby, ed.), pp. 624–629. Williams & Wilkins, Baltimore, Maryland.
Huang, H. T., Seto, T. A., and Shull, G. M. (1963). *Appl. Microbiol.* **11**, 1–6.
Jeffery, J. D'A., Abraham, E. P., and Newton, G. G. F. (1961). *Biochem. J.* **81**, 591–596.
Kabins, S. A., Sweeney, M. S., and Cohen, S. (1966). *Ann. Intern. Med.* **65**, 1271–1277.
Kind, A. C., Kestle, D. G., Standiford, H. C., and Kirby, W. M. M. (1969). *In* "Antimicrobial Agents and Chemotherapy—1968" (G. L. Hobby, ed.), pp. 361–365. Williams & Wilkins, Baltimore, Maryland.
Kukolja, S. (1968). *J. Med. Chem.* **11**, 1067–1069.
Kurita, M., Atashari, S., Hattori, K., and Takano, T. (1966). *J. Antibiot. Ser.* **A19**, 243–249.
Lederle Laboratories Division of American Cyanamid Co., Inc., Unpublished data of R. G. Shepherd, A. Lewis, and M. L. Sassiver.
Lee, C-C., Herr, E. B., Jr., and Anderson, R. C. (1963). *Clin. Med.* **70**, 1123–1138.
Lewis, A., Sassiver, M. L., and Shepherd, R. G. (1969). *In* "Antimicrobial Agents and Chemotherapy—1968" (G. L. Hobby, ed.), pp. 109–114. Williams & Wilkins, Baltimore, Maryland.
Lilly (1965a). *Farmdoc Complete Spec. Book* **423**, 621–632 (Belg. Pat. 652,148; Derwent Basic No. 18,222).
Lilly (1965b). *Farmdoc Complete Spec. Book* **453**, 97–99 (U.S. Pat. 3,225,038; Derwent Basic No. 19,441).
Lilly (1969a). *Farmdoc Complete Spec. Book* **879**, 653–658 (U.S. Pat. 3,449,338; Derwent Basic No. 38,076).
Lilly (1969b). *Farmdoc Complete Spec. Book* **898**, 181–183 (U.S. Pat. 3,459,746; Derwent Basic No. 39,001).
Lilly (1969c). *Farmdoc Complete Spec. Book* **898**, 195–197 (U.S. Pat. 3,459,749; Derwent Basic No. 39,004).
Loder, B., Newton, G. G. F., and Abraham, E. P. (1961). *Biochem. J.* **79**, 408–416.
Morin, R. B., Jackson, B. G., Flynn, E. H., and Roeske, R. W. (1962). *J. Amer. Chem. Soc.* **84**, 3400–3401.
Morin, R. B., Jackson, B. G., Mueller, R. A., Lavagnino, E. R., Scanlon, W. B., and Andrews, S. L. (1963). *J. Amer. Chem. Soc.* **85**, 1896–1897.
Morin, R. B., Jackson, B. G., Flynn, E. H., Roeske, R. W., and Andrews, S. L. (1969a). *J. Amer. Chem. Soc.* **91**, 1396–1400.
Morin, R. B., Jackson, B. G., Mueller, R. A., Lavagnino, E. R., Scanlon, W. B., and Andrews, S. L. (1969b). *J. Amer. Chem. Soc.* **91**, 1401–1407.

Muggleton, P. W., O'Callaghan, C. H., and Stevens, W. K. (1964). *Brit. Med. J.* **2**, 1234–1237.
Muggleton, P. W., O'Callaghan, C. H., Foord, R. D., Kirby, S. M., and Ryan, D. M. (1969). *In* "Antimicrobial Agents and Chemotherapy – 1968" (G. L. Hobby, ed.), pp. 353–360. Williams & Wilkins, Baltimore, Maryland.
Naito, T., Nakagawa, S., Takahashi, K., Fujisawa, K-I., and Kawaguchi, H. (1968). *J. Antibiot.* **21**, 300–305.
Newton, G. G. F., and Abraham, E. P. (1954). *Biochem. J.* **58**, 103–111.
Newton, G. G. F., and Abraham, E. P. (1955). *Nature (London)* **175**, 548.
Newton, G. G. F., and Abraham, E. P. (1956). *Biochem. J.* **62**, 651–658.
Newton, G. G. F., Abraham, E. P., and Kuwabara, S. (1968). *In* "Antimicrobial Agents and Chemotherapy – 1967" (G. L. Hobby, ed.), pp. 449–455. Williams & Wilkins, Baltimore, Maryland.
Nishida, M., Yokota, Y., Okui, M., Mine, Y., and Matsubara, T. (1968). *J. Antibiot.* **21**, 165–169.
O'Callaghan, C. H., Muggleton, P. W., Kirby, S. M., and Ryan, D. M. (1967). *In* "Antimicrobial Agents and Chemotherapy – 1966" (G. L. Hobby, ed.), pp. 337–343. Williams & Wilkins, Baltimore, Maryland.
O'Callaghan, C. H., Kirby, S. M., and Wishart, D. R. (1968). *In* "Antimicrobial Agents and Chemotherapy – 1967" (G. L. Hobby, ed.), pp. 716–722. Williams & Wilkins, Baltimore, Maryland.
O'Callaghan, C. H., Muggleton, P. W., and Ross, G. W. (1969). *In* "Antimicrobial Agents and Chemotherapy – 1968" (G. L. Hobby, ed.), pp. 57–63. Williams & Wilkins, Baltimore, Maryland.
Okui, M., Hattori, K., and Nishida, M. (1967). *J. Antibiot. Ser.* **A20**, 287–292.
Parke, Davis and Co. (1969). *Farmdoc Complete Spec. Book* **893**, 313–314 (U.S. Pat. 3,457,257; Derwent Basic No. 38,755).
Patchett, A. A., Assignor to Merck and Co., Inc. (1966). *Farmdoc Complete Spec. Book* **458**, 7–11 (U.S. Patent 3,227,712; Derwent Basic No. 19,591).
Pitt, J., Siasoco, R., Kaplan, K., and Weinstein, L. (1968). *In* "Antimicrobial Agents and Chemotherapy – 1967" (G. L. Hobby, ed.), pp. 630–635. Williams & Wilkins, Baltimore, Maryland.
Pollock, M. R. (1965). *In* "Antimicrobial Agents and Chemotherapy – 1964" (J. C. Sylvester, ed.), pp. 292–301. Williams & Wilkins, Baltimore, Maryland.
Rapp, R., and Micetich, R. G. (1968). *J. Med. Chem.* **11**, 70–73.
Rapp, R., Lemieux, R. U., and Micetich, R. G., Assignors to R. and L Molecular Res. Ltd. (1966). *Farmdoc Complete Spec. Book* **524**, 87–98 (U.S. Pat. 3,268,523; Derwent Basic No. 22,561).
Ridley, M., and Phillips, I. (1965). *Nature (London)* **208**, 1076–1078.
Ryan, C. W., Simon, R. L., and Van Heyningen, E. M. (1969). *J. Med. Chem.* **12**, 310–313.
Sabath, L. D. (1968). *In* "Antimicrobial Agents and Chemotherapy – 1967" (G. L. Hobby, ed.), pp. 210–217. Williams & Wilkins, Baltimore, Maryland.
Sabath, L. D., and Abraham, E. P. (1966). *In* "Antimicrobial Agents and Chemotherapy – 1965" (G. L. Hobby, ed.), pp. 392–397. Williams & Wilkins, Baltimore, Maryland.
Sabath, L. D., Jago, M., and Abraham, E. P. (1965). *Biochem. J.* **96**, 739–751.
Sassiver, M. L., Lewis, A., and Shepherd, R. G. (1969). *In* "Antimicrobial Agents and Chemotherapy – 1968" (G. L. Hobby, ed.), pp. 101–109. Williams & Wilkins, Baltimore, Maryland.
Sjöberg, B., Nathorst-Westfelt, L., and Örtengren, B. (1967). *Acta Chem. Scand.* **21**, 547–551.

Spencer, J. L., Flynn, E. H., Roeske, R. W., Siu, F. Y., and Chauvette, R. R. (1966). *J. Med. Chem.* **9**, 746–750.

Spencer, J. L., Siu, F. Y., Flynn, E. H., Jackson, B. G., Sigal, M. V., Higgins, H. M., Chauvette, R. R., Andrews, S. L., and Bloch, D. E. (1967a). *In* "Antimicrobial Agents and Chemotherapy – 1966" (G. L. Hobby, ed.), pp. 573–580. Williams & Wilkins, Baltimore, Maryland.

Spencer, J. L., Siu, F. Y., Jackson, B. G., Higgins, H. M., and Flynn, E. H. (1967b). *J. Org. Chem.* **32**, 500–501.

Stedman, R. J., Swered, K., and Hoover, J. R. E. (1964). *J. Med. Chem.* **7**, 117–119.

Stedman, R. J., Swift, A. C., Miller, L. S., Dolan, M. M., and Hoover, J. R. E. (1967). *J. Med. Chem.* **10**, 363–366.

Strominger, J. L., Izaki, K., Matsuhashi, M., and Tipper, D. J. (1967). *Fed. Proc.* **26**, 9–22.

Sullivan, H. R., and McMahon, R. E. (1967). *Biochem. J.* **102**, 976–982.

Sullivan, H. R., Billings, R. E., and McMahon, R. E. (1969a). *J. Antibiot.* **22**, 27–33.

Sullivan, H. R., Billings, R. E., and McMahon, R. E. (1969b). *J. Antibiot.* **22**, 195–200.

Takano, T., Hattori, K., Kurita, M., Ararashi, S., and Horibe, S. (1967). *Yakugaku Zasshi* **87**, 1141–1145.

Taylor, A. B. (1965). *J. Chem. Soc.*, pp. 7020–7029.

Tipper, D. J., and Strominger, J. L. (1965). *Proc. Nat. Acad. Sci. U.S.A.* **54**, 1133–1141.

Van Heyningen, E. (1965). *J. Med. Chem.* **8**, 22–25.

Van Heyningen, E. (1967). *Advan. Drug. Res.* **4**, 1–70.

Van Heyningen, E., and Ahern, L. K. (1968). *J. Med. Chem.* **11**, 933–936.

Van Heyningen, E., and Brown, C. N. (1965). *J. Med. Chem.* **8**, 174–181.

Wick, W. E. (1966). *In* "Antimicrobial Agents and Chemotherapy – 1965" (G. L. Hobby, ed.), pp. 870–875. Williams & Wilkins, Baltimore, Maryland.

Wick, W. E. (1967). *Appl. Microbiol.* **15**, 765–769.

Wick, W. E., and Boniece, W. S. (1965). *Appl. Microbiol.* **13**, 248–253.

Structure–Activity Relationships in the Tetracycline Series

ROBERT K. BLACKWOOD AND ARTHUR R. ENGLISH

Medical Research Laboratories
Chas. Pfizer & Co., Inc.,
Groton, Connecticut

I.	Introduction	237
II.	Basic Structural Requirements for Tetracycline Activity	239
	A. Skeletal and Chromophoric Requirements for Activity	241
	B. Modifications at the C-2 Position	242
	C. Modifications at the C-4 Position	245
	D. 5a(11a)-Dehydrotetracyclines and 5a-Epitetracyclines	246
	E. Modifications at the C-6 Position	248
	F. Modifications at the C-7 and C-9 Positions (Aromatic Substitution)	253
	G. 11a-Substituted Tetracyclines	256
	H. Modifications at the C-12a Position (Esters)	258
III.	Electronic and Lipophilic Effects on Tetracycline Activity	258
	A. *In Vitro* Gram-Positive Activity (Tetracycline Sensitive)	261
	B. *In Vitro* Gram-Positive Activity (Tetracycline Resistant)	261
	C. *In Vitro* Gram-Negative Activity	261
	D. *In Vivo* *Staphylococcus aureus* Activity	264
	References	265

I. Introduction

The tetracyclines may be defined as a family of broad-spectrum antibiotics which have a perhydronaphthacene skeleton in common. Although discovered later than chlortetracycline (**1**) and oxytetracycline (**2**), tetracycline (**3**) itself is considered as the parent compound for

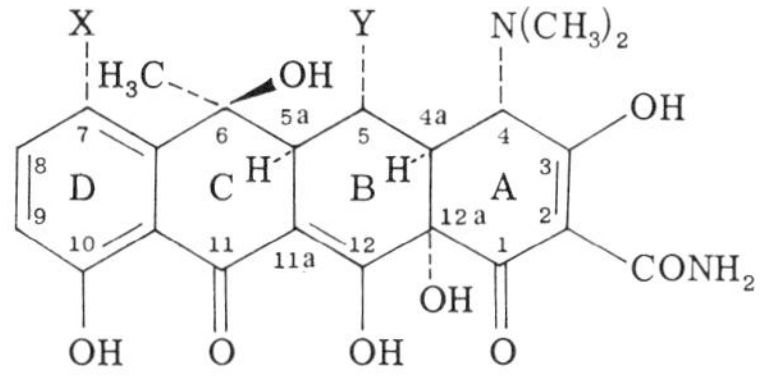

(1) X = Cl, Y = H
(2) X = H, Y = OH
(3) X = Y = H

nomenclature purposes. In addition to these three tetracyclines, other tetracyclines which have found extensive clinical use are demethylchlortetracycline (4), rolitetracycline (5), methacycline (6) and, most recently, doxycycline (7). Some common trade names and the typical oral, daily dosage of the tetracyclines are summarized in Table I. The present article deals only with the very earliest stages of the

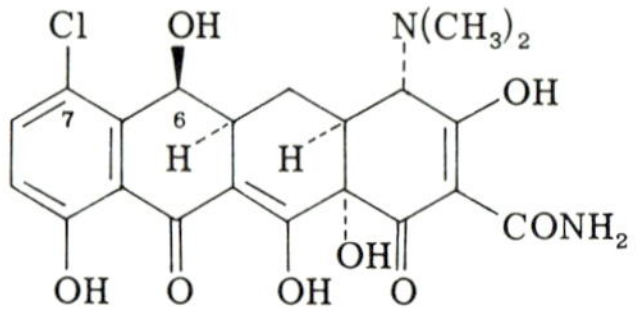

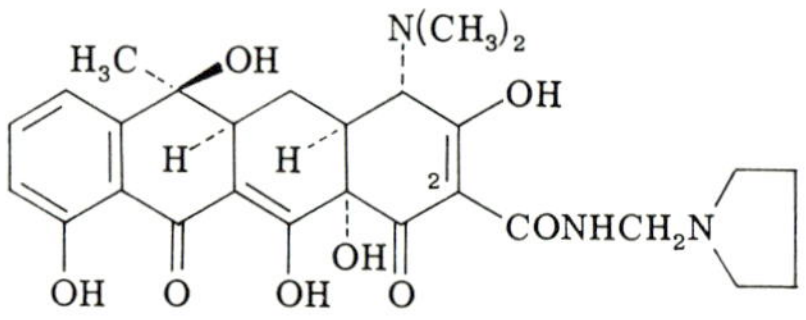

TABLE I
Typical Oral Dosage Regimen for Tetracyclines Marketed in the United States[a]

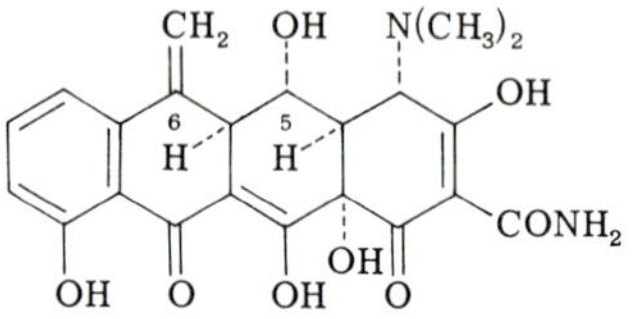

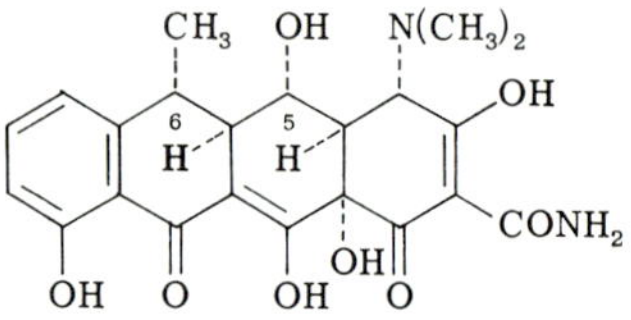

Structure	Chemical name	Generic name	Total daily dosage	No. of doses
1	7-Chlorotetracycline	Chlortetracycline	1 gm	4(250 mg qid)
2	5-Hydroxytetracycline	Oxytetracycline	1 gm	4(250 mg qid)
3	Tetracycline	Tetracycline	1 gm	4(250 mg qid)
4	6-Demethyl-7-chlorotetracycline	Demethylchlortetracycline	600 mg	4(150 mg qid) or 2(300 mg bid)
6	6-Methylenetetracycline	Methacycline	600 mg	4(150 mg qid) or 2(300 mg bid)
7	α-6-Deoxyoxytetracycline	Doxycycline	200 mg first day 100 mg subsequent days	2(100 mg bid) 1(100 mg id) or 2(50 mg bid)

[a]September, 1969.

route of a tetracycline from discovery to clinical use. This early stage is concerned with biological screening for *in vitro* and *in vivo* antibacterial activity. It is only at this level that a large number of derivatives have been studied so as to permit a detailed discussion of structure–activity relationships.

In the past decade a number of review articles have appeared concerning the tetracycline series. These have stressed either their chemistry (Blackwood, 1969a; Clive, 1968), biosynthesis (McCormick, 1968), or general pharmacological properties (Schach von Wittenau and Yeary, 1963; Schach von Wittenau and Delahunt, 1965; Schach von Wittenau, 1969). A single review (Barrett, 1963) includes an extensive compilation of biological data. Except for a few thoroughly studied tetracyclines, such literature data is quite limited, making detailed correlation of structure and activity difficult. We, therefore, plan to examine the contribution of the study of analogs to our current understanding of the structural features and physical properties essential for the characteristic antibacterial activity of the tetracycline antibiotics by presenting bacteriological data, for the most part previously unpublished, which will permit a rationale to be developed. We will first review, as systematically as possible, bacteriological data on analogs, in order to define the basic structural requirements for tetracycline activity. As a standard of reference, bacteriological data for clinically utilized tetracyclines are shown in Table II. Included in this and later tables are bioassay data (activities are expressed relative to 1.0 for tetracyclines), minimum inhibitory concentrations (MIC) against a representative cross section of bacteria, and animal protection results. Data in the latter two categories, unless otherwise specified, have been developed in our own laboratory, using previously published methods (English, 1966). Footnotes following names in the tables refer to chemical syntheses and are given whenever this has not been made clear in the text.

This detailed section is followed by a more general one in which electronic and lipophilic effects on *in vitro* and *in vivo* activity against various types of bacterial organisms, including tetracycline-resistant staphylococci, are considered.

II. Basic Structural Requirement for Tetracycline Activity

Although a very large number of tetracycline derivatives have been studied over the past 20 years, an exacting definition of the basic structural requirement remains elusive. The simplest known tetracycline with *in vitro* activity is 6-demethyl-6-deoxy-4-dedimethyl-

TABLE II
Antibacterial Activity of Tetracyclines Marketed in the United States

Structure	Name	BioAssay (relative to tetracycline)		Minimum inhibitory concentrations (μg/ml)							Protective dose 50% values (mg/kg)					
											S. pyogenes		S. aureus 5		P. multocida	
		K. pneumoniae	S. aureus 209 P	S. aureus 5	S. aureus (tetracycline resistant)	S. pyogenes	E. coli	A. aerogenes	S. typhosa	K. pneumoniae	Oral	Subcut.	Oral	Subcut.	Oral	Subcut.
1	Chlortetracycline	2.5	3.5	0.19	>100	0.04	1.56	1.56	1.56	0.39	8.5	2.1	7.6	3.0	19.6	4.5
2	Oxytetracycline	1.0	0.8	0.55	>100	0.07	1.09	2.91	3.12	2.5	8.2	2.3	7.2	2.6	18.2	4.1
3	Tetracycline	1.0	1.0	0.21	60	0.06	0.73	1.66	1.56	1.56	6.25	1.85	5.81	1.23	22.00	3.60
4	Demethylchlortetracycline	—	3.0	0.11	35	0.04	0.33	0.15	0.55	0.55	6.7	1.33	6.00	1.63	9.80	3.00
5	Rolitetracycline	1.0[a]	—	(b)	(b)	(b)	(b)	(b)	(b)	(b)	—	—	—	—	—	—
6	Methacycline	2.3	—	0.13	50	0.03	0.43	0.94	1.40	1.56	3.90	.66	4.50	1.19	13.00	2.60
7	Doxycycline	1.4	—	0.19	8.75	0.04	1.74	2.54	1.56	1.87	1.60	.98	2.55	1.00	6.60	3.90

[a] Corrected for molecular weight.
[b] *In vitro* activity equivalent to tetracycline has been reported (Seidel *et al.*, 1958).

aminotetracycline (8). However, this compound, which lacks a basic function at the C-4 position and which is relatively lipophilic in comparison to tetracycline itself, shows primarily gram-positive activity *in vitro*. Like other known dedimethylamino compounds it affords no significant *in vivo* activity (Table V). The simplest known tetracycline with broad-spectrum activity, both *in vitro* and *in vivo*, is 6-demethyl-6-deoxytetracycline (9) (Table VII). Consideration of all known

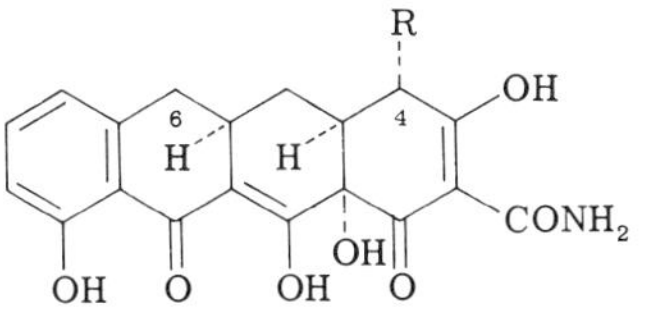

(8) R = H
(9) R = N(CH$_3$)$_2$

tetracycline varients would suggest that the minimum structural requirement for *in vitro* activity might be represented by structures **10** and **11**. It is presumed that a basic function (e.g., **12** and **13**) is required for useful *in vivo* activity. These generalizations may be made apparent from the systematic review of activity of analogs which is now presented.

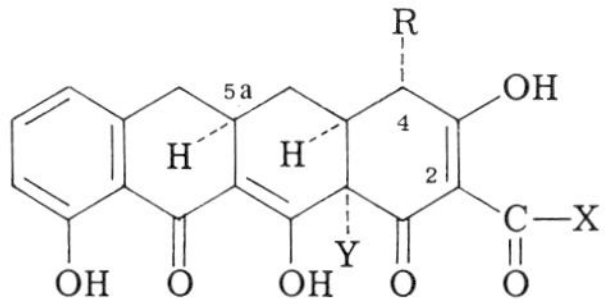

(10) R = H
(12) R = N(CH$_3$)$_2$

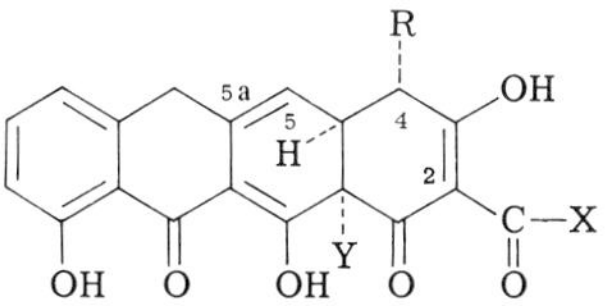

(11) R = H
(13) R = N(CH$_3$)$_2$

A. Skeletal and Chromophoric Requirements for Activity

Tetracycline possesses two ultraviolet-absorbing chromophoric groups, the A-ring and the BCD-rings, separated by the 12a-hydroxyl function (cf. structure **1**). Modification of these chromophoric groups through breaking of one of the rings (e.g., **14, 15**), aromatization of the

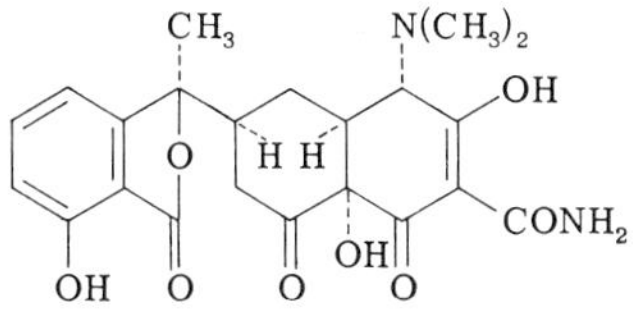

(14)

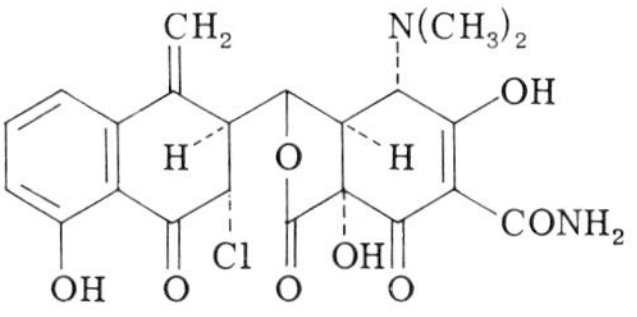

(15)

C or A ring (e.g., **16, 17**), extension of the chromophore (12a-deoxytetracyclines, e.g., **18**) or blockage of a chromophore (e.g., 11a-substitution, **19**) destroys tetracycline activity.[1] The antibacterial activities of compounds **14** to **18** are summarized in Table III. The activities of **19**, together with those of other 11a-halotetracyclines, are summarized in Table IX.

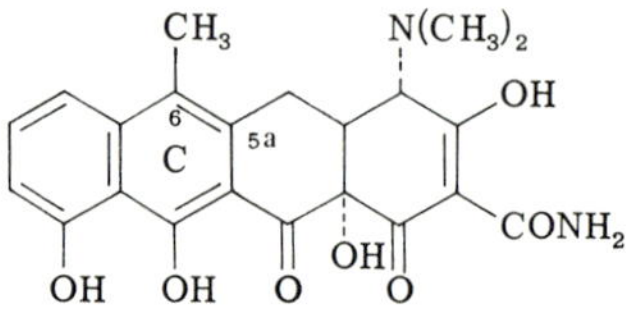

(16)

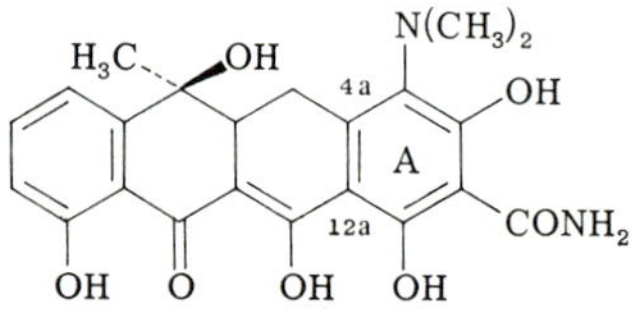

(17)

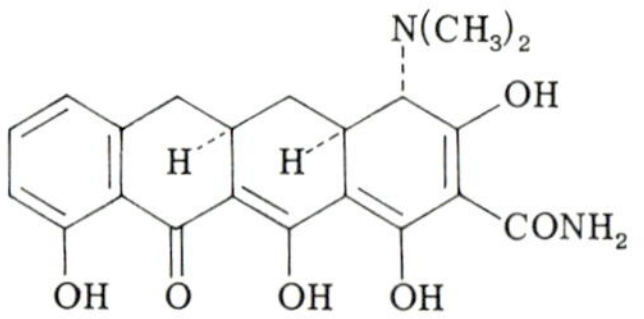

(18)

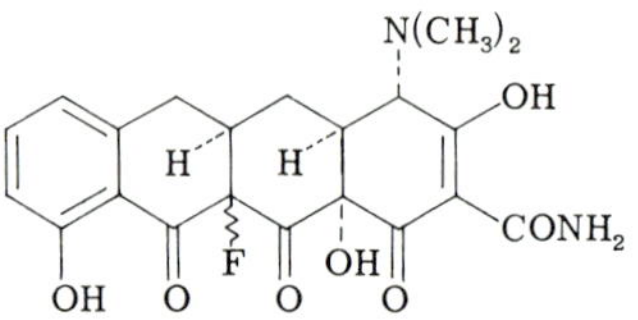

(19)

An overall stereochemical requirement is indicated by the fact that totally synthetic *dl*-6-demethyl-6-deoxytetracycline (**9** and its enantiomer) exhibits only half of the antibacterial activity of the pure optically active isomer derived by fermentation and subsequent chemical steps (Conover *et al.*, 1962).

B. Modifications at the C-2 Position

Modifications at the C-2 position of tetracyclines, as exemplified in Table IV, suggest that the attached carbonyl is essential for tetracycline activity, but that other functionality can replace the amine function of the amide. Thus nitriles (e.g., **20, 21**), prepared by dehydration of the amide (Stephens *et al.*, 1955) have no significant activity, but fermentation derived 2-acetyl-2-decarboxamidotetracyclines (e.g.,

---

[1]This broad statement cannot be made without reservation, since many of the compounds in these classes show weak antibacterial activity, which we judge to be unrelated to tetracycline activity. In the particular case of 5a,6-anhydro compounds (exemplified by **16**) this activity has been reported to result from a different mode of action (Koschel *et al.*, 1966). A lack of cross resistance with tetracycline is also noted (cf. **16** in Table III).

TABLE III
ANTIBACTERIAL ACTIVITY OF TETRACYCLINES WITH MODIFIED CHROMOPHORIC GROUPS[a]

		Minimum inhibitory concentrations (μg/ml)						
Structure	Name	S. aureus 5	S. aureus (tetracycline resistant)	S. pyogenes	E. coli	A. aerogenes	S. typhosa	K. pneumoniae
14	Isotetracycline[b]	>200	—	>200	>200	>200	>200	>200
15	Lactone degradation product of 11a-chloro-6-methyleneoxytetracycline[c]	>100	>100	>100	>100	>100	>100	>100
16	5a,6-Anhydrotetracycline[d]	0.7	0.4	6	12	50	6	12
17	4a,12a-Anhydrotetracycline[e]	50	50	50	>200	>200	>200	>200
18	12a-Deoxy-6-demethyl-6-deoxytetracycline[e,f]	3	6	6	6	6	12	12

[a] We have not observed *in vivo* activity with any of the compounds listed in this table.
[b] Booth *et al.*, 1953.
[c] Blackwood *et al.*, 1963.
[d] Waller *et al.*, 1952.
[e] Blackwood *et al.*, 1960.
[f] Green and Boothe, 1960.

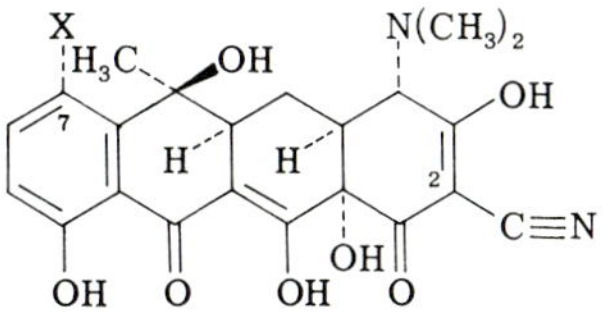

(20) X = Cl
(21) X = H

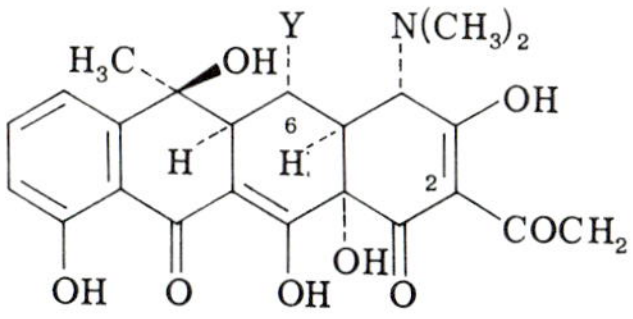

(22) Y = OH
(23) Y = H

TABLE IV
ANTIBACTERIAL ACTIVITY OF TETRACYCLINES MODIFIED AT THE C-2 POSITION

Structure	Name	Minimum inhibitory concentrations (μg/ml)							Protective dose 50% values (mg/kg) $S.\ aureus$ 5	
		$S.\ aureus$ 5	$S.\ aureus$ (tetracycline resistant)	$S.\ pyogenes$	$E.\ coli$	$A.\ aerogenes$	$S.\ typhosa$	$K.\ pneumoniae$	Oral	Subcut.
20	Chlorotetracyclinonitrile[a]	>100	—	>100	>100	>100	>100	50	—	—
21	Tetracyclinonitrile[b]	>100	—	>100	>100	>100	>100	>100	—	—
22	2-Acetyl-2-decarboxamidooxytetracycline[c]	3	>200	3	50	50	50	25	140	—
23	2-Acetyl-2-decarboxamidotetracycline[d]	3	—	0.7	50	50	50	50	>25	(e)
24	6-Demethyl-6-deoxy-N^2-t-butyltetracycline	0.7	1.5	1.0	>100	>100	>100	>100	>100	(f)

[a] Stephens et al., 1954.
[b] Stephens, 1962.
[c] Hochstein et al., 1960.
[d] Miller and Hochstein, 1962.
[e] 80% protected at 50 mg/kg.
[f] 30% protection at 50 mg/kg.

22, 23) and various derivatives substituted on the carboxamide nitrogen possess antibacterial activity.

N-*t*-Butyl-6-demethyl-6-deoxytetracycline (**24**) prepared by reaction of the corresponding nitrile with isobutylene under strongly acidic conditions (Stephens *et al.*, 1963) shows moderate *in vitro*

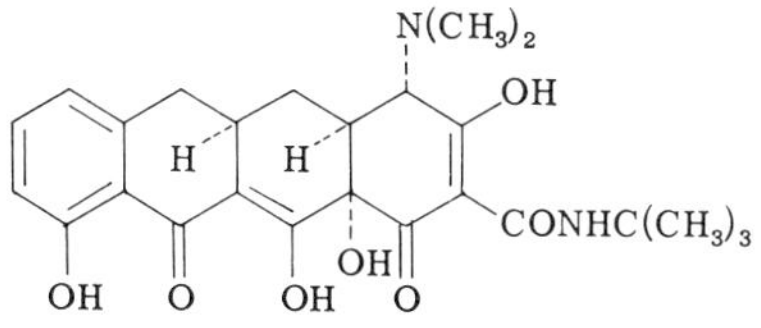

(24)

activity against gram-positive bacteria, including those resistant to tetracycline; however, gram-negative and *in vivo* activity are lost in this compound. As discussed in greater detail below, this is probably not the result of a purely structural effect, but due rather to the highly lipophilic character of the compound.

A common type of derivative of tetracyclines are the so-called Mannich bases, typified by pyrrolidinomethyltetracycline (rolitetracycline **5**), formed by the interaction of tetracycline with formaldehyde and pyrrolidine (Seidel *et al.*, 1958; Gottstein *et al.*, 1959). Because this reaction is reversible, these compounds in general show activity equivalent to that of the parent compound. Pharmacodynamic advantages are claimed for rolitetracycline and related compounds derived from formaldehyde and amines, particularly when used parenterally (Gradnik *et al.*, 1960). Several related compounds (e.g., **25, 26**) which are less prone to hydrolysis are reported to show depressed antibacterial activity (Martell *et al.*, 1967a; Tamorria and Esse, 1965).

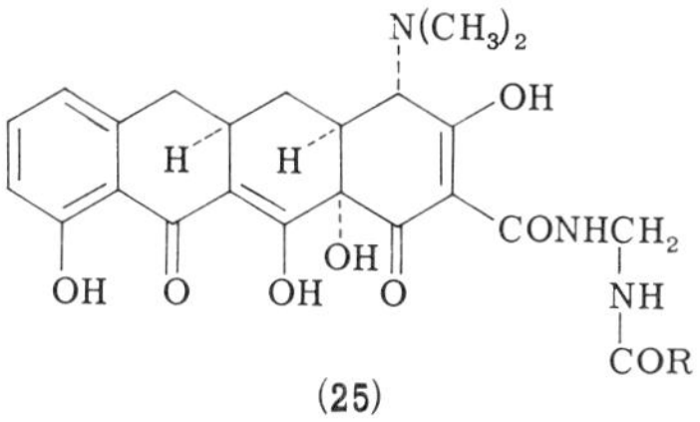

(25) 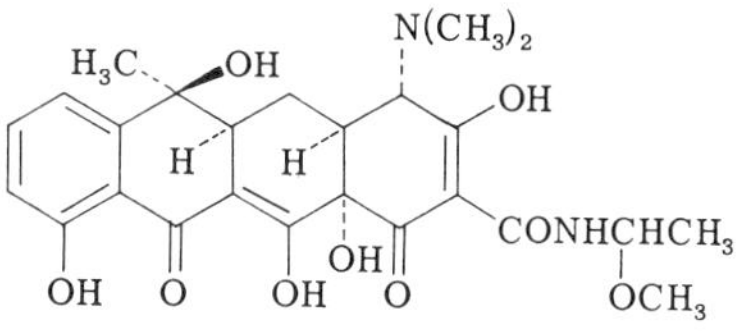 (26)

C. Modifications at the C-4 Position

The amine function at C-4 is not essential for antibacterial activity, as evidenced by the bacteriological results shown for 4-dedimethyl-

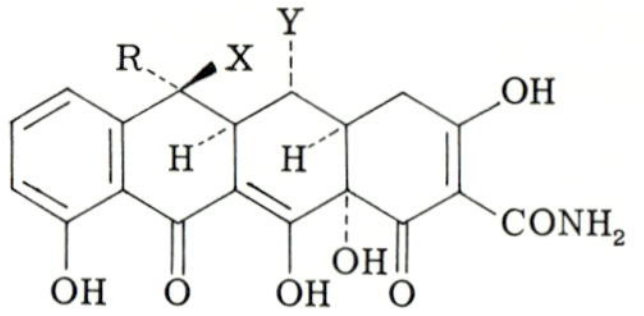

(27) R = CH$_3$, X = Y = OH
(28) R = CH$_3$, X = OH, Y = H
(8) R = X = Y = H

aminotetracyclines (**8, 27, 28**) in Table V. However, *in vivo* activity and gram-negative *in vitro* activity are borderline at best. 4-Epitetracyclines (e.g., **30–32**) also show reduced activity. A series of 6-demethyltetracyclines with modified amine functions at C-4 have

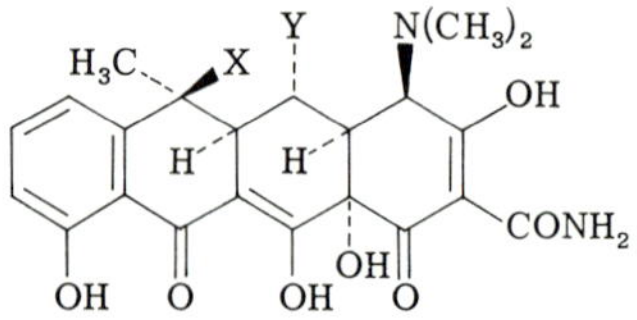

(30) X = OH, Y = H
(31) X = Y = OH
(32) X = H, Y = OH

been reported (Esse *et al.*, 1964). In the normal configuration these have depressed activity when compared to the parent dimethylamino compounds. It appears that the decrease in activity is correlated with increasing bulk of the dialkylamine function. The 4-epimers, as expected, are much less active. Several other tetracycline derivatives

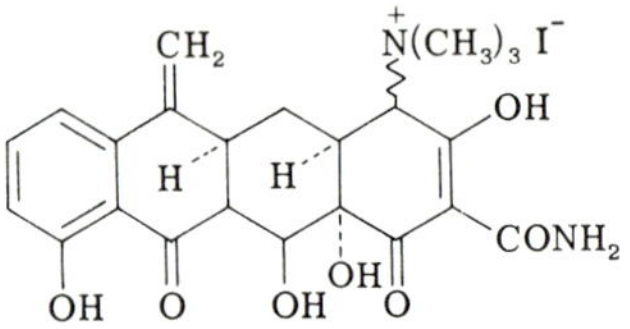

(33) 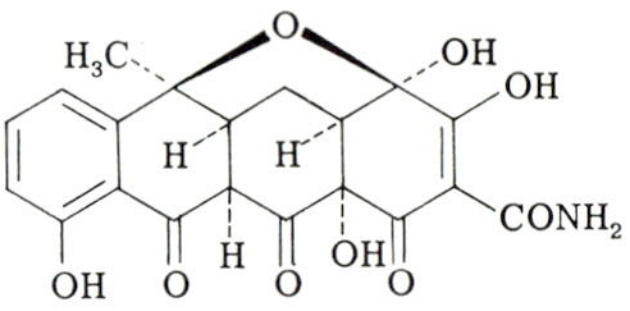   (34)

with modified substituents at C-4, methiodides (e.g., **33**), 4-oxo derivatives (e.g., **34**), the oxime and hydrazone derived therefrom (**35, 36**), and the alcohol (**37**) show little or no antibacterial activity.

D. 5A(11A)-DEHYDROTETRACYCLINES AND 5A-EPITETRACYCLINES

Interesting activity relationships have recently been reported in a series of 5a(11a)-dehydro- and 5a-epitetracyclines. The first reported

ANTIBACTERIAL ACTIVITY OF TETRACYCLINES MODIFIED AT THE C-4 POSITION

		Minimum inhibitory concentrations (μg/ml)							Protective dose 50% values (mg/kg)			
									S. pyogenes		S. aureus 5	
Structure	Name	S. aureus 5	S. aureus (tetracycline resistant)	S. pyogenes	E. coli	A. aerogenes	S. typhosa	K. pneumoniae	Oral	Subcut.	Oral	Subcut.
27	4-Dedimethylaminooxytetracycline[a]	25	>100	12	50	50	6	12	>155[d]	>38[e]	>200	>100
28	4-Dedimethylaminotetracycline[b]	12	25	6	100	50	12	12	—	—	72	39
8	4-Dedimethylamino-6-demethyl-6-deoxytetracycline[c]	0.8	1.5	1.5	25	50	50	50	—	—	>400	>200
30	4-Epitetracycline[f]	6	>100	3	12	12	12	12	—	—	—	—
31	4-Epioxytetracycline[g]	6	>100	3	12	25	12	12	8	0.5	—	—
32	4-Epi-α-6-deoxyoxytetracycline[h] (epidoxycycline)	3	>100	1.5	25	25	25	12	—	—	—	—
33	6-Methylenetetracycline methiodide[i]	12	>100	3	>100	>100	>100	50	>400	>400	—	—
34	4-Oxo-4-dedimethylaminotetracycline 4,6-hemiketal	>100	>100	>100	>100	>100	>100	>100	—	—	—	—
35	4-Oximino-4-dedimethylaminotetracycline[j]	>100	>100	>100	>100	>100	>100	>100	>400	>400	>200	>200
36	4-Hydrazono-4-dedimethylaminotetracycline[j]	50	>100	25	>100	>100	>100	>100	>25	>25	—	—
37	4-Hydroxy-4-dedimethylaminotetracycline[j]	>10	>100	5	>100	>100	>100	>100	—	—	—	—

[a] Hochstein et al., 1953.
[b] Boothe et al., 1958.
[c] Stephens et al., 1963.
[d] 15% survival at 155 mg/kg.
[e] 10% survival at 38 mg/kg.
[f] Stephens et al., 1956; Doerschuk et al., 1955.
[g] Conover, 1956.
[h] Blackwood, 1969b.
[i] Blackwood et al., 1963.
[j] Blackwood and Stephens, 1965.

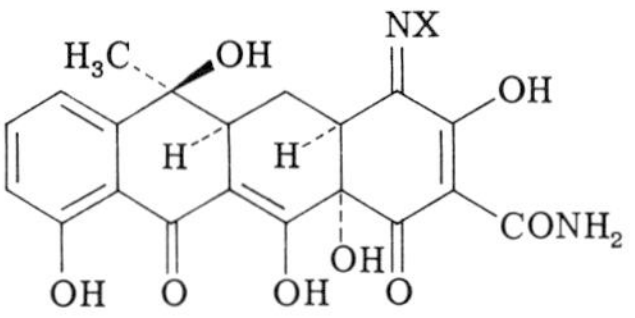

[Structures (35) X = OH, (36) X = NH₂, and (37) shown at top]

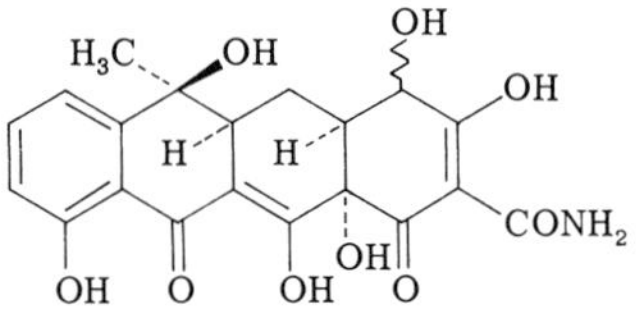

[Structures (38a) ⇌ (38b) shown]

[Structure (39) shown]

compounds (**38** and **39**), which possess the usual configuration at the C-6 position, were found to have insignificant activity (McCormick *et al.*, 1958). In marked contrast, the corresponding compounds possessing the epimeric configuration at the C-6 position show significant

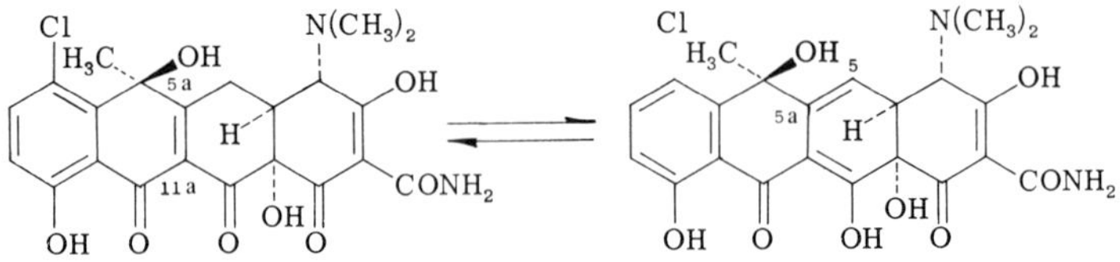

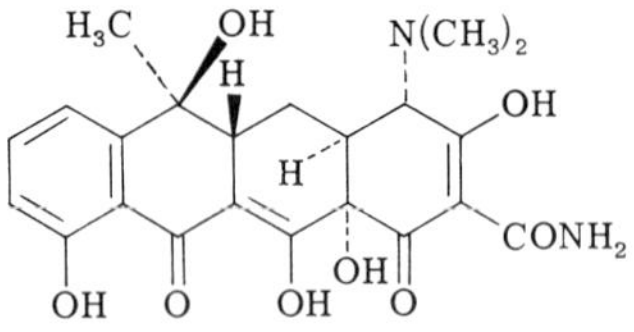

antibacterial activity (Martell *et al.*, 1967b). Literature data are summarized in Table VI. There is no obvious explanation for these results. At the very least it indicates the configuration at C-5a is not a critical feature for tetracycline activity. We would suggest, however, that the active form of the 5a(11a)-dehydro compounds is the tautomeric 5,5a-dehydro form (**38b**, cf. Schach von Wittenau *et al.*, 1963).

E. Modifications at the C-6 Position

A most effective approach to the discovery of highly active tetracycline derivatives has been studies yielding tetracyclines modified at the C-6 position. This has been all the more true because the acid stable derivatives which have generally resulted have permitted substitution reactions on the aromatic D-ring as discussed in the section

TABLE VI
RELATIVE ANTIBACTERIAL ACTIVITY OF 5a(11a)-DEHYDROTETRACYCLINES AND 5a-EPITETRACYCLINES AGAINST *Staphylococcus aureus*[a]

Structure	Name	Relative activity In vitro[b]	In vivo[c]
3	Tetracycline	1.0	1.0
	6-Epitetracycline	0.6	–
38	5a(11a)-Dehydrochlortetracycline		Inactive
39	5a-Epitetracycline		Inactive
40	5a(11a)-Dehydro-6-epichlortetracycline	1.5	0.5
41	5a-Epi-6-epitetracycline	0.4	–

[a] Data from Martell *et al.*, 1967b.
[b] Strain 209P.
[c] Strain Smith.

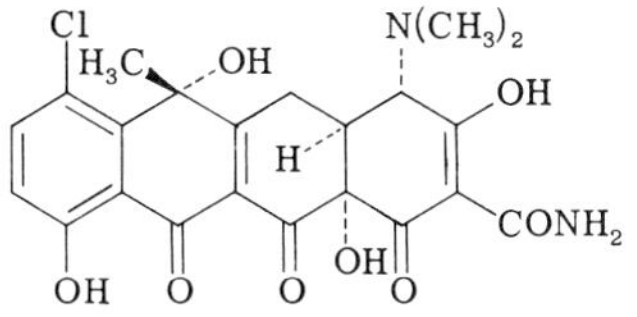

(40) 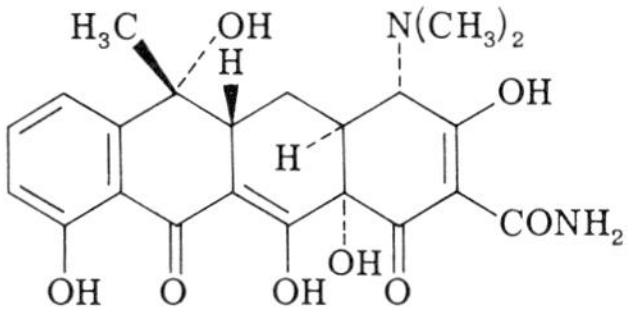 (41)

which follows. Neither the 6-methyl nor 6-hydroxyl is essential for activity (e.g., 6-demethyl-6-deoxytetracycline, **9**) and a variety of other functionality has been possible at the C-6 position with retention of antibacterial activity. Bacteriological data are summarized in Table VII.

Among the variations at C-6 are two fermentation-derived 6-demethyltetracyclines (**4, 42**). Catalytic hydrogenolysis (McCormick *et al.*, 1960; Stephens *et al.*, 1963) of fermentation tetracyclines provides β-6-deoxytetracyclines (**9, 43, 44**), compounds which show high retention of antibacterial activity. In contrast to the parent tetracyclines, which dehydrate to 5a,6-anhydrotetracyclines (e.g., **16**) these

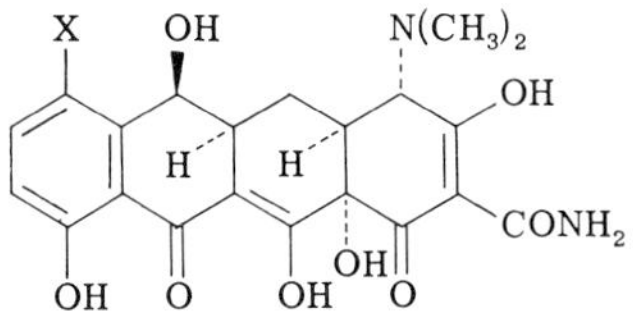

(4) X = Cl
(42) X = H

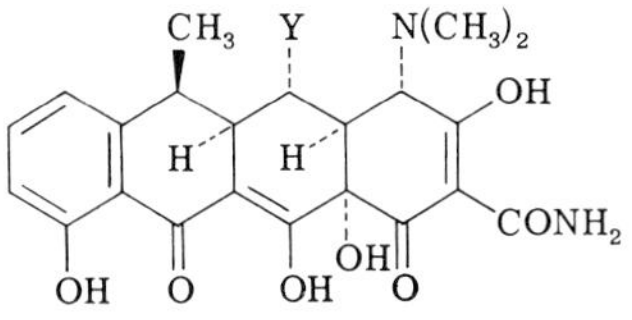

(43) Y = H
(44) Y = OH

TABLE VII

ANTIBACTERIAL ACTIVITY OF TETRACYCLINES AS MODIFIED AT THE C-6 POSITION

Structure	Name	Bioassay (relative to tetracycline)		Minimum inhibitory concentrations (µg/ml)							Protective dose 50% values (mg/kg) S. aureus 5	
		K. pneumoniae	S. aureus 209 P	S. aureus 5	S. aureus (tetracycline resistant)	S. pyogenes	E. coli	A. aerogenes	S. typhosa	K. pneumoniae	Oral	Subcut.
9	6-Demethyl-6-deoxytetracycline[a,b]	0.9[b]	1.6[a]	0.2	1.5	0.2	1.5	1.5	1.5	0.1	5	1.5
43	β-6-Deoxytetracycline[a,b]	0.5[b]	0.72[a]	0.8	9	0.4	6	6	6	6	25	8.5
44	β-6-Deoxyoxytetracycline[a,b]	0.4[b]	0.5[a]	0.8	200	0.1	12	12	6	6	22	—
6	6-Methyleneoxytetracycline[c]	2.3[c]	—	0.13	50	0.03	0.43	0.94	1.40	1.56	4.50	1.19
45	6-Methylenetetracycline[c]	1.2[c]	—	0.2	6	0.2	3	6	3	3	11	9.3
46	6-Methylene-7-chlorooxytetracycline[c]	6.3[b]	—	0.15	100	0.01	0.8	1.5	1.5	0.4	26	2.2
7	α-6-Deoxyoxytetracycline[b]	1.4[b]	—	0.19	8.8	0.04	1.74	2.54	1.56	1.87	2.55	1.00
47	α-6-Deoxytetracycline[b]	0.7[b]	—	0.6	3	0.1	6	12	6	3	12.5	7.5
48	13-Mercapto-α-6-deoxyoxytetracycline[c]	0.06[c]	—	3	>100	1.5	>100	>100	>100	>100	—	—
49	13-Phenylmercapto-α-6-deoxytetracycline[c]	0.01[c]	—	0.8	0.8	0.2	>100	>100	50	>100	>25	>25

	Compound									
50	13-Benzylmercapto-α-6-deoxytetracycline[c]	0.01[c]	—			>100	>100	>100	>100	—	—
51	13-Phenylmercapto-α-6-deoxyoxytetracycline[c]	0.2[c]	—	0.2	0.4	0.1	>100	>100	>100	—	—
52	13-Benzylmercapto-α-6-deoxyoxytetracycline[c]	0.26[c]	—	0.2	1.5	0.2	>100	>100	>100	(d)	50
53	13-(2-Hydroxymethylmercapto)-α-6-deoxyoxytetracycline[c]	0.06[c]	—	0.04	1.5	0.04	100	100	50	>12	>12
54	13-Acetylmercapto-α-6-deoxytetracycline[c]	0.6[c]	—	6	>100	0.4	>100	>100	>100	—	—
55	13-Acetylmercapto-α-6-deoxyoxytetracycline[c]	0.4[c]	—	0.4	12	0.4	50	—	25	—	—
56	13-Phenylmercapto-α-6-deoxytetracycline S-oxide[c]	0.4[c]	—	0.2	12	0.2	12	12	6	(e)	7.5
57	13-Benzylmercapto-α-6-deoxyoxytetracycline S-oxide[c]	0.24[c]	—	12.5	100	0.2	50	—	50	(f)	200
				3	>100	0.2	>100	>100	>100	—	12
58	7,13-Epithio-α-6-deoxytetracycline[c]	1.3[c]	—	0.4	3	0.2	3	6	6	(g)	
59	11a,13-Epithio-α-6-deoxyoxytetracycline[c]	<0.01[c]	—	100	>100	50	>100	>100	>100	>12.5	>12.5
										12.5,38	12.5

[a] McCormick et al., 1960.
[b] Stephens et al., 1963.
[c] Blackwood et al., 1963.
[d] 40% protection at 400 mg/kg.
[e] 20% protection at 25 mg/kg.
[f] 30% protection at 200 mg/kg.
[g] 10% protection at 12.5 mg/kg.

compounds are acid stable and suitable for aromatic substitution as summarized in the section which follows.

By protecting the 11a-position with halogen (see section below on modification at the 11a-position), exocyclic dehydration of tetracyclines has been accomplished (Blackwood et al., 1963), yielding 6-methylene derivatives (**6, 45, 46**). Excellent activity, as summarized in Table VII, is noted for these tetracyclines. The 6-methylenetetracyclines in turn have served as intermediates to other C-6 modified tetracyclines, most notably the α-6-deoxytetracyclines (**7, 47**), but also a series of sulfur-containing tetracyclines typified by compounds **48–59**.

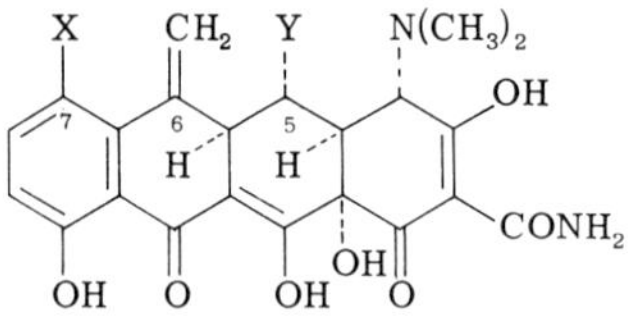

(6) X = H, Y = OH
(45) X = Y = H
(46) X = Cl, Y = OH

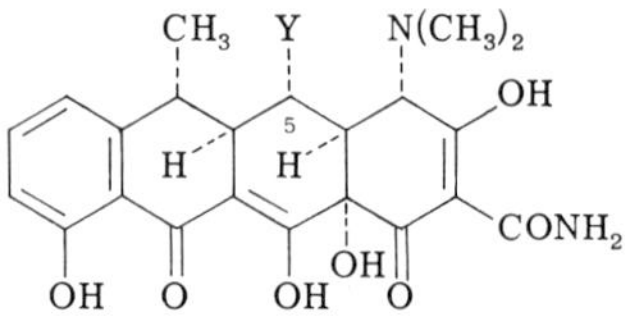

(7) Y = OH
(47) Y = H

The effect of lipophilicity on *in vitro* activity, a topic discussed more generally below, is readily seen by reference to the *Klebsiella pneumoniae* bioassay data in Table VII. Thus the highly lipophilic phenyl and benzyl mercaptan adducts of 6-methylenetetracycline (**49** and **50**, respectively) show markedly depressed activity, particularly against gram-negative bacteria. Consideration of *K. pneumoniae* bioassay data suggest that much activity is regained when these compounds are converted to their more polar sulfoxides (**56** and **57**). In line with this hypothesis is the greater bioassay activity of the more polar thiolacetic acid adducts (**54** and **55**) and the greater activity of

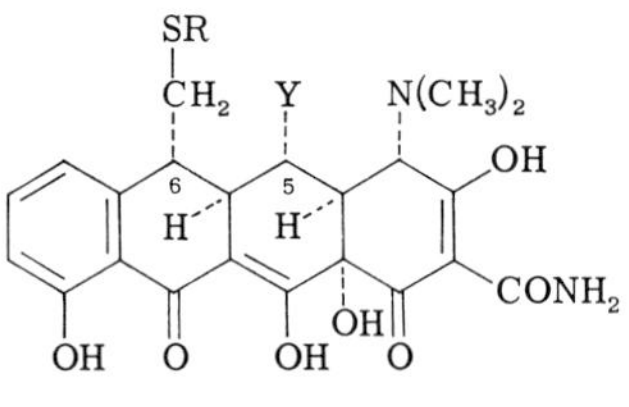

(48) R = H, Y = OH
(49) R = C_6H_5, Y = H
(50) R = $C_6H_5CH_2$, Y = H
(51) R = C_6H_5, Y = OH
(52) R = $C_6H_5CH_2$, Y = OH
(53) R = $HOCH_2CH_2$, Y = OH
(54) R = CH_3CO, Y = H
(55) R = CH_3CO, Y = OH

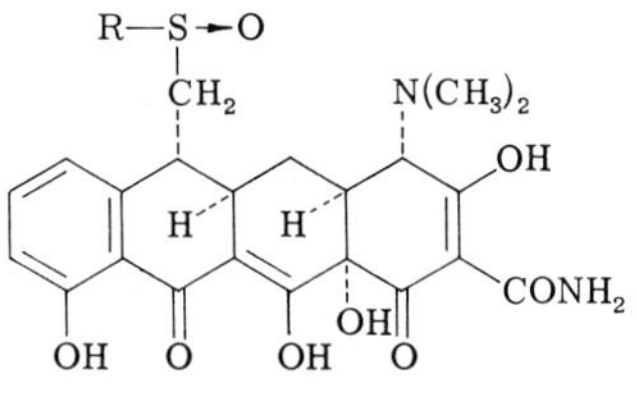

(56) R = C_6H_5
(57) R = $C_6H_5CH_2$

the phenyl and benzyl mercaptan adducts in the more polar oxytetracycline series (cf. **51** and **52** with **49** and **50**). The same correlation is not evident from a consideration of MIC data. Of special interest is the *in vivo* activity of the cyclic sulfur derivative **59**. Since this compound is inactive *in vitro*, it is apparent that a metabolic reaction must take place which cleaves the sulfur-11a-carbon bond to yield an active tetracyline of undetermined structure (compare section below on 11a-substituted tetracyclines). We can offer no explanation for the activity pattern of the isomeric sulfur derivative, **58**, which, in spite of excellent activity *in vitro*, shows no significant *in vivo* activity.

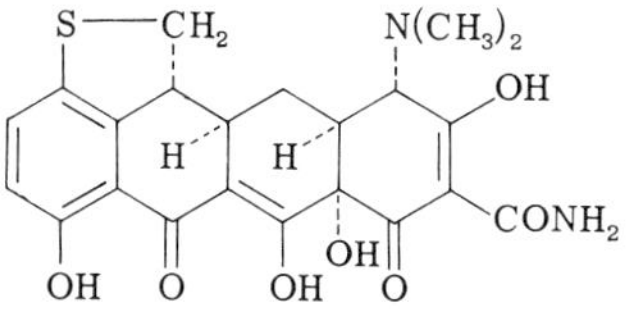

(58) 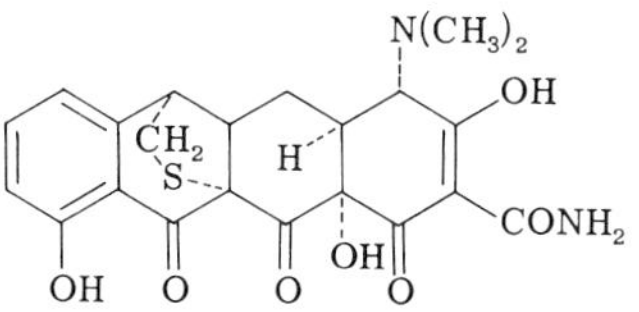   (59)

When the total series of C-6 modified tetracyclines described in this section are examined, other structure-activity generalizations emerge. One is that the 6-methylenetetracyclines are generally more active than the parent tetracyclines. This may be primarily an electronic effect—one which is roughly equivalent to that of a C-7 chlorine (see the section immediately below). Second, it will be noted that the α-6-deoxytetracyclines (**7** and **46**) are more active than the corresponding β-deoxytetracyclines (**44** and **43**, respectively). Finally, introduction of large bulk at C-6 appears to have an overall depressing effect on gram-negative activity. Several of the tetracyclines discussed in this section are included in the more general section below on electronic and lipophilic effects on tetracycline activity.

F. Modifications at the C-7 and the C-9 Positions (Aromatic Substitution)

The acid instability of the fermentation tetracyclines render impractical direct acid-catalyzed aromatic substitution reactions. However, availability of the 6-deoxytetracyclines, no longer possessing the acid-labile 6-hydroxyl function, made such reactions feasible, and extensive series of derivatives substituted at C-7 and/or C-9 have been described in recent years. With few exceptions, only the *in vitro* activity (as bioassays relative to tetracyclines) has been reported for these compounds. Both chemistry and this type of activity data have been reviewed (Blackwood, 1969a) and need not be repeated here.

We shall therefore limit our discussion to illustrating examples, stressing those compounds where *in vivo* data are available in our own files. Pertinent results are summarized in Table VIII.

Consider first substition at C-7. Electron withdrawing groups at this position clearly enhance *in vitro* activity. Thus compare the *in vitro* activity of chlortetracycline (**1**) with tetracycline (**3**) in Table II, the activity of methacycline (**6**) with its 7-chloro derivative (**46**) in Table VII, and 7-nitro-, 7-chloro-, and 7-bromo-6-methyl-6-deoxytetracyclines (**60–62**) with that of the unsubstituted analog (**9**) in Table VIII. Amine functions (e.g., **63, 64**) can be either strongly electron withdrawing or donating, depending upon whether or not the amine is protonated. At the pH employed for bioassays and determination of minimum inhibitory concentrations, pK_a measurements indicate that the C-7 amine should not be protonated. One might, therefore, expect an effect opposite to that seen with electron-withdrawing substituents, i.e., depressed *in vitro* activity. This is certainly not the case, particularly in the 7-dimethylamino-6-demethyl-6-deoxytetracycline (minocycline, **68**). A possible explanation is that near the surface of a

(60) X = NO_2
(61) X = Cl
(62) X = Br
(9) X = H
(63) X = NH_2
(64) X = $N(CH_3)_2$

(65) X = NO_2
(66) X = Cl
(67) X = NH_2
(68) X = $N(CH_3)_2$

bacterium there is a microregion of lower pH (as would be the case if the bacterium carried a negative charge, creating a region near the cell wall with an excess of positively charged ions; (cf. Weibull, 1960). In this case the amine at C-7 would be protonated and become strongly electron withdrawing, with resultant enhanced *in vitro* activity. The simple 7-amino substituent being a weaker base, is less apt to be protonated under the same conditions.

In vivo data for the same 7-substituted tetracyclines are also shown in Table VIII. Correlations seen *in vitro* do not appear to carry over to *in vivo* activity. The interplay of electronic and lipophilic factors as they may affect *in vivo* activity are considered for several of these compounds in the more general section below.

TABLE VIII
ANTIBACTERIAL ACTIVITY OF TETRACYCLINES MODIFIED AT THE C-7 AND C-9 POSITIONS

Structure	Name	Bioassay (relative to tetracycline)			Minimum inhibitory concentration (μg/ml)						Protective dose 50% values (mg/kg) S. aureus 5	
		K. pneumoniae	S. aureus 209 P	S. aureus 5	S. aureus (tetracycline resistant)	S. pyogenes	E. coli	A. aerogenes	S. typhosa	K. pneumoniae	Oral	Subcut.
60	7-Nitro-6-demethyl-6-deoxytetracycline[a]	4.6[b]	6.4[a]	0.2	12.5	0.04	0.15	0.07	0.39	0.78	18	4
61	7-Chloro-6-demethyl-6-deoxytetracycline[c]	—	3.0[c]	0.03	0.8	0.1	1.5	1.5	1.5	3.0	4.0	1.5
62	7-Bromo-6-demethyl-6-deoxytetracycline[b,d]	1.3[b]	2.0[a]	0.3	0.8	0.2	3	1	6	3	(g)	(h)
9	6-Demethyl-6-deoxytetracycline	0.9	1.6	0.2	1.5	0.2	1.5	1.5	1.5	1.5	5	1.5
63	7-Amino-6-demethyl-6-deoxytetracycline[a]	1.0[b]	0.4[a]	1.5	>100	0.2	0.8	3	6	3	20	3
64	7-Dimethylamino-6-demethyl-6-deoxytetracycline[e]	—	>1.0[e]	0.1	0.8	0.02	3	3	3	1.5	3.5	—[i]
65	9-Nitro-6-demethyl-6-deoxytetracycline[a]	0.2[b]	0.12[a]	1.5	25	0.8	1.5	1.5	1.5	1.5	—	—
66	9-Chloro-6-demethyl-6-deoxytetracycline[f]	—	—	1.5	12	0.8	12	12	200	25	6.2	2.0
67	9-Amino-6-demethyl-6-deoxytetracycline[a]	0.76[b]	1.6[a]	1.5	25	0.1	1.5	6	6	6	1	0.4
68	9-Dimethylamino-6-demethyl-6-deoxytetracycline[e]	—	—	6	50	3	12	25	50	50	4.3	2.5

[a] Petisi et al., 1962.
[b] Beereboom et al., 1960.
[c] Hlavka et al., 1962a.
[d] Hlavka et al., 1962b.
[e] Martell and Boothe, 1967.
[f] Stephens et al., 1963.
[g] 10% survival at 50 mg/kg.
[h] 25% survival at 12.5 mg/kg.
[i] Also shows in vivo activity against tetracycline-resistant Staphylococcus aureus (Redin, 1966).

The activities of tetracyclines substituted at C-9 do not parallel the activities of tetracyclines analogously substituted at C-7, even although electronic factors should be similar. Thus 9-nitro- and 9-chloro-6-demethyl-6-deoxytetracyclines (**65** and **66**) show depressed activity relative to corresponding 7-substituted analogs (**60** and **61**, respectively). Perhaps this is a result of hydrogen bonding with the phenolic hydroxyl at C-10 disrupting an active site in the tetracycline molecule. 9-Amino-6-demethyl-6-deoxytetracycline (**67**) shows activity somewhat better than that of the 7-analog (**63**), while 9-dimethylamino-6-demethyl-6-deoxytetracycline (**66**) shows depressed *in vitro* activity (perhaps a steric factor). However, in spite of this poor *in vitro* activity of the latter derivative, surprisingly good *in vivo* activity is noted, suggesting the same favorable electronic/lipophilic factors operating *in vivo* as may operate for minocycline (**64**).

G. 11a-Substituted Tetracyclines

Reference to the requirement of the 11,12-β-diketone system (BCD chromophore) for tetracycline activity has been made above. Consistent with this view is the poor *in vitro* activity of several 11a-halogenated tetracyclines (**19, 69–76**) and the sulfuric acid ester (**78**) as summarized in Table IX. However, *in vivo* activity will be noted for several of these compounds. 11a-Bromo and 11a-chloro compounds are readily reduced, both biologically and chemically, to the parent compounds. As a result those which are stable to degradation show significant *in vitro* activity, as well as *in vivo* activity comparable to the unsubstituted compounds. Thus compare the activities of 11a-bromo-6-demethyl-6-deoxytetracycline (**69**) and 11a-chloro-6-demethyl-6-deoxytetracycline (**70**) in Table IX with the activity of 6-demethyl-6-deoxytetracycline (**9**) in Table VIII; the activity of 7,11a-dichloro-6-demethyl-6-deoxytetracycline in Table IX with the activity of 7-chloro-6-demethyl-6-deoxytetracycline (**61**) in Table VIII; and the activity 11a-chloro-6-methylenetetracycline (**72**) in Table IX with the activity of 6-methylenetetracycline in Table VII. Those 11a-fluoro compounds which are stable to degradation (e.g., **19**, Table IX) show little *in vitro* activity, but some *in vivo* activity. The remaining 11a-halo compounds in Table IX (**73–75**) are very unstable, especially in neutral or basic solution. They are therefore devoid of *in vitro* activity or oral *in vivo* activity. It may be noted, however, that subcutaneous activity is seen with at least one of these (**73**). The latter compound degrades rapidly in neutral solution to a lactone (**15**) discussed earlier. Finally, it will be noted that the sulfuric acid ester of oxytetracycline (**78**) possesses neither *in vitro* nor *in vivo* activity.

TABLE IX
ANTIBACTERIAL ACTIVITY OF 11a-SUBSTITUTED TETRACYCLINES

Structure	Name	Minimum inhibitory concentration (μg/ml)							Protective dose 50% values (mg/kg) S. aureus 5	
		S. aureus 5	S. aureus (tetracycline resistant)	S. pyogenes	E. coli	A. aerogenes	S. typhosa	K. pneumoniae	Oral	Subcut.
69	11a-Bromo-6-demethyl-6-deoxytetracycline[b]	3	12	1.5	6	6	6	6	14	1.5[a]
70	11a-Chloro-6-demethyl-6-deoxytetracycline[b]	6	50	3	12	25	25	12	5.5	5.2
19	11a-Fluoro-6-demethyl-6-deoxytetracycline[b]	100	>100	50	>100	>100	>100	>100	75	28
71	7, 11a-dichloro-6-demethyl-6-deoxytetracycline[b]	12	—	6	50	100	100	100	>50	4.5
72	11a-Chloro-6-methylenetetracycline[c]	50	50	12	>100	>100	>100	>100	6.9	3.0
73	11a-Chloro-6-methyleneoxytetracycline[c]	100	—	100	>100	>100	>100	>100	>25	19
74	11a-Chlorotetracycline 6,12-hemiketal[c]	100	>100	50	>100	>100	>100	>100	>25[a]	>6.2[a]
75	11a-Chloroxytetracycline 6,12-hemiketal[c]	100	100	25	>100	>100	>100	>100	>25	>6.2
76	11a-Fluorotetracycline 6,12-hemiketal[c]	50	100	25	>100	>100	>100	>100	>25	>6.2
77	11a-Fluorooxytetracycline 6,12-hemiketal[c]	>100	>100	50	>100	>100	>100	>100	>200	50
78	Oxytetracycline sulfuric acid ester	50	>200	12	200	200	200	200	>100	>100

[a] 20% protection at these levels.
[b] Stephens et al., 1963.
[c] Blackwood et al., 1963.
[d] IV dosage.

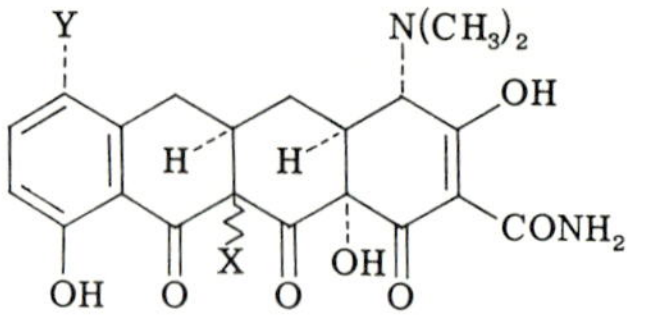

(19) X = F, Y = H
(69) X = Br, Y = H
(70) X = Cl, Y = H
(71) X = Y = Cl

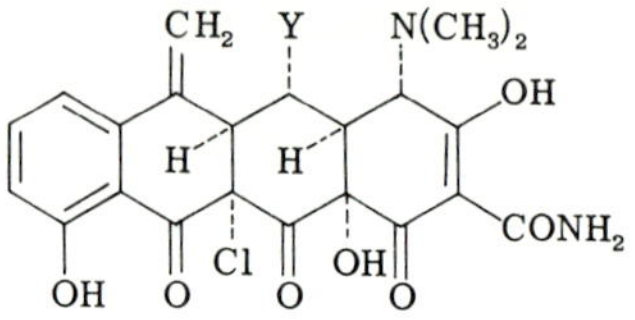

(72) Y = H
(73) Y = OH

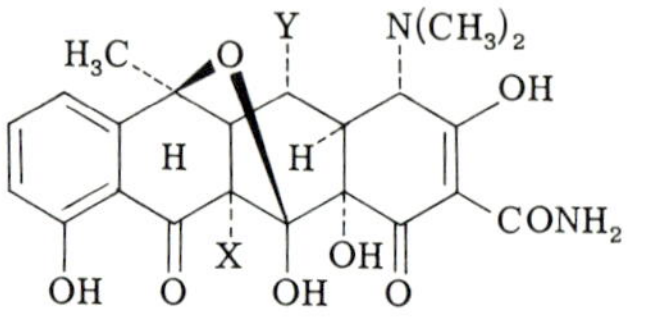

(74) X = Cl, Y = H
(75) X = Cl, Y = OH
(76) X = F, Y = H
(77) X = F, Y = OH

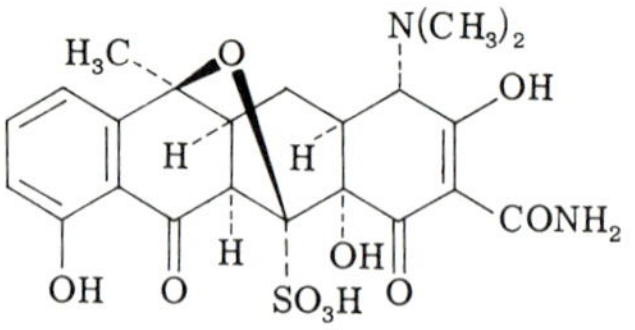

(78)

H. Modifications at the C-12a Position (Esters)

The activity of 12a-deoxytetracycline (**18**) and 4a,12a-anhydrotetracycline (**17**) have already been discussed in connection with a chromophoric requirements for activity (Table III). The activity of tetracyclines with other modifications at the C-12a-position are summarized in Table X. The O^{12a}-formate of tetracycline (**79**) is readily hydrolyzed to parent tetracycline, and so is found to have activity approximately equivalent to that of tetracycline (cf. 3 in Table II). A series of mono and diesters of oxytetracycline (**80–83**) show greatly depressed activity. The minor activity noted may well be due to partial hydrolysis of the esters. The requirement for C-12a stereochemistry is indicated by the lack of activity of the 12a-epidedimethylamino derivative (**84**) in comparison to corresponding compound retaining the normal configuration (**27**, cf. Table V).

III. Electronic and Lipophilic Effects on Tetracycline Activity

The effect of lipophilicity of tetracyclines on their pharmacodynamic properties in dogs has been discussed in detail, with several parameters correlated with the distribution coefficients (K_D) of various tetracyclines between aqueous buffers and chloroform (Schach von Wittenau and Yeary, 1963; Schach von Wittenau and Delahunt,

TABLE X
ANTIBACTERIAL ACTIVITY OF TETRACYCLINES MODIFIED AT THE C-12a POSITION

| | | Minimum inhibitory concentration (μg/ml) | | | | | | | Protective dose 50% values (mg/kg) | | | | | |
| | | | | | | | | | S. pyogenes | | S. aureus 5 | | P. multocida | |
Structure	Name	S. aureus 5	S. aureus (tetracycline resistant)	S. pyogenes	E. coli	A. aerogenes	S. typhosa	K. pneumoniae	Oral	Subcut.	Oral	Subcut.	Oral	Subcut.
79	O^{12a}-Formyltetracycline[a]	0.2	—	0.2	1.5	1.5	0.8	0.8	3	3	3	3	—	—
80	O^{12a}-Acetyloxytetracycline[b]	3	>200	3	25	50	25	25	36	12	—	>200[e]	>50	>50
81	O^{12a}-Propionyloxytetracycline[b]	3	>100	0.8	100	25	25	6	>50	11	>50 >	12.5	>50	>50
82	O^5,O^{12a}-Diacetyloxytetracycline[c]	3	>200	3	25	50	25	25	>100	>100	—	—	—	—
83	O^{10},O^{12a}-Diacetyloxytetracycline[b]	3	>100	0.8	25	100	25	6	>50	15	—	—	>50	>50
84	12a-Epi-4-dedimethylaminooxytetracycline[d]	>100	—	—	>100	>100	>100	>100	—	—	—	—	—	—

[a] Blackwood et al., 1960.
[b] Blackwood et al., 1962.
[c] Hochstein et al., 1953.
[d] M. Schach von Wittenau et al., 1965.
[e] 40% protection at this level.

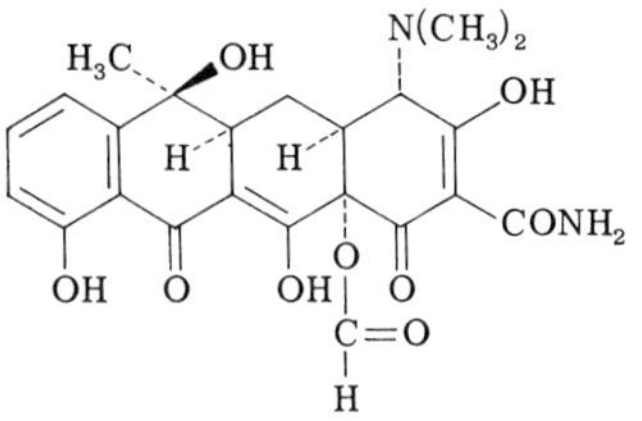

(79)

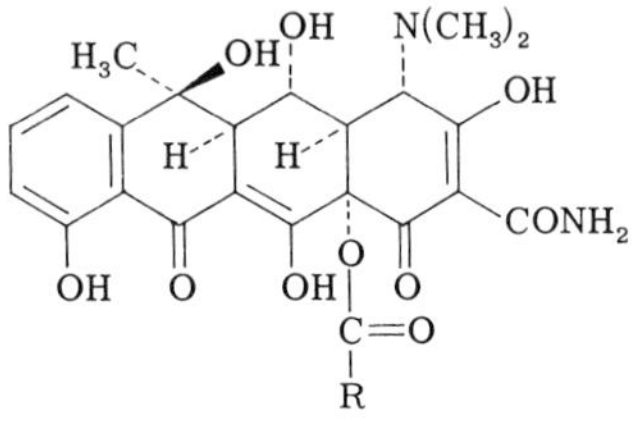

(80) R = CH₃
(81) R = C₂H₅

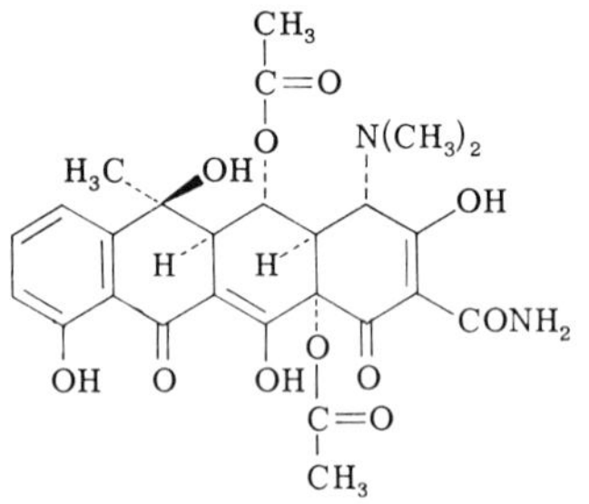

(82)

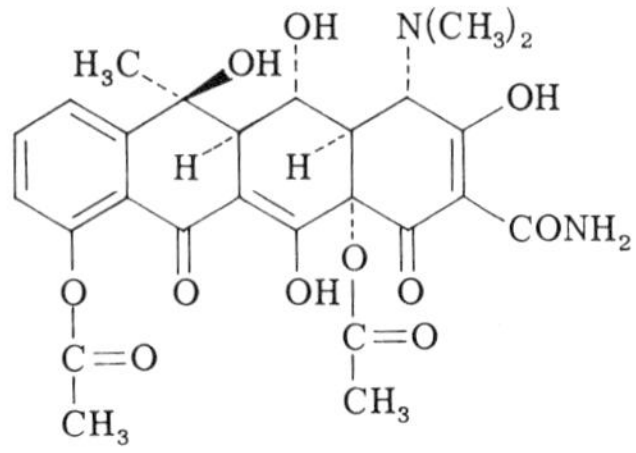

(83)

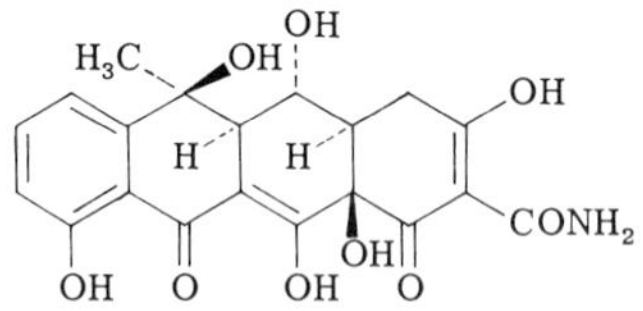

(84)

1965; Schach von Wittenau, 1969). Distribution coefficients reported there have been extended, by the same methods, to additional tetracycline derivatives at pH 7.0. The data are summarized as log K_D values in Table XI. The tetracyclines in this table are placed in the order from lowest lipophilicity (most polar) to highest lipophilicity (least polar).

Frequent reference has been made to the effect of electron-withdrawing groups at the C-7 position on enhancement of *in vitro* activity. We suggest that the enhanced *in vitro* activity of 6-methylenetetracycline may result from a similar electronic effect. In Table XI, tetracyclines having a electronic factor equivalent to tetracycline are listed as zero. 7-Chloro or 6-methylene substitution is listed as −1 (electron withdrawing). 7-Nitro or combined 7-chloro and 6-methylene are listed as −2 (strongly electron withdrawing). 7-Amino substitution is listed as +1 (electron donating) and 7-dimethylamino substitution listed as +2 (strongly electron donating). These relative assignments of electronic factor for halogen and amine substitution are fully consistent with substituent constants employed in the Hammett equation (for review, see Barlin and Perrin, 1966). For the sake of clarity, tetracyclines having a zero electronic factor are underlined and the factor is repeated twice in the table. Various aspects of *in vitro* and *in vivo* activity are now considered in light of these parameters.

A. *In Vitro* GRAM-POSITIVE ACTIVITY (TETRACYCLINE SENSITIVE)

In Table XI, *in vitro* activity against tetracycline sensitive gram-positive bacteria is exemplified by bioassay and MIC values against *Staphylococcus aureus* and *Streptococcus pyogenes*. It will be noted that over the entire spectrum of lipophilicity, both bioassay and MIC values stay within a relatively narrow range, indicating that lipophilicity is of no great importance as a factor effecting the *in vitro* activity of tetracyclines against sensitive gram-positive bacteria. An electronic factor is noted, however, in particular enhancement of *in vitro* activity by electron-withdrawing substituents. Activity is depressed with the electron-donating 7-amino substituent (63). However, 7-dimethylamino substitution (64) is an exception. This phenomenon, which may relate to charge on bacteria, has already been discussed above in connection with tetracyclines modified at the C-7 and C-9 positions.

B. *In Vitro* GRAM-POSITIVE ACTIVITY (TETRACYCLINE RESISTANT)

In Table XI, *in vitro* activity against tetracycline resistant bacteria is exemplified by MIC values against a tetracycline-resistant *S. aureus*. The large factor of lipophilicity is evident; as lipophilicity is increased, the MIC against resistant *Staphylococcus* approaches and becomes equal to the MIC against sensitive *Staphylococcus*. Unique *in vivo* activity against tetracycline resistant staphylococci has been reported for minocycline (64, Martell and Boothe, 1967).

C. *In Vitro* GRAM-NEGATIVE ACTIVITY

In contrast to tetracycline-sensitive gram-positive bacteria, gram-negative activity is greatly influenced by the lipophilicity of a tetracycline. The gram-negative bacteria are exemplified in Table XI, as well as earlier tables, by *Escherichia coli, Aerobacter aerogenes, Salmonella typhosa,* and *Klebsiella pneumoniae*. A general trend to

TABLE XI
Electronic and Lipophilic Effects on the Antibacterial Activity of Tetracycline

Partial structure[a]	CHCl$_3$; pH, 7.0 Buffer	Electronic factor	Bioassay (relative to tetracycline) K. pneumoniae	S. aureus	S. aureus	Minimum inhibitory concentrations (μg/ml) S. aureus (Tetracycline resistant)	S. pyogenes	E. coli	A. aerogenes	S. typhosa	K. pneumoniae	Protective dose % values (mg/kg) S. aureus Oral	Subcut.	Electronic factor
2	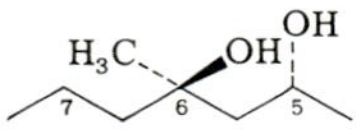 1.91[b]	0	1.0	0.8	0.55	>100	0.07	1.09	2.91	3.12	2.5	7.2	2.6	0
44	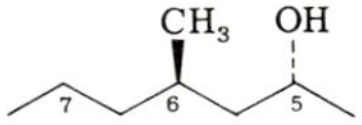 1.37[c]	0	0.4	0.5	0.8	200	0.1	12	12	6	6	22	—	0
3	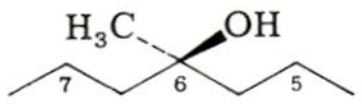 1.32[b]	0	1.0	1.0	0.21	60	0.06	0.73	1.66	1.56	1.56	5.81	1.23	0
6	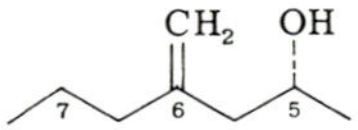 0.93[b]	−1	2.3	—	0.13	50	0.03	0.43	0.94	1.40	1.56	4.50	1.19	−1
4	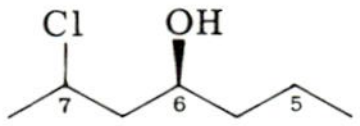 0.83[b]	−1	—	3.0	0.11	35	0.04	0.33	0.1	0.55	0.55	6.00	1.63	−1
46	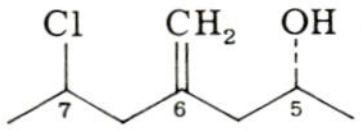 0.69[c]	−2	6.3	—	0.15	100	0.01	0.8	1.5	1.5	0.4	26	2.2	−2

STRUCTURE–ACTIVITY RELATIONSHIPS IN THE TETRACYCLINE SERIES 263

No.	Substituent													
63	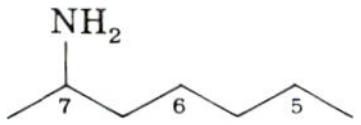 NH$_2$ at 7	0.29^d	+1	1.0	0.4	1.5	>100	0.2	0.8	3	6	3	20	3	+1
60	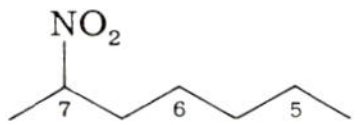 NO$_2$ at 7	0.20^d	−2	4.6	6.4	0.2	12.5	0.04	0.15	0.07	0.39	0.76	18	4	−2
7	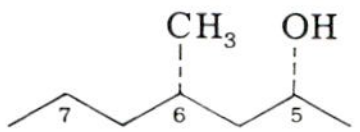 CH$_3$, OH at 6	0.20^b	0	1.4	—	0.19	8.75	0.04	1.74	2.54	1.56	1.87	2.55	1.00	0
9	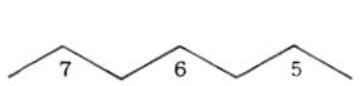 (6-deoxy)	−0.89^b	0	0.9	1.6	0.2	1.5	0.2	1.5	1.5	1.5	0.1	5	1.5	0
45	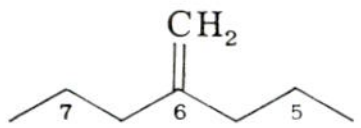 =CH$_2$ at 6	−1.22^c	+1	1.2	—	0.2	6.3	0.2	3	6	3	3	11	9.3	+1
64	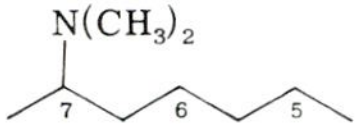 N(CH$_3$)$_2$	−1.60^d	+2	—	>1.0	0.1	0.8	0.02	3	3	3	1.5	3.5	—	+2
49	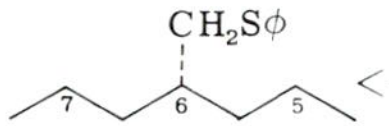 CH$_2$Sϕ	<−2.0^e	0	0.01	—	0.8	0.8	0.2	>100	>100	30	>100	>25	>25	0
24	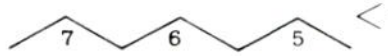 (N^2-t-butyl)	<−2.0^e	0	—	—	0.7	1.5	1	>100	>100	>100	>100	>100	>50	0

^{a} Except where specified (24), substitution is otherwise identical to tetracycline.
^{b} Schach von Wittenau and Yeary, 1963.
^{c} Schach von Wittenau, 1969.
^{d} Determined by the method of Schach von Wittenau and Yeary, 1963.
^{e} The high distribution of these tetracyclines into CHCl$_3$ precludes accurate determination of K_D.

decreased gram-negative activity, as lipophilicity increases, can be seen by examination of Table IX. In the extremely lipophilic tetracyclines (**49** and **24**), gram-negative activity is essentially lost. Superimposed is an electronic effect on *in vitro* activity, such that electron-withdrawing substituents (factor = −1 or −2) enhance activity, while electron-withdrawing substituents (factor = +1 or +2) reduce *in vitro* activity.

D. *In Vivo Staphylococcus aureus* ACTIVITY

In vivo activity of tetracyclines is not readily predictable from *in vitro* activity. As a result of important pharmacodynamic factors, *in vivo* activity correlates very differently with electronic and lipophilic factors. Examination of the *in vivo* data in Table XI demonstrates that an optimum lipophilicity exists for *in vivo* activity. Thus compare the activity of doxycycline (**7**, log K_D 0.20) with other tetracyclines having zero electronic factor. This conclusion is fully consistent with pharmacodynamic studies in dogs (Schach von Wittenau, 1969; Schach von Wittenau and Yeary, 1963; Schach von Wittenau and Delahunt, 1965) in which the same lipophilic factor has been correlated with tissue binding, free drug levels, rate of excretion, serum half-lives, etc. The same phenomenon is also reflected in more detailed mouse studies (English and Lynch, 1967; English, 1967, 1968). For example, in a series of four tetracyclines of intermediate lipophilicity, with a protocol involving drug dosage at various times prior to infection, (doxycycline, **7**) protected mice even when dosed up to 8 hours before infection.

In sharp contrast to the electronic effect on *in vitro* activity is a possible electronic effect on *in vivo* activity. Thus strongly electron withdrawing substituents (factor = −2, **46** and **60**) appear to depress *in vivo* activity, while a strongly electron-donating subsituent (factor = +2, **64**) enhances *in vivo* activity. These generalizations are particularly true of oral activity. 7-Amino-6-demethyl-6-deoxytetracycline (**63**, factor = +1) appears exceptional. We might speculate that its poor *in vivo* activity reflects metabolism to an inactive compound (e.g., oxidation to a quinone in the D-ring).

ACKNOWLEDGMENT

The authors wish to thank Drs. C. R. Stephens, L. H. Conover, M. Schach von Wittenau, J. J. Beereboom, H. H. Rennhard, J. J. Korst, and K. Butler for many stimulating discussions of structure–activity relationships in the tetracycline series. We are especially appreciative for permission from Dr. M. Schach von Wittenau to publish additional chloroform-pH 7.0 buffer distribution data from his files and to Mr. N. Belcher and his co-workers for having maintained excellent central files of biological screening information.

REFERENCES

Barlin, G. B., and Perrin, D. D. (1966). *Quart. Rev.* **20**, 75.
Barrett, G. C. (1963). *J. Pharm. Sci.* **52**, 309.
Beereboom, J. J., Ursprung, J. J., Rennhard, H. H., and Stephens, C. R. (1960). *J. Amer. Chem. Soc.* **82**, 1003.
Blackwood, R. K. (1969a). *In* "Kirk-Othmer Encyclopedia of Chemical Technology," 2nd ed., Vol. 18 or 19, in press. Wiley (Interscience), New York.
Blackwood, R. K. (1969b). To be published.
Blackwood, R. K., and Stephens, C. R. (1962). U. S. Patent 3,047,617 (assigned to Chas. Pfizer and Co., Inc.).
Blackwood, R. K., and Stephens, C. R. (1965). *Can. J. Chem.* **43**, 1382.
Blackwood, R. K., Rennhard, H. H., and Stephens, C. R. (1960). *J. Amer. Chem. Soc.* **82**, 5194.
Blackwood, R. K., Beereboom, J. J., Rennhard, H. H., Schach von Wittenau, M., and Stephens, C. R. (1963). *J. Amer. Chem. Soc.* **85**, 3943.
Boothe, J. H., Morton, J., Petisi, J. P., Wilkinson, R. G., and Williams, J. H. (1953). "Antibiotics Annual, 1953–1954," p. 47. Medical Encyclopedia, Inc., New York.
Boothe, J. H., Bonvicino, G. E., Waller, C. W., Petisi, J. P., Wilkinson, R. W., and Broshard, R. B. (1958). *J. Amer. Chem. Soc.* **80**, 1654.
Clive, D. L. (1968). *Quart. Rev.* **23**, 435.
Conover, L. H. (1956). *Symp. on Antibiot. Mold Metabolites, Spec. Publ. No.* **5**. The Chemical Society, London.
Conover, L. H., Butler, K., Johnson, J. D., Korst, J. J., and Woodard, R. B. (1962). *J. Amer. Chem. Soc.* **84**, 3222.
Doerschuk, A. P., Bitler, B. A., and McCormick, J. R. D. (1955). *J. Amer. Chem. Soc.* **77**, 4687.
English, A. R. (1966). *Proc. Soc. Exp. Biol. Med.* **122**, 1107.
English, A. R. (1967). *Proc. Soc. Exp. Biol. Med.* **126**, 487.
English, A. R. (1968). *Proc. Soc. Exp. Biol. Med.* **128**, 333.
English, A. R., and Lynch, J. E. (1967). *Proc. Soc. Exp. Biol. Med.* **124**, 586.
Esse, R. C., Lowery, J. A., Tamorria, C. R., and Sieger, G. M. (1964). *J. Amer. Chem. Soc.* **86**, 3877.
Gottstein, W. J., Minor, W. F., and Cheney, L. C. (1959). *J. Amer. Chem. Soc.* **81**, 1198.
Gradnik, B., Pedrazzoli, A., and Ferrero, E. (1960). *Pharm. Acta Helv.* **35**, 529.
Green, A., and Boothe, J. H. (1960). *J. Amer. Chem. Soc.* **82**, 3950.
Hlavka, J. J., Krazinski, H., and Boothe, J. H. (1962a). *J. Org. Chem.* **27**, 3674.
Hlavka, J. J., Schneller, A., Krazinski, H., and Boothe, J. H. (1962b). *J. Amer. Chem. Soc.* **84**, 1426.
Hochstein, F. A., Stephens, C. R., Conover, L. H., Regna, P. P., Pasternack, R., Gordon, P. N., Pilgrim, F. J., Brunings, K. J., and Woodward, R. B. (1953). *J. Amer. Chem. Soc.* **75**, 5455.
Hochstein, F. A., Schach von Wittenau, M., Tanner, F. W., and Murai, K. (1960). *J. Amer. Chem. Soc.* **82**, 5934.
Koschel, K., Hartmann, G., Kersten, W., and Kersten, H. (1966). *Biochem. Z.* **344**, 761.

McCormick, J. R. D. (1968). *In* "The Biogenesis of Antibiotic Substances," (Z. Vanek and Z. Hostalek, eds.), p. 83. Academic Press, New York.
McCormick, J. R. D., Miller, P. A., Growich, J. A., Sjolander, N. O., and Doerschuk, A. P. (1958). *J. Amer. Chem. Soc.* **80,** 5572.
McCormick, J. R. D., Jensen, E. R., Miller, P. A., and Doerschuk, A. P. (1960). *J. Amer. Chem. Soc.* **82,** 3381.
Martell, M. J., and Boothe, J. H. (1967). *J. Med. Chem.* **10,** 44.
Martell, M. J., Ross, A. S., and Boothe, J. H. (1967a). *J. Med. Chem.* **10,** 485.
Martell, M. J., Ross, A. S., and Boothe, J. H. (1967b). *J. Amer. Chem. Soc.* **89,** 6780.
Miller, M. W., and Hochstein, F. A. (1962). *J. Org. Chem.* **27,** 2525.
Petisi, J., Spencer, J. L., Hlavka, J. J., and Boothe, J. H. (1962). *J. Med. Pharm. Chem.* **5,** 538.
Redin, G. S. (1966). "Antibiotics and Chemotherapy—1966," (G. L. Hobby, ed.), p. 371. American Society for Microbiology, Ann Arbor, Michigan.
Schach von Wittenau, M. (1969). *Proc. 4th Int. Conf. Pharmacol., July, 1969, Basel, Switzerland*, in press.
Schach von Wittenau, M., and Delahunt, C. S. (1965). *J. Pharmacol. Exp. Therap.* **152,** 164.
Schach von Wittenau, M., and Yeary, R. (1963). *J. Pharmacol. Exp. Therap.* **140,** 258.
Schach von Wittenau, M., Hochstein, F. A., and Stephens, C. R. (1963). *J. Org. Chem.* **28,** 2454.
Schach von Wittenau, M., Blackwood, R. K., Conover, L. H., Glauert, R. H., and Woodward, R. B. (1965). *J. Amer. Chem. Soc.* **87,** 134.
Seidel, W., Soder, A., and Linder, F. (1958). *Muench. Med. Wochenschr.* **17,** 661.
Stephens, C. R. (1962). U. S. Patent 3,028,409 (assigned to Chas. Pfizer and Co., Inc.).
Stephens, C. R., Conover, L. H., Pasternack, R., Hochstein, F. A., Moreland, W. T., Regna, P. P., Pilgrim, F. J., Brunings, K. J., and Woodward, R. B. (1954). *J. Amer. Chem. Soc.* **76,** 3568.
Stephens, C. R., Bianco, E. J., and Pilgrim, F. J. (1955). *J. Amer. Chem. Soc.* **77,** 1701.
Stephens, C. R., Conover, L. H., Gordon, P. N., Pennington, F. C., Wagner, R. L., Brunings, K. J., and Pilgrim, F. J. (1956). *J. Amer. Chem. Soc.* **78,** 1515.
Stephens, C. R., Beereboom, J. J., Rennhard, H. H., Gordon, P. N., Murai, K., Blackwood, R. K., and Schach von Wittenau, M. (1963). *J. Amer. Chem. Soc.* **85,** 2643.
Tamorria, C. R., and Esse, R. C. (1965). *J. Med. Chem.* **8,** 870.
Waller, C. W., Hutchings, B. L., Broschard, R. W., Goldman, A. A., Stein, W. J., Wolf, C. F., and Williams, J. H. (1952). *J. Amer. Chem. Soc.* **74,** 4981.
Weibull, C. (1960). *In* "The Bacteria: A Treatise on Structure and Function" (I. C. Gunsalas and R. Y. Stanier, eds.), Vol. I. p. 158. Academic Press, New York.

Microbial Production of Phenazines

J. M. INGRAM AND A. C. BLACKWOOD

Department of Microbiology,
Macdonald College of McGill University,
Quebec, Canada

I.	Introduction	267
II.	Pyocyanine	267
III.	Chlororaphine	270
IV.	Phenazine-1-Carboxylic Acid	272
V.	Iodinin	274
VI.	Other Phenazines	275
VII.	Discussion and Possible Future Studies	279
	References	280

I. Introduction

The natural occurrence of phenazine pigments has been known since late in the nineteenth century. All of these pigments contain the same basic structure whose nucleus and numbering system is outlined in Fig. 1. The description of increased numbers of these phena-

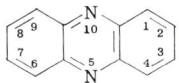

FIG. 1. Structure and numbering system of the phenazine nucleus.

zine structures and of microorganisms known to produce these derivatives is the product of relatively recent research. The present review is an attempt to collate the available information concerning the production and the biosynthetic pathways of the various phenazine pigments.

II. Pyocyanine

Pyocyanine is known widely as the blue pigment produced by some strains of *Pseudomonas aeruginosa* under certain cultural conditions and may be regarded as the parent of phenazine pigments since it was described first and undoubtedly is the most thoroughly studied. Pyocyanine has been studied also because of its antibiotic properties and because of its occurrence in pathogenic strains of the organism. Recent reviews by Caltrider (1967) and MacDonald (1967) on the mechanism of action and on the biosynthesis of pyocyanine, respectively, summarize most of the data found in the literature. The pigment was first isolated by Gessard (1890) following the growth of *P. aeruginosa* on complex media. The original structure proposed by Wrede and Strack (1929) was later shown by Hilleman (1938) to be double the now accepted formula which is illustrated in Fig. 2.

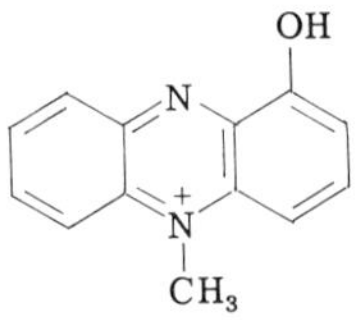

FIG. 2. Structure of pyocyanine.

The first documented evidence concerning the production of pyocyanine in a chemically defined medium was that of Jordan (1899). He demonstrated the ability of some cultures to produce the pigment when grown on a mixture of salts plus asparagine or plus the ammonium salts of succinic, lactic, acetic, or citric acids. The description of additional chemically defined media was initiated by the studies of Burton et al. (1947). Since that time many media, usually modifications of that mentioned above, have been described in an effort to optimize the amount of pigment produced. The media usually contain glycerol as a carbon source, leucine and/or alanine as nitrogen sources, and Fe^{+2}, Mg^{+2}, SO_4^{-2}, and PO_4^{-3} although the level of phosphate is critical and must be kept low to ensure maximum production of pyocyanine. A summary of the specific media used for pyocyanine production as well as various cultural conditions has been tabulated by MacDonald (1967).

Recent reports which appeared subsequent to those discussed above confirmed and extended observations previously noted. Studies by Vallette and co-workers (1966) with strain A-237 of *P. aeruginosa* show that optimum pyocyanine production occurs at a concentration of 0.08 M phosphate when the organism is cultured on a synthetic medium of succinic acid and ammonium chloride. This phosphate concentration appears to be higher than that found optimal for the ATCC 9027 strain (MacDonald, 1967). In addition, these workers found that methyl phosphate, as the phosphorus source, yielded approximately twice the amount of pyocyanine as when orthophosphate was supplied; however, the absolute concentrations of the pigment produced were not stated. The fact that methyl phosphate yields higher concentrations of pyocyanine seems to substantiate the conclusions of Ingledew and Campbell (1969b) who state that phosphate deficiency may be the trigger for phenazine synthesis. These authors developed a resuspension medium to follow the competition of shikimic and quinic acids on the incorporation of ^{14}C-labeled 2-ketogluconate into pyocyanine by a mutant which was unable to use these acids for growth. In a resuspension medium devoid of organic phosphate, pyocyanine production by the mutant was increased approximately 30% over that of the parent strain when 2-ketogluconate was the carbon source. In the later stages of pigment production free inorganic phosphate appeared in the resuspension medium. This was attributed to ribosome degradation which is known to occur when *P. aeruginosa* is subjected to starvation conditions (Hou *et al.*, 1966).

The previous discussions concerning the media for the production of pyocyanine are based upon the fact that cell growth occurs to some extent. The production of pyocyanine in resting cell suspensions has been studied by Grossowicz *et al.* (1957); Frank and De Moss (1959); Halpern *et al.* (1962); and Kurachi (1959); these studies are discussed and evaluated by MacDonald (1967). Very little has been done with resting cells or near-resting cell suspensions since 1962. Even in the resuspension system of Ingledew and Campbell (1969b), growth, as measured by optical density at 490 mμ, occurred while the production of pyocyanine proceeded.

The effect of inhibitors upon the production of pyocyanine has been studied by various workers. Frank and De Moss (1959) state that those inhibitors which produce useful inhibition of growth also inhibit the production of pigment. These workers studied the effects of potassium cyanide, sodium azide, sodium arsenate, and chloramphenicol upon growth and pigment production. Schneierson and co-workers (1960), however, found that of four strains of *P. aeruginosa* treated with subinhibitory concentrations of chloramphenicol, three exhibited a complete and permanent loss in the capacity to produce pyocyanine and no effect on growth was noted in these tests. Erythromycin at 50 μg per ml also eliminated pigment production in all four strains while apparently having no effect upon growth.

The biosynthesis of pyocyanine from labeled substrates has been thoroughly reviewed by MacDonald (1967). In general, glycerol is one of the most efficient sources of pyocyanine carbon. In most of these studies, however, the percentage of carbon label finally incorporated into the pigment was only a small proportion of the total carbon utilized. Additional studies by Millican (1962) and MacDonald

(1963) using a direct label incorporation, or isotope dilution showed that quinic or shikimic acids serve as good sources of pyocyanine carbon. However, in each study cell growth occurred and presumably the competitive carbon source was metabolized subsequent to incorporation into pyocyanine.

The problem of precursor metabolism prior to incorporation into pigments was circumvented as described in an elegant study by Ingledew and Campbell (1969a). These authors prepared a mutant Q^- which was unable to oxidize and, hence, unable to metabolize shikimic and quinic acids. When cultures of the wild type W^+ were added to a resuspension medium containing 2-ketogluconate-^{14}C as the initial sole source of carbon, it was found by isotope dilution that 9% of the pyocyanine carbon was derived from shikimic acid. However, when the same experiment was repeated, with the Q^- strain, 98% of the pyocyanine ring carbon was derived from shikimic acid whereas 40% was derived from quinic acid. These authors also reported studies to rule out the possibility that quinic acid is a direct precursor of pyocyanine carbon. Their data strongly suggest that quinic acid is converted to shikimic acid which is then incorporated into the pyocyanine molecule. Their conclusions were tempered by the statement that anthranilic or chorismic acids cannot entirely be ruled out as precursors by these competition experiments alone since these intermediates may be enzyme-bound *in vivo* and unavailable for equilibration with the exogenously added substrates. In any event, the nature of the steps involved in nitrogen donation to shikimic acid remains to be established.

III. Chlororaphine

Guignard and Sauvageau (1894) isolated and described two pigments produced by *Bacillus chlororaphis (Pseudomonas chlororaphis)*. Later studies by Kögl and Postowsky (1930) showed the two pigments to be reduced chlororaphine (green) and oxidized oxychlororaphine (yellow). The structure of the latter is given in Fig. 3. The exact structure of chlororaphine remains in dispute but it is

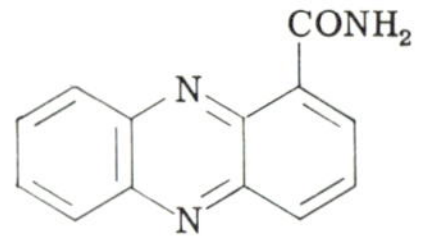

FIG. 3. Structure of oxychlororaphine.

generally agreed that it is the dihydroderivative of oxychlororaphine. Oxychlororaphine is produced by certain cultures of *P. chlororaphis* and *P. aeruginosa*. The pigment was first produced in the synthetic medium of glycerol and asparagine plus salts as described by Mercier and Lasseur (1911), but later workers reverted to a glycerol plus peptone culture medium. In studies with *P. aeruginosa*, Takeda and Nakanishi (1959) followed production in a glycerol, urea, and salts medium. In a comparative study concerning the simultaneous production of pyocyanine, phenazine-1-carboxylic acid, and oxychlororaphine by *P. aeruginosa*, strain Mac 436, Chang and Blackwood (1969) evaluated several media for optimum production of the three respective pigments. Maximum production of all three as well as maximum production of oxychlororaphine occurred in the medium of Frank and De Moss (1959).

Chlororaphine is obtained from culture filtrates by filtration, centrifugation, or by ether extraction. Mixtures of chlororaphine and oxychlororaphine are resolved by fractional crystallization from acetone in the absence of atmospheric oxygen since under these conditions, oxychlororaphine is much less soluble than chlororaphine. The further purification of oxychlororaphine is achieved by column chromatography on alumina, sublimation at about 210°C, or by repeated crystallization (Swan and Felton, 1957).

Studies on the biosynthesis of oxychlororaphine at present are less numerous than those available for pyocyanine. Early work by Kögl *et al.* (1932) demonstrated that the addition of phenazine-1-carboxylic acid, picolinic acid, pyrocatechol or pyrocatechol-3-carboxylic acid stimulated the production of oxychlororaphine by cultures of *P. chlororaphis*. Takeda and Nakanishi (1959) later showed that the addition of D,L-γ-aminobutyric or L-aspartic acids increased the yield of oxychlororaphine by cultures of *P. aeruginosa*. P. Bechard and A. C. Blackwood (unpublished observations, 1970) have also shown that the addition of phenazine-1-carboxylic acid to cultures of *P. aeruginosa* Mac 436 stimulates the production of oxychlororaphine and the increase in oxychlororaphine is proportional to the amount of phenazine-1-carboxylic acid added. In addition, phenazine-1-carboxylic-^{14}C acid is converted to oxychlororaphine-^{14}C whose specific activity is approximately equal to the specific activity of the added phenazine-1-carboxylic acid. In a study by Carter and Richards (1961) on the biogenesis of oxychlororaphine, carboxy-labeled anthranilic acid was incorporated to the extent of 0.002%. Of this quantity, approximately 80% was found in the carboxamide group while 20%

appeared in the phenazine moiety. On the basis of the incorporation of alanine-1-^{14}C and alanine-2-^{14}C, as well as the incorporation pattern of anthranilic acid carboxyl-^{14}C, these authors concluded that the shikimic acid pathway was responsible for the synthesis of the phenazine ring. Chang and Blackwood (1968) studied the simultaneous production of pyocyanine and oxychlororaphine by *P. aeruginosa* Mac 436 and found that ^{14}C was incorporated maximally into both pigments from glycerol-1,3-^{14}C and glycerol-2-^{14}C. Shikimic acid, although not as efficient a carbon donor as glycerol, was superior to glucose-^{14}C, succinate-^{14}C, or acetate-^{14}C. The specific activity of the two pigments following synthesis in the presence of glycerol-1,3-^{14}C, glycerol-2-^{14}C, or shikimic-1,6-^{14}C acid was essentially the same. These authors concluded that the route of synthesis of the two pigments was similar and that the shikimic acid rather than the acetate pathway was involved.

IV. Phenazine-1-Carboxylic Acid

Phenazine-1-carboxylic acid, whose structure is illustrated in Fig. 4, was first described in simultaneous publications by Kluyver (1956) and Haynes *et al.* (1956). The organism, *Pseudomonas aureofaciens*,

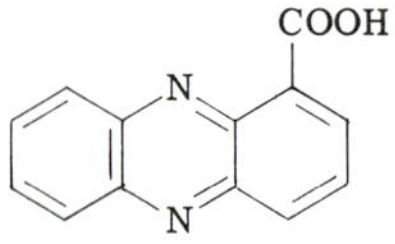

FIG. 4. Structure of phenazine-1-carboxylic acid.

was isolated from clay and observed to produce a yellow pigment. Takeda (1958) later demonstrated production of phenazine-1-carboxylic acid by *P. aeruginosa*. Isono *et al.* (1958) reported the production of the pigment by cultures of *Streptomyces misakiensis*. Toohey *et al.* (1965) isolated and characterized phenazine-1-carboxylic acid as one of the causative antibiotic substances of "barren ring." *Pseudomonas aureofaciens* was cultured on a complex medium containing glycerol, peptone or yeast extract, and salts by Kluyver (1956), Haynes *et al.* (1956) and Toohey *et al.* (1965). Takeda (1958) later simplified the medium by substituting urea as a nitrogen source. Levitch and Stadtman (1964) in a biosynthetic study of phenazine-1-carboxylic acid, described an incubation mixture which was composed of 0.2 M glycerol, 0.02 M lysine, and dimethyl glutarate buffer. In a study on the simultaneous production of pyocyanine, phenazine-1-carboxylic acid, and oxychlororaphine, Chang and Blackwood (1969) found that phenazine-1-carboxylic acid was produced in good yield by *P. aeruginosa* Mac 436 in the medium of Frank and De Moss (1959).

Phenazine-1-carboxylic acid is extractable into organic solvents following acidification of the fermentation broth. The crude acid was further purified by alumina column chromatography by Kluyver (1956) and Isono *et al.* (1958) or by paper chromatography by Toohey *et al.* (1965). Haynes *et al.* (1956) isolated the acid following carbon dioxide sparging. Levitch and Stadtman (1964) purified phenazine carboxylic acid following repeated crystallizations from 95% ethyl alcohol. Chang and Blackwood (1968) isolated pure phenazine-1-carboxylic acid free of pyocyanine and oxychlororaphine following extraction into $CHCl_3$ and subsequent silica gel G thin-layer chromatography employing $CHCl_3$, methanol, and acetone as the developing solvents.

Studies on the biosynthesis of phenazine carboxylic acid by Takeda and Nakanishi (1959) showed that pigment production is stimulated on the addition of L-cysteine, D,L-tryptophan or D,L-tyrosine to the fermentation medium. Levitch and Stadtman (1964) employing an incubation mixture for the production of acid by *P. aureofaciens* showed that glycerol was the carbon source of preference as glucose, fructose, ribose, erythrose, lactic, pyruvic, succinic, and gluconic acids yielded little or no pigment. Pigment production in the presence of labeled substrates showed that glycerol-1,3-^{14}C or glycerol-2-^{14}C were at least 300% more efficient in the donation of carbon than shikimic acid or threonine. It was assumed that the incorporation of shikimic-U-^{3}H, as used in this study, reflected the incorporation of shikimic acid carbon. The carbon contributions of acetic, succinic, and pyruvic acids as well as alanine or histidine were less than 10% of that contributed by glycerol. Isotopic competition experiments between unlabeled shikimic acid and glycerol-1,3-^{14}C showed that shikimic acid depressed the radioactivity contribution from glycerol-^{14}C by 70%. These authors suggested from these experiments that perhaps one or both rings of phenazine-1-carboxylic acid were derived from the shikimic acid pathway. Studies on the simultaneous production of pyocyanine, phenazine-1-carboxylic acid, and oxychlororaphine by *P. aeruginosa* Mac 436 (Chang and Blackwood, 1968) showed that the specific activity of the phenazine carboxylic acid was the same as that of pyocyanine or oxychlororaphine when glycerol-1,3-^{14}C, glycerol-2-^{14}C or shikimic-1,6-^{14}C acid were tested as carbon donors for phenazine ring synthesis. These authors concluded that all three pigments were synthesized via the same route which most probably involved the shikimic acid pathway.

V. Iodinin

Iodinin (1,6-phenazinediol 5,10-dioxide) may be regarded as the parent compound of this particular class of phenazines. The structures of the three diolphenazines: iodinin, 1,6-phenazinediol 5-oxide, and 1,6-phenazinediol are shown in Fig. 5. The pigment,

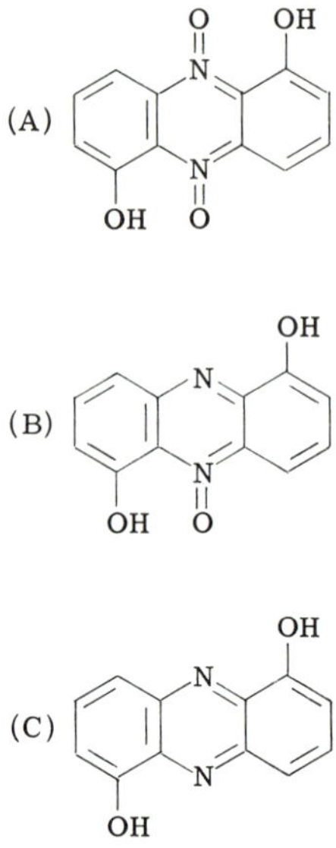

FIG. 5. Structures of (A) 1,6-phenazinediol 5,10-dioxide (iodinin); (B) 1,6-phenazinediol 5-oxide; and (C) 1,6-phenazinediol.

iodinin, was first observed in milk culture and shown by Davis (1939) to be produced by *Chromobacterium iodinum (Pseudomonas iodimum)*. The pigment was later purified by Clemo and McIlwain (1938) and the name iodinin was proposed by McIlwain (1937). Iodinin is known to be produced by various organisms including *Streptomyces thioluteus, Brevibacterium crystalloiodinum, Waksmania aerata, Microbispora aerata*, and the "Malloch" strain of some Norcardiacaeae. The pigment appears as large, deep purple crystals in fermentation broths or in solid agar cultures. The pigment may be washed from the surface of solid cultures and extracted with $CHCl_3$. Repeated crystallizations from $CHCl_3$ solution yields a compound which melts at 236°C (Swan and Felton, 1957). Gerber and Lechevalier (1965) isolated pure iodinin, 1,6-phenazinediol 5-oxide, and 1,6-phenazinediol following paper chromatography of the $CHCl_3$ extract of the fermentation liquor with a developing solvent composed of toluene-ethanol-water. The three diol pigments are produced by cultures of *P. iodinum* when grown on yeast extract, cerelose, and tap water. Following growth on a rotary shaker for 68 hours, 165 mg/liter of iodinin, 12 mg/liter 1,6-phenazinediol 5-oxide, and 3.8 mg/liter 1,6-phenazinediol were isolated. The three pigments were obtained in yields of 1 mg/liter for iodinin and 0.3 mg/liter for both 1,6-phenazinediol 5-oxide and 1,6-phenazinediol following growth of *Streptomyces thioluteus* on a 6% Pablum medium (Gerber and Lechevalier, 1964). Certain strains of *Waksmania aerata* produced iodinin in yields of 100–200 mg/liter following growth for 14 days on aqueous Pablum medium (Gerber and Lechevalier, 1964) and trace amounts of 1,6-phenazinediol were also detected. Similarly, iodinin (no yield reported) and 1,6-phenazinediol 5-oxide in yields of 50 μg/liter were isolated from *Waksmania aerata* following growth on aqueous Pablum.

The biosynthesis of iodinin by *Brevibacterium iodinum* has been studied in some detail by Podojil and Gerber (1967). These authors showed that washed cells added to either saline phosphate or the salts E solution of Frank and De Moss (1959) synthesized iodinin from L-glutamine, L-asparagine, L-glutamic acid, and fumaric acid. D,L-glutamic-3-^{3}H acid, when added to washed cell suspensions, was incorporated into iodinin at a level of 0.11% whereas the addition to a 36-hour growing culture gave incorporations to a level of 0.01%. Uniformly labeled shikimic-^{14}C acid, on the other hand, was incorporated to an extent of approximately 4% of the initial label added. Tritium-labeled 1,6-phenazinediol and 1,6-phenazinediol 5-oxide yielded iodinin labeled to 10 and 15%, respectively, of the initial tritium added. Isotope dilution experiments demonstrated that unlabeled phenazine, 1,6-phenazinediol, phenazine-1,6-dicarboxylic acid, or fumaric acid suppressed the incorporation of tritium from labeled D,L-glutamic acid to degrees varying between 70 and 40%. Similar experiments showed that unlabeled L-phenylalanine depressed the radioactivity incorporated from uniformly labeled shikimic-^{14}C by 58%. The authors concluded that the shikimic acid pathway is most probably involved in the biosynthesis of iodinin and that 1,6-phenazinediol and the 5-oxide are direct precursors. These conclusions corroborated and extended the views put forth in previous studies (Gerber and Lechevalier, 1965; Irie *et al.*, 1960).

VI. Other Phenazines

The groups of pigments just discussed account for most of the information available concerning phenazine production and possible routes of biosynthesis. However, other phenazines have been isolated, characterized, and studied in lesser detail. The monohydroxy phenazines, whose structures are illustrated in Fig. 6, have, in general, been obtained as minor products following cultural and isolation pro-

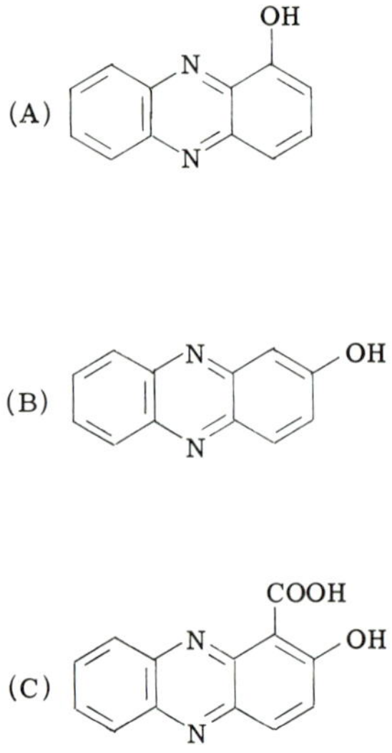

FIG. 6. Structures of (A) 1-hydroxyphenazine; (B) 2-hydroxyphenazine; and (C) 2-hydroxyphenazine-1-carboxylic acid.

cedures of other pigments. It is claimed (Schoental, 1941), that certain cultures of *P. aeruginosa* also produce 1-hydroxyphenazine, the product of pyocyanine demethylation. Toohey and co-workers (1965) described the occurrence of a pigment 2-hydroxyphenazine-1-carboxylic acid which was synthesized by a strain of *P. aureofaciens* during the course of phenazine-1-carboxylic acid production. These workers provided convincing chemical and biological evidence as proof of structure. It was also noted that a mixed culture (composition unspecified) of soil microorganisms converted 2-hydroxyphenazine-1-carboxylic acid to phenazine-1-carboxylic acid. Levitch and Rietz (1966), on the other hand, noted the production of a contaminating pigment following the synthesis of phenazine-1-carboxylic acid by *P. aureofaciens*. Isolation of the pigment from large-scale fermentations and subsequent comparison of its properties with an authentic sample showed it to be identical in all respects to 2-hydroxyphenazine. Preliminary studies on the incorporation of labeled substrates were performed with this pigment. Approximately 0.7% of added glycerol-2-^{14}C carbon was incorporated into the pigment while 2.15% of the added shikimic-U-^{14}C acid was converted to 2-hydroxyphenazine. As suggested by the authors, the results were consistent with those obtained with phenazine-1-carboxylic acid and suggested that the shikimic acid pathway was responsible for phenazine ring synthesis.

Holliman (1957) described the occurrence of two water-soluble pigments, aeruginosins, which were produced by a red strain of *P. aeruginosa*. In a subsequent report (Holliman, 1961) a procedure employing Celite and charcoal chromatography followed by pyridine gradient elution was described for the separation of the two pigments. The preliminary evidence indicated that the two structures were closely related and were probably derivatives of 2-aminophenazine. Herbert and Holliman (1969) subsequently suggested the formula for aeruginosin B as 2-amino-6-carboxy-10-methyl-8-sulfophenazinium betaine and Holliman (1969) established the structure of aeruginosin A as 2-amino-5-methyl-6-carboxyphenazine (see Fig. 7).

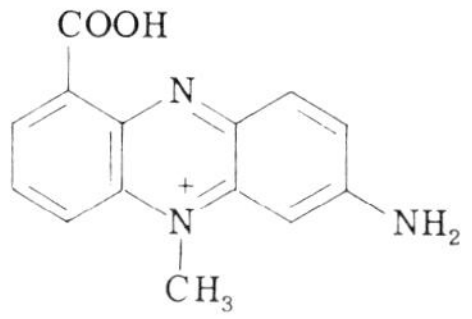

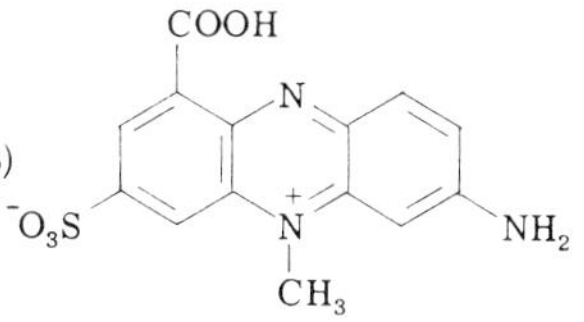

FIG. 7. Structures of (A) aeruginosin A and (B) aeruginosin B.

The production of antibiotic substances by *Streptomyces griseoluteus* was first reported by Umezawa *et al.* (1950) and Okami (1952). The chemical structures of the two antibiotics, griseolutein A and B were later reported by Nakamura *et al.* (1959) and Nakamura *et al.* (1964). Griseolutein A was shown to be 1-methoxy-4-[(hydroxyacetoxy)methyl]-9-carboxyphenazine, and griseolutein B, 1-methoxy-4,6,7,12-tetrahydro-6-hydroxy-6-hydroxymethyl-2H-oxazino [5,4,3-d,e] phenazine-11-carboxylic acid. Yagashita (1960) isolated griseoluteic acid, 1-methoxy-4-hydroxymethyl-9-carboxyphenazine from culture filtrates of *S. griseoluteus*. The structures of the griseoluteins are shown in Fig. 8. This report (Yagashita, 1960) demonstrated

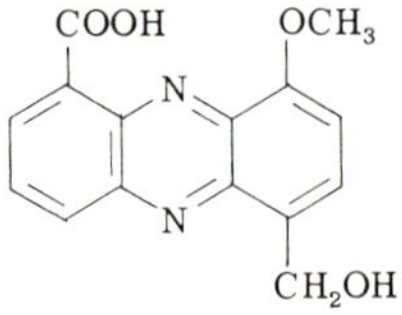

FIG. 8. Structures of (A) griseolutein A; and (B) griseoluteic acid.

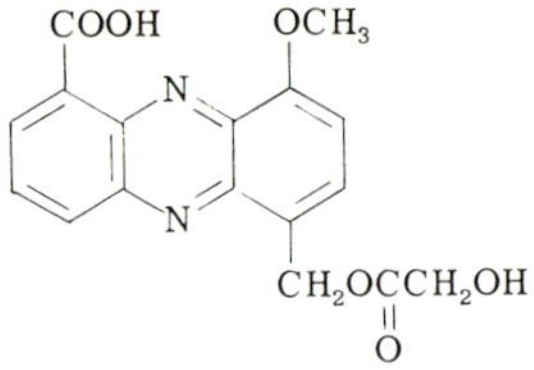

the ability of whole mycelium preparations and sonic extracts to convert griseoluteic acid and glycolic acid to griseolutein A.

One of the most recently described phenazine pigments is myxin. This wide-spectrum antibiotic was first isolated from a strain of *Sorangium* by Peterson *et al.* (1966). The structure of the pigment, shown in Fig. 9, was established as 1-hydroxy-6-methoxyphenazine

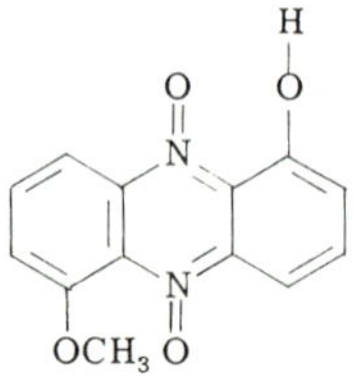

FIG. 9. Structure of myxin.

10-dioxide by Edwards and Gillespie (1966), but later studies by Weigele and Leimgruber (1967) and Hanson (1968) produced a revised structure, 1-hydroxy-6-methoxyphenazine 5,10-dioxide. Myxin is produced by fermentation on glucose, tryptone, and salts and subsequently purified by absorption and elution from Amberlite CG-50 resin. Following acidification of the resin eluate, the pigment was extracted with ethyl ether (Cook *et al.*, 1968). Radioactive myxin has been prepared (Lesley *et al.*, 1967) following growth of *Sorangium* on uniformly labeled glucose. Lesley and Behki (1967) tested the effect of myxin on the synthesis of DNA, RNA, and protein by *Escherichia coli*. The authors concluded that the primary site of myxin action was on cellular DNA followed by a secondary inhibition of RNA and protein synthesis. No studies on the biosynthesis of myxin are available at the present time.

The preceding discussions have, in general, been concerned with chemical descriptions, production procedures, and interconversions of the phenazine pigments. Recent studies, however, have increased the number of phenazine derivatives known to be produced in nature. The descriptions are limited to chemical identifications, and conditions for optimum production are lacking. Additional phenazine derivatives produced by *S. thioluteus* have been described by Gerber (1967), and those produced by an unidentified bacterium by Gerber (1969).

VII. Discussion and Possible Future Studies

A critical assessment of the current state of information regarding the production and biological synthesis of the phenazine pigments is premature at the present time. Specific data are available for individual pigments but generalities encompassing all phenazines for the most part, are unjustified. From the discussions presented, glycerol appears as the carbon source in a chemically defined growth medium that increases the production of most of the pigments although oxychlororaphine synthesis seems to be depressed (Chang and Blackwood, 1969). The production of pyocyanine occurs equally well with 2-ketogluconate as the carbon source under the proper conditions (Ingledew and Campbell, 1969a). With regard to the requirement for inorganic ions, the concentration of phosphate ion appears critical for the optimum production of some phenazine pigments (Frank and De Moss, 1959; Ingram and Blackwood, 1962; Levitch and Stadtman, 1964; Ingledew and Campbell, 1969b). Ingledew and Campbell (1969b) state that phosphate deficiency may be the trigger for pyocyanine synthesis by *P. aeruginosa*. Similar conclusions have also been reached by P. Bechard and A. C. Blackwood (unpublished observations, 1970). These latter studies indicate that the production of pyocyanine commences when the phosphate concentration reaches a critical minimum level and terminates as the phosphate level rises. The increase in phosphate in the medium presumably occurs in a manner similar to that described by Hou *et al.* (1966). P. Bechard and A. C. Blackwood (unpublished observations, 1970) have also shown that pyocyanine production by *P. aeruginosa* occurs under growth-limiting conditions which may be correlated to phosphate deficiency, that is, if the phosphate level is increased at the appropriate time, an increase in cell mass is observed. If the subsequent cell mass is again subjected to similar phosphate deprivation, the quantity of pyocyanine synthesized is increased. Similar results relating increased cell numbers and pigment synthesis have been found by Ingledew and Campbell (1969b). If however, too much phosphate is added so that the culture medium is not depleted beyond the critical level, the production of additional pigment is not achieved. On the basis of these and similar studies, the reviewers believe that a complete understanding of the phosphate effect may explain the reason for pyocyanine, and perhaps phenazine pigment, synthesis and may also reveal additional detailed information on the biosynthetic pathways involved.

The collective results concerning the biosynthetic pathway for phenazine pigments indicate that shikimic acid is for the most part involved. The recent studies by Ingledew and Campbell (1969a) with a mutant of *P. aeruginosa* show conclusively that pyocyanine is probably synthesized *in toto* from shikimic acid. Production and isolation of additional mutations of the shikimic acid pathway are therefore necessary and warranted. Studies with other phenazines strongly suggest that the conversion of one pigment to another probably occurs. Podojil and Gerber (1967) reported that for *Brevibacterium iodinum* 1,6-phenazinediol and 1,6-phenazinediol 5-oxide are precursors of 1,6-phenazinediol 5,10-dioxide (iodinin). In addition, the early experiments of Kögl et al. (1932) and the recent work of P. Bechard and A. C. Blackwood (unpublished observations, 1970) suggests that phenazine-1-carboxylic acid may be a precursor of phenazine-1-carboxamide (oxychlororaphine). In view of these probable interconversions, it appears that more detailed studies of this nature would be desirable.

REFERENCES

Burton, M. O., Eagles, B. A., and Campbell, J. J. R. (1947). *Can. J. Res.* **25C**, 121–128.
Caltrider, P. G. (1967). *In* "Antibiotics" (D. Gottlieb and P. D. Shaw, eds.), Vol. II, p. 117. Springer-Verlag, Berlin.
Carter, R. E., and Richards, J. H. (1961). *J. Amer. Chem. Soc.* **83**, 495–496.
Chang, P. C., and Blackwood, A. C. (1968). *Can. J. Biochem.* **46**, 925–929.
Chang, P. C., and Blackwood, A. C. (1969). *Can. J. Microbiol.* **15**, 439–444.
Clemo, G. R., and McIlwain, H. (1938). *J. Chem. Soc.* pp. 479–483.
Cook, F. D., Peterson, E. A., and Gillespie, D. C. (1968). Can. Patent 784,213; *Chem. Abstr.* **69**, 18039d.

Davis, J. G. (1939). *Chem. Abstr.* **33**, 6383.
Edwards, O. E., and Gillespie, D. C. (1966). *Tetrahedron Lett.* **40**, 4867–4870.
Frank, L. H., and De Moss, R. D. (1959). *J. Bacteriol.* **77**, 776–782.
Gerber, N. N. (1967). *J. Org. Chem.* **32**, 4055–4057.
Gerber, N. N. (1969). *J. Het. Chem.* **6**, 297–300.
Gerber, N. N., and Lechevalier, M. P. (1964). *Biochemistry* **3**, 598–602.
Gerber, N. N., and Lechevalier, M. P. (1965). *Biochemistry* **4**, 176–180.
Gessard, C. (1890). *Ann. Inst. Pasteur* **4**, 88.
Grossowicz, N., Hayat, P., and Halpern, Y. S. (1957). *J. Gen. Microbiol.* **16**, 576–583.
Guignard and Sauvageau (1894). From D. G. Swan, and D. G. I. Felton, (1957). "The Chemistry of Heterocyclic Compounds, Phenazines." p. 180. Wiley (Interscience), New York.
Halpern, Y. S., Teneth, M., and Grossowicz, N. (1962). *J. Bacteriol.* **83**, 935–936.
Hanson, A. W. (1968). *Acta Cryst.* **B24**, 1084–1087.
Haynes, W. C., Stodola, F. H., Locke, J. M., Pridham, T. C., Conway, H. F., Sohns, V. E., and Jackson, R. W. (1956). *J. Bacteriol.* **72**, 412–417.
Herbert, R. B., and Holliman, F. G. (1969). *J. Chem. Soc.* pp. 2517–2520.
Hilleman, H. (1938). *Ber. Deut. Chem. Ges.* **B71**, 46–52.
Holliman, F. G. (1957). *Chem. Ind.* p. 1668.
Holliman, F. G. (1961). *S. African J. Lab. Clin. Med.* **8**, 81–82.
Holliman, F. G. (1969). *J. Chem. Soc.* pp. 2514–2516.
Hou, C. I., Gronlund, A. F., and Campbell, J. J. R. (1966). *J. Bacteriol.* **92**, 851–855.
Ingledew, W. M., and Campbell, J. J. R. (1969a). *Can. J. Microbiol.* **15**, 535–541.
Ingledew, W. M., and Campbell, J. J. R. (1969b). *Can. J. Microbiol.* **15**, 595–598.
Ingram, J., and Blackwood, A. C. (1962). *Can. J. Microbiol.* **8**, 49–56.
Irie, T., Kurosawa, E., and Nagaoka, I. (1960). *Bull. Chem. Soc. Jap.* **33**, 1057–1059.
Isono, K., Anzai, K., and Suzuki, S. (1958). *J. Antibiot.* **11A**, 264–267.
Jordan, E. O. (1899). *J. Exp. Med.* **4**, 627.
Kluyver, A. J. (1956). *J. Bacteriol.* **72**, 406–411.
Kögl, F., and Postowsky, J. J. (1930). *Ann.* **480**, 280–297.
Kögl, F., Tönnis, B., and Groenewegen, H. J. (1932). *Ann.* **497**, 265.
Kurachi, M. (1959). *Bull. Inst. Chem. Res. Kyoto Univ.* **37**, 101–111.
Lesley, S. M., and Behki, R. M. (1967). *J. Bacteriol.* **94**, 1837–1845.
Lesley, S. M., Behki, R. M., and Gillespie, D. C. (1967). *Can. J. Microbiol.* **13**, 1251–1257.
Levitch, M. E., and Rietz, P. (1966). *Biochemistry* **5**, 689–692.
Levitch, M. E., and Stadtman, E. R. (1964). *Arch. Biochem. Biophys.* **106**, 194–199.
MacDonald, J. C. (1963). *Can. J. Microbiol.* **9**, 809–819.
MacDonald, J. C. (1967). In "Antibiotics" (D. Gottlieb and P. D. Shaw, eds.), Vol. II, p. 52. Springer-Verlag, Berlin.
McIlwain, H. (1937). *J. Chem. Soc.* pp. 1704–1711.
Mercier, L., and Lasseur, P. (1911). *C. R. Acad. Sci.* **152**, 1415.
Millican, R. C. (1962). *Biochem. Biophys. Acta* **57**, 407–409.
Nakamura, S., Lin Wang, E., Murase, M., Maeda, K., and Umezawa, H. (1959). *J. Antibiot.* **12A**, 55–58.
Nakamura, S., Maeda, K., and Umezawa, H. (1964). *J. Antibiot.* **17**, 33–36.
Okami, Y. (1952). *J. Antibiot.* **5**, 477–480.
Peterson, E. A., Gillespie, D. C., and Cook, F. D. (1966). *Can. J. Microbiol.* **12**, 221–230.
Podojil, M., and Gerber, N. N. (1967). *Biochemistry* **6**, 2701–2705.

Schneierson, S. S., Amsterdam, D., and Perlman, E. (1960). *Antibiot. Chemother.* **10,** 30–33.
Schoental, R. (1941). *Brit. J. Exp. Pathol.* **22,** 137–147.
Swan, G. A., and Felton, D. G. I. (1957). "The Chemistry of Heterocyclic Compounds, Phenazines," pp. 174–192. Wiley (Interscience), New York.
Takeda, R. (1958). *Hakko Kogaku Zasshi* **36,** 286–290.
Takeda, R., and Nakanishi, I. (1959). *Hakko Kogaku Zasshi* **38,** 9–19.
Toohey, J. I., Nelson, C. D., and Krotkov, O. (1965). *Can. J. Bot.* **43,** 1055–1062.
Umezawa, H., Hayano, S., Maeda, K., Ogata, Y., and Okami, Y. (1950). *Jap. Med. J.* **3,** 111–117.
Vallette, J. P., Lacoste, A. M., Labeyrie, S., and Neuzil, E. (1966). *C. R. Soc. Biol.* **160,** 1562–1567.
Weigele, M., and Leimgruber, W. (1967). *Tetrahedron Lett.* pp. 715–718.
Wrede, F., and Strack, E. (1929). *Z. Physiol. Chem.* **181,** 58–76.
Yagashita, K. (1960). *J. Antibiot.* **13A,** 83–96.

The Gibberellin Fermentation

E. G. JEFFERYS

Imperial Chemical Industries Limited, Pharmaceuticals Division, Alderley Park, Macclesfield, Cheshire, Great Britain

I.	Introduction and Scope of Review	283
II.	Gibberellins—Structure and Terminology	284
III.	Other Compounds Isolated from Gibberellin Fermentations	284
IV.	Organisms, Nomenclature, and Strains	284
V.	Assay, Extraction, and Purification Methods	291
VI.	The Biosynthesis of Gibberellins	294
VII.	The Course of Fermentation	296
VIII.	The Effects of Environmental Changes on the Course of Fermentation	299
	A. The Effects of Varied Temperature	299
	B. The Effects of Varied pH	301
	C. The Effects of Varied Nutrient Conditions	302
	D. The Effects of Light	307
IX.	The Production of Gibberellins	307
X.	Present and Potential Applications	309
	References	310

I. Introduction and Scope of Review

The purpose of this review is to collate and discuss fermentation aspects of the production of gibberellins. It will include sections on the organisms that produce gibberellins, the structure of gibberellins, their biosynthesis, and production. A brief review of the applications to which their plant growth-promoting properties have been put is also included.

Several excellent reviews and literature surveys of historical and chemical aspects have been published, among which the "Source Book on Gibberellin" by F. H. Stodola (Stodola, 1958) is an essential key to the early literature, which was already impressively large by 1957. Other surveys which should be consulted for information and views of these early days include those of Stowe and Yamaki (1957), Wakagi (1958), Grove (1961, 1963), and Stowe *et al.* (1961).

Early work was, of course, based on surface culture fermentations (Kurosawa, 1934; Yabuta *et al.*, 1939, 1940; Yabuta and Hayashi, 1936, 1937, 1939; Curtis and Cross, 1954). From these studies a wide range of products were reported, usually in a low yield, after prolonged periods of incubation. First publications dealing with the submerged culture production of gibberellins include those of Kitamura *et al.* (1953), Borrow *et al.* (1955), Stodola *et al.* (1955, 1957), Darken *et al.* (1959), and Fuska *et al.* (1960). The literature survey upon which this review is based necessarily overlaps previous reviews.

II. Gibberellins — Structure and Terminology

The fungal gibberellins form part of a larger family of compounds, the other members being found in higher plants. It is possible that gibberellins, or closely related compounds, are also produced by bacteria. To date[1] twenty-seven gibberellins are known, of which four, A_1, A_3, A_4, and A_7, have been isolated from both fungi and higher plants, while gibberellins A_2, A_{9-15}, and A_{24-25} have been isolated only from fungi (Cavell *et al.*, 1967; Cross, 1968; D. M. Harrison, personal communication, 1969). Gibberellic acid (gibberellin A_3) has received the greatest attention so far, but gibberellins A_4 and A_7 (and mixtures of A_4 and A_7) are now becoming more available and have commercially attractive properties. The structures of the fungal gibberellins are shown in Fig. 1, and comments on the conditions of production will be made later.

III. Other Compounds Isolated from Gibberellin Fermentations

The intense studies which these fermentations have attracted have yielded a large number of compounds. Some of these are clearly implicated in the biosynthesis and breakdown of gibberellins (Table I) while others are either "shunt" products or compounds, the presence of which would be expected on general metabolic considerations (Tables II, III, and IV). Several enzyme systems have also been detected (Table V).

IV. Organisms, Nomenclature, and Strains

Considerable confusion is evident in the literature in the naming of fungal cultures which produce gibberellins, and an authoritative review of strains and nomenclature would be helpful. Names which have been attributed to producing organisms include:

Fusarium heterosporum	Nees
Fusarium moniliforme	Sheldon
Fusarium oxysporum	Schlechtendahl
Gibberella fujikuroi	(Sawada) Wollenweber
Gibberella moniliformis	(Sheldon) Wineland

[1] Mid 1969.

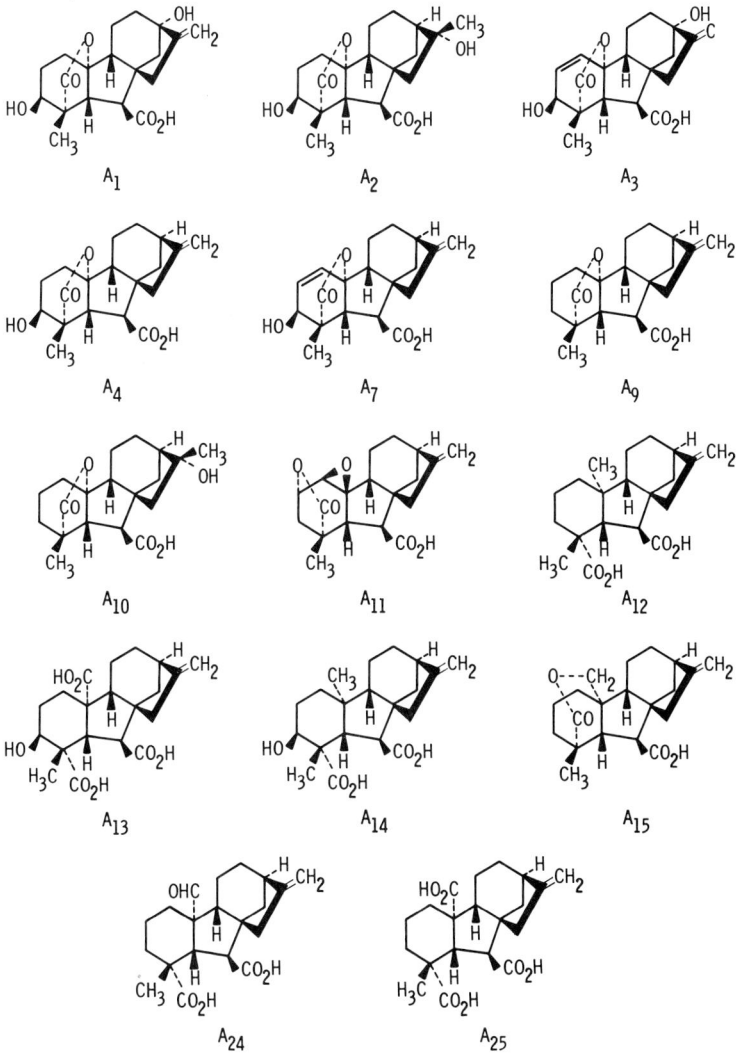

FIG. 1. The structures of the fungal gibberellins. Gibberellins A_2, A_{9-15}, and A_{24-25} have been isolated only from fungi. The remainder have also been isolated from higher plants.

In a brief review Borrow et al. (1955) indicated that *F. moniliforme* Sheld. should be regarded as the imperfect stage of *G. fujikuroi* (Saw.) Wr., and that *F. heterosporum* Nees as used by Yabuta et al. (1934) is probably incorrect. The organisms used in fermentation

TABLE I
COMPOUNDS RELATED TO GIBBERELLINS ISOLATED FROM FERMENTATIONS

Gibberic acid[a]	Kuhr, 1962
Dehydroallogibberic acid[a]	Cross et al., 1963
Allogibberic acid[a]	Kuhr, 1962
Gibberellenic acid[a]	Kuhr, 1962
	Gerzon et al., 1957
Gibberellin X	Schreiber et al., 1966
2,3,7-Trihydroxy-1-methyl-8-methylenegibb-4-ene-1,10-dicarboxylic acid, 1,3-lactone[a]	Kuhr, 1962 Muromtsev et al., 1966
2-O-acetyl gibberellic acid	Schreiber et al., 1966
Fujenal[a]	Cavell and MacMillan, 1967
	Cross et al., 1963
	Brown et al., 1967
Fujenoic acid	Cross et al., 1963
(−)-Kaurene	Cross et al., 1962a, 1963
(−)-Kauranol	Cross et al., 1962a, 1963
(−)-Kaur-16-en-19-oic acid	Cavell and MacMillan, 1967
7,Hydroxykaurenolide	Cross et al., 1962a, 1963
	Cavell and MacMillan, 1967
7,18-Dihydroxykaurenolide	Cross et al., 1963
	I.C.I., 1964b
7,16,18-Trihydroxykaurenolide	Cross et al., 1963
	I.C.I., 1964b
13-Epi-(−)-manoyl oxide (olearyl acid)	Cross et al., 1963
	Serebryakov et al., 1966
$C_{19}H_{26}O_4$ oxolactone	Wierzchowski and Wierzchowska, 1961
$C_{20}H_{26}O_4$ oxolactone	Wierzchowski and Wierzchowska, 1961

[a] These products may be extraction artifacts.

TABLE II
ORGANIC ACIDS DETECTED IN GIBBERELLIN-PRODUCING ORGANISMS

Acetic acid	Harhash, 1967a
	Sandhu, 1960
Behenic acid	Cavell and MacMillan, 1967
Cerotic acid	Cavell and MacMillan, 1967
Citric acid	Harhash, 1967a
Fujic acid	Sternberg, 1962
Fumaric acid	Harhash, 1967a
5-Hydroxymethylfuran-2-carboxylic acid	Kawarada et al., 1955
5-Hydroxymethyl-2-furoic acid	Cross et al., 1963
Fusaric acid	Yabuta et al., 1934
	Stoll, 1954
	Stoll and Renz, 1957
	Sandhu, 1960
	Hill et al., 1966
Dehydrofusaric acid	Stoll, 1954
	Stoll and Renz, 1957
Gluconic acid	Bilai et al., 1961a
α-Ketoglutaric acid	Harhash, 1966, 1967a
Malic acid	Harhash, 1967a
Malonic acid	Nakmura et al., 1958
Orsellenic acid	Bentley et al., 1965
Phthalic acid	Cross et al., 1963
Pyruvic acid	Harhash, 1966, 1967a
Succinic acid	Cross et al., 1963
	Harhash, 1967a
Tetracosanic acid	Cavell and MacMillan, 1967
$C_{20}H_{28}O_5$ (dibasic)	Cross et al., 1963

TABLE III
AMINO ACIDS DETECTED IN GIBBERELLIN-PRODUCING ORGANISMS

Alanine	Bilai and Zakordonets, 1965[a,b]
	Carito and Pisano, 1966[a,c]
α-Alanine	Sandhu, 1960
	Harhash 1967a[d]
γ-Aminobutyric acid	Sandhu, 1960
Aspartic acid	Bilai and Zakordonets, 1965[a,b]
	Harhash 1967a[d]
Glutamic acid	Bilai and Zakordonets, 1965[a,b]
	Carito and Pisano, 1966
	Harhash, 1967a[d]
Glycine	Harhash, 1967a
Leucine	Harhash, 1967a[d]
Lysine	Bilai and Zakordonets, 1965[a,b]
Phenylalanine	Harhash, 1967a[d]
Serine	Sandhu, 1960
Tryptophan	Harhash, 1967a[d]
Tyrosine	Bilai and Zakordonets, 1965[a,b]
Valine	Bilai and Zakordonets, 1965[a]

[a] From cells.
[b] From filtrate.
[c] 2.6 gm/liter.
[d] It seems likely that racemization occurred during the isolation of these products since this author describes the presence of the DL form.

TABLE IV
OTHER COMPOUNDS DETECTED IN GIBBERELLIN-PRODUCING ORGANISMS

Adenine	Yabuta et al., 1941
D-Arabitol	Borrow et al., 1964b
Betaine	Yabuta et al., 1941
Ergosterol	Yabuta et al., 1941
Fungisterol	Yabuta et al., 1941
Fusigen (a sideramine)	Diekmann, 1967
Gibberine[a]	Sternberg, 1962
D-Mannitol	Borrow et al., 1964b
Dimethylphthalate	Cross et al., 1963
Phenyl ethyl alcohol	Cross et al., 1963
Tyrosol	Cross et al., 1963
Vasinfuscarin	Stoll, 1954
Pigment ($C_{14}H_{10}O_7$, ? anthraquinone)	Nakamura et al., 1957
Polysaccharide of	
Glucuronic acid	Martin and Adams, 1956
Glucose	Siddiqui and Adams, 1961
Galactose	
Mannose	
Mucilage based on	
Uronic acid	Yabuta et al., 1941
Ketose	
Galactose	
Pentose	
$C_{20}H_{36}O_8$ (unidentified liquid)	Cavell and MacMillan, 1967
Sitostanol	Serebryakov et al., 1966
β-Sitosterol	Serebryakov et al., 1966

[a] This product may be an extraction artifact.

TABLE V
ENZYME SYSTEMS AND COENZYMES DETECTED IN
GIBBERELLIN-PRODUCING ORGANISMS

Amylase	Harhash, 1967b
Aspartic deaminase	Harhash, 1967b
Butyrase	Harhash, 1967b
Catalase	Harhash, 1967b
Cellulolytic enzyme	Capellini and Peterson, 1966
Coenzyme Q_8	Lavate and Bentley, 1964
Coenzyme Q_9	Lavate and Bentley, 1964
Coenzyme Q_{10} (H-10)	Merck and Co., 1966
Dihydro coenzyme Q_{10}	Lavate and Bentley, 1964
Cutinase	Heinen, 1960
	Heinen and Linskens, 1960
Fructosanase	Loewenburg and Reese, 1957
Galactose oxidase	Gancendo et al., 1967
3-Indole acetaldoxime hydrolyase	Kumar and Mahadevan, 1963
Invertase	Harhash, 1967b
Laccase	Harhash, 1967b
Lipase	Harhash, 1967b
Maltase	Harhash, 1967b
Oxalate decomposing enzyme	Nagata and Hayashi, 1957
Pectolytic enzyme	Capellini and Peterson, 1966
Pectin transeliminase	Kathirvelu and Mahadevan, 1967
Phosphatase (acid)	Harhash, 1967b
Phosphorylase	Harhash, 1967b
Polygalacturonate transeliminase	Kathirvelu and Mahadevan, 1967
Proteolytic enzyme	Harhash, 1967b
Renninlike enzyme	Knight, 1966
RNase	Harhash, 1967b
Serine deaminase	Harhash, 1967b
Urease	Harhash, 1967b

practice can be referred to *Gibberella fujikuroi* (Saw.) Wr. (*stat. conid. Fusarium moniliforme* Sheld.).

Bilai et al. (1961b), comparing strains showed that nine named as *G. fujikuroi* produced plant growth stimulators, but that only occasional strains named as *F. oxysporum* did so.

Borrow et al. (1955) commented on the correlation between the ability of isolates to produce gibberellins, and the origin of the strains from infected rice, rather than from other species of plants in the host range of species; but Gordon (1960) showed that isolates from maize and several other host plants produced gibberellinlike activity.

Strain improvements by conventional mutation processes have been made by many workers (e.g., Imshenetsky and Ulyanova, 1961,

1962b). Erokhina and Sokolova (1966) showed increases in titer of up to 60% resulting from ultraviolet radiation, fast neutrons, and γ-ray treatments. They also demonstrated (Sokolova and Erokhina, 1966) that mutants could be produced from nonsporing mycelial organisms. Imshenetsky and Ulyanova (1961, 1962a) showed that mutant strains which gave increased yields of gibberellins used more glucose, produced more acetic acid and ethanol, and produced less citric and gluconic acids than did the parent strain.

Screening for strain improvements is normally done in shake-flask cultures. Good strains could be missed in such a system unless the medium is selected to match the gas-transfer efficiency of the shake-flask system, and flasks are sampled sufficiently frequently for the rate of production to be used as the primary criterion for selection, rather than the final titer after a prescribed period.

Work has been reported on many strains, both natural and upgraded by selection, mutation, and breeding. It is clear that strains vary greatly, not only in their inherent ability to produce gibberellins, but also in the balance of their products and their response to environmental changes. This wide variation between strains invalidates the comparison of fermentation conditions, when the strains also vary.

V. Assay, Extraction, and Purification Methods

Assay systems initially used in studies of the gibberellins (Stodola, 1958) were inevitably based on their effects on extensions of plant tissues. Such systems may still have their uses where high levels of specificity and of sensitivity are demanded, but they are not relevant in fermentation studies.

For routine control purposes, colorimetric methods based on Folin-type reagents have been used, but have the disadvantage that many interfering substances must be removed and the gibberellins purified prior to measurement (Graham and Henderson, 1961; Muromtsev and Nestyuk, 1962; Udagawa and Kinoshita, 1961; Sternberg and Voinescu, 1961). Head (1959) described a polarographic method. Infrared (IR), and UV spectrophotometric, or fluorescence methods are probably the best for control purposes, while for the determination of product purity, gas-liquid or thin-layer chromatographic methods are recommended; tracer techniques would also seem to be useful. Various methods used for gibberellin studies are collated in Table VI.

While it has been suggested (Bolgarev et al., 1962) that lyophilized culture filtrates may be useful in agricultural applications, it seems safer to expect that products should be selectively extracted, standard-

TABLE VI
Assay Techniques Used in Gibberellin Fermentation Studies and Production Control

IR methods	
Carbonyl-stretching—on extracts	Borrow et al., 1964a
Absorbance at 12.86 and 10.85 μ	Washburn et al., 1959
UV methods	
Conversion to gibberellenic acid	Holbrook et al., 1961
(measure at 25.4 mμ)	
Fluorescence methods	
Cold H_2SO_4 on extracts in solvent	Kavanagh and Kuzel, 1958
	Theriault et al., 1961
	Borrow et al., 1964a
Gas chromatographic methods	
As methyl esters	Ikekawa et al., 1963
As methyl and trimethylsilyl esters	Cavell et al., 1967
Paper chromatography	
$KMnO_4$ spray	Bird and Pugh, 1958
Thin-layer chromatography/UV fluorescence	
No pretreatment	Ikekawa et al., 1963
	Kagawa et al., 1963
After treatment with H_2SO_4	Aseeva, 1963
	Ikekawa et al., 1963
	Kagawa et al., 1963
H_2SO_4/methyl alcohol	Podojil and Ševčik, 1960
H_2SO_4/ethyl alcohol	Cavell et al., 1967
	MacMillan and Suter, 1963
Tracer techniques	
Mass isotope dilution	Arison et al., 1958
Derivative labeling	Baumgartner et al., 1959
Tritiated derivatives	Baumgartner et al., 1963

ized, and reasonably pure. It should, however, be noted that the dilution required to bring the concentration of gibberellic acid in a good fermentation broth, down to the level required for application to plants, would dilute out any phytotoxicity due to fusaric acid, should any be present.

The original process for the isolation of "gibberellin" (Yabuta et al., 1939, 1940) used adsorption by charcoal followed by solvent elution, usually with acetone. A number of processes describe the use of ion-exchange resins with both aqueous and solvent-based eluants; several solvent extraction systems have also been proposed.

Adsorption on charcoal was originally used by Imperial Chemical Industries, Ltd. (I.C.I.) (1957) for the extraction of gibberellin A_3. This was followed by elution with acetone of the air-dried (20–30%

moisture) charcoal, in the presence of some water, which is essential for efficient extraction. They used a subsequent purification process involving a buffer–ethyl acetate system. Borrow *et al.* (1955) and Stodola *et al.* (1955) both showed that elution of carbon with ammoniacal methanol was useful, while Nestyouk *et al.* (1961) used ammoniacal *n*-butanol.

A patent filed by Société d'Études et d'Applications Biochimiques (1963) described an ion-exchange resin process by which acidic, nongibberellinlike components were first removed by treatment with alkaline-earth metal hydroxides, or their salts, followed by the use of anionic resins and elution with ammonia or buffered solutions of alkaline-earth metal salts; examples quoted claimed 89–100% extraction. Other ion-exchange processes describe the use of strongly basic resins (Merck and Co., 1960a). Using the chloride forms of Dowex 1-X2 and of Amberlite IRA 401 gave extraction efficiencies of 94 and 98%, respectively, when eluted with methanol acidified with HCl, and of 87 and 93%, respectively, when eluted with methanol acidified with H_2SO_4. Another patent (Merck and Co., 1960b) provided for elution of resins with NH_4Cl in aqueous methanol. Other resin processes for extraction are described by Podojil *et al.* (1961), Ezhov (1965), and by Instytut Antibiotykow (1965), the latter process provided concentrates of the potassium salt of gibberellin A_3, with recoveries of ca. 64%.

Solvent extraction processes were developed as yields were increased. Most claim the use of esters, of which ethyl and butyl acetates are frequently featured and appear to be the solvent of choice for commercial-scale work. However, ketones, alcohols, and ethers have also been used either alone, or in mixtures. All solvent processes rely on the extraction in acid (pH 2–4) conditions, the subsequent recovery being either by adsorption on solid sodium or potassium bicarbonate, or by buffer-solvent processes in which the relative solubilities of the free acid in solvent and of the sodium or potassium salt in an aqueous phase are used to effect separation, concentration, and purification of the product. Both Pfizer and Co. (1959) and I.C.I. (1959) showed that ethyl acetate and methyl isobutyl ketone (MIBK) were effective. However, in addition Pfizer claimed methyl ethyl ketone and I.C.I. claimed a wider range of esters, ketones, alcohols, and ethers, with a stated preference for ethyl acetate, isoamyl acetate, methyl *n*-propyl ketone or MIBK, with extraction efficiencies on plant-scale examples of 70–80%. The yield in the Pfizer patent was ca. 55% overall. No useful comparison of the solvent stages can however be made from the data, which are derived from the whole process. Other examples of the use of esters and, in particular, of ethyl acetate are seen in other I.C.I. patents (1960a, b) in which yields of 90–100% and 40–60%, respectively, were obtained. Probst (1961) also described the use of MIBK but extraction efficiencies cannot be calculated from the data given.

Alcohols, expecially n-butanol, isobutanol, and benzyl alcohol have been claimed to be inefficient or to give an impure product (Pfizer and Co., 1959), but Muromtsev et al. (1962, 1963) recommended the use of butanol as a primary extracting solvent.

As previously noted, commercial-scale purification stages may be based on buffer–solvent systems, but other processes have also been described, e.g., that of De Rose (1962) for the preparation of crystalline potassium salt of gibberellin A_3 by solution in alcohol, and that of Probst (1961) by purification on sodium or potassium bicarbonate columns. Stodola et al. (1957) described the separation of gibberellins A_1, A_2, and A_3 on buffered partition columns and Kucherov et al. (1965) described the separation of gibberellins A_1, A_3, A_4, and A_9 on silica gel–phosphate systems.

VI. The Biosynthesis of Gibberellins

Most of these studies have been aimed at elucidating the metabolic pathways by which gibberellin A_3 is synthesized, and have been based on the isolation and characterization of products from fermentations, followed by studies using labeled precursors to validate the pathways suggested as a result of the primary isolations.

The main outlines of the biosynthetic pathways are now known, but many finer details await further work; the biosynthetic interrelations between gibberellins are by no means clear. The most recent review of the subject is by Cross (1968), and a more detailed treatise covering this and other aspects of the chemistry and mode of action of the gibberellins is being prepared (MacMillan, 1969). The early stage in the formation of the gibberellin molecule is common to the biosynthesis of diterpenes, in general, and consists of the condensation of acetate units, via acetyl coenzyme A, through mevalonate to geranyl pyrophosphate and then via farnesyl pyrophosphate to geranylgeranyl pyrophosphate. The second stage involves the cyclization of the geranylgeranyl pyrophosphate to an intermediate, from which the kaurene-based compounds are subsequently formed. Kaurenelike compounds are then modified by ring contraction and this stage is followed by various hydroxylations to the gibberellin range of compounds.

Confirmation of this general outline comes from tracer studies. Cross et al. (1964) showed that (−)-kaurene could act as a precursor for gibberellin A_3. Geissman et al. showed that (−)-kaur-16-en-19-oic acid is converted to gibberellin A_3, and Cavell and MacMillan (1967) showed it to be present in the mycelium, suggesting an "obligate role" for the compound in the biosynthesis of gibberellin A_3. It is possible that some hydroxylation may occur in the "kaurene" stage of biosynthesis and this is yet to be resolved.

On the basis of studies of mutants of *F. moniliforme* Phinney and Spector (1967) suggested that from a pool of precursors, at least three separate lines diverge, one providing gibberellin A_9 and then A_{10}, a second providing gibberellins A_4 and A_7, with subsequent further hydroxylation to A_1 and A_3, and a third line providing the other C19 gibberellins. In later work Spector and Phinney (1968) identified two genes which control different steps in the biosynthetic pathway. One gene (g_1) controls the production of all the gibberellins, and the second (g_2) controls the production of gibberellins A_1 and A_3 only. It has been shown (Cross and Norton, 1966; Cross et al., 1968) that gibberellin A_{13} does not lie on the direct pathway to gibberellin A_3.

These outlines are in accord with suggestions made by Geissman et al. (1966) and Verbiscar et al. (1967) that the sequences in the biosynthetic pathways suggest that there is an allover change toward higher oxidation levels during the biosynthesis of gibberellins. Extrapolation of this thesis provides a biochemical basis for the indications that the gas-transfer efficiency of the fermentation system is a very significant factor in the commercial production of gibberellins.

Claims are substantiated in the Patent literature that some of the compounds implicated in the biosynthesis of gibberellins will increase yields of gibberellin A_3. It seems unlikely that the increases thus obtained have any economic interest. For instance, Redemann (1959) described an increase in yield of gibberellin (as measured by a plant assay) from 22.5 mg/liter to 50 mg/liter on the addition of 122 mg/liter of the sodium salt of senecioic (β,β-dimethylacrylic) acid while I.C.I. (1960d) showed that on the 80-liter scale, the addition of 15 gm mevalonic acid every 12 hours (465 gm added in all) increased the yield of gibberellic acid from 27 gm (ca. 340 mg/liter) to 42 gm (ca. 525 mg/liter). Later I.C.I. (1964a) claimed the improvement in yield of gibberellin A_3 by the addition of (−)-kaurene, (−)-isokaurene, (−)-kauranol or of (−)-13-epimanoyl oxide (olearyl alcohol).

Studies have been made of the effects of plant growth inhibitors on the gibberellin fermentation, since some of these compounds have been implicated in affecting the biosynthesis of gibberellins in higher plants. Although their role is not clear, it is probable (Cross and Myers, 1969) that "CCC" [(2-chloroethyl)trimethylammonium chloride] and "Amo 1618" [2'-isopropyl-4'-(trimethylammonium chloride)-5'-methylphenyl-piperidine-1-carboxylate] act before the "kaurene" stage in biosynthesis. Adamiec (1966) also reported the inhibition of gibberellin biosynthesis by CCC and TIBA (2,3,5-triiodobenzoic acid) (both at 50–100 µg/ml); but he claimed a stimulation of biosynthesis in the presence of 0.1 µg/ml of TIBA, and also by the same concentration of kinetin, indoleacetic acid, or CCC. Pigment production was, however, inhibited by both concentrations of TIBA. None of these compounds, as used, inhibited respiration.

VII. The Course of Fermentation

The gibberellin A_3 fermentation has all the attributes of the now classical "shunt metabolism." Initially, exponential proliferation of cells occurs in the presence of all nutrients and qualitative and quantitative changes of metabolism result from the exhaustion of nutrients (conditions which will be termed *nutrient limitation*) or from the development of conditions in which the rate of the supply of a nutrient becomes restricting (which will be termed *nutrient restriction*). These definitions are obviously simplifications of the true situation, since in nutrient limitation there will be some turnover of components resulting in the continued, but relatively slow, supply of the exhausted nutrient.

The production of gibberellin A_3 (and probably other gibberellins) becomes rapid with the onset of nitrogen-limited conditions and continues in the presence of sufficient energy supplies, sometimes at a constant rate, sometimes with a late reduction in rate, until a condition, as yet unidentified, develops which rapidly inhibits production. All known nutrients being available in excess, the total amount of gibberellin A_3 produced is directly proportional to the initial concentration of assimilable nitrogen in the medium (Borrow et al., 1964a).

The earliest of the submerged fermentations (Kitamura et al., 1953) was based on the use of the Japanese surface culture medium, as was also that of Stodola et al. (1955). Borrow et al. (1955) used a simplified Raulin–Thom medium which gave higher yields, and subsequently (Borrow et al., 1961) investigated the effects of varied physical and nutrient conditions on aspects of the fermentation. Media were designed so that either glucose, nitrogen, phosphate, or magnesium were exhausted at a selected dry weight and remaining nutrients were provided so that they might be exhausted in selected sequences, thereafter, except in the case of glucose-limited fermentations in which the lack of reserves in the mycelium meant that the fermentation ceased soon after the exhaustion of glucose.

Distinct phases of growth were defined by the results of this study. The phases were described in rather more simple terms than subsequent work suggests to be justifiable, but the concepts are still valid. They provide a useful approach to fermentation studies and were used in subsequent kinetic studies (Borrow *et al.*, 1964a, b). Borrow *et al.* (1961) also introduced the term "contribution" as a measure of the amount of nutrient used in the accumulation of a unit of cell mass. This concept had the advantage that it gave an immediate measure of the cell content for most nutrients, but with glucose and oxygen no such interpretation was possible. In view of its now widespread use, the term "yield constant" (the reciprocal of the contribution), will be used here, and the "contribution" abandoned.

Culture ACC 917 being a nonsporing strain, mycelial inocula were used in all the studies of Borrow *et al.* A *lag phase* was noted only on ammonium acetate media, or on media containing high initial concentrations of glucose (greater than 30 gm/100 ml). The *balanced phase* was a period of proliferation in the presence of all nutrients, during which a unit increase in dry weight was accompanied by approximately constant uptakes of glucose, nitrogen, phosphate, magnesium, and potassium; the increase in dry weight was exponential, the morphology of the hyphae remained essentially constant, and the composition of the hyphae also remained remarkably constant in terms of its fat, carbohydrate, DNA, RNA, and polyol contents. In conditions which supported high cell densities in the presence of all nutrients, exponential growth was shown to give way first to linear and then to decelerating growth. Exponential growth ceased at a dry weight which was shown to be dependent on the rate of agitation of the fermentor, and it has subsequently been shown that the change arises from oxygen restriction, the linear characteristics reflecting the rate of oxygen transfer and the subsequent deceleration probably reflecting feedback of metabolic changes resulting from conditions of severe oxygen restriction.

In both these situations the composition of the mycelium changed. It seems probable that the dry weight at which exponential growth ceases in unlimited or otherwise unrestricting conditions, can be used as an integrated measure of the gas-transfer efficiency of a fermentor.

Where nutrient limitation occurred before the onset of oxygen restriction then in phosphate- and magnesium-limited fermentations a *transition phase* was defined during which it was shown that the continued proliferation resulted from the use of intracellular acid-soluble phosphate in phosphate-limited conditions, and probably of some reserve form of magnesium in magnesium-limited fermentations.

When nitrogen was exhausted, exponential or linear proliferation ceased. The dry weight continued to increase, but growth, in the sense of an increase in the level of nitrogen-containing components or of DNA-phosphorus ceased, the increase in dry weight being a measure of the increase in the cell-bound fat and carbohydrate. The *storage phase* was initiated by the exhaustion of nitrogen in the presence of glucose, and it is with this phase that the commercial production of gibberellins is associated. The phase can be continued by glucose feeds, the cell mass increasing until the fat and carbohydrate contents reach maximal values of ca. 45 and 32% of the dry weight, respectively; the production of gibberellin A_3 can likewise be extended by feeds — often until after the time at which the maximum dry weight is achieved. This point is taken as the end of the storage phase. In the continued presence of external supplies of glucose, the *maintenance phase* ensues and continues until external glucose is exhausted and, thereafter, until internal reserves of fat, but not carbohydrate, are used. If glucose is exhausted after nitrogen then the production of gibberellins may continue for a period, the duration of which depends upon the duration of the storage phase and, hence, upon the amount of reserve fat available; but when this has been exhausted to about the level present in balanced phase mycelium then the *terminal phase* begins. This is characterized by hyphal breakdown, decreasing dry weight, and the liberation of mycelial components, phosphate, and ammonia into the medium.

As already noted, the production of gibberellin A_3 on nitrogen-limited synthetic media is initiated at the time of exhaustion of nitrogen. The course of production is linear (Borrow *et al.*, 1964a) and, within limits, the rate of production and the amount produced are proportional to the initial concentration of nitrogen supplied. That is to say, that within these limits, the productivity (the rate of production per unit of nitrogen used) remains constant. If assimilable carbon is exhausted from the medium before maximum production has been achieved then some further production continues at the expense of stored fat; but, of course, in this situation the maximum titer will not necessarily reflect the initial concentration of assimilable nitrogen.

Bu'Lock *et al.* (1965) modified the approach of Borrow *et al.* (1961), describing two main phases of activity. Their *tropophase* is equivalent to the exponential part of the balanced phase, and their *idiophase* includes the transition, storage, and maintenance phases. It is not clear how the linear and decelerating stages of the balanced phase would fit into their scheme. Bu'Lock *et al.* studied respirometric, enzymatic, and secondary metabolic activities in the work which led to the development of their concepts, but they did not routinely attempt to correlate the observed changes with the nutritional status of the organism, especially in relation to the availability of oxygen. The onset of the tropophase in the presence of conventional nutrients could have resulted from oxygen restriction in the shake-flask conditions used. It is clearly desirable that some synthesis should be attempted of the varied views expressed by several "schools." Such a synthesis should result from studies in which all the various criteria are examined and correlated in a small range of different types of fermentation.

VIII. The Effects of Environmental Changes on the Course of Fermentation

Changes may be imposed upon fermentations, and subsequently maintained by continued analysis and control, or else initial conditions may be selected and the fermentation allowed to develop thereafter. Variations of temperature, pH, and in some instances, nutrient concentration fall into the first category, but without continuous monitoring and control, fluctuations of nutrient levels will arise. It is, however, difficult in batch culture to select and maintain standard agitation and gas-transfer conditions, and it is at present more usual to prescribe initially the composition and flow rate of the gas, and the rate of agitation. Likewise it is not possible to measure directly the feedback of product on its own rate of production in batch culture. Terui and Kagawa (1958) reported that the growth of *G. fujikuroi* was "considerably inhibited" by 10 mg % gibberellin, but such concentrations would not normally be present during growth.

A. The Effects of Varied Temperature

Most published work refers to temperatures over the range 25°–33°C but Stoll (1954) studied growth on agar over the range 3°–36°C, and Borrow *et al.* (1964b) over the range 8°–40°C in submerged culture conditions. The work of Borrow *et al.* (1964b) showed that most optimum temperatures were in the range of 29°–32°C. The minimum temperature for growth of strain ACC 917 was below

8°C (the lowest temperature achievable under their conditions). Stoll (1954) found that the minimum temperature for growth was 3°C. At 38°C growth ceased when the dry weight had reached ca. 2 mg/gm; no growth occurred at 40°C (Borrow et al., 1964b). The attributes studied showed the typical "skew" curve with a "tail-off" at lower temperatures when plotted as a function of temperature, but some attributes showed a marked discontinuity in the range 17°–20°C above which the increase with increasing temperature was less than at lower temperatures.

In this study, two nitrogen-limited media were used, differing only in the initial concentration of ammonium tartrate. No significant differences were noted between the two media early in fermentation in respect of the specific growth rates, yield constants, or mycelial analyses. During the storage and maintenance phases however all specific rates, i.e., rates per unit of nitrogen supplied, were lower on the more concentrated medium, the strength of which (100mM of nitrogen/liter) had been chosen to bring the fermentations into conditions of oxygen restriction at the usual (28°C) temperature of fermentation, and, hence, provide conditions in which a period of linear growth would be experienced before the exhaustion of nitrogen.

The yield constant for nitrogen was independent of temperature. With increasing temperature, that for glucose increased slightly while that for magnesium increased markedly. The nature of the mycelium changed little during the balanced phase, but the changes in the storage phase were markedly different with differing temperatures and, as previously noted, also between the two media.

The production of gibberellin A_3 was maximal at about 29°C (I.C.I., 1960e). There is a suggestion (Sumiki et al., 1966) that a temperature of ca. 32°C favors the production of gibberellins A_4 and A_7, but these authors also used a medium initially at pH 6.8 and this was probably the more relevant factor. Cross et al. (1963) isolated phthalic acid from one fermentation at 12°C, but in no other conditions.

Variations of the temperature of fermentation therefore has, as would be expected, both quantitative and qualitative effects on the fermentation. The optimum temperature for growth is between 31°–32°C, while for the production of gibberellin A_3 it is ca. 29°C, with a rapid fall-off above this temperature. No evidence is available from the literature to suggest that fermentations have been run with the balanced phase at 31°–32°C and the production phase at about 29°C, but there might be marginal advantages to be gained from this manipulation.

B. The Effects of Varied pH

It is not always possible to disentangle the effects of varied pH from effects caused by varied sources of nitrogen, nor is it yet common to have data available from experiments where the pH has been rigidly maintained at the initially prescribed value. More often, the effects of varied pH have to be deduced from conditions arising from variations in both nutrient and prescribed situation. Moreover, the nature of the source of nitrogen is important in determining the reaction of the organism to varied levels of pH. Borrow et al. (1964a) showed that on ammonium tartrate-based media the specific growth rate is fairly constant over the range pH 3.5–6.5, but that decreases occur away from this range. However, it should be noted that in this experiment the pH drifted down, making the significance of these results rather difficult to assess. It has since been realized (A. Borrow, unpublished) that glycine-based media are useful for studies on the effects of pH, since the nitrogen is assimilated without the concurrent release of excess acid radicals. Hence, prescribed conditions of pH are more readily maintained.

When ammonium nitrate was used as a source of nitrogen a complex "hunting" developed between pH and growth described (Borrow et al., 1961, 1964a) as the development of "steps." NH_3— N was assimilated more rapidly than NO_3— N, but in the presence of NO_3— N the assimilation of ammonia was inhibited at pH 2.8–3.0. Rapid growth, assimilation of ammonia and nitrate, and decrease in pH to the critical level were followed by periods of very little growth; however, nitrate assimilation continued until the pH had increased enough to restart the ammonia assimilating mechanism. The cycle was repeated several times, the number and duration of which varied in a manner which suggested that the hypothesis was correct.

The yield constant also varied with varied initial pH. On ammonium tartrate or urea (Borrow et al., 1964a), the yield constant for nitrogen increased with increasing initial pH, but that of glucose was independent of the initial pH.

Both quantitative and qualitative effects of pH on the production of gibberellins can be seen, but again these effects are usually confounded by variations in the source of nitrogen and by lack of control throughout the experiment. However, in one experiment a reliable quantitative result is available which showed that the production of gibberellin A_3 was independent of pH over range 3.0–5.5, but that both the rate of production and the productivity (rate per unit of nitrogen supplied) decreased with changing pH when the pH was controlled outside this range.

Stodola et al. (1955), using a medium with a low initial pH and a strong tendency to become more acidic (since the nitrogen was supplied as ammonium chloride), produced gibberellins A_1 and A_3. This was confirmed by Kuhr et al. (1961) and by Fuska et al. (1962) and shown to be a reflection of the low pH — usually in the range of pH 2–3; this production of mixtures of A_1 and A_3 could occur on media with varied sources of nitrogen.

At the other end of the range it has been demonstrated that when the pH of the brew was increased to, and maintained at, values around pH 7, increased production of gibberellins A_4, A_7, A_9, A_{12}, A_{14}, and A_{16} could be detected, and additionally small quantities of a dibasic acid ($C_{20}H_{28}O_5$) were isolated (Cross, 1966; Cross and Norton, 1965; Cross et al., 1960a,b, 1962a; Galt, 1968). As previously noted Sumiki et al. (1966) also obtained gibberellins A_4 and A_7 on a medium with an initial pH of 6.8.

C. The Effects of Varied Nutrient Conditions

The work of Ricicova et al. (1960) listed and reviewed many of the media used in studies of the gibberellin fermentation; but, as noted by these authors, "It is advisable to exercise the utmost caution in interpreting the results of fermentation procedures as the production of gibberellic acid depends considerably on the strain used." Production also depends upon the interaction of conditions resulting from the presence and uptake of nutrients. At one extreme of conditions it is possible to measure with fair confidence the effects of varied concentrations of glucose on gibberellin productivity in otherwise similar media. It is less easy unequivocally to measure the effects of varied concentrations of even a simple source of nitrogen, due to the simultaneous, *pro rata* uptake of other nutrients. This results in the varied levels having also different excesses of these other nutrients and therefore different pH conditions, in addition to the conditions resulting from the varied cell concentrations. The situation becomes further complicated when natural substrates which combine nitrogen and carbon are investigated.

Over the range 4–40% glucose, Borrow et al. (1964a) showed that the specific growth rate of strain ACC 917 decreased with increasing glucose concentration, and that the form of the relation varied with the nature of the source of nitrogen. The range tested is, of course, several orders of magnitude greater than that over which the Michaelis–Menten type of relation operates. In an experiment in which the initial supply of glucose was so calculated as to provide a range from 5–30% at the time of exhaustion of nitrogen, they showed that with increasing concentration of glucose the rate of production of gibberellin A_3 and the productivity (the cell concentration being the same throughout) decreased. The measured maxima also decreased.

With varied sources of nitrogen their results were less straightforward, but showed a logical pattern. With each source of nitrogen both the rate of production of gibberellin A_3 and the productivity first increased with increasing concentration of nitrogen. Further increases in concentration of nitrogen led to a decrease in productivity, while the overall rate could still show an increase, but at the higher concentrations tested, both the productivity and the rate of production decreased. These results were interpreted as reflecting interactions between the increasing demand for oxygen by the organism, and the decreasing efficiency of gas-transfer due to the presence of the increasing concentration of cells. The form of the relation between productivity and initial concentration of nitrogen was similar on glycine, urea, ammonium tartrate, ammonium nitrate, and ammonium acetate media, but the values for maximal productivity varied, those for glycine and urea being highest (ca. 0.070 mg/hour mmole nitrogen) and that for ammonium acetate being lowest (ca 0.034 mg/hour mmole).

These results emphasize the importance of developing media recipes to obtain the best performance from particular fermentors, the "pay-off" being a power function of initial nitrogen supplied up to a critical point, thereafter becoming linear before decreasing.

Many energy sources have been investigated for use in the gibberellin fermentation. Glucose and sucrose have frequently been used and combinations of sucrose with dextrin recommended (Abbott Laboratories, 1963). Glycerol (Kitamura et al., 1953; Darken et al., 1959), and glycerol–starch, glycerol–glucose, and glycerol–glucose–lactose mixtures were used by Darken et al. (1959), the latter providing the highest titers in their work (880 mg/liter). Shklyar and Globus (1961, 1963) attempted to replace the carbon sources used by Darken et al. (1959) by molasses. They found that decreased, but economically useful yields of gibberellin resulted. Dietrich (1960) showed that gibberellins could be produced on molasses residues, sulfite liquors, whey, or skimmed milk, but only low titers are recorded (50 mg/liter in 3–5 days). Russian workers have successfully used natural oils as carbon sources, e.g., sunflower oil (Agnistikova et al., 1966; Erokhina,

1967; Muromtsev et al., 1968). Muromtsev and Dubovaya (1964) compared production of gibberellin on a range of oils with that on sucrose and found, using the same initial concentration (4%) of carbon source, that yields were increased by linseed oil (559% of sucrose yield), sunflower oil (435%), olive oil (382%), cotton seed oil (336%), and ethyl palmitate (300%), while stearic acid depressed the yield to 49%.

Hydrocarbons have also been tested for their ability to support the growth of F. moniliforme (Flippin et al., 1964; Hitzman and Mills, 1963; Zaichenko and Koval, 1966), but no data on growth rates or productivity are given.

In view of the inhibitory effect which high concentrations of glucose exert on productivity, it is not surprising that feed processes have been described (Borrow et al., 1964a; I.C.I., 1960b; Abbott, 1963; Serzedello and Whitaker, 1960); and, of course, a form of feed process arises from the use of carbohydrate polymers such as plant meals, e.g., soya flour, soya meal, cotton seed meal (I.C.I., 1960c). These natural sources of carbohydrate also provide nitrogen compounds and can affect the buffering capacity of media sufficiently to make it impossible to attribute precise effects to particular components. Some attempts to trace the source of improvements resulting from the use of natural sources of nitrogen were made by the Czechoslovakian school, who investigated the effects of fractions of soya bean flour (Fuska et al., 1964; Podojil and Ricicova, 1965), and showed differential stimulation of production of gibberellins A_1 and A_3 by some fractions. However, differences in pH in the different experiments may have complicated their results.

Many sources of nitrogen have been tested. Nitrates, and inorganic and organic ammonium salts have been used mainly for research purposes, while process development has centered around the less expensive natural sources. Harhash (1966) reported that ammonia was assimilated in preference to nitrate, confirming the observations of Borrow et al. (1961, 1964a); but he was not concerned with gibberellin production. The original Japanese media were based on the use of ammonium chloride, which was also used by Stodola et al. (1955). This produced very low pH levels and, together with the use of glycerol, resulted in the production of both gibberellins A_1 and A_3. Podojil and Ricicova (1965) showed that by providing as the nitrogen source corn steep liquor, only gibberellin A_3 was produced, but when soya bean flour was substituted, both A_1 and A_3 were produced. Ammonium acetate was shown by Borrow et al. (1964a) to support poor productivity, but this source is claimed by Nestyouk et al. (1961) to provide good yields. Their process yielded 4.18 gm from 36 liters of broth; a yield of 116 mg/liter. Productivity cannot be calculated from their data.

Cross et al. (1963) commented that glycine-based fermentations provided a rich source of metabolites and reported the isolation of fujenal, the mono-, di-, and trihydroxykaurenolides and of fujenoic acid, as well as gibberellins A_4 and A_9.

Fuska et al. (1960, 1962) studied a wide range of sources of nitrogen and recommended a medium based on peanut or soya bean meal as an improvement on their previous corn steep liquor medium. They concluded that "the increase in gibberellic acid is thus evidently related to the specific effect of the nitrogenous substance and not to the percentual concentration of total nitrogen." It seems possible that the stimulation in productivity afforded by natural sources of nitrogen (and also carbohydrate) could be attributable to the presence in these seed meals of substances which act as precursors for the gibberellins. Information on the presence of such compounds would be of interest and perhaps of considerable economic significance.

Little work has been published on the effects of trace elements on the fermentation, most workers adding excess trace elements, or relying on impurity levels. However, Krasilnikov et al. (1963) showed that the response to trace elements varied between cultures. Zinc, as a single trace element, was more effective than others in stimulating production by a strain named as G. fujikuroi but copper was the single most effective element for a *Fusarium* culture. In combination, G. fujikuroi responded best to zinc, molybdenum, and copper, while the *Fusarium* produced best with copper, cobalt, and boron.

The production of gibberellin A_3 requires a continuous supply of oxygen. At some stages of fermentation a prolonged period of oxygen-limitation can prove to be a shock from which recovery is impossible. As quoted earlier from Geissman (1966), the biosynthetic pathway would appear to progress through compounds of increasing levels of oxidation, and Cross et al. (1963) showed that at low rates of supply of oxygen, a low yield of acidic compounds was obtained, metabolism having been diverted to produce a new range of compounds. *Fusarium moniliforme* fermentations run under conditions of oxygen restriction develop a very characteristic sharp "estery" smell—probably due to the production of ethyl acetate.

The kinetics of the effect of varied rates of supply of oxygen on growth have not been elucidated. Borrow et al. (1964a) showed that over the range 570–830 rpm in "Hoover" (washing machine) fermentors the dry weight at which exponential growth ceased, (X_b), increased from 14.8 mg/gm to ca. 37.9 mg/gm. This experiment was performed using ammonium nitrate as the source of nitrogen, and consequently the specific growth rate was low. Therefore, the overall demand rate for oxygen would have been lower than in the experiment quoted in which glycine was used as source of nitrogen when X_b in the "Hoover" was 5.7 mg/gm, while it was greater than 26.8 mg/gm in the more conventional fermentor. X_b was also sensitive to changes in air flow rate at constant rates of agitation. As might be expected, the yield constant for nitrogen was unaffected by these changes, but that for glucose always decreased at the onset of linear growth, suggesting that this phase, being associated with oxygen restriction, is characterized by the less efficient utilization of glucose associated with partial anaerobiosis.

No information appears to have been published which directly relates the rate of production of gibberellins to aeration–agitation regimes, but it was suggested (Borrow et al., 1964a) that the high productivity achieved by Darken et al. (1959) was associated with the use of a conventional fermentor, and that the lower productivities then reported by other workers were associated with the use of fermentation systems with poorer gas-transfer characteristics. It is for these reasons that it was suggested earlier in this chapter that shake-flask screening processes and production media should be developed to match the gas-transfer characteristics of individual pieces of equipment. It is known that the Q_{O_2} of *F. moniliforme* decreases after the balanced phase (J. C. Swait, unpublished). The media must therefore be designed so that the maximum rate of demand for oxygen can be met by the equipment. If a protracted period of the fermentation is run in conditions of severe oxygen restriction then a reduction in productivity can be expected.

In addition to supplying oxygen, gas-transfer processes can affect the CO_2 content of the brew. In conditions where low levels of inoculum were provided, and when CO_2 was removed from the incoming air to the fermentor, then a protracted "lag" occurred (I.C.I., 1958). Conversely, the supplementation of the air supply by CO_2 in these conditions can greatly reduce the time between inoculation and the onset of production, partly by eliminating the "lag" and partly by increasing the overall growth rate. The yield of gibberellin A_3 was also shown to be increased in these conditions. Where two-stage processes are involved (I.C.I., 1960a) and vigorous inocula are used, then it is probable that sufficient CO_2 is evolved from the organism to meet its own demands.

D. THE EFFECTS OF LIGHT

Observations made by Zweig and Devay (1959), and followed up by Mertz and Henson (1967a) suggested that light can increase the amount of growth and the production of gibberellins in *F. moniliforme* cultures. Furthermore, the inhibitors AMO 1618 and CCC acted differentially on light- and dark-grown cultures (Mertz and Henson, 1967b).

The fact that gibberellin A_3 was not detected in dark-grown cells is hard to reconcile with general experience, and the presentation of plant bioassay results in terms of plant growth per unit of mycelial dry weight makes impossible any quantitative comparison with other work, but there would seem to be a phenomenon here worthy of further, and more quantitative study.

IX. The Production of Gibberellins

1. Gibberellin A_1

Gibberellin A_1 was a component of the original Japanese "gibberellin" and is produced by protracted fermentations held in very acid conditions (less than pH 3.5). The original Japanese media, and that used by Stodola *et al.* (1955) had no magnesium added, although traces would have been present in the water and as impurity, but this condition might also stimulate the production of gibberellin A_1.

2. Gibberellin A_2

Takahashi *et al.* (1955) described the isolation of this compound, together with that of A_1 and A_3. They noted that the ratio between the three products could be varied by varying the growth conditions, and also showed variations between strains.

3. Gibberellin A_3

This is usually referred to as gibberellic acid, and has been the main subject of study. Using an unimproved strain, yields of ca. 1.6 gm/liter have been recorded (Borrow *et al.*, 1964a). Mutation and selection have been predominant factors in improving yields. Production can be maximized as follows:

Temperature
    for growth   31°–32°C
    for production   29°C

Airflow/agitation regime	To be as vigorous as possible
Nitrogen source	(a) Ammonia nitrogen preferred
	(b) Natural plant meals are suitable additional sources
	(c) Concentration to be adjusted to match gas-transfer characteristics of fermentation equipment
Carbon source	(a) Sucrose better than glucose
	(b) Natural plant meals and oils very good. (Note the possibility of these containing precursors.)
	(c) Feed process, or slow release desirable
Salts	(a) Magnesium, potassium, phosphate, and sulfate also required.
	(b) Trace element requirement will be met by impurities in commercial fermentations but must be considered in synthetic media
pH	Range to be pH 3.5–5.5.

Holme and Zacharias (1965) described the production of gibberellin A_3 in continuous culture conditions and reported specific production rates as high as those obtainable in comparable batch culture conditions.

Radioactive gibberellin A_3 has been prepared by McComb (1964) who obtained material with a good specific activity (ca. 4 μc/mg), but in low yield, by a replacement culture technique using sodium acetate-1-^{14}C.

4. Gibberellins A_4, A_7, A_9, A_{12}, A_{14}, A_{16}, and A_{24}

All members of this group have been isolated from fermentations run in conditions in which the pH was increased to greater than pH 5.5 at the time of exhaustion of nitrogen.

Gibberellin A_3-producing strains can be switched over to increased production of gibberellins A_4 and A_7 by increasing the pH to the range pH 6–7.5 after the time of exhaustion of nitrogen, but Takahashi et al. (1957) described the production of gibberellin A_4 on a medium which is unlikely to have run to neutral conditions. Light (Mertz and Henson, 1967a,b) and temperatures in the range of 30°–34°C (Kagawa et al., 1965; Sumiki et al., 1966) have also been suggested as predisposing strains to produce gibberellin A_7. It is unlikely that the high (30%) initial concentration of glucose used by I.C.I. (1963) did more than slow down the production of gibberellin A_7, the primary stimulus being the increased pH (Cross et al., 1960a,b). Gibberellins A_9, A_{12}, and A_{14} are produced in similar conditions to those described for A_7 (Cross et al., 1960b, 1962b; Cross and Norton, 1965; Cross, 1966); but gibberellins A_{16} and A_{24}, while being produced in similar conditions, result from the growth of mutants ACC 917/B47 (Galt, 1968) and M419 (Harrison et al., 1968), respectively.

5. Gibberellins A_{10}, A_{13}, and A_{15}

These compounds were isolated in varying, but usually low yields from the mother liquors of gibberellin A_3 production batches of strain ACC 917 (Hanson, 1966, 1967; Galt, 1965), but the yield of A_{13} was higher from the mutant strain ACC 917/B47, and especially when triparanol (MER 29), an inhibitor of cholesterol biosynthesis, was added to the fermentation.

6. Gibberellins from Other Organisms

Gibberellinlike activity has been reported from many widely different organisms, but none has been shown to produce commercially interesting levels of activity, indeed, the identity of the products as gibberellins is, in many instances, somewhat equivocal (Aubé and Sackston, 1965; Brown and Burlingham, 1968; Curtis, 1957; Galsky and Lippincott, 1967; Katznelson and Cole, 1965; Krasilnikov et al., 1958; Montuelle and Cheminais, 1964; Netien and Oddoux, 1961; Panosyan and Babayan, 1966; Vancura, 1961; Zarnescu and Nita, 1964).

X. Present and Potential Applications

The gibberellins are a widely distributed and fundamentally fascinating group of compounds, the discovery of which has led to great advances in our knowledge of the control of growth in higher plants. The present state of this work has been reviewed by Brian (1966) and Cleland (1969). Commercial interest in the compounds is centered around their use in malting, horticulture, and agriculture (Amdal Co., 1969; Plant Protection Ltd., 1969). Additional applications have also been claimed for their use in protein deficiency and stress conditions (Laboratories Laroche Navarron, 1964), and in the use of waste mycelium as an animal feedstuff supplement (Commercial Solvents and Co., 1961).

In the malting process, gibberellin A_3 added to the steep water reduces the time needed for germination (Briggs, 1963) and improves the quality of the malt.

The use of gibberellin A_3 (gibberellic acid) is established on a commercial scale for the treatment of artichokes, celery, cucumbers, lettuce, winter spinach, cherries, grapes, lemons, oranges, pears, and rhubarb. The response of fruit trees to treatment usually results in the improved quality of the product or increased fruit "set," whereas, with vegetables the rewards are in improved yields and earlier marketing.

One attribute of gibberellin A_3 is to shorten the period of development of biennial plants. Seed crops may therefore be obtained from lettuce and from sugar beet in 1 year instead of 2.

Proposed uses for gibberellin A_3 in apples are to promote the establishment and early bearing of new trees and perhaps to alter fruit shape—an effect more marked with gibberellins A_4 and A_7 (Williams and Stahly, 1969). New applications to other fruit crops are envisaged, and it is probable that in field crops, grass, hops, and sugar cane may be the first to benefit from their use. Interest is increasing in the treatment of many crops with gibberellic acid as a means of facilitating mechanical harvesting. By altering growth habit and structure, or by compressing the harvesting period, losses normally inherent in machine harvesting may be greatly reduced.

Results from animal tests indicate that gibberellin A_3 has very low mammalian toxicity and that it occurs naturally in many, if not all, vegetable foodstuffs, in some cases in amounts greater than would occur from extraneous addition. It is to be expected that increasing applications will be found for this and other members of this interesting group of compounds.

Acknowledgments

The information on the commercial application of gibberellins was kindly provided by Amdal Company and by Plant Protection Ltd.

Thanks are due to Professor P. W. Brian, F. R. S., and Dr. C. T. Calam for making available translations of Japanese and Russian papers; and Dr. J. MacMillan for a preview of chapters from his forthcoming book.

I also would like to thank Professor Brian and my colleagues for helpful criticism.

References

Abbott Laboratories. (1963). Brit. Pat. No. 919,186.
Adamiec, A. (1966). *Acta Soc. Botan. Pol.* **35**, 487–510.
Agnistikova, V. N., Dubovaja, L. P., Lekareva, T. A., Lupova, L. M., Muromtsev, G. S., Kucherov, V. F., and Serebryakov, E. P. (1966). *Mikrobiologia* **35**, 1037–1043.
Amdal Company. (1969). "Pro-Gibb" Promotional Literature.
Arison, B. H., Speth, O. C., and Trenner, N. R. (1958). *Anal. Chem.* **30**, 1083–1085.

Aseeva, I. V. (1963). *Gibberelliny Ikh Deistvie Rast.*, pp. 73-75.
Aubé, C., and Sackston, W. E. (1965). *Can. J. Bot.* **43**, 1335-1342.
Baumgartner, W. E., Lazer, L. S., Dalziel, A. M., Cardinal, E. V., and Varner, E. L. (1959). *J. Agr. Food Chem.* **7**, 422-425.
Baumgartner, W. E., Lazer, L. S., and Dalziel, A. (1963). *Advan. Tracer Methodol.* **1**, 257-262.
Bentley, R., Ghaphery, J. A., and Keil, J. G. (1965). *Arch. Biochem. Biophys.* **111**, 80-87.
Bilai, V. I., and Zakordonets, L. A. (1965). *Mikrobiol. Zh. (Kiev)* **27**, 3-6.
Bilai, V. I., Verner, D. O., Bondarchuk, A. O., V'yun, A. A., and Dymovich, V. O. (1961a). *Mikrobiol. Zh. (Kiev)* **22**, 39-47.
Bilai, V. I., Zanevich, V. E., and Malashenko, Yu.R. (1961b). *Mikrobiol. Zh. (Kiev)* **23**, 34-38.
Bird, H. L., Jr., and Pugh, C. T. (1958). *Plant Physiol.* **33**, 45-46.
Bolgarev, P. T., Muromtsev, G. S., and Nestyouk, M. N. (1962). Bull. Inventions No. 23. Certificate No. 762055/23-24.
Borrow, A., Brian, P. W., Chester, V. E., Curtis, P. J., Hemming, H. G., Henehan, Catherine, Jefferys, E. G., Lloyd, P. B., Nixon, I. S., Norris, G. L. F., and Radley, Margaret. (1955). *J. Sci. Food Agr.* **6**, 340-348.
Borrow, A., Jefferys, E. G., Kessell, R. H. J., Lloyd, Eithne C., Lloyd, P. B., and Nixon, I. S. (1961). *Can. J. Microbiol.* **7**, 227-276.
Borrow, A., Brown, Sheila, Jefferys, E. G., Kessell, R. H. J., Lloyd, Eithne C., Lloyd, P. B., Rothwell, A., Rothwell, B., and Swait, J. C. (1964a). *Can. J. Microbiol.* **10**, 407-444.
Borrow, A., Brown, Sheila, Jefferys, E. G., Kessell, R. H. J., Lloyd, Eithne C., Lloyd, P. B., Rothwell, A., Rothwell, B., and Swait, J. C. (1964b). *Can. J. Microbiol.* **10**, 445-466.
Brian, P. W. (1966). *Int. Rev. Cytol.* **19**, 229-266.
Briggs, D. E. (1963). *J. Inst. Brewing (London)* **69**, 244-248.
Brown, M. E., and Burlingham, S. K. (1968). *J. Gen. Microbiol.* **53**, 135-144.
Brown, J. C., Cross, B. E., and Hanson, J. R. (1967). *Tetrahedron* **23**, 4095-4103.
Bu'Lock, J. D., Hamilton, D., Hulme, M. A., Powell, A. J., Smalley, H. M., Shepherd, D., and Smith, G. N. (1965). *Can. J. Microbiol.* **11**, 765-778.
Cappellini, R. A., and Peterson, J. L. (1966). *Bull. Torrey Bot. Club* **93**, 52-55.
Carito, S. L., and Pisano, M. A. (1966). *Appl. Microbiol.* **14**, 39-44.
Cavell, B. D., and MacMillan, J. (1967). *Phytochemistry* **6**, 1151-1154.
Cavell, B. D., MacMillan, J., Pryce, R. J., and Sheppard, A. C. (1967). *Phytochemistry* **6**, 867-874.
Cleland, R. E. (1969). In "The Physiology of Plant Growth and Development" (M. B. Wilkins, ed.) pp. 49-77. McGraw Hill, London.
Commercial Solvents Co. (1961). Brit. Pat. No. 860, 765.
Cross, B. E. (1966). *J. Chem. Soc., C.* 501-504.
Cross, B. E. (1968). In "Progress in Phytochemistry" (L. Reinhold and Y. Liwschitz, eds.) pp. 195-222. Wiley (Interscience), New York.
Cross, B. E., and Myers, P. L. (1969). *Phytochemistry* **8**, 79-83.
Cross, B. E., and Norton, K. (1965). *J. Chem. Soc.* 1570-1572.
Cross, B. E., and Norton, K. (1966). *Tetrahedron Lett.* **48**, 6003-6007.
Cross, B. E., Galt, R. H. B., and Hanson, J. R. (1960a). *Tetrahedron Lett.* **15**, 18-22.
Cross, B. E., Galt, R. H. B., and Hanson, J. R. (1960b). *Tetrahedron Lett.* **23**, 22-24.

Cross, B. E., Galt, R. H. B., Hanson, J. R., and Klyne, W. (1962a). *Tetrahedron Lett.* **4**, 145–150.
Cross, B. E., Galt, R. H. B., and Hanson, J. R. (1962b). *Tetrahedron* **18**, 451–459.
Cross, B. E., Galt, R. H. B., Hanson, J. R., and (in Part) Curtis, P. J., Grove, John Frederick, and Morrison, A. (1963). *J. Chem. Soc.* 2937–2943.
Cross, B. E., Galt, R. H. B., and Hanson, J. R. (1964). *J. Chem. Soc.* 295–300.
Cross, B. E., Norton, K., and (in Part) Stewart, J. C. (1968). *J. Chem. Soc. C.*, 1054–1063.
Curtis, R. W. (1957). *Science* **125**, 646.
Curtis, P. J., and Cross, B. E. (1954). *Chem. Ind. (London)*, 1066.
Darken, Marjorie A., Jensen, A. L., and Shu, P. (1959). *Appl. Microbiol.* **7**, 310–303.
De Rose, A. F. (1962). U. S. Pat. No. 3, 057, 878.
Diekmann, H. (1967). *Arch. Mikrobiol.* **58**, 1–5.
Dietrich, K. R. (1960). Ger. Pat. No. 1, 081, 402.
Erokhina, L. I. (1967). *Genetika* **7**, 77–82.
Erokhina, L. I., and Sokolova, E. V. (1966). *Genetika* **1**, 109–115.
Ezhov, V. A. (1965). *Uch. Zap. Mord. Gos. Univ.* **46**, 81–85.
Flippin, R. S., Smith, C., and Mickelson, M. N. (1964). *Appl. Microbiol.* **12**, 93–95.
Fuska, J., Kuhr, I., Podojil, M., andŠevčík, V. (1960). *Folia Microbiol. (Prague)* **6**, 18–21.
Fuska, J., Kuhr, I., Ševčík, V., Musilek, V., and Podojil, M. (1962). Czech. Pat. No. 104,329.
Fuska, J., Kuhr, I., Zajicek, I. (1964). *Mikrobiologiya* **33**, 783–786.
Galsky, A. G., and Lippincott, J. A. (1967). *Plant Physiol. Suppl.* **42**, S–29.
Galt, R. H. B. (1965). *J. Chem. Soc.*, 3143–3151.
Galt, R. H. B. (1968). *Tetrahedron* **24**, 1337–1339.
Gancendo, J. M., Gancendo, C., and Asensio, C. (1967). *Arch. Biochem. Biophys.* **119**, 588–590.
Geissman, T. A., Verbiscar, A. J., Phinney, B. O., and Cragg, G. (1966). *Phytochemistry* **5**, 933–947.
Gerzon, K., Bird, H. L. Jr., and Woolf, D. O. (1957). *Experientia* **13**, 487–489.
Gordon, W. L. (1960). *Nature* **186**, 698–700.
Graham, H. D., and Henderson, J. H. M. (1961). *Plant Physiol.* **36**, 405–408.
Grove, John Frederick (1961). *Quart. Rev.* **15**, 56–70.
Grove, John Frederick (1963). In "Biochemistry of Industrial Microorganisms" (C. Rainbow and A. H. Rose, eds.) pp. 320–340. Academic Press, New York.
Hanson, J. R. (1966). *Tetrahedron* **22**, 701–703.
Hanson, J. R. (1967). *Tetrahedron* **23**, 733–735.
Harhash, A. W. (1966). *Acta Biol. Med. Ger.* **17**, 8–16.
Harhash, A. W. (1967a). *Acta Biol. Med. Ger.* **18**, 1–7.
Harhash, A. W. (1967b). *Acta Biol. Med. Ger.* **18**, 121–124.
Harrison, D. M., MacMillan, J., and Galt, R. H. B. (1968). *Tetrahedron Lett.* **27**, 3137–3139.
Head, W. F. (1959). *J. Amer. Pharm. Assoc., Sci. Ed.* **48**, 631–634.
Heinen, W. (1960). *Acta Bot. Neer.* **9**, 167–190.
Heinen, W., and Linskens, H. F. (1960). *Naturwissenschaften* **47**, 18.
Hill, R. D., Unrau, A. M., and Canvin, D. T. (1966). *Can. J. Chem.* **44**, 2077–2082.
Hitzman, D. O., and Mills, A. M. (1963). U. S. Pat. No. 3, 984, 106.
Holbrook, A. H., Edge, W. J., and Bailey, F. (1961). *Advan. Chem. Ser.* **28**, 159–167.
Holme, T., and Zacharias, B. (1965). *Biotechnol. Bioeng.* **7**, 405–415.
Ikekawa, N., Kagawa, T., and Sumiki, Y. (1963). *Proc. Jap. Acad.* **39**, 507–512.

Imperial Chemical Industries Ltd. (1957). Brit. Pat. No. 783,611. (= U. S. Pat. No. 2,842,051; Belg. Pat. No. 570,005).
Imperial Chemical Industries Ltd. (1958). Brit. Pat. No. 803,591. (= U. S. Pat. No. 2,865,812; Australian Patent No. 24,274/56).
Imperial Chemical Industries Ltd. (1959). British Pat. No. 821,733. (= U. S. Patent No. 2,950,288; Aust. Pat. No. 36,374/58; Belg. Pat. No. 566,541; Ger. Pat. No. 1,088,911).
Imperial Chemical Industries Ltd. (1960a). Brit. Pat. No. 838,032. (= U. S. Pat. No. 2,906,670; Aust. Pat. No. 35,014/58; Belg. Pat. No. 567,954; Ger. Pat. No. 1,155,413).
Imperial Chemical Industries Ltd. (1960b). Brit. Pat. No. 838,033. (= U. S. Pat. No. 2,906,671; Aust. Pat. No. 35,015/58; Belg. Pat. No. 567,956; Ger. Pat. No. 1,150,348).
Imperial Chemical Industries Ltd. (1960c). Brit. Pat. No. 839,652. (= U. S. Pat. No. 2,990,337; Aust. Pat. No. 40,951/48; Belg. Pat. No. 571,077).
Imperial Chemical Industries Ltd. (1960d). Brit. Pat. No. 844,341. (= U. S. Pat. No. 2,977,285).
Imperial Chemical Industries Ltd. (1960e). Brit. Pat. No. 850,018. (= U. S. Pat. No. 2,906,673; Aust. Pat. No. 34,926/58; Belg. Pat. No. 567,955; Ger. Pat. No. 1,150,347).
Imperial Chemical Industries Ltd. (1963). Brit. Pat. No. 914,893.
Imperial Chemical Industries Ltd. (1964a). Brit. Pat. No. 957,634. (= Fr. Pat. No. 3564M).
Imperial Chemical Industries Ltd. (1964b). Brit. Pat. No. 967,596.
Imshenetsky, A. A., and Ulyanova, O. M. (1961). *Dokl. Akad. Nauk Akad. Sci. SSSR.* **138**, 1204–1207.
Imshenetsky, A. A., and Ulyanova, O. M. (1962a). *Mikrobiologia* **31**, 832–837.
Imshenetsky, A. A., and Ulyanova, O. M. (1962b). *Nature* **195**, 62–63.
Instytut Antybiotyków (1965). Pol. Pat. No. 49,464.
Kagawa, T., Fukinbara, T., and Sumiki, Y. (1963). *Agr. Biol. Chem.* **27**, 598–599.
Kagawa, T., Fukinbara, T., and Sumiki, Y. (1965). *Agr. Biol. Chem.* **29**, 285–291.
Kathirvelu, R., and Mahadevan, A. (1967). *Curr. Sci.* **36**, 396.
Katznelson, H., and Cole, S. E. (1965). *Can. J. Microbiol.* **11**, 733–741.
Kavanagh, F., and Kuzel, N. R. (1958). *J. Agr. Food Chem.* **6**, 459–463.
Kawarada, A., Takahashi, N., Kitamura, H., Seta, Y., Takai, M., and Tamura, S. (1955). *Bull. Agri. Chem. Soc. Jap.* **19**, 84–86.
Kitamura, H., Kawarada, A., Seta, Y., Takahashi, N., Otsuki, T., and Sumiki, Y. (1953). *Bull. Agri. Chem. Soc. Jap.* **27**, 545–549.
Knight, S. G. (1966). *Can. J. Microbiol.* **12**, 420–422.
Krasilnikov, N. A., Chailakhyan, M. K., Asseva, I. V., and Khlopenkova, L. P. (1958). *Dokl. Akad. Nauk SSSR* **123**, 1124–1127.
Krasilnikov, N. A., Shirokov, O. G., and Kuchaeva, A. G. (1963). *Gibberelliny Ikh Deistvie Rast.*, pp. 39–44.
Kucherov, V. F., Gurvich, I. A., Simolin, A. V., and Milshtein, I. M. (1965). *Dokl. Akad. Nauk SSSR* **163**, 765–767.
Kuhr, I. (1962). *Folia Microbiol. (Prague)* **7**, 358–363.
Kuhr, K., Fuska, J., Podojil, M., and Ševčík, V. (1961). *Folia Microbiol. (Prague)* **6**, 179–185.
Kumar, S. A., and Mahadevan, S. (1963). *Arch. Biochem. Biophys.* **103**, 516–518.
Kurosawa, E. (1934). *Ann. Phytopathol. Soc. Jap.* **4**, 65–66.
Laboratories Laroche Navarron (1964). Fr. Pat. No. M2480.
Lavate, W. V., and Bentley, T. (1964). *Arch. Biochem. Biophys.* **108**, 287–291.
Loewenburg, J. R., and Reese, E. T. (1957). *Can. J. Microbiol.* **3**, 643–650.
McComb, A. J. (1964). *J. Gen. Microbiol.* **34**, 401–411.

MacMillan, J. (1969). Personal preview from "Gibberellins, their Action and Chemistry." In press Academic Press, London and New York.
MacMillan, J., and Suter, P. J. (1963). Nature 197, 790.
Martin, S. M., and Adams, G. A. (1956). Can. J. Microbiol. 2, 715–721.
Merck and Co. Inc. (1960a). Brit. Pat. No. 847,435.
Merck and Co. Inc. (1960b). Ger. Pat. No. 1,085,878.
Merck and Co. Inc. (1966). Brit. Pat. No. 1,049,034.
Mertz, D., and Henson, W. (1967a). Nature 214, 844–846.
Mertz, D., and Henson, W. (1967b). Physiol. Plant. 20, 187–199.
Montuelle, B., and Cheminais, L. (1964). C. R. Acad. Sci. (24) 258, 6016–6017.
Muromtsev, G. S., and Dubovaya, L. P. (1964). Mikrobiologia 33, 1048–1055.
Muromtsev, G. S., and Nestyuk, M. N. (1962). Izv. Akad. Nauk SSSR. Ser. Biol. 27, 825–831.
Muromtsev, G. S., Dendze-Pletman, B. B., Kleiner, G. I., and Nestyuk, M. N. (1962). USSR Certificate No. 147,294.
Muromtsev, G. S., Nestyuk, M. N., Kleiner, G. I., Ionova, I. V., Dendze-Pletman, B. B., and Krutova, R. L. (1963). Gibberelliny Ikh Deistvie Rast., pp. 54–59.
Muromtsev, G. S., Agnistikova, V. N., Lupova, L. M., Dubovaya, L. P., Lekareva, T. A., Serebryakov, E. P., and Kucherov, V. F. (1966). Khim. Prirodn. Soedin. 2, 114–120.
Muromtsev, G. S., Rakovskii, Y. S., Dubovaya, L. P., Temnikova, T. V., and Fedchenko, A. N. (1968). Prikl. Biokhim. Mikrobiol. 4, 398–407.
Nagata, Y., and Hayashi, K. (1957). Nippon Nogei Kagaku Kaishi 31, 575–578.
Nakamura, Y., Shimomura, T., and Ono, J. (1957). Nippon Nogei Kagaku Kaishi 31, 669–672.
Nakamura, Y., Shimomura, T., and Ono, J. (1958). Nippon Nogei Kagaku Kaishi 32, 800–802.
Nestyouk, M. N., Dendze-Pletman, B. B., Ionova, N. B., Iofo, R. N., Kleiver, G. I., Kravchenko, B. F., Krutova, R. L., Muromtsev, G. C., and Rusanova, N. V. (1961). Bull. Invent. No. 18. USSR Certificate No. 141,352.
Netien, G., and Oddoux, L. (1961). C. R. Acad. Sci. 253, 520–522.
Panosyan, A. K., and Babayan, G. S. (1966). Biol. Zh. Arm. 19, 78–84.
Pfizer and Co. Inc. (1959). Brit. Pat. No. 819,110.
Phinney, B. O., and Spector, C. (1967). Ann. N. Y. Acad. Sci. 144, 204–210.
Plant Protection Ltd. (1969). "Berelex." Technical Service Department, Yalding, Kent, Great Britain.
Podojil, M., and Ricicova, A. (1965). Folia Microbiol. (Prague) 10, 55–59.
Podojil, M., and Ševčík, V. (1960). Folia Microbiol. (Prague) 5, 192–197.
Podojil, M., Ševčík, V., Kuhr, I., and Fuska, J. (1961). Folia Microbiol. (Prague) 6, 273–276.
Probst, G. W. (1961). U. S. Pat. No. 2,980,700.
Redemann, C. T. (1959). U. S. Pat. No. 2,918,413.
Ricicova, A., Podojil, M., Musilek, V., and Ševčík, V. (1960). Folia Microbiol. (Prague) 5, 181–191.
Sandhu, R. S. (1960). Phytopathol. Z. 37, 33–60.
Schreiber, K., Schneider, G., Sembdner, G., and Focke, I. (1966). Phytochemistry 5, 1221–1225.
Serebryakov, E. P., Kucherov, V. F., and Muromtsev, G. S. (1966). Khim. Prir. Soedin. 2, 55–58.

Serzedello, A., and Whitaker, N. (1960). *Rev. Agr. (São Paulo)* **35**, 15–24.
Shklyar, M. Z., and Globus, G. A. (1961). *Byull. Nauchn. Tekhn. Inform. Sel'skokhoz. Mikrobiol.* **10**, 6–8.
Shklyar, M. Z., and Globus, G. A. (1963). *Gibberelliny Ikh Deistvie Rast.*, 50–53.
Siddiqui, I. R., and Adams, G. A. (1961). *Can. J. Chem.* **39**, 1683–1694.
Société d'Études et d'Applications Biochimiques. (1963). Brit. Pat. No. 936,548. (= U. S. Pat. No. 3,118,908; Fr. Pat. No. 1,278,673; Belg. Pat. No. 609,332).
Sokolova, E. V., and Erokhina, L. I. (1966). *Tr. Mosk. Obshchest. Ispyt. Prir.* **22**, 253–256.
Spector, C., and Phinney, B. O. (1968). *Physiol. Plant.* **21**, 127–136.
Sternberg, M. (1962). *Arch. Biochem. Biophys.* **98**, 299–304.
Sternberg, M., and Voinescu, R. (1961). *Folia Microbiol. (Prague)* **6**, 189–191.
Stodola, F. H. (1958). "Source book on Gibberellin 1828–1957," Agr. Res. Service, U. S. Dept. Agr. (ARS-71-11).
Stodola, F. H., Raper, K. B., Fennell, Dorothy I., Conway, H. F., Sohns, V. E., Langford, C. T., and Jackson, R. W. (1955). *Arch. Biochem. Biophys.* **54**, 240–245.
Stodola, F. H., Nelson, G. E. N., and Spence, D. J. (1957). *Arch. Biochem. Biophys.* **66**, 438–444.
Stoll, C. (1954). *Phytopathol. Z.* **22**, 233–274.
Stoll, C., and Renz, J. (1957). *Phytopathol. Z.* **29**, 380–387.
Stowe, B. B., and Yamaki, T. (1957). *Annu. Rev. Plant Physiol.* **8**, 181–216.
Stowe, B. B., Stodola, F. H., Hayashi, T., and Brian, P. W. (1961). *In* "Plant Growth Regulation" (R. M. Klein, ed.) pp. 465–471. Iowa State Univ. Press, Ames, Iowa.
Sumiki, Y., Kagawa, T., and Fukanbara. (1966). Jap. Pat. No. 16558(66).
Takahashi, N., Kitamura, H., Kawarada, A., Seta, Y., Takai, M., Tamura, S., and Sumiki, Y. (1955). *Bull. Agr. Chem. Soc. Jap.* **19**, 267–277.
Takahashi, N., Seta, Y., Kitamura, H., and Sumiki, Y. (1957). *Bull. Agr. Chem. Soc. Jap.* **21**, 396–398.
Terui, M., and Kagawa, H. (1958). *Hirosaki Daigaku Nogakubu Gakujutsu Hokoku* **4**, 88–91.
Theriault, R. J., Friedland, W. C., Peterson, M. H., and Sylvester, J. C. (1961). *J. Agr. Food Chem.* **9**, 21–23.
Udagawa, K., and Kinoshita, S. (1961). *Nippon Nogei Kagaku Kaishi* **35**, 219–223.
Vancura, V. (1961). *Nature* **192**, 88–89.
Verbiscar, A. J., Cragg, G., Geissmann, T. A., and Phinney, B. O. (1967). *Phytochemistry* **6**, 807–814.
Wakagi, S. (1958). *J. Fermentation Ass. Jap.* **16**, 150–164.
Washburn, W. H., Scheske, F. A., and Schenck, J. R. (1959). *J. Agr. Food Chem.* **7**, 420–422.
Wierzchowski, P., and Wierzchowska, Z. (1961). *Naturwissenschaften* **48**, 653.
Williams, M. W., and Stahly, E. A. (1969). *J. Amer. Soc. Hort. Sci.* **94**, 17–19.
Yabuta, T., and Hayashi, T. (1936). *Agr. Hort. (Japan)* **11**, 27–33.
Yabuta, T., and Hayashi, T. (1937). *Agr. Hort. (Japan)* **12**, 1073–1083.
Yabuta, T., and Hayashi, T. (1939). *J. Agr. Chem. Soc. Jap.* **15**, 257–266.
Yabuta, T., Kambe, K., and Hayashi, T. (1934). *J. Agric. Chem. Soc. Jap.* **10**, 1059–1068.
Yabuta, T., Sumiki, Y., and Uno, S. (1939). *J. Agric. Chem. Soc. Jap.* **15**, 1209–1220.

Yabuta, T., Sumiki, Y., Katayama, E., and Motoyama, H. (1940). *J. Agr. Chem. Soc. Jap.* **16,** 1157–1158.
Yabuta, T., Sumiki, Y., Tamura, T., and Murayama, N. (1941). *J. Agric. Chem. Soc. Jap.* **17,** 673–676.
Zaichenko, A. M., and Koval, E. Z. (1966). *Mikrobiol. Zh. (Kiev)* **28,** 8–14.
Zarnescu, A., and Nita, L. (1964). *An. Inst. Cercet. Cereale Plante Deh Fundulea, Inst. Cent. Cercet. Agr. Ser.* **B32,** 443–445.
Zweig, G., and Devay, J. E. (1959). *Mycologia* **51,** 877–886.

Metabolism of Acylanilide Herbicides[1]

RICHARD BARTHA AND DAVID PRAMER

Department of Biochemistry and Microbiology,
College of Agriculture and Environmental Science,
Rutgers — The State University, New Brunswick, New Jersey

I. Introduction	317
II. Metabolism of Acylanilides	319
A. Liberation and Degradation of the Aliphatic Moiety	319
B. Transformations of the Aromatic Moiety	324
C. Proposed Pathways of Acylanilide Metabolism	332
III. Biological Activity of Metabolites	336
IV. Concluding Comments	338
References	339

> There is relatively little information about the ultimate fate of persistent pesticides in soil or in other parts of any ecosystem, or about the sequence in which the degradation processes take place. For some chemicals of interest, the initial products formed are known and are measurable in monitoring programs; but for these chemicals, the products formed next and the sequence in which they are formed are not known or are known for only a few types of natural habitats. Until these products are identified and their potential biological activities are ascertained, it is impossible to assess meaningfully their toxicity to man or to the biota or their residence times in nature.
>
> *Report of the Committee on Persistent*
> *Pesticides of the National Research Council,*
> *May, 1969*

I. Introduction

It has been estimated that every year the foliage and soils of the United States are doused with approximately one billion pounds of synthetic organic pesticides. In this process more than 800 substances are used in different registered formulations that now exceed 60,000 in number (Minter *et al.*, 1969). Acylanilides are a relatively recent addition to the arsenal of chemical weapons employed by man in his war against pests. They are effective weed killers displaying a number of attractive features that include selective toxicity and low cost. These and other attributes have focused attention on the acylanilides and their use as herbicides has increased. The names, sources, and structures of some acylanilide herbicides are presented in Fig. 1.

[1]The review of the pertinent literature was closed in December, 1969.

CHEMICAL NAME	STRUCTURE	SOURCE	COMMON and TRADE NAME
N-(3-CHLORO-4-METHYLPHENYL) -2-METHYLPENTANAMIDE	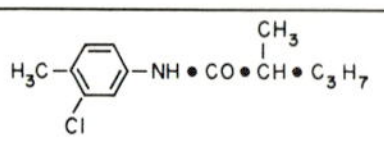 H$_3$C-⟨⟩-NH•CO•CH(CH$_3$)•C$_3$H$_7$; Cl	NIAGARA-FMC	solan Solan
N-(3,4-DICHLOROPHENYL) CYCLOPROPANCARBOXAMIDE*	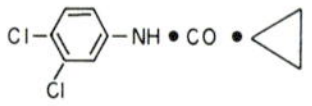 Cl-⟨⟩-NH•CO•◁ ; Cl	GULF OIL	cypromid Clobber, S-6000
N-(3,4-DICHLOROPHENYL) METHACRYLAMIDE	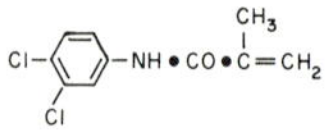 Cl-⟨⟩-NH•CO•C(CH$_3$)=CH$_2$; Cl	NIAGARA-FMC	dicryl Dicryl
N-(3,4-DICHLOROPHENYL)* -2-METHYLPENTANAMIDE	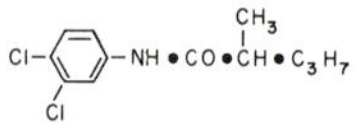 Cl-⟨⟩-NH•CO•CH(CH$_3$)•C$_3$H$_7$; Cl	NIAGARA-FMC	karsil Karsil
N-(3,4-DICHLOROPHENYL) PROPIONAMIDE	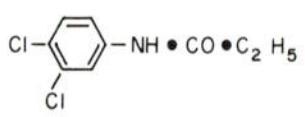 Cl-⟨⟩-NH•CO•C$_2$H$_5$; Cl	MONSANTO ROHM & HAAS	propanil Rogue STAM F34
N-ISOPROPYL-2 -CHLOROACETANILIDE	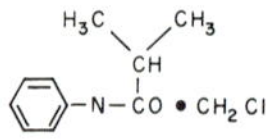 ⟨⟩-N(CH(CH$_3$)$_2$)-CO•CH$_2$Cl	MONSANTO	propachlor Ramrod

FIG. 1. Acylanilide herbicides. Compounds marked by an asterisk are not marketed at present in the United States.

Solan, dicryl, and propanil are employed at levels of 1.0–6 lb/acre for postemergence control of weeds. Solan is recommended for use in cultivating tomato, dicryl for cotton, and propanil for both tomato and rice. Propachlor (3–6 lb/acre) is applied for preemergence control of weeds in corn and soybean, but cypromid and karsil, which were developed for postemergence control of weeds in corn and other crops, are not marketed in the United States at this time (Weed Society of America, 1967). Acylanilides that are the more effective herbicides have in common certain molecular dimensions (length of about 13 Å extended and 11 Å collapsed) and substituents, particularly chlorine in positions 3 and 4 of the aromatic moiety (Huffman and Allen, 1960). From an economic point of view propanil is by far the most important acylanilide. Its ability to selectively control barnyard grass *(Echinocloa crusgalli)* in rice *(Oryza sativa)* fields dramatically increases yields (Smith, 1961; Syrbu, 1967; Oelke and Morse, 1968). Sales of propanil were valued at $14 million in 1967 (Anonymous, 1968) and this figure is likely to have increased and will continue to do so, considering the use of rice as a dietary staple by a major portion of the world population.

The mode of herbicidal action of acylanilides is complex and not as yet clearly defined. They are able at low levels to inhibit the Hill reaction displayed by isolated chloroplasts (Moreland and Hill, 1963). The phytotoxicity of propanil is manifested by reductions in plant growth, respiration, and oxidative phosphorylation (Hofstra and Switzer, 1968); propachlor interferes with protein synthesis (Duke *et al.*, 1967; Jaworski, 1969a). The selective action of acylanilides depends on their stability in plants. Resistant plants are able to metabolize and detoxify them, but sensitive plants are unable to do so and are destroyed. The biochemistry of the detoxification reactions in resistant plants will be described subsequently but the mode of action of acylanilide herbicides will not be considered further.

Attention here will be concentrated on the transformations of acylanilides, as mediated by microorganisms and other biological systems, and results of studies at the enzyme level also will be reviewed. Evidence for an unexpected type of metabolic product will be presented and used to support generalizations concerning the influence of molecular configuration on the biochemical transformation of acylanilides and related herbicides. The pathways proposed here for acylanilide metabolism may apply in part to the metabolism of other compounds that also contain a substituted aniline moiety. This is true of some phenylcarbamate and phenylurea herbicides. Present understanding of the degradation of these substances was recently reviewed by Herrett (1969) and Geissbühler (1969), respectively.

II. Metabolism of Acylanilides

The commercial and agricultural literature describes acylanilide herbicides as, "completely metabolized by the crop plant, e.g., rice" .... "degraded in soil" .... and, as a result of microbial action, "broken down quite rapidly" (Weed Society of America, 1967). The remainder of this review can serve as a basis for a decision by the reader as to the extent to which these statements are valid.

A. LIBERATION AND DEGRADATION OF THE ALIPHATIC MOIETY

1. Liberation and Degradation in Soil

Soil samples treated with high concentrations (approximately 500 ppm) of dicryl, karsil, or propanil displayed an initial increase and subsequent decrease in carbon dioxide production. This respiratory pattern was interpreted as evidence that the herbicides were incompletely oxidized and transformed in part to a relatively stable and inhibitory product (Bartha *et al.*, 1966, 1967). A detailed study using propanil as a model system demonstrated that carbon dioxide production was increased when soil samples were amended with propionate, the aliphatic moiety of the herbicide, and decreased when 3,4-dichloraniline (DCA), the aromatic moiety of the herbicide, was the soil supplement. The algebraic sum of these two effects corresponded to the result obtained with propanil, and Bartha et al. (1966, 1967) suggested that in soil the herbicide molecule was hydrolyzed to propionate and DCA. The propionate was oxidized in part to carbon dioxide, but the DCA persisted and caused a decrease in soil respiration. Since no comparable effects were observed in tests with sterilized soil, it was concluded (Bartha et al., 1967; Bartha and Pramer, 1967) that the transformation of propanil was biochemical and mediated by microorganisms.

Gas chromatographic evidence for the hydrolytic cleavage of acylanilide herbicides in soil was obtained by Bartha (1968), and he also observed that degradation rate was a function of chain length of the aliphatic moiety of 3,4-dichloroacylanilides. Propanil (C_3), dicryl (C_4), and karsil (C_6) were decomposed in the order listed. The aliphatic moieties of the solan and karsil molecules are identical, and, although the aromatic moiety of the former compound is 3-chloro-4-methylaniline and that of the latter is DCA, there was no significant difference in decomposition rate of these compounds in soil. An alkyl substituent on the nitrogen atom of the unchlorinated aniline moiety of the propachlor endowed the compound with considerable stability in soil (Bartha, 1968; 1969a).

2. Liberation and Degradation by Microbial Cultures and Enzymes

Various microorganisms are equipped biochemically to cleave the C — N bond and liberate the side chain of acylanilides. The N-formyl-L-kynurenine hydrolase partially purified from *Neurospora crassa* by Jacoby (1954) was highly specific. It catalyzed hydrolysis of formyl, but not acetyl compounds. Crude and partially purified cell-free preparations from a strain of *Pseudomonas striata* grown on a phenylcarbamate herbicide, as a sole source of energy, and organic carbon were active on both propanil and its acetic acid analog. The former compound was hydrolyzed more rapidly than the latter (Kearney, 1965).

Bartha et al. (1967) reported the isolation of a fungus from propanil-treated soil able to develop in a medium containing the herbicide as the only organic carbon and energy source. The organism was identified as *Fusarium solani* (Lanzilotta, 1968). It hydrolyzed propanil, forming propionate and DCA. The propionate was used for growth but the DCA accumulated in the medium to levels that inhibited growth and, eventually, caused cell leakage and lysis. Kaufman and Miller (1969) also isolated *F. solani* from propanil-enriched soil. Bacteria, particularly *Pseudomonas* species, can utilize propanil, but develop poorly on laboratory media that contain the herbicide as the sole organic constituent (unpublished results).

Lanzilotta (1968) had difficulty obtaining extracts of *F. solani* that would hydrolyze propanil, however, enzyme activity was detected when acetanilide was used to replace propanil as substrate, and the fungal acylamidase was concentrated by salt precipitation and characterized. The relation of substrate concentration to reaction velocity was described by Michaelis-Menten kinetics, and the K_m for acetanilide was estimated at 0.195 mM. Propachlor acted as a competitive inhibitor of acetanilide hydrolysis and had a K_i of 0.165 mM. Hydrolysis rates were decreased by various *para* substitutions of acetanilide. 3,4-Dichloroacetanilide was less susceptible than the unchlorinated parent compound to enzyme action, but propanil was hydrolyzed much more rapidly than unchlorinated propionanilide. Chloro substitution in the acyl moiety of acetanilide reduced its ability to serve as substrate for the enzyme.

Acyl chain length had a marked effect on enzyme activity. The fungal acylamidase was highly specific for *N*-acetylarylamines. It did not catalyze hydrolysis of formanilide or butyranilide, and acetanilide was hydrolyzed at 42 times the rate of propionanilide. The enzyme produced by *F. solani* did not catalyze the hydrolysis of dicryl, karsil, propachlor, phenylcarbamates, or phenylureas (Lanzilotta, 1968).

Sharabi and Bordeleau (1969) isolated from enriched soil two species of *Penicillium* and one species of *Pulullaria* able to decompose karsil. Primary products of karsil hydrolysis by cells and cell-free extracts of a *Penicillium* species were identified as 2-methylvaleric acid and DCA. The former compound was metabolized to formic and acetic acids when used by the fungus as a carbon source. The specificity of the partially purified acylamidase was measured using 23 acylanilides. Activity increased with increasing chain length to four carbons. Substitution or branching of the *N*-acyl group affected enzyme activity and the effect was related to both the nature and location of the substituent. Deacylation was also influenced by the type and position of substituents on the phenyl ring. The enzyme had no activity on tested phenylcarbamate and phenylurea herbicides.

Apparently, different microorganisms produce acylamidases with substantially different activities and substrate specificities. A major difference between the *Penicillium* and *F. solani* acylamidases is the inability of the latter to hydrolyze karsil and dicryl.

3. *Liberation and Degradation by Other Biological Systems*

The metabolism of acylanilides by plants has been studied in an effort to identify the biochemical basis of their selective action as herbicides.

McRae *et al.* (1964) employed propanil-^{14}C to demonstrate that the herbicide was detoxified by hydrolysis in rice and other resistant plants, but sensitive plants, such as barnyard grass, were unable to transform the compound. This report is consistent with more recent investigations (Still and Kuzirian, 1967) in which tissue homogenates and partially purified enzyme preparations from leaves of resistant and sensitive plants were compared, and the former were observed to have much greater acylamidase activity than the latter. The rice enzyme was associated with particles and could not be solubilized. Radioautography established that a number of different products resulted from propanil degradation, but, of these, only DCA was identified (Still and Kuzirian, 1967).

Frear and Still (1968) used an enzyme concentrate to establish the substrate range of rice acylamidase. Propanil was hydrolyzed more rapidly than other acylanilides with longer or shorter side chains. Alkyl branching at position α or β to the carbonyl rendered molecules resistant to enzyme action. This observation explains, at least in part, the report of McRae *et al.* (1964) that rice is sensitive to karsil. The acyl moiety of karsil branches at the β-position and would, therefore, not be detoxified in the plant by deacylation. Enzyme action was influenced also by the number and location of chlorine substitutions. The susceptibility of compounds to hydrolysis decreased in the following order: 2,3-dichloro, 2,4-dichloro, 2-chloro, 3-chloro, 3,4-dichloro, 3,5-dichloro, 2,5-dichloro, 4-chloro, and 2,6-dichloro. These differences appeared to result from steric and electronic effects.

Of particular interest were the observations of Bowling and Hudgins (1966) and McRae *et al.* (1964) that certain carbamate and organophosphate insecticides were inhibitors of rice acylamidase at levels as low as 10^{-6} M (Frear and Still, 1968). If these compounds are applied in combination with propanil to rice fields, they interfere with detoxification of the herbicide and severe damage to the crop can result (Unger *et al.*, 1964; Matsunaka, 1968). Yih *et al.* (1968b) detected an accumulation of N-(3,4-dichlorophenyl)lactamide in barnyard grass and in rice that was pretreated with an acylamidase inhibitor, and suggested conversion to this intermediate as a step that immediately precedes propanil detoxification by hydrolysis.

To determine the fate of the aliphatic moiety of the herbicide after liberation by enzymatic hydrolysis, Still (1968) treated pea and rice plants with propanil-^{14}C labeled on either C-1 or C-3 of the propionate moiety. Radioactivity was detected throughout the tissue of root-treated plants. Time-course studies in which $^{14}CO_2$ was recovered, indicated that an intact C_3 acyl moiety was cleaved from the herbicide and metabolized in plants to CO_2 by β-oxidation. The appearance of radioactivity in shoot tissue was attributed to assimilation of products of propionate metabolism. Both the susceptible pea plants and the tolerant rice plants oxidized a large part of propanil-^{14}C to $^{14}CO_2$, and Still (1968) noted that this result is not consistent with the hypothesis that the basis of selective herbicidal action by propanil is detoxification by cleavage of the amide bond in rice and other resistant plants. An explanation for this inconsistency may be the fact that in Still's 1968 studies plants were root treated. It is known that roots of rice and barnyard grass contain the same amount of acylamidase activity, but rice leaves display 60 times more acylamidase activity than barnyard grass leaves (Frear and Still, 1968). Apparently, note must be taken of differences in enzyme distribution in determining the relative phytotoxicity of propanil to different plant species and in studies of the biochemical basis in plants of herbicide resistance and sensitivity.

There have been only two studies of the metabolism by plants of acylanilide herbicides other than propanil. The report by McRae et al. (1964) that rice was unable to degrade karsil was described and discussed previously, and only the work of Jaworski (1969b) remains to be presented. He investigated the transformation of propachlor in plants such as corn and soybean that are resistant to the herbicide. Except for the elimination of a chlorine atom from the acyl chain, propachlor remained intact and reacted with a normal but unknown plant metabolite to form a polar product that was not identified.

Acylanilide metabolism by animal systems is the subject of only three reports. Mehler and Knox (1950) described a N-formyl-L-kynurenine hydrolase from mammalian liver, but it was highly specific for o-formamido compounds and unlikely to act on acylanilide herbicides. A mitochondrial preparation from chick kidney rapidly deacylated a variety of acylanilides, and hydrolysis was influenced by ring substitution and side-chain length (Nimmo-Smith, 1960). Substitution in the *para* position increased the relative rate of deacylation but *ortho* substitution caused a great decrease in rate. When the effect of increasing acyl chain length from formyl to hexanoyl was tested, enzyme action was greatest for the acetyl and propionyl compounds. Williams and Jacobson (1966) studied the metabolism of propanil in mice, rats, rabbits, and dogs. The livers of all of these animals exhibited substantial acylamidase activity and hydrolyzed propanil with the release of DCA. The bulk of the enzyme activity was associated with the microsomal fraction of liver homogenates and, like the rice enzyme, it was subject to inhibition by carbamate and organophosphate insecticides.

B. Transformations of the Aromatic Moiety

1. Transformations In Soil

Once released, the aliphatic side chain of acylanilide herbicides is rapidly metabolized via conventional pathways to carbon dioxide, water, and cell substance, but the fate of the aromatic moiety has proved to be less usual and more intriguing. Evidence that the arylamine was transformed biologically was first obtained from soil studies. Macrae and Bautista (1966) noted a change with time in absorption of ultraviolet radiation (248 mμ) by supernatants from suspensions of propanil-treated soil. Since no change was obtained if the soil was sterilized before treatment, the microbial population was credited with an ability to degrade the herbicide. However, carbon balances performed by Bartha *et al.* (1967) indicated that extensive oxidation of the aromatic moiety of propanil should not be anticipated, and more recent, but previously unpublished, results obtained in the laboratory indicate that less than 1.0% of radiocarbon added to soil as ring labeled propanil-^{14}C is oxidized to $^{14}CO_2$ in 3 weeks. Continued study demonstrated that DCA released by hydrolysis of propanil in soil was unstable and transformed biochemically to a variety of products. Progress toward the identification of these substances was hampered by the notion that DCA would be metabolized by established mechanisms, including ring hydroxylation and cleavage to an aliphatic product that would be further oxidized. This preconception has proved erroneous. DCA is oxidized in soil by reactions that are polymerizing rather than degradative. Products of DCA metabolism in soil have a greater molecular weight than the parent compound, so the process of change is better described as a transformation rather than a degradation.

A major product of DCA transformation in soil was identified by Bartha and Pramer (1967) as 3,3′,4,4′-tetrachloroazobenzene (TCAB). The compound was extracted and crystallized from propanil-treated soil, but insight into its identity was provided by its mass spectrum and molecular weight. Microbial activity was required for TCAB synthesis since the compound was not detected in sterilized soil that received propanil or DCA.

Recognition of the fact that DCA and other anilines are products of the degradation of various phenylcarbamate (Herrett, 1969) and phenylurea (Geissbühler, 1969), as well as acylanilide herbicides, suggested that condensation reactions and the formation of TCAB or other azo compounds may widely occur in soil. Tests have shown that TCAB is indeed produced in soils supplemented with the acylanilides dicryl or karsil (Bartha, 1968), or the phenylcarbamate swep (Bartha and Pramer, 1969); each of which contains a DCA moiety. Solan, which contains a 3-chloro-4-methylaniline moiety, was transformed similarly, but the yield of 3,3′-dichloro-4,4′-dimethylazobenzene was not as great as that of TCAB produced from DCA based herbicides (Bartha, 1969a). In the laboratory where losses by leaching and volatilization are prevented and high concentrations (500 ppm) of the herbicide are used, 20–30% of the aromatic moiety can be converted to TCAB.

Kearney et al. (1969a) have investigated the influence of soil type and propanil concentration on TCAB formation. Three different response curves were observed using four soils and the herbicide at levels ranging from 1.0–1000 ppm: (1) a linear relation between amount of propanil added and TCAB produced was noted in only one soil; (2) a sharp decrease in TCAB at the highest rate of propanil used in two soils; and (3) a large accumulation of TCAB at the highest rate of propanil addition in the remaining soil. The second type of response was interpreted as product inhibition and the third type suggested two possible competing routes of propanil metabolism: one favoring ring oxidation and the other favoring ring condensation. Previously unpublished results obtained by Bartha indicate that some DCA produced during propanil metabolism in soil is lost by volatilization, some undergoes polymerization to TCAB and other substances that can be extracted from soil, and a portion is bound tenaciously and cannot be extracted from soil. This immobilization may be due to chemical bonding of DCA to appropriate sites on humus material. When ring-labeled propanil-^{14}C was added to soil at levels of 5, 100, and 500 ppm, some of the activity was volatilized as DCA but less than 1.0% was oxidized to $^{14}CO_2$ during 3 weeks of observation. The treated soils were finally extracted with solvent and recoveries were 20, 28, and 42% of the added activities, respectively. Acid digestion of the soil after solvent extraction accounted for an additional 73, 69, and 54%, respectively, of the added radiocarbon. The suggestion is that low levels of aniline intermediates will favor the formation of soil-bound residues, whereas high levels of aniline intermediates will predominantly produce azo compounds and other residues that can be extracted from soil by organic solvents.

Thin-layer chromatography reveals the presence of many different products of acylanilide herbicide metabolism that can be extracted from soil. TCAB and others are readily detected because they are highly colored compounds. Plimmer and Kearney (1969) reported results of mass spectral studies which indicate that several of these products are high molecular weight compounds formed by condensations of DCA in soil. Rosen et al. (1970) identified 4-(3,4-dichloroanilino)-3,3',4'-trichloroazobenzene as a photochemical transformation product of DCA and the same compound has been detected in solvent extracts of soil treated with DCA or propanil (H. A. B. Linke, personal communication). Further studies are needed to elucidate the nature of the unextractable residue that represents as much as 73% of the aniline moiety of a herbicide added to soil.

Evidence now exists that TCAB is produced under field as well as laboratory conditions. Kearney et al. (1970) surveyed rice-producing soils in Arkansas with known histories of crop rotation and propanil application to determine the influence of rate of treatment, depth, and time on TCAB formation. When propanil was applied at 3 lb/acre, TCAB was detected in 2 of 47 soil samples, but when the application rate was doubled, TCAB was found in all samples analyzed. The maximum TCAB concentration was measured at 0.18 ppm and the average TCAB residue in soils that received 6 lb/acre propanil was 0.09 ppm. The latter value is equivalent to a 4.5% conversion of propanil-derived arylamine to TCAB in soil. It is not surprising that this figure is substantially less than the 20–30% conversion measured in the laboratory, since in the field, leaching, adsorption, and volatilization will each reduce the amount of propanil-derived DCA that is available in soil for TCAB synthesis. In general, concentration and occurrence of TCAB decreased with increasing time and depth in soil. Of particular note, however, was the detection of residual propanil and TCAB in the soils treated with the herbicide 2 and 3 years prior to sampling (Kearney et al., 1970). The persistence of propanil was unexpected but Bartha (1968) had previously observed TCAB to resist change for 3 weeks in soil.

The influence of molecular configuration on the ability of anilines to be transformed to azo compounds in soil was described by Bartha et al. (1968), and evidence was presented that the reaction is catalyzed by the enzyme peroxidase. For these studies it was necessary to synthesize a series of chlorinated azobenzenes as analytical standards. A total of thirteen different compounds was prepared by reduction of chloronitrobenzenes or by oxidation of their corresponding chloroanilines (Linke et al., 1969), and their spectral characteristics were described and compared (Linke and Pramer, 1969). Figure 2 illustrates the anilines tested and the azo compounds formed from them in soil.

Aniline itself was transformed to unidentified polymeric products such as "aniline black," but no azo compound was formed. All monochloro- and dichloroanilines were converted in part to their corresponding dichloro- and tetrachloroazobenzenes, and in part to compounds of higher molecular weight that were polyaromatic but unidentified. The exception to this statement was 2,3-dichloroaniline. Dimerization of this substance failed to produce the expected 2,2′,3,3′-tetrachloroazobenzene but resulted instead in the partially dechlorinated 3,3′-dichloroazobenzene. The same change occurred in soil (Bartha et al., 1968) and during chemical synthesis (Linke et al., 1968). Two dichloroanilines and the two trichloroanilines tested remained unchanged for 30 days, and it was concluded that chlorination of the 2,5- or 2,6-positions stabilized the molecule and protected aniline from biochemical transformation. Alternatively, chloroaniline unsubstituted in both *ortho* (2,6-) positions, or in adjacent *ortho* and *meta* (2,3- or 5,6-) positions, are likely to give rise to azo compounds in soil (Bartha et al., 1968). In the two cases examined, 4-methyl substituted anilines were also transformed in soil to azo compounds (Fig. 2), but greater proportions of intensely colored high molecular weight polyaromatic compounds were produced than from the corresponding chloro substituted anilines.

Of particular interest are recent reports (Bordeleau et al., 1969; Kearney et al., 1969b; Bordeleau and Bartha, 1970) that two different anilines added simultaneously to soil give rise to asymmetric as well as symmetric azo compounds. The suggestion here that hybrid residues may be produced in field soils treated with more than one aniline-based herbicide was tested by Bartha (1969b). When propanil and solan, licensed for agricultural use on the same crop, were added to a single sample of soil, it was demonstrated that they do indeed undergo

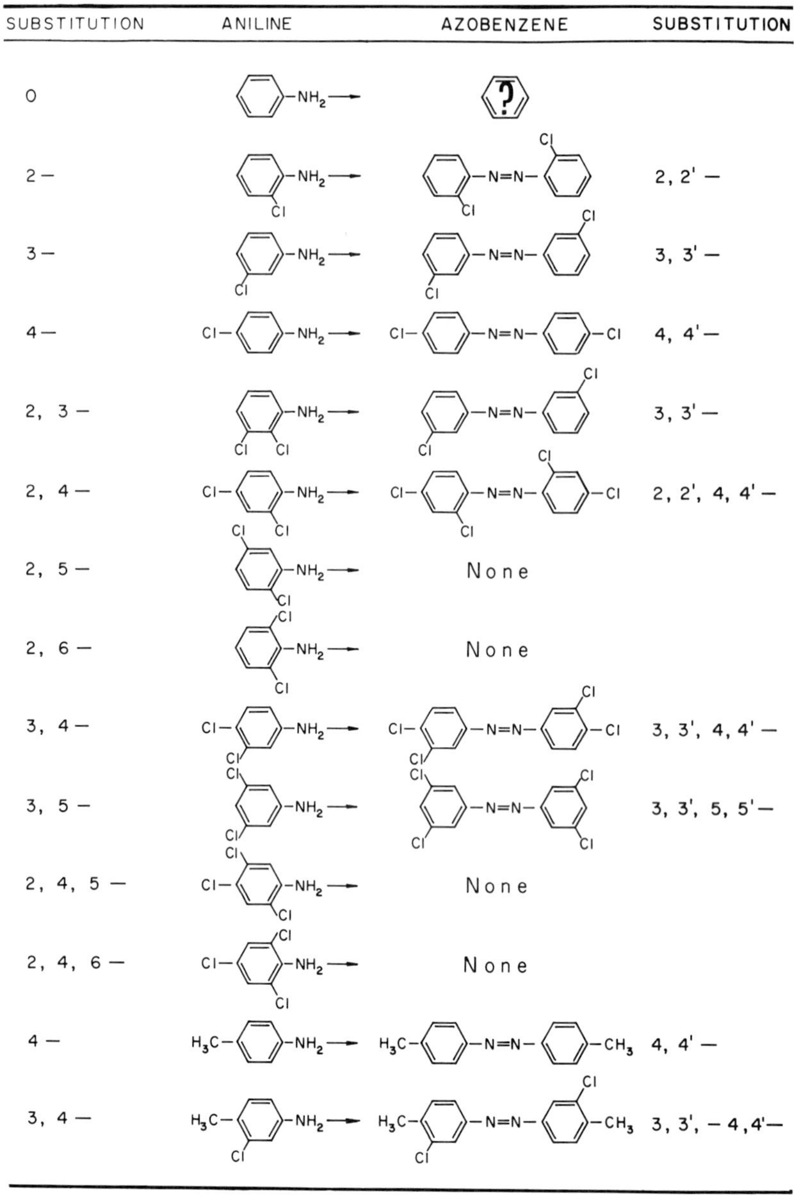

SUBSTITUTION	ANILINE	AZOBENZENE	SUBSTITUTION
0	⌬-NH₂ →	⌬(?)	
2 —	⌬(2-Cl)-NH₂ →	⌬(2-Cl)-N=N-⌬(2'-Cl)	2, 2' —
3 —	⌬(3-Cl)-NH₂ →	⌬(3-Cl)-N=N-⌬(3'-Cl)	3, 3' —
4 —	Cl-⌬-NH₂ →	Cl-⌬-N=N-⌬-Cl	4, 4' —
2, 3 —	⌬(2,3-Cl₂)-NH₂ →	⌬(2-Cl)-N=N-⌬(3'-Cl)	3, 3' —
2, 4 —	Cl-⌬(2-Cl)-NH₂ →	Cl-⌬(2-Cl)-N=N-⌬(2'-Cl)-Cl	2, 2', 4, 4' —
2, 5 —	⌬(2,5-Cl₂)-NH₂ →	None	
2, 6 —	⌬(2,6-Cl₂)-NH₂ →	None	
3, 4 —	Cl-⌬(3-Cl)-NH₂ →	Cl-⌬(3-Cl)-N=N-⌬(3'-Cl)-Cl	3, 3', 4, 4' —
3, 5 —	⌬(3,5-Cl₂)-NH₂ →	⌬(3,5-Cl₂)-N=N-⌬(3',5'-Cl₂)	3, 3', 5, 5' —
2, 4, 5 —	Cl-⌬(2,5-Cl₂)-NH₂ →	None	
2, 4, 6 —	Cl-⌬(2,6-Cl₂)-NH₂ →	None	
4 —	H₃C-⌬-NH₂ →	H₃C-⌬-N=N-⌬-CH₃	4, 4' —
3, 4 —	H₃C-⌬(3-Cl)-NH₂ →	H₃C-⌬(3-Cl)-N=N-⌬(3'-Cl)-CH₃	3, 3', — 4, 4' —

FIG. 2. Influence of substitution on the transformation of anilines to azobenzenes. A question mark within a benzene ring indicates that products were unidentified polyaromatics but not azobenzene, and "none" designates an absence of products extractable with organic solvents.

transformation in part to symmetric (TCAB and 3,3'-dichloro-4,4'-dimethylazobenzene), as well as the asymmetric hybrid (3,3',4-trichloro-4'-methylazobenzene) residues.

The four-electron transfer required for the conversion of two aniline molecules to one of azobenzene was anticipated to be complex and to involve various possible intermediates (Bartha and Pramer, 1967). Since unmodified aniline molecules cannot interact directly, azo compounds are formed by the random reactions of intermediates either with each other or with excess aniline. An attempt was made by Bordeleau et al. (1969) and Bordeleau and Bartha (1970) to identify the key intermediates in the conversions of 4-chloroaniline and DCA to azo compounds. Their results are consistent with a proposal that anilines are converted in part to phenylhydroxylamines which reacted with their aniline precursors to form azo compounds. Since azo compounds were not produced from anilines in sterilized soils but were formed from phenylhydroxylamines and anilines in both sterilized and natural soil, it was concluded that the former transformation was biochemical and the latter occurred spontaneously. Unpublished experiments using trapping agents indicate that the phenylhydroxylamine intermediate is generated via the anilino free radical.

Belasco and Pease (1969) have recently challenged the role of DCA as precursor of TCAB on grounds that in their experiments more TCAB was produced from propanil than from DCA. We feel that these results were due to the use of biologically less active air-dried soil and to losses of DCA by volatility. Using fresh soils and closed incubation systems, we consistently obtained higher and more rapid TCAB production from free DCA than from propanil.

2. Transformations by Microbial Cultures and Enzymes

Although the conversion of anilines to azo compounds was mediated by soil microorganisms, conventional enrichment procedures were not successful in providing pure cultures that were able to perform the transformation. Apparently azo compound formation offers little selective advantage and is incidental to the metabolic activity of peroxidase-producing organisms: the reaction provides little energy and no carbon for microbial growth. A method for detecting and enumerating peroxidase-positive microorganisms was developed by Bordeleau and Bartha (1969), and was used to demonstrate that the difference in capacity of two soils to convert anilines to azobenzenes was positively correlated with numbers of peroxidase-producing inhabitants. A list of microorganisms that display peroxidase activity has been compiled by Saunders et al. (1964); many of these are indigenous to soil. Various peroxidase-positive bacteria, actinomycetes, and fungi have been isolated and are under investigation in this laboratory.

The reported (Daniels and Saunders, 1953) catalysis by horseradish peroxidase of the synthesis of 4,4'-dichloroazobenzene from 4-chloroaniline suggested that a peroxidatic mechanism may be responsible for production of azo compounds in soil. This was noted by Bartha et al. (1968) who concluded that the utilization of different aniline substrates by soil organisms and by peroxidase was highly correlated. Galstyan (1958) described the peroxidation of pyrogallol by soil extracts but the method he employed was found by Bartha and Bordeleau (1969a,b) to be inadequate, and, therefore, they developed and applied a new procedure for measuring peroxidases in soil. No difficulty was encountered in demonstrating substantial amounts of cell-free enzyme in two different soils, and the amounts of activity varied directly with the ability of the soils to transform added DCA to TCAB.

Buffered solutions (0.05 M phosphate at pH 6) extracted peroxidase from soil, and the solutions were filter sterilized so the enzyme was cell free (Bartha and Bordeleau, 1969a,b). When supplied with H_2O_2 these extracts catalyzed the conversion of chloroanilines to chloroazobenzenes and the activity was heat labile. Azo compounds are formed rapidly from anilines in Nixon sandy loam. A phosphate buffer extract of this soil was treated with $(NH_4)_2SO_4$ to 80% of saturation, the precipitated protein was concentrated by centrifugation, and after dialysis was lyophilized. Analyses demonstrated that 500 gm of soil yielded 63 mg of protein that had a peroxidase specific activity of 0.18 units/mg. Unpublished results of subsequent studies established that the optimum buffer concentration was 0.2 M phosphate and that acetate was not an effective extractant. In the range of pH 6-8, recoveries of peroxidase were independent of hydrogen ion concentration.

3. Transformations by Other Biological Systems

The metabolic fate of DCA in rice plants was investigated by Still (1967). The plants were exposed to propanil in liquid culture for 5 days and their roots and shoots were then separated, lyophilized, and extracted with methanol. Chromatographic analyses of the extracts detected four aniline-positive metabolites. One of these was identified as DCA. Qualitative tests of the remaining three compounds indicated that each had a reducing sugar component, and one was identified as N-(3,4-dichlorophenyl)glucosylamine. Efforts to characterize the remaining two metabolites were unsuccessful, but one is now known to be a conjugate of DCA with glucose, xylose, and fructose (Yih *et al.*, 1968a). Time-course studies demonstrated a precursor-product relation between propanil, DCA, and N-(3,4-dichlorophenyl)glucosylamine, but no evidence was obtained that TCAB is a metabolite of either propanil or DCA in rice. Yih *et al.* (1968a) confirmed the absence of TCAB and the presence of complexed DCA in propanil-treated rice, however, they used ring-labeled propanil-^{14}C and established that a considerable amount of activity remained in the residue after extraction of plant tissue with methanol. The distribution of this residual activity between plant celluloses and lignin was measured at 25.7 and 42.4%, respectively. Thus, soluble aniline-carbohydrate complexes are minor products of the transformation of DCA in rice. A major portion of the aromatic moiety of propanil is complexed with polymeric plant constituents, primarily lignin. Comparative tests with ^{14}C-ring-labeled and carbonyl-^{14}C-labeled propanil demonstrated clearly that in plants aniline is lignin bound as DCA and not as the unhydrolyzed herbicide.

The distribution of the aniline moiety of propanil in rice has been traced as far as the grain. Still and Mansager (1969) identified DCA in caustic hydrolyzates of rice grain obtained from experimental plots that were treated with propanil and in rice grain purchased on the local consumer market. Moreover, the grain of rice plants treated with N-(3-chlorophenyl)carbamate (chlorpropham) contained 3-chloroaniline. The destructive nature of the analytical methods employed in these investigations made it impossible to characterize the parent compounds, but it is clear that anilines are incorporated into the marketed product of rice grown on soils treated with aniline-based herbicides.

Different investigators (Still, 1967; Yih *et al.*, 1968a; Still, 1969) have failed to identify TCAB as a product of propanil metabolism in rice plants. This is surprising since rice tissue displays acylamidase and peroxidase activities. The hydrolysis of propanil to DCA occurs in both plants and soil. In plants the aniline is converted to several compounds including sugar and lignin conjugates. In soil it is transformed in part to TCAB. The absence of TCAB in propanil-treated rice may result from sugars and lignin having a greater affinity than peroxidase for DCA, or from differences in the specificity of peroxidases from different sources. Evidence to support the latter possibility was published by Lieb and Still (1969). They compared the ability of crystalline horseradish peroxidase and extracts of horseradish root, barnyard grass, and rice, to oxidize a series of structurally related anilines. The two horseradish preparations were identical in their substrate specificity but differed from the barnyard grass and rice enzymes. The latter were more limited than the former and unable to use DCA as an electron donor.

Although TCAB is not generated in propanil-treated rice it is produced in soil that receives the herbicide (Bartha and Pramer, 1967). Still (1969) inquired into the extent to which the compound may be absorbed by plant roots and translocated to shoots. He treated the roots of rice in nutrient solution with TCAB-^{14}C for 12 days. Analyses indicated that 5.6% of the supplied azo compound was absorbed by roots and 3.2% of this amount was translocated to shoots. No attempt was made to measure the persistence of TCAB in rice tissue, and it was not determined if the compound is translocated to the edible grain of rice.

Mention has been made previously of the studies of Bartha et al. (1968) that noted a similarity in aniline oxidation to azo compounds as it occurred in soil and in solutions of peroxidase. The preparation employed was horseradish peroxidase type II with an activity of 135 purpurogallin units/mg (Sigma Chemical Co., St. Louis, Missouri). Horseradish peroxidase, much as the soil enzyme system, converts anilines to a number of intensely colored polyaromatic compounds in addition to azobenzenes (Bartha et al., 1968; Knowles et al., 1969; Lieb and Still, 1969). Saunders et al. (1964) isolated and identified several three-, four-, and five-ring compounds as products of peroxidase-mediated oxidations of anilines with H_2O_2. The three-ring compound 4-(3,4-dichloroanilino)-3,3',4'-trichloroazobenzene has been demonstrated to be a product of the transformation of aniline by ultraviolet radiation (Rosen et al., 1970), in soil (H. A. B. Linke, personal communication), and by horseradish peroxidase (unpublished results).

C. Proposed Pathways of Acylanilide Metabolism

An attempt has been made here to use the foregoing detailed information and construct summary schemes that illustrate the most probable biochemical pathways of acylanilide metabolism by microorganisms in soil and by plants. Diagrammed also is an integrated view that relates the different possible routes of movement of an acylanilide herbicide in the environment. Illustrated compounds have been assigned Roman numerals to expedite their identification as an aid to discussion. Intermediates that are postulated on the basis of indirect evidence but not as yet identified by accepted chemical criteria are enclosed in brackets.

Current evidence for the transformation of propanil in soil is consistent with the sequence of chemical events illustrated in Fig. 3. The

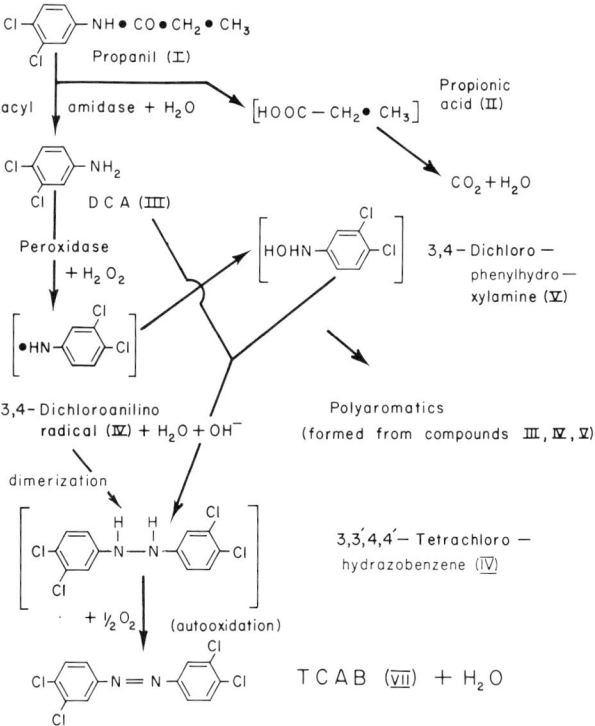

FIG. 3. Proposed pathways of propanil transformation in soil. For explanation see text.

initial step in the proposed pathway is cleavage of the herbicide molecule (I) to propionic acid (II) and DCA (III). It is possible that propanil in soil may be activated by oxidative change prior to hydrolysis, but the hydrolytic reaction is catalyzed by acylamidase that is presumably of microbial origin. Propionic acid is metabolized by soil microorganisms to carbon dioxide, water, and cell substance. DCA is peroxidatically converted via the 3,4-dichloroanilino free radical (IV) to 3,4-dichlorophenylhydroxylamine (V). Compounds III and V react to generate 3,3′,4,4′-tetrachlorohydrazobenzene (VI), which is atmospherically oxidized to TCAB (VII). Compound VI could arise by dimerization of the anilino radical, but this is energetically a less likely route than that which is illustrated. The spectrum of polyaromatic products of DCA transformation in soil appears to result from complex reactions and interactions involving compounds III, IV, and V. Compounds IV, V, and VI are extremely labile and have not been isolated and characterized chemically. Their proposed existence and role as intermediates are based on indirect evidence, including the results of kinetic studies and tests which free radical trapping agents that remain in part unpublished. Oxidation and cleavage of the DCA ring is a biochemically possible alternative to TCAB formation that as yet has not been demonstrated to occur. Moreover, TCAB may not be recalcitrant. The azo configuration is capable of being oxidized or reduced, and it is likely that further investigation will extend understanding of propanil metabolism by identifying products of the degradation of TCAB by microorganisms in soil.

The metabolic fate of propanil in rice plants is illustrated by Fig. 4.

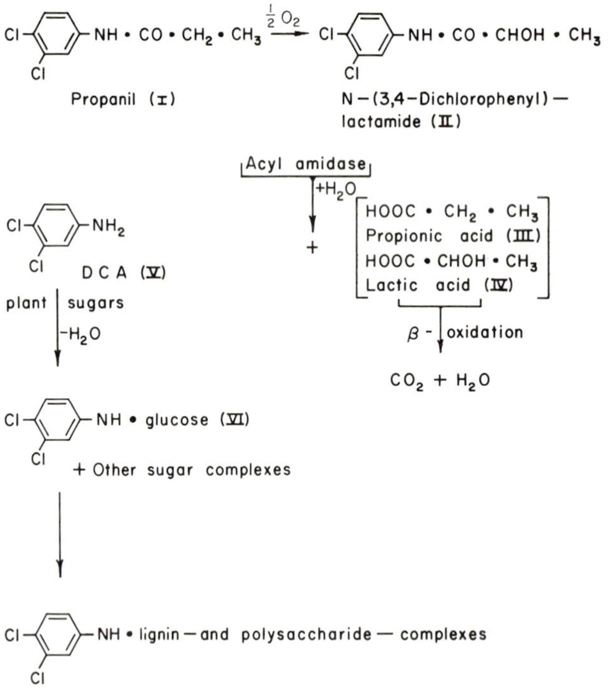

$$Cl\text{-}C_6H_3Cl\text{-}NH \cdot CO \cdot CH_2 \cdot CH_3 \xrightarrow{\frac{1}{2}O_2} Cl\text{-}C_6H_3Cl\text{-}NH \cdot CO \cdot CHOH \cdot CH_3$$

Propanil (I) \quad\quad N-(3,4-Dichlorophenyl)-lactamide (II)

Acyl amidase, $+H_2O$

$$\begin{bmatrix} HOOC \cdot CH_2 \cdot CH_3 & \text{Propionic acid (III)} \\ HOOC \cdot CHOH \cdot CH_3 & \text{Lactic acid (IV)} \end{bmatrix}$$

$Cl\text{-}C_6H_3Cl\text{-}NH_2$ \quad DCA (V)

plant sugars | $-H_2O$

β-oxidation

$CO_2 + H_2O$

$Cl\text{-}C_6H_3Cl\text{-}NH \cdot$ glucose (VI)
+ Other sugar complexes

$Cl\text{-}C_6H_3Cl\text{-}NH \cdot$ lignin— and polysaccharide— complexes

FIG. 4. Proposed pathways of propanil transformation in the rice plant (*Oryza sativa*). For explanation see text.

The herbicide enters susceptible and resistant plants at the same rate, but only the latter have the ability to detoxify the compound. This is accomplished in rice by rapid hydrolysis of propanil (I) to DCA (V) and an aliphatic acid (III or IV). The reaction is catalyzed by an acylamidase, but the identities of the substrate and the acid product are uncertain. Propionic acid (III) will be formed from propanil, but if the report of Yih *et al.* (1968b) that propanil is oxidized to N-(3,4-dichlorophenyl)lactamide (II) prior to hydrolysis is confirmed, lactic (IV) and not propionic will be the acid formed. In either case, DCA is produced and the aliphatic acid will ultimately be oxidized to carbon dioxide and water. DCA is transformed to N-(3,4-dichlorophenyl)-glucosylamine (VI) and other more complex substances that contain DCA, glucose, xylose, and fructose. However, the major portion of the DCA is found associated with polymeric plant constituents, mainly lignin. TCAB has not been identified as a product of DCA metabolism in rice plants.

Propanil is frequently applied as an aerial spray. Herbicide released from airplanes will settle on the surface of plants, but some will be deposited directly on the soil, and more will reach soil as run-off from treated plants and in the tissue of susceptible plants that die due to propanil treatment. Figure 5 illustrates different possible routes of movement and types of transformation of propanil, and attempts also

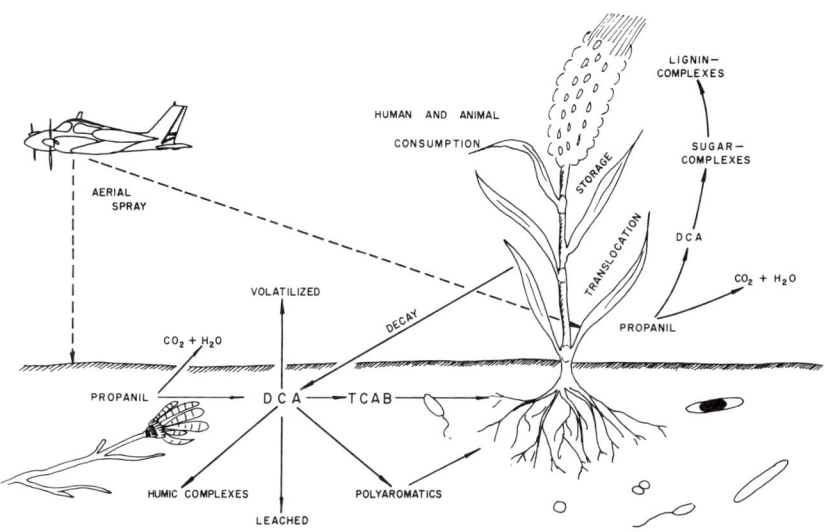

FIG. 5. Diagrammatic illustration of the transformations and movements of propanil and its metabolites in the environment.

to relate events that occur in the rice plant to those that take place in the soil environment.

Propanil is metabolized by some plants and by microorganisms in soil as described previously and illustrated in Figs. 4 and 3, respectively. In both systems, the herbicide molecule is hydrolytically cleaved to DCA and an aliphatic acid. Most of the latter compound is oxidized to carbon dioxide and water, but DCA is transformed via a number of different routes: in plants it is complexed by sugars and lignin; in soil it is converted in part to TCAB and polyaromatic compounds and is complexed by humus, but some is lost from soil by leaching and volatilization. DCA and TCAB produced in soil may be absorbed by plant roots, translocated to shoots and grain, and be consumed by animals or man. The ultimate fate of the various DCA derivatives that have been identified in soil and in rice plants remains to be determined. TCAB and higher molecular weight compounds resulting from the biochemical polymerization of DCA may be degraded with time in soil. Likewise, complexes of DCA with carbohydrates, lignin, and humus need not be irreversible. They may be metabolized with the release and/or destruction of DCA, but these possibilities have not been investigated. Likewise, the distribution and metabolic fate of DCA and TCAB in animal systems warrant attention but have not as yet been studied.

Acylanilide herbicides are certainly not "completely metabolized by the crop plant." This and other published statements (Weed Society of America, 1967) describing acylanilides as "degraded in soil" and, as a result of microbial action, "broken down quite rapidly," are generally misleading. They appear to be based on measurements of disappearance of the parent compound without regard for the formation and accumulation of products that may pollute the environment and be of concern to public health and welfare.

III. Biological Activity of Metabolites

There is a paucity of published information concerning the biological activity of metabolites of acylanilide herbicides. Toxicities of some pure substances have been measured, but little is known of synergistic effects and much remains to be done to alleviate concern about longterm exposures to low levels, and genetic, neoplastic, and teratogenic effects. Present knowledge relating in turn to plants, microorganisms, and animals is summarized below.

N-(3,4-Dichlorophenyl)lactamide, suggested by Yih et al. (1968b) to be the initial transformation product of propanil in rice displayed herbicidal properties identical to the parent compound. It was hydrolyzed, however, and the products formed, lactic acid was metabolized to carbon dioxide and water, and DCA was complexed by various plant constituents. DCA and TCAB enter roots and are translocated to shoots without any reported phytotoxicity, but Prasad and Pramer (1969) observed some cytogenetic effects of propanil and its metabolites. The herbicide was more toxic than DCA or TCAB and caused curling, inhibited division, and deformed nuclei in root cells of onion (*Allium cepa*).

Under certain conditions, metabolites of acylanilide herbicides depress the activities of microorganisms in soil. Bartha et al. (1967) described respiratory effects and noted some inhibition of nitrification by dicryl, karsil, and propanil in soil. Thompson and Corke (1969) examined the relation of chloro substitution to ability of anilines to inhibit nitrification. Ammonium oxidation was more sensitive than nitrite oxidation and toxicity was directly correlated to the lipophilic nature of the compounds tested. When molecular configuration of different anilide herbicides was related to their ability to inhibit growth of the alga *Chlorococcum aplanosporum* in solution medium, Sharabi (1969) reported the following order of effectiveness: karsil, solan, propanil, dicryl, and propachlor. DCA suppressed growth and chlorophyll synthesis by *C. aplanosporum* in solution culture, but TCAB did not. Propanil and DCA were toxic to the alga in soil as well as in solution medium, but for comparable effects higher levels of the compounds were required in soil than in solution. Previously unpublished results obtained in this laboratory indicate that propanil and DCA, but not TCAB, decrease the viability of spores of *Aspergillus nidulans*, however, DCA and TCAB, but not propanil, are mutagenic. DCA was the most genetically active of a series of structurally related anilines tested for their ability to induce reversion of a methionine requirement by a stable auxotrophic strain of *A. nidulans*.

Although some literature (Weed Society of America, 1967) lists a low order of mammalian toxicity among the properties that recommend acylanilide herbicides for agricultural use, Williams and Jacobson (1966) reported that high levels of propanil caused acute distress in dogs and other animals. The livers of these animals displayed acylamidase activity and hydrolyzed the herbicide to propionic acid and DCA. Anilines adversely affect red blood cells and can cause anemia (Stecher, 1968), but the acute oral toxicity of propanil was not altered by selective inhibition of liver acylamidase, and it appears that the observed effects were due to the parent compound and not to the aniline metabolite (Williams and Jacobson, 1966). Two of a bewildering number of products of DCA metabolism have been identified as azo compounds [TCAB and 4-(3,4-dichloroanilino(-3,3′,4′-trichloroazobenzene]. Most azo compounds are innocuous but a few are carcinogenic in animals, mostly to the liver (Weisburger and Weisburger, 1966). TCAB, the azo product of propanil metabolism by soil microorganisms, does not appear to have the molecular geometry required for chemical carcinogenicity (Arcos and Arcos, 1962), but it was synthesized in quantity and supplied to the National Cancer Institute for test. The compound was fed in peanut oil to rats at a level of 4 mg/week for an initial 3 week period, and at 10 mg/week for an additional 37 weeks. Weight gains were good and, when the animals were sacrificed at the 60th week, there was a small amount of fatty degeneration of the liver, but no tumors among the TCAB-treated rats (E. B. Mattheis, personal communication). Mammalian liver contains azo reductases (Fouts *et al.*, 1957; Mascitelli-Coriandoli, 1959), and it is possible that TCAB in the rat was cleaved enzymatically and metabolized.

IV. Concluding Comments

We have attempted here to describe and compare present understanding of the metabolism of acylanilide herbicides by soil microorganisms, plants, and animals. Biochemical and ecological pathways are proposed for the production and movement of metabolites of the herbicides in various environments, and their biological activities are described and discussed. Some novel reactions and unexpected products have been observed. Of particular note in this regard is the enzymatic hydrolysis of acylanilide herbicides with the release and subsequent peroxidatic polymerization of substituted anilines to azo compounds and other polyaromatic residues in soil. Some of these substances persist in soil and at least one is absorbed by plant roots and translocated to the shoots. These events are a matter of concern because some azo compounds are carcinogens. It has been possible to identify and synthesize a sufficient quantity of only one azo substance for test, and fortunately, it proved to have little toxicity and no carcinogenicity in rats. Much remains to be done, however, because many products are as yet unidentified and untested.

The pathways of acylanilide metabolism are still incompletely defined, but according to present knowledge, they terminate with various aniline complexes and azo compounds. Since some anilines and azo compounds have undesirable biological effects and may constitute a health hazard, recognition of their existence as environmental pollutants constitutes an obligation to further investigation and to a clarification of their ultimate fate. In our opinion the need to meet this obligation is not nullified by the argument that concentrations of pesticides and their metabolites employed in laboratory tests greatly exceed those recommended or detected in the field. It is well established that damage may result from long-term exposure to low levels of certain chemicals, and that pesticide concentrations may be magnified as residues are consumed and transported along a food chain. What has been learned of the biochemistry and microbiology of acylanilide herbicides is in principle directly applicable to other compounds constructed with an aniline moiety. Since chloroanilines are present in acylanilide, phenylcarbamate, and phenylurea herbicides, and these represent a large portion of the agricultural chemicals now in use, the conclusions reached here have rather broad and significant implications.

ACKNOWLEDGMENTS

This is a paper of the Journal Series, New Jersey Agricultural Experiment Station, New Brunswick, New Jersey. The research performed by the authors and their associates was supported in part by USPHS Research Grant ES-16. We are grateful to several colleagues and collaborators for their permission to cite unpublished results and review manuscripts that were in press.

REFERENCES

Anonymous. (1968). *Chem. Eng. News* **46** (22), 23–24.
Arcos, J. C., and Arcos, M. (1962). In "Fortschritte der Arzneimittelforschung" (E. Jucker, ed.), pp. 407–581. Berghauser Verlag, Stuttgart.
Bartha, R. (1968). *J. Agr. Food Chem.* **16**, 602–604.
Bartha, R. (1969a). *Weed Sci.* **17**, 471–478.
Bartha, R. (1969b). *Science* **166**, 1299–1300.
Bartha, R., and Bordeleau, L. M. (1969a). *Bacteriol. Proc.* p. 4.
Bartha, R., and Bordeleau, L. M. (1969b). *Soil Biol. Biochem.* **1**, 139–143.
Bartha, R., and Pramer, D. (1967). *Science* **156**, 1617–1618.
Bartha, R., and Pramer, D. (1969). *Bull. Environ. Contam. Toxicol.* **4**, 240–245.
Bartha, R., Lanzilotta, R. P., and Pramer, D. (1966). *Bacteriol. Proc.* p. 5.
Bartha, R., Lanzilotta, R. P., and Pramer, D. (1967). *Appl. Microbiol.* **15**, 67–75.
Bartha, R., Linke, H. A. B., and Pramer, D. (1968). *Science* **161**, 582–583.
Belasco, I. J., and Pease, H. L. (1969). *J. Agr. Food Chem.* **17**, 1414–1417.
Bordeleau, L. M., and Bartha, R. (1969). *Appl. Microbiol.* **18**, 274–275.
Bordeleau, L. M., and Bartha, R. (1970). *Bull. Environ. Contam. Toxicol.* **5**, 34–37.
Bordeleau, L. M., Linke, H. A. B., and Bartha, R. (1969). *Bacteriol. Proc.* p. 4.
Bowling, C. C., and Hudgins, H. R. (1966). *Weeds* **14**, 94–95.
Daniels, D. G. H., and Saunders, B. C. (1953). *J. Chem. Soc. London*, pp. 822–826.
Duke, W. B., Slife, F. W., and Hanson, J. B. (1967). *Weed Sci. Soc. Amer. Nat. Meeting Washington, D. C., Abstr. No.* **50**.

Fouts, J. R., Kamm, J. J., and Brodie, B. B. (1957). *J. Pharmacol. Exp. Ther.* **120**, 291–300.
Frear, D. S., and Still, G. G. (1968). *Phytochemistry* **7**, 913–920.
Galstyan, A. S. (1958). *Dokl. Akad. Nauk Arm. SSR* **26**, 285–288.
Geissbühler, H. (1969). *In* "Degradation of Herbicides" (P. C. Kearney and D. D. Kaufman, eds.), pp. 79–111. Dekker, New York.
Herrett, R. A. (1969). *In* "Degradation of Herbicides" (P. C. Kearney and D. D. Kaufman, eds.), pp. 113–145. Dekker, New York.
Hofstra, G., and Switzer, C. M. (1968). *Weed Sci.* **16**, 23–28.
Huffman, C. W., and Allen, S. E. (1960). *J. Agr. Food Chem.* **8**, 298–302.
Jacoby, W. B. (1954). *J. Biol. Chem.* **207**, 657–663.
Jaworski, E. G. (1969a). *J. Agr. Food Chem.* **17**, 165–170.
Jaworski, E. G. (1969b). *In* "Degradation of Herbicides" (P. C. Kearney and D. D. Kaufman, eds.), pp. 165–185. Dekker, New York.
Kaufman, D. D., and Miller, D. E. (1969). *Weed Sci. Soc. Amer. Nat. Meeting, Las Vegas, Nevada, Abstr. No.* **235**.
Kearney, P. C. (1965). *J. Agr. Food Chem.* **13**, 561–564.
Kearney, P. C., Plimmer, J. R., and Guardia, F. S. (1969a). *158th A.C.S. Nat. Meeting, New York, AGFD Abstr. No.* **31**.
Kearney, P. C., Plimmer, J. R., and Guardia, F. S. (1969b). *J. Agr. Food Chem.* **17**, 1418–1419.
Kearney, P. C., Smith, R. J., Jr., Plimmer, J. R., and Guardia, F. S. (1970). *Weed Sci.*, in press.
Knowles, C. O., Gupta, A. K. S., and Hassan, T. K. (1969). *J. Econ. Entomol.* **62**, 411–414.
Lanzilotta, R. P. (1968). Ph.D. Thesis, Rutgers – The State University, New Brunswick, New Jersey.
Lanzilotta, R. P., Bartha, R., and Pramer, D. (1967). *Bacteriol. Proc.* p. 8.
Lieb, H. B., and Still, C. C. (1969). *Plant Physiol.* **44**, 1672–1673.
Linke, H. A. B., and Pramer, D. (1969). *Z. Naturforsch.* **24b**, 997–999.
Linke, H. A. B., Bartha, R., and Pramer, D. (1968). *Naturwissenschaften* **55**, 444.
Linke, H. A. B., Bartha, R., and Pramer, D. (1969). *Z. Naturforsch.* **24b**, 994–996.
McRae, D. H., Yih, R. Y., and Wilson, H. F. (1964). *Weed Sci. Soc. Amer. Nat. Meeting, Chicago, Illinois, Abstr.* p. 87.
Macrae, I. C., and Bautista, E. M. (1966). *Ann. Rep. Int. Rice Res. Inst., Manila, Philippines,* p. 127.
Mehler, A. H., and Knox, W. E. (1950). *J. Biol. Chem.* **187**, 431–438.
Mascitelli-Coriandoli, E. (1959). *Z. Naturforsch.* **14**, 70–71.
Matsunaka, S. (1968). *Science* **160**, 1360–1361.
Minter, P. C., Hayes, W. J., and Caras, G. J. (1969). *J. Chem. Doc.* **9**, 73–75.
Moreland, D. E., and Hill, K. L. (1963). *Weeds* **11**, 55–60.
Nimmo-Smith, R. H. (1960). *Biochem. J.* **75**, 284–293.
Oelke, E. A., and Morse, M. D. (1968). *Weed Sci.* **16**, 235–239.
Plimmer, J. R., and Kearney, P. C. (1969). *158th A. C. S. Nat. Meeting, New York, AGFD Abstr. No.* **29**.
Prasad, I., and Pramer, D. (1969). *Cytologia* **34**, 351–352.
Rosen, J. D., Siewierski, M., and Winnett, G. (1970). *J. Agr. Food Chem.* **18**, 494–496.
Saunders, B. C., Holmes-Siedle, A. G., and Stark, B. P. (1964). "Peroxidase." Butterworths, Washington, D. C. and London.
Sharabi, N. E. (1969). Ph.D. Thesis, Rutgers – The State University, New Brunswick, New Jersey.

Sharabi, N. E., and Bordeleau, L. M. (1969). *Appl. Microbiol.* **18**, 369–375.
Smith, R. J. (1961). *Weeds* **9**, 318–322.
Stecher, P. G., ed. (1968). "The Merck Index." Merck & Co., Inc., Rahway, New Jersey.
Still, C. C., and Kuzirian, O. (1967). *Nature* **216**, 799–800.
Still, G. G. (1967). *Science* **159**, 992–993.
Still, G. G. (1968). *Plant Physiol.* **43**, 543–546.
Still, G. G. (1969). *Weed Res.* **9**, 211–217.
Still, G. G., and Mansager, E. R. (1969). *Weed Res.* **9**, 218–223.
Syrbu, G. A. (1967). *Izv. Akad. Nauk Kaz. SSR. Ser. Biol.* **1**, 38–42.
Thompson, F. R., and Corke, C. T. (1969). *Can. J. Microbiol.* **15**, 791–796.
Unger, V. H., McRae, D. H., and Wilson, H. F. (1964). *Weed Sci. Soc. Amer. Nat. Meeting, Chicago, Illinois, Abstr.* pp. 86–87.
Weed Society of America. (1967). "Herbicide Handbook," pp. 62–63, 64–72, 93–94, and 117–119. Humphrey Press, Geneva, New York.
Weisburger, J. H., and Weisburger, E. K. (1966). *Chem. Eng. News* **44** (6), 124–142.
Williams, C. H., and Jacobson, K. H. (1966). *Toxicol. Appl. Pharmacol.* **9**, 495–500.
Yih, R. Y., McRae, D. H., and Wilson, H. F. (1968a). *Science* **161**, 376–377.
Yih, R. Y., McRae, D. H., and Wilson, H. F. (1968b). *Plant Physiol.* **43**, 1291–1296.

Therapeutic Dentifrices

J. K. PETERSON

*Division of Dental Health,
North Dakota State Department of Health,
Bismarck, North Dakota*

I.	Introduction	343
II.	Plaque- and Calculus-Inhibiting Dentifrices	344
III.	Desensitizing Dentifrices	344
IV.	Caries-Inhibitory Dentifrices	345
	A. Ammonia and Urea Compound Dentifrices	345
	B. Dentifrices Incorporating Antibiotics	346
	C. Antienyzme Dentifrices	347
	D. Fluoride Dentifrices	348
V.	Sodium Fluoride Dentifrices	350
VI.	An Amine Fluoride Dentifrice	350
VII.	Stannous Fluoride Dentifrices	350
VIII.	Phosphate Fluoride Dentifrices	355
IX.	Summary	358
	References	359

I. Introduction

Wholesale dentifrice sales totaled about $255 million in the United States in 1969. Each 1% of the market is worth over $2½ million in sales to the manufacturer. Before the American Dental Association Council on Dental Therapeutics' critical recognition of the anticaries properties of a dentifrice, the extent of each manufacturer's share of the market was almost entirely determined by the size and success of its advertising campaign. They actually had a formula telling them how much needed to be spent on advertising to get and hold a certain percentage of the market. Evidence that this formula still works in a predictable manner is the recent years success of two "nontherapeutic" dentifrices — Ultra Brite and Macleans.

At the time of Crest's recognition by the ADA Council on Dental Therapeutics in 1960, Crest had 12% of the market, compared to 23% for Gleem and 35% for Colgate. Crest sales rapidly climbed with recognition to 35%. Their sales increase of over $50 million per year was due both to the ADA recognition and an increased advertising effort. Since Crest's ADA recognition, three additional stannous fluoride dentifrices received qualified recognition: Cue, Fact, and Super Stripe. None of these last three were still being produced in late 1969.

In October 1969, Colgate with MFP became the first nonstannous fluoride dentifrice to obtain ADA recognition as an effective therapeutic dentifrice. Since the Colgate Company earmarked over $20 million a year in 1969 and 1970 to promote Colgate with MFP, it may well regain some of the market lost to Crest after 1960. In the fall of 1969, before the ADA recognition of Colgate with MFP, Crest still had 35% of the market, Colgate, 25%; Ultra Brite, 12%; and followed by Gleem, Macleans, and Pepsodent.

Most tests have been with those dentifrices formulated with the hope of inhibiting caries. This discussion will be concerned primarily with these. However, there are reports of dentifrices designed to treat gingivitis, to inhibit plaque and calculus formation, and to relieve hypersensitive teeth.

II. Plaque- and Calculus-Inhibiting Dentifrices

Several investigators are working with antibiotic and enzyme formulations in dentifrices with the hope of favorably influencing the health of the periodontal tissues by inhibiting plaque and calculus formation. Harrisson et al. (1963) reported the results of testing three toothpastes to which different enzymes had been added. After 6 months use, 60-70% of the subjects using toothpaste containing a predominately proteolytic or a predominately amylolytic enzyme exhibited a decrease in the accumulation of soft accretions, calculus, and stain. Only 20% of the individuals using a cellulose-containing toothpaste exhibited similar effects. Molle (1967) reported a 54% decrease in plaque production in a double-blind crossover study with a dentifrice incorporating a *Bacillus subtilis* proteolytic–amylolytic enzyme.

This type of dentifrice formulation has great potential but much more work needs to be completed and published before any conclusions can be drawn on its practical value.

III. Desensitizing Dentifrices

Several dentifrice formulations have been developed, tested, and marketed in the hopes of relieving hypersensitivities. One, named Thermodent, incorporates 1.4% formalin. (Emoform is the British and Swiss version.) Another, Sensodyne, incorporates a solution of strontium chloride as 10% of the dentifrice mixture. Also the new Colgate MFP dentifrice has been tested for desensitizing properties.

A British assessment by Forrest (1963) of Thermodent (Emoform) reports frequent success in treating generalized areas of moderate sensitivity and exposed occlusal dentin. He was not successful in relieving severe cervical sensitivity. Kimmelman et al. (1969) in a double-blind study reported that Thermodent relieved or alleviated sensitivity 66% of the time while a placebo gave 39% success. Smith and Ash (1964a) reported no significant alteration or cervical hypersensitivity to mechanical or thermal stimuli after the use of Thermodent for 30 or 60 days.

The results from the strontium chloride (Sensodyne) tests are similarly conflicting. Ross (1961) reported 83% of his patients had good to complete relief with results showing progressive improvement over a period of several months. Meffert and Hoskins (1964) reported complete relief in 49% of his cases, very good relief in 29%, and only 7% secured little or no relief.

Again in another double-blind study, Smith and Ash (1964b) were unable to measure any significant alteration in cervical hypersensitivity to either mechanical or thermal stimuli after the use of Sensodyne for 30 or 60 days.

Bolden et al. (1968) recently have reported a 66% improvement in sensitivity after brushing with the new Colgate with MFP (0.76% monofluorophosphate) dentifrice. This compares with a 51% improvement with a 1.4% formalin dentifrice, 42% improvement with a stannous fluoride dentifrice, and a 46% improvement with the non-MFP placebo dentifrice (Table I).

IV. Caries-Inhibitory Dentifrices

The investigations into possible therapeutic effects resulting from the addition of anticaries agents to dentifrices can be divided into several categories:
1. Ammonia and urea compounds.
2. Antibiotics.
3. Antienzyme compounds.
4. Flouride compounds.

Investigations that have been made in each of these categories will be reviewed. The more recent and more significant investigations will be considered in greater detail.

A. AMMONIA AND UREA COMPOUND DENTIFRICES

There have been no recent field tests reported of dentifrices incorporating ammonia and urea compounds. The Council on Dental Therapeutics in April 1951 classified this category of dentifrices in Group C "needing further study."

TABLE I
Tests of Desensitizing Dentifrices

Investigator	Percentage reduction in sensitivity	Remarks
Thermodent (Emoform—1.4% formaldehyde)		
Forrest (1963)	Frequent relief	No relief in cervical areas
Kimmelmann et al. (1969)	66	39% Relieved with placebo
Smith and Ash (1964a)	No relief	Cervical areas—double blind
Bolden et al. (1968)	51	46% Relieved with placebo
Sensodyne (10% strontium chloride)		
Ross (1961)	83	Good to complete relief
Meffert and Hoskins (1964)	49	Complete relief
	29	Very good relief
	7	No relief
Smith and Ash (1964b)	No relief	Cervical areas—double blind
Colgate with MFP (0.76% monofluorophosphate)		
Bolden et al. (1968)	66	46% Improved with placebo
		42% Improved with 0.4% stannous fluoride—IMP

Kerr and Kesel (1951), Cohen and Donzanti (1954), and Vogel and Hess (1957) have reported positive tests. Other studies report a drop in lactobacillus counts and were thus considered to indicate positive tests. However, there is now considerable doubt that *Lactobacillus acidophilus* has a part in caries etiology. A recent animal study at the National Institutes of Health showed no caries activity with just lactobacilli present. It has been postulated that lactobacilli are present because of the environment provided by open carious lesions and have little or no part in the actual caries process.

At any rate ammonia and urea compound dentifrices have little status in the current therapeutic dentifrice picture.

B. Dentifrices Incorporating Antibiotics

The Council on Dental Therapeutics classified dentifrices incorporating penicillin in Group C twice (1950, 1952) after considering evidence relating to their effectiveness and safety. Each time the one positive study led to the Council's interest and consideration. This was the study published by Zander (1950) in which he reported 50–60% reduction in caries over 2 years. Hill et al. (1953) followed with a 1-year report of 16% caries reduction. Negative results were reported by Walsh and Smart (1951) and Lunin and Mandel (1955).

Since Hill and others reported some increase in the development of penicillin-resistant organisms and some mild sensitivities with the use of the penicillin dentifrice, the Council expressed concern over (1) the possibility of sensitizing a significant portion of the population to penicillin, and (2) the development of penicillin-resistant microorganisms in those using the dentifrice. These concerns led to the termination of serious consideration of penicillin dentifrices.

Mention was made recently by Keyes of the possibility of trying antibiotics again in dentifrices. Many antibiotics have been discovered, which have little or no significance in treating general illnesses, that may have value in the control of microorganisms related to the caries process. Shiere (1957) has reported a 26% reduction of caries with a tyrothricin dentifrice.

C. Antienzyme Dentifrices

Burnett (1957) reported that the only antienzyme agents to have been given adequate but limited clinical trials in dentifrices were sodium dehydroacetate and sodium N-lauroyl sarcosinate.

Sulser et al. (1958) reported a 51% decrease in decayed and filled surfaces (DFS) in a 2-year, young adult study using a dehydroacetate-oxylate dentifrice. This was a promising beginning and should have been investigated further.

Several studies have been completed with the sodium N-lauroyl sarcosinate dentifrice (the Colgate Gardol formulation). Four separate clinical tests (Science, 1956) were carried out under the auspices of Northwestern Dental School which showed a combined reduction of 47% in decayed and filled surfaces. Another study reported by Frasher and Hein (1958) in a fluoridated area showed a 48% reduction in 27 months. Hayden and Glass (1959) reported a study showing a 38% reduction in interproximal caries at the end of 1 year. In a British bite-wing radiograph study of dental and medical students, Emslie (1963) reported an apparent, but not significant, caries reduction of 17.5% with palmitoyl sarcosinate dentifrice. Backer-Dirks et al. (1960) reported no significant difference in caries increment in a Belgian study with a sarcosinate formulation containing a carbonate abrasive.

The most recent sarcosinate dentifrice field test was reported by

Finn and Jamison (1963). In this study, the Colgate sarcosinate dentifrice was matched against Crest and the monofluorophosphate formula. Similar results were recorded for the sarcosinate and stannous fluoride dentifrices. However, without an inactive control group, a reliable assessment of the sarcosinate dentifrice results was impossible.

D. Fluoride Dentifrices

The mode of topical action of the fluoride ion in dentifrices may differ from that of fluoride in water. Jenkins (1962) reported fluoride concentrations in saliva and tissue fluids are perhaps a thousand times greater when fluoride dentifrices are used than with fluoridated water. McCann and Bullock (1955) determined that fluoride can react with enamel in five different ways:

1. At low concentrations as with fluoridation, the reaction is mostly an ionic exchange with the hydroxyl ion resulting in fluorapatite formation.
2-3. At dentifrice and topical concentrations, additional reactions occur including the precipitation of calcium and magnesium fluorides. A low pH increases these precipitations.
4. Exchange with carbonate.
5. Adsorption.

Increasing the fluoride content of the enamel decreases its solubility in acids and other decalcifying agents and thus increases its caries resistance. Fluoride-containing dentifrices will raise the fluoride concentration of the surface enamel by fluorapatite formation (which some regard as irreversible) and by calcium fluoride precipitation which is readily lost through dissolution. One of the advantages of dentifrices with available fluoride is that frequent use will restore the calcium fluoride which tends to be dissolved off by the oral fluids.

Most laymen and many dentists have assumed that it is just necessary to add fluorides to a conventional dentifrice, the use of which will prevent caries. Unfortunately, it has turned out to be difficult to prepare a dentifrice formula where fluoride ions remain stable and available to the tooth surface at the time of brushing.

The conventional dentifrice abrasive, dicalcium phosphate, ties up the fluoride ion completely, making it unavailable to the tooth surface. Heat-treated calcium pyrophosphate is used as the abrasive in Crest. It is somewhat more compatible with stannous fluoride, but since it slowly combines with the stannous fluoride, stannous pyrophosphate is added to increase the shelf life. Cue has insoluble metaphosphate, also known as IMP, as an abrasive, as does Fact and the Pepsodent stannous fluoride dentifrices. It is claimed that stannous fluoride retains its stability longer in this mixture. Wachtel and Strange (1965) at the National Naval Medical Center made *in vitro* measurements of enamel solubility reduction on etched enamel with three of the stannous fluoride dentifrices. He found protection from all three of about the same magnitude with fresh dentifrice and a much reduced magnitude of protection from all three with aged dentifrice.

He also noted that the protection against acid solubility afforded by the three dentifrices was the same as was obtained by a water solution containing one-tenth the amount of stannous fluoride found in the dentifrices. In other words, the other constituents in all three formulas reduced the effectiveness of the stannous fluoride by nine-tenths. In fact, this led Wachtel to wonder if the caries protection afforded by the stannous fluoride dentifrices was due to reduction in enamel solubility. He suggests as an alternative possibility Jenkins' (1962) evidence that stannous fluoride will produce an "antienzymatic effect" on plaque bacteria. This suggests that stannous fluoride-containing dentifrices may exert an antibacterial action. He acknowledges, however, that perhaps the action of these dentifrices *in situ* is one that cannot be fully judged under *in vitro* conditions.

Other chemical changes that tend to decrease the effectiveness of fluoride dentifrices are:
1. Oxidation of the stannous ion to stannic when tin is included.
2. Hydrolysis.
3. pH change—a rise in the pH toward neutral occurs in dentifrice mixtures with time; this decreases the enamel uptake of the fluoride ion.

The lower pH of the stannous fluoride dentifrices has in some quarters been considered the sole reason for their observed superiority over natural pH sodium fluoride formulations. Unfortunately, when the pH of sodium fluoride solutions are lowered to the point where the fluoride uptake is appreciably greater, the enamel also becomes etched or decalcified by the acid. Apparently, little or no decalcification occurs with the low pH stannous fluoride dentifrices.

Brudevold *et al.* (1963a) determined that solutions of sodium fluoride in acid sodium phosphate increased the amount of fluoride available for uptake by the enamel while causing minimal demineralization. This buffering protection by the phosphate has been disputed in some quarters. However, topical studies by Brudevold *et al.* (1963b), Wellock and Brudevold (1963), and Pameijer *et al.* (1963)

show caries reduction in the neighborhood of 70% in addition to positive 1- and 2-year results with an acidulated phosphate fluoride dentifrice (Brudevold et al., 1964; Brudevold and Chilton, 1966).

Investigations concerning four types of fluoride dentifrices will be discussed; those incorporating sodium fluoride at the natural pH, those using amine fluoride, those using stannous fluoride, and those using phosphate fluorides.

V. Sodium Fluoride Dentifrices

There have been few recent studies of dentifrices involving near-neutral pH sodium fluoride dentifrices. One study used a sodium fluoride plus N-lauroyl sarcosinate formulation along with Crest by Admiral Kyes et al. (1961); the results were negative. Another used one of the five dentifrices studied by Brudevold et al. (1964), and Brudevold and Chilton (1966). The results for the neutral sodium fluoride dentifrice were negative at the end of 1 year and not significantly positive at the end of 2 years.

Wallace (1962) reported the completion of eleven controlled investigations. Two showed positive results and nine showed negative or insignificant results. After sponsoring a series of controlled investigations, Radike (1956) summarized the results as essentially negative. These conclusions were disputed by Bibby (1956) who felt that with a compatible dentifrice formulation, consistent positive results could be obtained. Some recent laboratory tests with more compatible abrasives would seem to indicate that perhaps Bibby may be correct. Gron and Brudevold (1967) have stated that the early failures of sodium fluoride dentifrices have been due to incompatible abrasives, low fluoride content, neutral pH, and inadequacies in clinical trials.

VI. An Amine Fluoride Dentifrice

Marthaler (1965) has recently reported the results of a 3-year unsupervised brushing study in which he started with first- and sixth-grade children. Two amine fluoride dentifrice formulas were used. He found caries reductions of approximately 32% for the younger age group and 25% for the older.

VII. Stannous Fluoride Dentifrices

Wallace (1962) reported that of ten tests of dentifrices containing stannous fluoride, only one showed negative results. These were all Crest studies. The test generally considered to be negative was made by the Kyes group (1961) at the Naval Academy when a sodium fluoride plus N-lauroyl sarcosinate dentifrice was also tested. Neither the 8% reduction measured for Crest nor the sodium fluoride sarcosinate results were significant. It is interesting that when Colgate's sodium N-lauroyl sarcosinate dentifrice (without sodium fluoride) was pitted against Crest and a sodium monofluorophosphate formula (Finn and Jamison, 1963) (but with no control dentifrice), the essentially similar results in this case were interpreted to mean that "Colgate is unsurpassed in preventing dental caries." This is particularly deceiving when the third formula—the monofluorophosphate—showed 25% fewer new decayed and filled surfaces than either of the other two. Without an admitted inactive control, there is no way of knowing whether this was a positive or negative Crest or sarcosinate test, or how effective the monofluorophosphate dentifrice really was. Additional monofluorophosphate tests will be discussed later in the section on Phosphate Fluoride Dentifrices.

We are now in a new era of dentifrice tests. Instead of a conventional control, which contains everything but the caries-inhibition component of the test dentifrice, we see tests with an active and an inactive control. Crest is being used as the *active control* in several studies. Since Crest is conceded to have therapeutic value, a Type II error possibility may be inferred if both Crest and the test dentifrice in question are reported equally ineffective. A Type II error occurs when no difference is found when a true difference really exists (ADA, 1955).

The *inactive control* in these same studies is a commercially available dentifrice which is conceded to have no therapeutic properties.

The Council on Dental Therapeutics of the American Dental Association studied the results of Crest tests and gave Crest a Group B classification (Council, 1960). This recognized that the dentifrice had some anticaries properties not before demonstrated to the Council's satisfaction by any other dentifrice. The Council, at that time, suggested additional investigation in certain areas would be desirable. The results of additional investigations were subsequently presented to the Council which ultimately led to the Group A classification (1964a).

The Council based its decision on data from laboratory, animal, and clinical studies, the results of which are listed in Tables II and III. Additional studies involving multiple uses of topical application and Crest by Scola and Ostrom (1966) and Bixler and Muhler (1966a) suggest that the additional procedures generally complement each other.

TABLE II
Crest Studies Considered by Council on Dental Therapeutics — August 1964[a]

Investigators	Percentage reduction in caries incidence	Remarks
Weisenstein[b]	12	2 Years — normal home use plus supervised brushing once a day
Jordan and Peterson (1959)	21	2 Years — normal home use plus supervised brushing once a day
Bixler and Muhler[b]	32	2 Years — supervised brushing three times daily
Peffley and Muhler (1960)	46	2 Years — supervised brushing three times daily
Muhler and Bixler (1966b)	54	2 Years — supervised brushing three times daily

[a] Supervised brushing.
[b] Unpublished data.

The results of an additional independent 3-year study were reported by Horowitz et al. of the Public Health Service (1966). This study showed caries reductions of 10, 12, and 21% at the end of 1, 2, and 3 years. Only the 3-year result was statistically significant.

Three additional Crest studies were reported from England by Slack et al. (1967b), James and Anderson (1967), and Jackson and Sutcliffe (1967). These were unsupervised brushing tests. Results generally showed very modest benefits for the test groups by visual examination and considerably better results with the use of radiographic examinations.

Laboratory and clinical data regarding their stannous fluoride dentifrice, Cue, was submitted by Colgate to the Council on Dental Therapeutics. This led to a Group B classification for Cue (1964b). The main difference in this formulation is that insoluble metaphosphate is used as the abrasive agent, while Crest uses calcium pyrophosphate. The clinical studies that aided the Council in its decision were investigations by Thomas and Jamison (1966), and unpublished reports by Mergele and Henriques (Table IV).

Although the results of each of these studies were not statistically significant at the same level of confidence, each did show less caries

TABLE III
CREST STUDIES CONSIDERED BY COUNCIL ON DENTAL THERAPEUTICS—
AUGUST 1964

Investigators	Percentage reduction in caries incidence	Remarks
Muhler (1962)[a]	21	3 Years—normal home use
Zacherl (1964)[a]	40	18 Months—normal home use
Gish and Muhler (1966)[b]	32	2 Years—Normal home use in water fluoridated area
Kyes et al. (1961)[c]	8	2 Years—Annapolis young adults unsupervised brushing. All carious surfaces filled at beginning of study
Muhler and Radike (1957)[c]	32	2 Years—young adults unsupervised brushing
Subsequent Crest studies		
Horowitz et al. (1966)	21	3 Years—unsupervised brushing
Slack et al. (1967b)	0	3 Years—unsupervised brushing for visual examination only
	36	Radiographic examination of posterior surfaces
James and Anderson (1967)		Unsupervised brushing
	16% Boys	Visual examination
	24% Girls	
	35% Boys	Radiographic examinations of posterior proximals
	49% Girls	
Jackson and Sutcliffe (1967)	13	3 Years—unsupervised brushing proximal surfaces of posterior teeth

[a] Unsupervised brushing.
[b] Flouride areas.
[c] Adults.

incidence in the test group when compared to the control. Also, in each of the three studies, Cue was compared to Crest with similar degrees of caries inhibition reported. Additional positive Cue studies have been reported by Naylor and Emslie (1967) from England and Fanning et al. (1968) from Australia.

Bristol-Myers submitted laboratory data and the 1- and 2-year results of Brudevold's study (Brudevold and Chilton, 1966) to gain the

TABLE IV
CUE STUDIES CONSIDERED BY THE COUNCIL ON DENTAL THERAPEUTICS – AUGUST 1964

Investigators	Percentage reduction in caries incidence	Remarks
Thomas and Jamison (1966)	37	2 Years – supervised brushing three times daily prophylaxis each 6 months
Mergele et al.[a]	16	2 Years – supervised brushing twice a day – fluoridated community
Henriques et al.[a]	17	2 Years – normal home use
Subsequent Cue studies		
Naylor and Emslie (1967)	15	Unsupervised brushing
Fanning et al. (1968)	21	2 Years – unsupervised brushing

[a] Unpublished data.

TABLE V
FACT TEST CONSIDERED BY COUNCIL ON DENTAL THERAPEUTICS – OCTOBER 1965[a]

| Dentifrice | Length of test (years) | Percentage reduction | |
		DFT[b]	DFS[c]
Commercially available (fluoride free)		Control	
Stannous fluoride	1	6	−3
Calcium pyrophosphate (Crest)	2	11	8
Stannous fluoride	1	24	12
Insoluble metaphosphate (Fact)	2	32	30

[a] Unsupervised brushing.
[b] DFT, decayed and filled teeth.
[c] DFS, decayed and filled surfaces.

Council on Dental Therapeutics' Group B classification for Fact (1965). Approximate reductions of 12 and 30% were reported for 1 and 2 years when compared with a nonfluoride dentifrice. Fact also compared favorably with Crest in the same study (Table V).

Lever Brothers received the Council's Group B classification for

Super Stripe (1966). Clinical evidence was obtained from a 2-year study by Segal et al. (1966). A 19% decayed and filled surface reduction was reported for all subjects and a 31% DFS reduction in a subgroup which eliminated those subjects who had a history of rampant caries and those who appeared highly resistant to caries or were caries immune (Table VI).

TABLE VI
SUPER STRIPE TEST CONSIDERED BY COUNCIL ON DENTAL THERAPEUTICS—
JUNE 1966

Investigators	Percentage reduction in caries incidence	Remarks
Segal et al. (1966)	19	Daily supervised brushing 2-year results
Subsequent Super Stripe tests		
Slack and Martin (1964)	No difference	2 Years—unsupervised brushing—boys and girls
Slack et al. (1967a)		3 Years—unsupervised brushing—girls only
	29	Proximal surfaces—radiographic examination
	30	Total surfaces—visual examination
	35	Total surfaces—radiographic examination

There are two additional reports of studies of an insoluble metaphosphate stannous fluoride dentifrice using the Unilever formula by Slack and Martin (1964); Slack et al. (1967a) in England. No differences between control and study groups were reported after 2 years in one test. In the second test, after 3 years of unsupervised brushing, Slack reported significantly fewer (28.5%) proximal surfaces becoming carious when measured by radiographic diagnosis alone. Thirty to thirty-five percent fewer total decayed and filled surfaces were reported with visual and radiographic measurements.

VIII. Phosphate Fluoride Dentifrices

At least two phosphate fluoride dentifrice types have received field tests. One incorporates monofluorophosphate as the fluoride vehicle and the other adds acid orthophosphate to sodium fluoride. They both employ calcium-free abrasive systems using insoluble sodium metaphosphate (IMP).

Enamel solubility reduction was determined *in vivo* for a sodium monofluorophosphate toothpaste by Herd and Overell (1964). This was accomplished by clamping silicone rubber cells to the lingual surfaces of upper central incisors. The rate of acid dissolution then was measured before and after brushing with the monofluorophosphate toothpaste. Statistically significant lesser amounts of tooth structure were dissolved after brushing with the paste.

The first published field test report on a monofluorophosphate dentifrice was the one made by Finn and Jamison (1963) and which also included Crest and the Colgate 2% sodium N-lauroyl sarcosinate (Gardol) formula. The monofluorophosphate formula also contains 1% sodium N-lauroyl sarcosinate, but has a lower pH and uses insoluble sodium metaphosphate (IMP) as the abrasive. As noted earlier, the monofluorophosphate dentifrice permitted approximately 25% fewer new decayed and filled surfaces than either Crest or Colgate's Gardol formula.

Further tests of Colgate's monofluorophosphate dentifrice have recently led to consideration by the ADA Council on Dental Therapeutics and a Class A rating (1969). The Council cited eight different studies from Australia, Great Britain, and the United States in which caries reductions in decayed, missing, and filled teeth ranged from 17–34%. Included were both supervised and unsupervised brushing tests, and were conducted in both fluoridated and low fluoride areas. They usually employed a stannous fluoride dentifrice as an active control and a nonfluoride dentifrice as an inactive control. The reductions were reported by the Council as "similar, in general, to those that have been reported for stannous fluoride-containing dentifrices." The data from these investigations are summarized in Table VII.

Two-year results of Brudevold and Chilton (1966) are available for the sodium fluoride-acid orthophosphate dentifrice developed by Bristol-Myers. In this study, five dentifrices were used:

1. A fluoride-free dentifrice as an inactive control.
2. A neutral sodium fluoride dentifrice.
3. A stannous fluoride-insoluble metaphosphate (IMP) dentifrice (Fact).
4. A stannous fluoride-calcium pyrophosphate dentifrice (Crest).
5. The sodium fluoride acid orthophosphate and insoluble metaphosphate (IMP) dentifrice (Ipana).

TABLE VII
Colgate Monofluorophosphate Dentifrice Tests Considered by Council on Dental Therapeutics — October 1969

Investigators	Percentage reduction in caries increment	Remarks
Finn and Jamison (1963)	25	When compared with Crest and Colgate without MFP
Naylor and Emslie (1967)	18	3 Years — unsupervised brushing
Fanning et al. (1968)	20	2 Years — unsupervised brushing
Moller (Council, 1969)	19	Supervised brushing once a day — 30 months
Thomas and Jamison (Council, 1969)	34	Supervised brushing twice a day for 2 years
Mergele (Council, 1969)	21	Supervised brushing three times a day — 22 months
Mergele (Council, 1969)	17	3 Years — unsupervised brushing — fluoridated community

A comparison of these group caries increments expressed as percentages are illustrated in Table VIII. The four fluoride dentifrices are compared with the nonfluoride dentifrice (inactive control).

Three-year data from a second independent field test of the same dentifrice were reported by Peterson and Williamson (1967, 1968)

TABLE VIII
Sodium Fluoride — Acid Orthophospate Dentifrice Tests

Investigators	Percentage reduction in caries increment	Remarks
Brudevold et al. (1964; Brudevold and Chilton, 1966)	24	Unsupervised brushing 1 Year
	26	2 Years
		(15% for NaF dentifrice)
		(30% for Fact)
		(8% for Crest)
Peterson and Williamson (1967, 1968)		Unsupervised brushing
	12	1 Year
	13	2 Years
	20	3 Years

(Table VIII). Crest was used for the active control and a third fluoride-free dentifrice was used for the inactive control. Statistically significant less caries increment was also found in the phosphate-fluoride dentifrice group in this test. However, the reduction was somewhat less than that found by Brudevold. On the other hand, the performance of the active control was better in this test with no significant difference occurring between the two therapeutic dentifrices. Parallel results to Brudevold's were obtained in an enamel solubility reduction test by Gwinnet and Buonocore (1965) in which least demineralization was obtained after treatment with Ipana, intermediate demineralization after Crest and Cue, and the most demineralization after the nonfluoride Gleem.

IX. Summary

Dental scientists have tested and added various therapeutic agents to dentifrices with the hopes of improving or maintaining the health of the gingival tissues, inhibiting plaque and calculus formation, relieving hypersensitive teeth, or inhibiting the caries process.

As of late 1969, few tests of dentifrices incorporating agents to improve or maintain the health of the periodontal tissues have been completed. Although frequent optimistic rumors are heard, little has been published.

Dentifrices incorporating 1.4% formaldehyde, 10% strontium chloride, and 0.76% monofluorophosphate have been specifically tested on patients with hypersensitive teeth. Results have been equivocal with the first two formulations. When control groups were used and the double-blind procedure followed, often little or no benefit was demonstrated. Significant improvement was reported in the one test with monofluorophosphate formulation.

Caries inhibition to a moderate degree has been obtained by adding therapeutic agents to dentifrices. This has been largely from formulations which incorporate specific fluoride compounds with compatible dentifrice abrasives. There appears to be a possibility of antibiotic or antienzyme agents also being used effectively in the future to control plaque and cariogenic organisms and thus inhibit caries. There does not appear to be any therapeutic value in the addition of ammonia compounds or detergents in dentifrices.

Five fluoride dentifrices have received recognition for their caries-inhibition properties by the Council on Dental Therapeutics of the American Dental Association. These are four stannous fluoride dentifrices: Procter and Gamble's *Crest*, Colgate's *Cue*, Bristol-

Myers' *Fact*, and Lever Brothers' *Super Stripe;* and one monofluorophosphate dentifrice, Colgate's *Colgate with MFP*. Crest was first to receive recognition and holds the largest share of the present market. Only Colgate with MFP is as extensively advertised and competes effectively for sales with Crest.

There are those that deplore the American Dental Association's Council on Dental Therapeutic's decision to classify therapeutic dentifrices. In rebuttal, an editorial in the August 1964 *Journal of the American Dental Association* enumerates the benefits that have accrued because of this policy: "First and most important, the health, of the public has benefited. Next the Council's program has been strengthened by the improved cooperation of commercial firms and by the further recognition of its opinions in matters of dental health. Also, commercial firms have stepped up their dental research efforts and are demonstrating new attainments of excellence. The efficacy of dentifrices is being steadily improved, and the likelihood of discoveries of other products which benefit the health of the public is good."

Doubt is expressed by some investigators that therapeutic dentifrices will ever contribute significantly to caries inhibition. Conventional tooth brushing is more often than not infrequently and haphazardly carried out, when even at best the drugs are diluted greatly by water and saliva making only momentary contact with the tooth surface. Undoubtedly, frequent and conscientious brushing with a therapeutic dentifrice is necessary to obtain satisfactory caries inhibition. This is emphasized by the American Dental Association pronouncement on each tube of toothpaste bearing the Association's approved label: ". . . can be of significant value when used in a conscientiously applied program of oral hygiene and regular professional care."

References

American Dental Association (1955). "Clinical Testing of Dental Caries Preventives," p. 15, 37.
American Dental Association (1964). "Editorial," *J. Amer. Dent. Ass.* **69,** 212.
Backer-Dirks, O., Kwant, G. W., and Starmons, J. L. E. M. (1960): *Dent. Abstr.* **5,** 371.
Bibby, B. G. (1956). *J. Amer. Dent. Ass.* **52,** 243, 755.
Bixler, D., and Muhler, J. C. (1966a). *J. Amer. Dent. Ass.* **72,** 392–396.
Bixler, D., and Muhler, J. C. (1966b). *J. Amer. Dent. Ass.* **72,** 653–658.
Bolden, T. E., Volpe, A. R., and King, W. J. (1968). *J. Periodontol.* **6,** 112–114.
Brudevold, F., and Chilton, N. W. (1966). *J. Amer. Dent. Ass.* **72,** 889–894.
Brudevold, F., Gardner, D. E., Spinelli, M., and Speirs, R. (1963a). *Arch. Oral Biol.* **8,** 167–177.

Brudevold, F., Wellock, W. D., and Pameijer, J. H. N. (1963b). *Int. Ass. Dent. Res. Meeting Abstr.* 332
Brudevold, F., Chilton, N. W., and Wellock, W. D. (1964). *J. Oral Ther. Pharmacol.* **1**, 1.
Burnett, G. W. (1957). *Dent. Abstr.* **2**, 550.
Cohen, A., and Donzanti, A. (1954). *J. Amer. Dent. Ass.* **49**, 185–190.
Council on Dental Therapeutics (1950). *J. Amer. Dent. Ass.* **40**, 619–622.
Council on Dental Therapeutics (1952). *J. Amer. Dent. Ass.* **45**, 466–468.
Council on Dental Therapeutics (1960). *J. Amer. Dent. Ass.* **61**, 272–274.
Council on Dental Therapeutics (1964a). *J. Amer. Dent. Ass.* **69**, 195–196.
Council on Dental Therapeutics (1964b). *J. Amer. Dent. Ass.* **69**, 197–198.
Council on Dental Therapeutics (1965). *J. Amer. Dent. Ass.* **71**, 930–931.
Council on Dental Therapeutics (1966). *J. Amer. Dent. Ass.* **72**, 1515.
Council on Dental Therapeutics (1969). *J. Amer. Dent. Ass.* **79**, 937–938.
Emslie, R. D. (1963). *J. Dent. Res.* **42**, 1079.
Fanning, E. A., Goljamanos, T., and Vowles, N. J. (1968). *Australian Dent. J.* **13**, 201–206.
Finn, S. B., and Jamison, H. C. (1963). *J. Dent. Child.* **30**, 17–25.
Forrest, J. O. (1963). *Brit. Dent. J.* **114**, 103–106.
Frasher, L. A., and Hein, J. W. (1958). *J. Dent. Res.* **37**, 75–76.
Gish, C. W., and Muhler, J. C. (1966). *J. Amer. Dent. Ass.* **73**, 853–855.
Gron, P., and Brudevold, F. (1967). *J. Dent. Child.* **34**, 122–127.
Gwinnett, A. J., and Buonocore, M. G. (1965). *Int. Ass. Dent. Res. Meeting Abstr.* 123.
Harrisson, J. W. E., Salisbury, G. B., Abbott, D. D., and Packman, E. W. (1963). *J. Periodontol.* **34**, 334–337.
Hayden, J., and Glass, R. L. (1959). *J. Dent. Res.* **38**, 671–672.
Herd, J. K., and Overell, B. G. (1964). *Brit. Dent. J.* **117**, 286.
Hill, T. J., *et al.* (1953). *J. Dent. Res.* **32**, 448–452.
Horowitz, H. S., *et al.* (1966). *J. Amer. Dent. Ass.* **72**, 408–422.
Jackson, D., and Sutcliffe, P. (1967). *Brit. Dent. J.* **123**, 40–48.
James, P. M. C., and Anderson, R. J. (1967). *Brit. Dent. J.* **123**, 33–39.
Jenkins, G. N. (1962). "Caries Symposium Zurich" p. 104–111. Huber, Bern.
Jordan, W. A., and Peterson, J. K. (1959). *J. Amer. Dent. Ass.* **58**, 42–44.
Kerr, D. W., and Kesel, R. G. (1951). *J. Amer. Dent. Ass.* **42**, 180–188.
Kimmelmann, B. B., King, R. J., Lorenzo, J., Solomon M., Zafran, J. N., and Zayon, G. M. (1969). *J. New Jersey State Dent. Ass.* **40**, 279–288.
Kyes, F. M., Overton, N. F., and McKean, T. W. (1961). *J. Amer. Dent. Ass.* **63**, 189–193.
Lunin, M., and Mandel, I. (1955). *J. Amer. Dent. Ass.* **51**, 696–702.
Marthaler, T. M. (1965). *Brit. Dent. J.* **119**, 153–163.
McCann, H. G., and Bullock, F. A. (1955). *J. Dent. Res.* **34**, 59–67.
Meffert, R. M., and Hoskins, S. W. (1964). *J. Periodontol.* **35**, 232–235.
Molle, W. H. (1967). *J. S. Calif. Dent. Ass.* **35**, 391–395.
Muhler, J. C. (1962). *J. Amer. Dent. Ass.* **64**, 216–224.
Muhler, J. C., and Radike, A. W. (1957). *J. Amer. Dent. Ass.* **55**, 196–198.
Naylor, M. N., and Emslie, R. D. (1967). *Brit. Dent. J.* **123**, 17–23.
Pameijer, J. H. N., Brudevold, F., and Hunt, E. E. (1963). *Arch. Oral. Biol.* **8**, 183–185.
Peffley, G. E., and Muhler, J. C. (1960). *J. Dent. Res.* **39**, 871–874.
Peterson, J. K., and Williamson, L. (1967). *J. Oral Ther. Pharmacol.* **4**, 1–4.
Peterson, J. K., and Williamson, L. (1968). *Int. Ass. Dent. Res. Meeting Abstr.* 255.
Radike, A. W. (1956). *J. Amer. Dent. Ass.* **52**, 244.

Ross, M. (1961). *J. Periodontol.* **32**, 49–53.
Science (1956). Vol. 123, p. 988.
Scola, F. P., and Ostrom, C. A. (1966). *J. Amer. Dent. Ass.* **73**, 1306–1311.
Segal, A. H. et al. (1966). *Int. Ass. Dent. Res. Meeting Abstr.* 250.
Shiere, F. R. (1957). *J. Dent. Res.* **36**, 237–244.
Slack, G. L., and Martin, W. J. (1964). *Brit. Dent. J.* **117**, 275–280.
Slack, G. L., Berman, D. S., Martin, W. J., and Young, J. (1967a). *Brit. Dent. J.* **123**, 9–15.
Slack, G. L., Berman, D. S., Martin, W. J., and Hardie, J. M. (1967b). *Brit. Dent. J.* **123**, 26–33.
Smith, B. A., and Ash, M. M. (1964a). *J. Amer. Dent. Ass.* **68**, 639–647.
Smith, B. A., and Ash, M. M. (1964b). *J. Periodontol.* **35**, 222–231.
Sulser, G. F., Fosket, R. R., and Fosdick, L. S. (1958). *J. Amer. Dent. Ass.* **56**, 368–375.
Thomas, A. E., and Jamison, H. C. (1966). *J. Amer. Dent. Ass.* **73**, 844–852.
Vogel, P., and Hess, W. (1957). *J. Dent. Child.* **24**, 237–242.
Wachtel, L. W., and Strange, C. G. (1965). *J. Dent. Res.* **44**, 3–9.
Wallace, D. (1962). *Dent. Progr.* **2**, 242–248.
Walsh, J. P., and Smart, R. S. (1951). *New Zealand Dent. J.* **47**, 118–122.
Wellock, W. D., and Brudevold, F. (1963). *Arch. Oral Biol.* **8**, 179–182.
Zacherl, W. (1964). Presented at the Canadian Dental Association Meeting, July, 1964.
Zander, H. A. (1950). *J. Amer. Dent. Ass.* **40**, 569–574.

Some Contributions of the U.S. Department of Agriculture to the Fermentation Industry

GEORGE E. WARD

Dawe's Laboratories, Inc.,
Chicago, Illinois

I.	Introduction	363
II.	Citric Acid Research of Dr. J. N. Currie	364
III.	Arlington Farms Research Period	364
	A. Appointment of Mr. H. T. Herrick and Assignment of Dr. O. E. May	364
	B. Gluconic Acid Surface Fermentation	366
	C. Staff Appointments	366
	D. Submerged Gluconic Acid Fermentation	367
	E. Rotary Aluminum Fermentors	367
	F. Kojic Acid Fermentation	368
	G. Citric Acid Fermentation	368
	H. Lactic Acid and Fumaric Acid Fermentation	368
	I. Sorbose Fermentation	369
	J. Ketogluconic Acid Fermentations	369
	K. Staff Changes: 1936–1940	370
	L. Equipment	370
IV.	Regional Research Laboratory Period: 1940–1945	371
	A. Establishment of the Regional Research Laboratories	371
	B. Staffing of the Fermentation Division at NRRL	371
	C. Itaconic Acid Fermentation	372
	D. Alcohol Production from Wheat	372
	E. 2,3-Butanediol Fermentation	373
	F. Penicillin Fermentation Research	373
V.	Regional Research Laboratory Period: 1945–1969	377
VI.	Vitamin B_{12} Fermentation	379
	References	379

I. Introduction

During the first two decades of this century the fermentation industry was limited mainly to the production of yeast, alcoholic beverages, vinegar, industrial alcohol, lactic acid, and a few enzymes. Today the industry has a much wider scope, producing a variety of antibiotics, vitamins, organic acids, enzymes, sugars, sugar alcohols, polysaccharides, amino acids, microbial insecticides, sterols, and modified steroids. Research conducted by the staff of the United States Department of Agriculture has played an important role in the early development of the fermentation industry during the past half century. This chapter will describe in some detail the contributions of Agriculture Department scientists between 1916 and 1945, and will then touch briefly on the numerous additional contributions made during the past quarter century.

II. Citric Acid Research of Dr. J. N. Currie

Scientific studies conducted about 55 years ago in the microbiological laboratory of the Bureau of Chemistry in Washington had a profound effect on the fermentation industry. This laboratory, headed by Dr. Charles Thom, was primarily responsible for the enforcement of certain aspects of the Pure Food and Drugs Act, but in the course of the work a large number of cultures of *Aspergillus* and *Penicillium* was collected. The oxalic acid-producing characteristics of these fungi were studied and reported by Dr. Thom and Dr. James N. Currie (1916), a dairy chemist of the Bureau of Animal Industry. It was observed that several of the strains of *Aspergillus niger* produced citric acid in addition to oxalic acid. In 1917 Dr. Currie described nutritional and environmental conditions conducive to citric acid formation by several *A. niger* strains which were grown in surface cultures in glass flasks and shallow pans. Soon after publication of this paper Dr. Currie left the Department to continue his studies at Chas. Pfizer and Co., Inc., and in 1923 this company began to produce citric acid in commercial quantities by a fermentation method believed to be based on Dr. Currie's experience. Citric acid production by fermentation has been in continuous operation at the Pfizer company since that time, and in recent years a second company, Miles Laboratories, Inc., has also produced this acid by fermentation. The estimated current annual production of citric acid in this country is about 125 million pounds.

III. Arlington Farm Research Period

A. Appointment of Mr. H. T. Herrick and Assignment of Dr. O. E. May

A second period of industrial fermentation research at the Agriculture Department began in 1925 with the appointment of Mr. Horace T. Herrick (Fig. 1) to be Chemist-in-Charge of the Color and Farm Waste Division of the Bureau of Chemistry and Soils, located at Arlington Farms, Virginia. The activities of that Division were conducted in a building known as the Color Laboratory, located on the Arlington Farms, situated between Arlington National Cemetery and

FIG. 1. Mr. H. T. Herrick.

the Potomac River. The farm was originally attached to Arlington House, the estate of Robert E. Lee. The Color Laboratory, a two-story, poured-concrete structure, received its name from the fact that it was built during World War I to house research projects aimed at developing methods for producing dyestuffs which, until then, had been largely a German monopoly. (The Color Laboratory was vacated in 1940 and the Arlington Farms site was completely altered by the construction of the Pentagon Building during the early years of World War II.)

Prior to his government service Mr. Herrick had been a chemical engineer with the Zinsser Company, which conducted a fermentation process for the conversion of tannin to gallic acid by *Aspergillus* species. In planning research projects for the industrial utilization of agricultural materials Mr. Herrick, undoubtedly influenced to some degree by Dr. Currie's successful production of citric acid and by his own experience at Zinsser, proposed that mold fermentations would be an area worthy of exploration. He assigned the task of organizing and developing a fermentation program to Orville E. May (Fig. 2), a young chemist who had entered the government service in 1923 as a "chemical laboratorian," and completed his university studies by attending evening classes at The George Washington University. To become familiar with the elements of mycology Dr. May spent several

Fig. 2. Dr. Orville E. May.

months at the Washington laboratories of Dr. Thom and his assistant, Dr. Margaret Church.

B. Gluconic Acid Surface Fermentation

Tartaric acid, widespread in nature and a commercially important acid, was considered to be a likely product of mold metabolism, and tests were made for the presence of this acid in many fermentation broths. Tartaric acid was not found, but the studies did show that several species of *Aspergillus* and *Penicillium* produced large amounts of gluconic acid. The *Penicillium luteum-purpurogenum* group was especially good for gluconic acid production in surface cultures and a semi-plant scale procedure for producing gluconic acid in aluminum trays was soon developed (May *et al.*, 1929).

C. Staff Appointments

The staffing of a fermentation research group was undertaken between 1927 and 1931. Dr. May's first technical assistant was Samuel M. Weisberg, who served 1927–1928. Andrew J. Moyer, mycologist, and Percy A. Wells, chemist, were added in 1929; George E. Ward, chemist, was employed in 1930, and in 1931 Lewis B. Lockwood joined the group of five which was the nucleus for the fermentation research conducted by the Industrial Fermentation Laboratory during the early 1930's.

D. SUBMERGED GLUCONIC ACID FERMENTATION

Gluconic acid production from glucose continued to be a prime research topic. Submerged fermentation was investigated and found to have advantages over surface culture methods. *Aspergillus niger* strains were found superior to *Penicillium* strains for submerged culture. Shake flask techniques were not yet developed, and the first submerged studies were made by aerating inoculated fermentation medium in sintered glass false-bottom gas washing bottles mounted in an autoclave which was placed under air pressure; the air flow through each bottle was controlled with a needle valve (May *et al.*, 1934). Encouraging results were obtained, although loss of medium by foaming and partial plugging of the porous disc were sometimes encountered.

E. ROTARY ALUMINUM FERMENTORS

The limitations of the gas washing bottle technique were appreciated, and with the assistance of Mr. R. Hellbach, the laboratory's very able instrument maker, a set of rotary fermentors was constructed of high purity aluminum which was found not to inhibit the gluconic acid fermentation (Herrick *et al.*, 1935). The cylindrical shell revolved horizontally on hollow trunnions through which sterile air could be admitted under controlled pressure, and effluent gas could be metered. The inner wall of the shell contained buckets and baffles which imparted violent mixing and exposure of the mass to the gaseous atmosphere. The fermentors could be operated at air pressure up to 75 psig, at speeds up to 13 rpm, and at variable aeration rates. Each fermentor had a total volume of about 18 liters and was usually charged with about 3.2 liters of fermentation broth.

The rotary aluminum fermentors permitted independent evaluation of aeration, agitation, and pressure; in a series of tests it was shown that the highly oxidative gluconate fermentation proceeded especially well at 2 to 3 atm of pressure, at aeration rates of about 0.5 volume of air per volume of medium per minute, and at about 13 revolutions of the drum per minute, which gave a high degree of agitation (Wells *et al.*, 1937a). A patent describing the method of carrying out fermentations of carbohydrates by submerged growths of fungi was obtained by May *et al.* (1935) and assigned to the Secretary of Agriculture, thereby conserving this technology for the American public.

The success of the small rotary fermentors in the research laboratory led to the construction of an aluminum fermentor with a working capacity of 140 gallons which was installed about 1936 at the Bureau's Agricultural By-Products Laboratory at Ames, Iowa (Wells et al., 1937c). Scale-up fermentations for the production of calcium gluconate and of sorbose in this fermentor verified the results in the small rotary fermentors (Gastrock et al., 1938; Wells et al., 1939).

It is of interest to note that all the submerged gluconate fermentations studied by the Arlington group during the 1930's used calcium carbonate as the neutralizing agent, thereby yielding calcium gluconate as the fermentation product. At that time calcium gluconate was the principal marketable salt of gluconic acid; the sequestering properties of strongly alkaline gluconate solutions were not recognized until about 1950; today the annual market for sodium gluconate is about 15 million pounds, for gluconic acid about 6 million pounds, and for calcium gluconate probably less than 1 million pounds, according to U.S. Tariff Commission statistics.

F. Kojic Acid Fermentation

The production of kojic acid by the surface cultivation of *Aspergillus flavus* on glucose nutrient medium was studied by the Arlington group (May et al., 1931, 1932) and a large number of samples was distributed to chemical and pharmaceutical companies in search of a market. Uses for kojic acid did not develop until the 1950's when it was found suitable for the production of maltol, a flavoring material.

G. Citric Acid Fermentation

A significant contribution to an understanding of the chemistry of the citric acid fermentation was the finding that the quantity of carbon dioxide produced by a high-yielding strain of *Aspergillus niger* was insufficient to validate proposed mechanisms which involved simple decarboxylation of pyruvic acid, a possible intermediate (Wells et al., 1936).

H. Lactic Acid and Fumaric Acid Fermentation

The acid-producing propensities of *Rhizopus* species were investigated and strains were selected which produced good yields of fumaric acid and L(+)-lactic acid (Ward et al., 1936). A submerged fermentation procedure for the production of L(+)-lactic acid was described by Ward et al. (1938), and this method is believed to be the basis for the present commercial production of this acid by Miles Laboratories, Inc.

I. SORBOSE FERMENTATION

Much research effort was devoted to studies of the fermentation propensities of the acetic acid bacteria. A commercially important product formed upon the oxidation of D-sorbitol by *Acetobacter suboxydans* is L-sorbose, which is used as an intermediate in the synthesis of ascorbic acid. An important contribution of the Arlington group to this fermentation was the demonstration that 30% solutions of D-sorbitol could be rapidly and efficiently oxidized to L-sorbose by *Acetobacter suboxydans* if increased air pressure (30 psig) and a high rate of aeration were employed, if corn steep water at low concentrations (3 gm per liter) was used as a nutrient supplement, and if sufficient calcium carbonate was added to the medium to neutralize the gluconic acid formed from the small quantities of glucose present in the sorbitol and in the corn steep water (Wells *et al.*, 1937b). Sorbose fermentations conducted previously had used dilute sorbitol solutions and had required expensive yeast extracts and peptones as nutrient supplements.

J. KETOGLUCONIC ACID FERMENTATIONS

An efficient method for the production of 5-ketogluconic acid from glucose by *Acetobacter suboxydans* was developed and the advantages of conducting this fermentation under highly aerobic conditions were described by Stubbs *et al.* (1940). This acid may be chemically converted to tartaric acid or reduced to a mixture of D-gluconic acid and L-idonic acid, the latter being of the proper configuration to form L-ascorbic acid.

Numerous *Acetobacter* strains were screened in an attempt to find a good producer of 2-ketogluconic acid, from which can be synthesized D-isoascorbic acid (now known as erythorbic acid). Although *Acetobacter* had been reported to produce 2-ketogluconic acid, no productive strains were recognized at that time by the Arlington group and the 2-ketogluconate project made little progress until one day when large quantities of calcium 2-ketogluconate were found in a contaminated fermentation. The contaminating organism was identified as a strain of *Pseudomonas fluorescens;* thereafter a survey of *Pseudomonas* cultures showed that 16 strains distributed among ten species produced 2-ketogluconic acid from glucose in submerged aerated culture (Lockwood *et al.*, 1941). *Pseudomonas* species have the advantage that they can utilize inorganic nitrogen salts and do not require complex nitrogen-containing organic compounds, as do *Acetobacter* species. The production of 2-ketogluconic acid in the rotary fermentors was described by Stubbs *et al.* (1940). The 2-ketogluconic acid fermentation using *Pseudomonas*, essentially as developed by the Arlington group, has been used commercially as a step in the production of erythorbic acid.

K. STAFF CHANGES: 1936–1940

Some important staff changes occurred in the late 1930's. In 1936 Dr. May left the group to become Director of the Department's Regional Soybean Industrial Products Laboratory at Urbana, Illinois, and Dr. Wells assumed the leadership of the Arlington fermentation research group. Also about this time E. A. Gastrock, N. Porges, and T. F. Clark were employed at the Agricultural By-Products Laboratory at Ames, Iowa to demonstrate fermentation processes in the 140-gallon aluminum rotary fermentor installed there. In 1937 Benjamin Tabenkin joined the Arlington group, and in the late 1930's J. J. Stubbs and E. T. Roe, already members of the scientific staff at the Color Laboratory, devoted the major portion of their time to fermentation research.

L. EQUIPMENT

It seems appropriate to mention here some of the equipment used for fermentation research at the Color Laboratory during the 1930's, the period of the Great Depression. Funds for the purchase of equipment were very limited, and by present standards many of the facilities were spartan. Factory-built pH meters were not yet commonly available and most acidity measurements were made colorimetrically or with a home-made quinhydrone electrode. In the early 1930's we did not have a source of compressed air and the author remembers several experiments in which a flow of air through a fermentation chamber was obtained by admitting water to a closed 55-gallon drum, thereby displacing air which could be conveyed through tubing to the chamber. Aeration rates were measured by home-made manometers which we calibrated by water displacement. Improvisation was the order of the day.

In the late 1930's two stirred aluminum fermentors, equipped with air spargers, were constructed at the laboratory. These were subsequently moved to the Northern Regional Research Laboratory and were the prototypes for some of the fermentors installed there.

The pilot-plant area of the Color Laboratory contained several serviceable pieces of equipment sent from Germany as part of World War I reparations. The items of particular usefulness to the fermentation research staff were two open glass-lined evaporating pans which operated at atmospheric pressure, and one single-effect vacuum evaporator of working capacity about 20 liters. This equipment was frequently used for concentrating and crystallizing fermentation products such as calcium gluconate, sorbose, calcium lactate, calcium 2-ketogluconate, and kojic acid.

IV. Regional Research Laboratory Period: 1940-1945

A. ESTABLISHMENT OF THE REGIONAL RESEARCH LABORATORIES

The outstanding accomplishments of the Regional Soybean Industrial Products Laboratory under Dr. May's direction resulted in a nationwide interest in establishing additional research laboratories to search for industrial uses for agricultural materials, and four such laboratories were established: at Peoria, Illinois; Philadelphia, Pennsylvania; New Orleans, Louisiana; and Albany, California. Dr. Wells was asked to supervise the planning for the Eastern Regional Research Laboratory and left the Arlington group; George Ward became acting head during 1938-1940. When the regional laboratories were opened the activities of the Color and Farm Waste Division at Arlington Farms were apportioned among the new laboratories, and the industrial fermentation work was assigned to the Northern Regional Research Laboratory at Peoria. Interestingly, Dr. May was named the first Director of the Northern Laboratory and thus again became intimately connected with the fermentation research program. He served as Director until 1942, when he returned to Washington to become Research Coordinator of the Agricultural Research Administration, and subsequently to become Chief of the Bureau of Agricultural and Industrial Chemistry. Mr. Herrick became the second Director of the Northern Laboratory in 1942 and served in this capacity until 1946, when he returned to Washington.

B. STAFFING OF THE FERMENTATION DIVISION AT NRRL

The Fermentation Division at the Northern Laboratory was a much larger operation than the Arlington effort. Dr. Robert D. Coghill, (Fig. 3), formerly Professor of Organic Chemistry at Yale University, was named Chief of the Fermentation Division, which consisted of three sections: the Culture Collection Section under Dr. Kenneth B. Raper, a protégé of Dr. Charles Thom (see Fig. 4); the Chemical Section under Dr. Frank H. Stodola who came to the laboratory from Columbia University; and the Survey and Development Section under George Ward. The last section included Dr. Andrew Moyer, Dr. Lewis Lockwood, and Mr. Benjamin Tabenkin, all of whom had been associated with the Arlington research group. Dr. Thom's collection

Fig. 3. Dr. Robert D. Coghill.

Fig. 4. Dr. Kenneth B. Raper (*left*) and Dr. Charles Thom (*right*).

of molds was brought to the Laboratory by Dr. Raper and it became the nucleus for an outstanding collection of microorganisms.

C. Itaconic Acid Fermentation

One of the first fermentation research projects at Peoria was the production of itaconic acid by *Aspergillus terreus* and *A. itaconicus*. *Aspergillus terreus* NRRL 1960, isolated by Dr. Raper from Texas soil, proved to be the best itaconate producer, and cultures now used commercially for itaconic acid production are descended from this strain. A submerged fermentation process for producing itaconic acid was developed (Lockwood and Ward, 1945; Lockwood and Nelson, 1946; Nelson *et al.*, 1952; Pfeifer *et al.*, 1952), and a good analytical method for itaconic acid assay was devised (Friedkin, 1945).

D. Alcohol Production from Wheat

World War II began shortly after the regional laboratories were activated, and research programs were modified to meet war needs. One of the strategic chemicals was ethyl alcohol, largely produced from the fermentation of blackstrap molasses under peacetime conditions. In the early war years shipping restrictions prevented the receipt of molasses from Cuba or Puerto Rico, and the alcohol plants were obliged to ferment grain mashes. Because of cost and starch content corn was the grain preferred for the alcohol fermentation in normal times, but war conditions required the consideration of wheat as an alternative or supplementary raw material. The distilleries had little experience with wheat fermentations and information was needed as quickly as possible on the alcohol yield obtainable from wheat varieties of different starch contents, on optimal malting procedures, on the suppression of foams sometimes encountered in wheat fermentations, on the choice of yeast culture, and on the recovery of fermentation residues for use in animal feeds. A committee composed of representatives of the U.S. Department of Agriculture, the War Production Board, and the Forest Products Research Laboratory was organized and Dr. May was appointed chairman. The Northern Regional Research Laboratory was the focal point for conferences with industry and for conducting fermentation experiments. The problems were solved within a short time, and wheat was used for alcohol production during the balance of the war. Wheat was the source of practically all of the alcohol used to produce butadiene for synthetic rubber manufacture during this period.

E. 2,3-BUTANEDIOL FERMENTATION

The production of synthetic rubber from butadiene was intensively studied by many research organizations during the early war years, and the Fermentation Division was asked to investigate the conversion of carbohydrates to 2,3-butanediol, a potential raw material for butadiene formation. Two fermentation methods for producing this diol were studied: (a) the conversion of sugar solutions by *Aerobacter aerogenes* to principally *meso*-2,3-butanediol plus minor quantities of *dextro*-2,3-butanediol and ethyl alcohol; and (b) the fermentation of unsaccharified starchy substrates by *Bacillus polymyxa* to *levo*-2,3-butanediol plus ethyl alcohol (Ward et al., 1944, 1945). Other Divisions at the Northern Laboratory studied methods for converting the 2,3-butanediol to butadiene. Eventually, the production of butadiene from ethyl alcohol and later from petroleum were judged to be superior to methods based on 2,3-butanediol, and the latter was not used industrially for synthetic rubber production.

F. PENICILLIN FERMENTATION RESEARCH

1. Historical Background

The contributions of the Fermentation Division of the Northern Regional Research Laboratory to the development of penicillin during the years of World War II are without question among the most constructive all-time accomplishments of the U.S. Department of Agriculture. Also, the development of penicillin from the stage of laboratory production of traces of a compound of only partially understood chemical structure to industrial production in many pharmaceutical and chemical plants of sufficient quantities to meet the needs of the armed forces and the civilian population, within a 3–4 year period, exemplifies the progress that can be made when there is unqualified cooperation between many competent scientific groups. Penicillin, discovered by Alexander Fleming in 1929, lay dormant and unappreciated during most of the 1930's. It was revived as a potential medical asset by British physicians near the beginning of the war and sufficient amounts of the antibiotic were laboriously produced to permit testing in a few selected patients, with very encouraging results. In July 1941 when penicillin research in England ceased because of bombing, Dr. Norman Heatley and Dr. Howard Florey were brought to this country by the Rockefeller Foundation to enlist the assistance of American research workers. The Washington offices of the Department of Agriculture received Drs. Heatley and Florey, evaluated the problem and sent them to Peoria to confer with the Fermentation Division of the Northern Regional Research Laboratory. Dr. Heatley remained at the Laboratory for several months to demonstrate the methods used in England for the production and assay of penicillin. The Oxford strain of *Penicillium notatum* was grown on the surface of sucrose nutrient salt medium to obtain a penicillin yield of 2 Oxford units per ml. The structure of penicillin was still unknown and it was then thought to contain only carbon, hydrogen, and oxygen, although subsequently it was found also to contain nitrogen and sulfur. The sample provided for an assay standard was thought to be upward of 90% pure, but knowledge obtained later showed that this working standard probably contained less than 3% active material.

2. *Improvements with Lactose and Corn Steep Water*

Within a few weeks of the visit of Drs. Florey and Heatley, the staff of the Fermentation Division demonstrated a significant increase in penicillin broth titers when the sucrose in the medium was replaced by lactose, which was consumed more slowly by the organism. A short time later additional yield increases were obtained when corn steep water was incorporated in the medium at levels as high as 4–8%, in contrast to the 0.3–0.5% levels which had been found adequate in the sorbose and 5-ketogluconate fermentations (Moyer and Coghill, 1946a). Apparently the corn steep water was functioning as a major raw material rather than as a source of trace nutrients. The provision of trace metals, amino acids, phenylacetate, and buffering capacity have been cited by various workers as desirable functions of corn steep water in this fermentation. Simultaneously with the nutritional studies, substrains of the Oxford *P. notatum* culture were isolated and tested, and substrain NRRL 1249.B21 was chosen. This substrain, grown in surface culture on medium containing lactose and corn steep water, produced broth titers of 150 to 200 units per ml. The first commercial production of penicillin used this substrain and medium, and employed metal trays, Roux bottles, or 2-quart milk bottles as culture vessels. Under the war conditions the operations were conducted with secrecy and all the penicillin produced was placed under allotment for the armed forces by the government.

Dr. Coghill, among his numerous other duties and responsibilities, was assigned the task of maintaining technical liaison with the British scientists engaged in penicillin research and production. Classified progress reports and several trans-Atlantic trips served to exchange important data promptly.

3. *Improved Cultures and Submerged Fermentation*

From the beginning of the penicillin work at the Northern Laboratory the disadvantages of the surface culture method were recognized and numerous unsuccessful attempts were made to produce good yields of penicillin with submerged growths of the Oxford strain of *P. notatum*. The Culture Collection Section of the Division screened all the *P. notatum* strains on hand and found one, designated *Penicillium notatum* NRRL 832, which produced fair yields of penicillin in submerged culture (Moyer and Coghill, 1946b). Industry used this strain to produce a large proportion of the antibiotic made in this country in 1943–1944.

A further improvement in cultures resulted from the isolation of a very productive strain of *Penicillium chrysogenum* from a moldy canteloupe found in a Peoria fruit market (Raper, 1946). A selected substrain of this organism, *P. chrysogenum* NRRL 1951.B25, was capable of producing broth titers of up to 250 units per ml and was provided to the industrial producers. Demerec of the Carnegie Institute, Cold Spring Harbor, New York, treated this substrain with X-rays to form mutation X-1612, capable of producing about 500 units per ml. Thereafter, a University of Wisconsin group treated X-1612 with ultraviolet irradiation to obtain a further mutation, designated as strain Q-176, which could produce up to 1000 units per ml.

Process improvement since about 1946 has been principally a concern of individual industrial research laboratories. It is understood that improvement in cultures, in medium, and in process control have brought present industrial penicillin yields to approximately 20,000 units per ml.

4. Improvements with Phenylacetic Acid and Related Compounds

As penicillin became available in modest quantities, small amounts were supplied to American and British chemists for studies of structure and composition. It was discovered that there were several different penicillins, which differed in the identity of a side chain. Penicillin F, having a Δ^2-pentenyl side chain, was the predominant product from surface fermentations conducted on medium lacking corn steep water. Penicillin G, having a benzyl side chain, was the predominant product from submerged fermentations of medium containing corn steep water. It was found by the Northern Laboratory group (Coghill and Moyer, 1947) that increases in penicillin yield and also in the proportion of Penicillin G would result if phenylacetic acid, phenylacetamide, or phenylethylamine were added to the fermentation medium. The addition of such precursors was especially important when the high-yielding strains were employed, so as to avoid the formation of large quantities of Penicillins F and K. Penicillin G (benzylpenicillin) was, by that time, the product preferred by the medical profession.

5. Assay and Reference Standards

During the penicillin project, Northern Regional Laboratory staff improved the microbiological assay procedures and made this information available to other research workers, regulatory agencies, and to industrial penicillin producers (Schmidt and Moyer, 1944; Schmidt et al., 1945; Schmidt, 1946; Reeves and Schmidt, 1945).

Also, a quantity of pure benzylpenicillin prepared by Dr. Stodola and his group was provided to the Health Organization of the League of Nations for use as an International Penicillin Standard (Coghill, 1946).

6. Comment

The successful commercial production of penicillin and its unquestioned value in medicine opened the door to the antibiotic age. Since 1945 hundreds of new antibiotics have been discovered and about twenty have had sufficient merit to justify their industrial production in this country. Corn steep water is used in most media and submerged culture methods similar to those developed for penicillin are usually employed.

The 1946 Lasker Group Award of the American Public Health Association was presented to the Northern Regional Research Laboratory for its studies leading to the mass production of penicillin. In 1947 the Penicillin Group (Fig. 5) was presented the United States Department of Agriculture Award for Distinguished Service for its "invaluable contribution to medical science by making the mass production of penicillin possible."

FIG. 5. The penicillin team at the Northern Regional Research Laboratory, 1945. Left to right: W. H. Schmidt, M. D. Reeves, M. Friedkin, J. L. Wachtel, D. Fennell, H. T. Herrick, F. H. Stodola, K. B. Raper, R. D. Coghill, G. E. Ward, A. J. Moyer, R. T. Milner, N. C. Schieltz, C. H. Van Etten, G. E. N. Nelson, L. J. Wickerham, and L. Smith.

V. Regional Research Laboratory Period: 1945–1969

Fermentation research and development at the Northern Regional Research Laboratory have continued since 1945 and many contributions of commercial value have resulted. The impossibility of listing here all of the contributions and workers will be apparent when it is realized that more than 600 scientific papers have been published by this research group between 1945 and 1969. The following list shows some of the more important contributions.

Riboflavin production by *Ashbya gossypii*, which is the basis for most commercial riboflavin production by fermentation today (Wickerham *et al.*, 1946; Tanner *et al.*, 1949; Pridham, 1952; Pfeifer *et al.*, 1950).

Fungal amylase production by *Aspergillus niger* (Corman and Langlykke, 1948; Corman and Tsuchiya, 1951; LeMense *et al.*, 1947, 1949; Tsuchiya *et al.*, 1950).

Amyloglucosidase (glucamylase) production by *Aspergillus awamori* (Smiley *et al.*, 1964; Cadmus *et al.*, 1966).

Scientific study of production of microbiologically derived foods such as soy sauce, miso, tempeh, sufu (Hesseltine and Shibasaki, 1961; Hesseltine *et al.*, 1963, 1967; Shibasaki and Hesseltine, 1961; Wang and Hesseltine, 1966).

Production of antibiotics, vitamin B_{12}, and animal growth factors by *Streptomyces* species (Benedict, 1953; Benedict *et al.*, 1954; Hall *et al.*, 1953; Stodola *et al.*, 1951).

Production of dextran and fructose by *Leuconostoc mesenteroides* and by the enzyme dextran sucrase derived therefrom (Tsuchiya *et al.*, 1952, 1953, 1955; Koepsell and Tsuchiya, 1952; Hellman *et al.*, 1955).

Production of sodium gluconate from glucose by *Aspergillus niger* (Blom *et al.*, 1952).

Production of gibberellins (Stodola *et al.*, 1955, 1957).

Production of invertase by yeasts (Wickerham and Dworschack, 1960; Wickerham, 1947; Wickerham and Kuehner, 1956; Dworschack and Wickerham, 1961).

Production of β-carotene by mated strains of *Blakeslea trispora* (Anderson *et al.*, 1958; Ciegler *et al.*, 1963; Ciegler, 1965).

Production of microbial polysaccharides, particularly of a gum by *Xanthomonas campestris* (Anderson *et al.*, 1960; Rogovin *et al.*, 1961; Smiley, 1966).

Production of microbial insecticides active against the Japanese beetle (Hall *et al.*, 1968; Haynes and Rhodes, 1969; Rhodes, 1965).

Production, analysis, and detoxification of aflatoxin (Ciegler *et al.*, 1966; Hesseltine *et al.*, 1966; Stubblefield *et al.*, 1968).

Production of mannitol from glucose by *Aspergillus candidus* (Smiley *et al.*, 1967; Strandberg, 1969).

An additional important contribution to the fermentation industry is the publication of the experience in maintaining the NRRL collection of yeasts, molds, and bacteria, principally in the lyophilized state, which was pioneered by the Culture Collection Section (Haynes et al., 1955; Hesseltine et al., 1960; Ellis and Roberson, 1968). The NRRL Culture Collection also serves as a depository for cultures used in processes for which United States patent applications are filed.

VI. Vitamin B_{12} Fermentation

A process for the production of vitamin B_{12} based on the cultivation of *Bacillus megaterium* was devised by staff of the Western Regional Research Laboratory and was used commercially until more productive procedures were developed (Lewis et al., 1949; Garibaldi et al., 1951, 1953).

Another Agriculture Department group which has made an important contribution to the fermentation industry is the Bureau of Dairy Industry in Washington. In the 1950's two members of their staff reported that propionic acid bacteria produce intracellular vitamin B_{12} when grown under microaerophilic conditions (Leviton and Hargrove, 1952; Hargrove and Leviton, 1955; Leviton, 1956). This finding has been investigated in great detail by several industrial fermentation companies, and a large part of the vitamin B_{12} now produced in the world is made by the cultivation of selected strains of *Propionibacterium freudenreichii* or *P. shermanii*.

REFERENCES

Anderson, R. F., Arnold, M., Nelson, G. E. N., and Ciegler, A. (1958). *J. Agr. Food Chem.* **6**, 543–545.
Anderson, R. F., Cadmus, M. C., Benedict, R. G., and Slodki, M. E. (1960). *Arch. Biochem. Biophys.* **89**, 289–292.
Benedict, R. G. (1953). *Bot. Rev.* **19**, 229–320.
Benedict, R. G., Shotwell, O. L., Pridham, T. G., Lindenfelser, L. A., and Haynes, W. C. (1954). *Antibiot. Chemother.* **4**, 653–656.
Blom, R. H., Pfeifer, V. F., Moyer, A. J., Traufler, D. H., Conway, H. F., Crocker, C. K., Farison, R. E., and Hannibal, D. V. (1952). *Ind. Eng. Chem.* **44**, 435–440.
Cadmus, M. C., Jayko, L. G., Hensley, D. E., Gasdorf, H., and Smiley, K. L. (1966). *Cereal Chem.* **43**, 658–669.
Ciegler, A. (1965). *Advan. Appl. Microbiol.* **7**, 1–34.
Ciegler, A., Lagoda, A. A., Sohns, V. E., Hall, H. H., and Jackson, R. W. (1963). *Biotechnol. Bioeng.* **5**, 109–121.
Ciegler, A., Lillehoj, E. B., Peterson, R. E., and Hall, H. H. (1966). *Appl. Microbiol.* **14**, 934–939.

Coghill, R. D. (1946). *Bull. Health Organ. League Nations* **12**, Extract No. 5.
Coghill, R. D., and Moyer, A. J. (1947). *J. Bacteriol.* **53**, 329–341.
Corman, J., and Langlykke, A. F. (1948). *Cereal Chem.* **25**, 190–201.
Corman, J. and Tsuchiya, H. M. (1951). *Cereal Chem.* **28**, 280–288.
Currie, J. N. (1917). *J. Biol. Chem.* **31**, 15–37.
Dworschack, R. G., and Wickerham, L. J. (1961). *Appl. Microbiol.* **9**, 291–294.
Ellis, J. J., and Roberson, J. A. (1968). *Mycologia* **60**, 399–405.
Friedkin, M. (1945). *Ind. Eng. Chem. Anal. Ed.* **17**, 637–638.
Garibaldi, J. A., Ijichi, K., Lewis, J. C., and McGinnis, J. (1951). U.S. Patent No. 2,576,932.
Garibaldi, J. A., Snell, N. S., and Lewis, J. C. (1953). *Ind. Eng. Chem.* **45**, 838–846.
Gastrock, E. A., Porges, N., Wells, P. A., and Moyer, A. J. (1938). *Ind. Eng. Chem.* **30**, 782–789.
Hall, H. H., Benedict, R. G., Wiesen, C. F., Smith, C. E., and Jackson, R. W. (1953). *Appl. Microbiol.* **1**, 124–129.
Hall, H. H., St. Julian, G., and Adams, G. L. (1968). *J. Econ. Entomol.* **61**, 840–843.
Hargrove, R. E., and Leviton, A. (1955). U.S. Patent No. 2,715, 602.
Haynes, W. C., and Rhodes, L. J. (1969). *J. Invertebr. Pathol.* **13**, 161–166.
Haynes, W. C., Wickerham, L. J., and Hesseltine, C. W. (1955). *Appl. Microbiol.* **3**, 361–368.
Hellman, N. N., Tsuchiya, H. M., Rogovin, S. P., Lamberts, B. L., Tobin, R., Glass, C. A., Stringer, C. S., Jackson, R. W., and Senti, F. R. (1955). *Ind. Eng. Chem.* **47**, 1593–1598.
Herrick, H. T., Hellbach, R., and May, O. E. (1935). *Ind. Eng. Chem.* **27**, 681–683.
Hesseltine, C. W., and Shibasaki, K. (1961). *Appl. Microbiol.* **9**, 515–518.
Hesseltine, C. W., Bradle, B. J., and Benjamin, C. R. (1960). *Mycologia* **52**, 762–774.
Hesseltine, C. W., Smith, M., Bradle, B., and Djien, K. S. (1963). *Develop. Ind. Microbiol.* **4**, 275–287.
Hesseltine, C. W., Shotwell, O. L., Ellis, J. J., and Stubblefield, R. D. (1966). *Bacteriol. Rev.* **30**, 795–805.
Hesseltine, C. W., Smith, M., and Wang, H. L. (1967). *Develop. Ind. Microbiol.* **8**, 179–186.
Koepsell, H. J., and Tsuchiya, H. M. (1952). *J. Bacteriol.* **63**, 293–295.
LeMense, E. H., Corman, J., Van Lanen, J. M., and Langlykke, A. F. (1947). *J. Bacteriol.* **54**, 149–159.
LeMense, E. H., Sohns, V. E., Corman, J., Blom, R. H., Van Lanen, J. M., and Langlykke, A. F. (1949). *Ind. Eng. Chem.* **41**, 100–103.
Leviton, A. (1956). U.S. Patent No. 2,753,289.
Leviton, A., and Hargrove, R. E. (1952). *Ind. Eng. Chem.* **44**, 2651–2655.
Lewis, J. C., Ijichi, K., Snell, N. S., and Garibaldi, J. A. (1949). *U.S. Dept. Agr., Bur. Agr. Ind. Chem., Mimeogr. Circ. Ser., AIC* **254**, October, 1949.
Lockwood, L. B., and Nelson, G. E. N. (1946). *Arch. Biochem. Biophys.* **10**, 365–374.
Lockwood, L. B., and Ward, G. E. (1945). *Ind. Eng. Chem.* **37**, 405–406.
Lockwood, L. B., Tabenkin, B., and Ward, G. E. (1941). *J. Bacteriol.* **42**, 51–61.
May, O. E., Herrick, H. T., Moyer, A. J., and Hellbach, R. (1929). *Ind. Eng. Chem.* **21**, 1198–1210.
May, O. E., Moyer, A. J., Wells, P. A., and Herrick, H. T. (1931). *J. Amer. Chem. Soc.* **53**, 774–782.
May, O. E., Ward, G. E., and Herrick, H. T. (1932). *Zentr. Bakteriol. Parasitenk. Infektionskr. Hyg. Abt. II* **86**, 129–134.

May, O. E., Herrick, H. T., Moyer, A. J., and Wells, P. A. (1934). *Ind. Eng. Chem.* **26**, 575–578.
May, O. E., Herrick, H. T., Moyer, A. J., and Wells, P. A. (1935). U.S. Patent No. 2,006,086.
Moyer, A. J., and Coghill, R. D. (1946a). *J. Bacteriol.* **51**, 57–78.
Moyer, A. J., and Coghill, R. D. (1946b). *J. Bacteriol.* **51**, 79–93.
Nelson, G. E. N., Traufler, D. H., Kelley, S. E., and Lockwood, L. B. (1952). *Ind. Eng. Chem.* **44**, 1166–1168.
Pfeifer, V. F., Tanner, F. W.; Jr., Vojnovich, C., and Traufler, D. H. (1950). *Ind. Eng. Chem.* **42**, 1776–1781.
Pfeifer, V. F., Vojnovich, C., and Heger, E. N. (1952). *Ind. Eng. Chem.* **44**, 2975–2980.
Pridham, T. G. (1952). *Econ. Bot.* **6**, 185–205.
Raper, K. B. (1946). *Ann. N.Y. Acad. Sci.* **48**, 41–56.
Reeves, M. D., and Schmidt, W. H. (1945). *J. Bacteriol.* **49**, 395–400.
Rhodes, R. A. (1965). *Bacteriol. Rev.* **29**, 373–381.
Rogovin, S. P., Anderson, R. F., and Cadmus, M. C. (1961). *J. Biochem. Microbiol. Technol. Eng.* **3**, 51–63.
Schmidt, W. H. (1946). *Bull. Health Organ. League of Nations* **12**, Extract No. 9, 260–267.
Schmidt, W. H., and Moyer, A. J. (1944). *J. Bacteriol.* **47**, 199–209.
Schmidt, W. H., Ward, G. E., and Coghill, R. D. (1945). *J. Bacteriol.* **49**, 411–412.
Shibasaki, K., and Hesseltine, C. W. (1961). *Develop. Ind. Microbiol.* **2**, 205–214.
Smiley, K. L. (1966). *Food Technol.* **20**, 112–116.
Smiley, K. L., Cadmus, M. C., Hensley, D. E., and Lagoda, A. A. (1964). *Appl. Microbiol.* **12**, 455.
Smiley, K. L., Cadmus, M. C., and Liepins, P. (1967). *Biotechnol. Bioeng.* **9**, 365–374.
Stodola, F. H., Shotwell, O. L., Borud, A. M., Benedict, R. G., and Riley, A. C., Jr. (1951). *J. Amer. Chem. Soc.* **73**, 2290–2293.
Stodola, F. H., Raper, K. B., Fennell, D. I., Conway, H. F., Sohns, V. E., Langford, C. T., and Jackson, R. W. (1955). *Arch. Biochem. Biophys.* **54**, 240–245.
Stodola, F. H., Nelson, G. E. N., and Spence, D. J. (1957). *Arch. Biochem. Biophys.* **66**, 438–443.
Strandberg, G. W. (1969). *J. Bacteriol.* **97**, 1305–1309.
Stubblefield, R. D., Shotwell, O. L., and Shannon, G. M. (1968). *J. Amer. Oil Chem. Soc.* **45**, 686–688.
Stubbs, J. J., Lockwood, L. B., Roe, E. T., Tabenkin, B., and Ward, G. E. (1940). *Ind. Eng. Chem.* **32**, 1626–1631.
Tanner, F. W., Jr., Vojnovich, C., and Van Lanen, J. M. (1949). *J. Bacteriol.* **58**, 737–745.
Thom, C., and Currie, J. N. (1916). *J. Agr. Res.* **7**, 1–15.
Tsuchiya, H. M., Corman, J., and Koepsell, H. J. (1950). *Cereal Chem.* **27**, 322–330.
Tsuchiya, H. M., Koepsell, H. J., Corman, J., Bryant, G., Bogard, M. O., Feger, V. H., and Jackson, R. W. (1952). *J. Bacteriol.* **64**, 521–526.
Tsuchiya, H. M., Hellman, N. N., and Koepsell, H. J. (1953). *J. Amer. Chem. Soc.* **75**, 757.
Tsuchiya, H. M., Hellman, N. N., Keopsell, H. J., Corman, J., Stringer, C. S., Rogovin, S. P., Bogard, M. O., Bryant, G., Feger, V. H., Hoffman, C. A., Senti, F. R., and Jackson, R. W. (1955). *J. Amer. Chem. Soc.* **77**, 2412–2419.
Wang, H. L., and Hesseltine, C. W. (1966). *Cereal Chem.* **43**, 563–570.
Ward, G. E., Lockwood, L. B., May, O. E., and Herrick, H. T. (1936). *J. Amer. Chem. Soc.* **58**, 1286–1288.

Ward, G. E., Lockwood, L. B., Tabenkin, B., and Wells, P. A. (1938). *Ind. Eng. Chem.* **30,** 1233-1235.
Ward, G. E., Pettijohn, O. G., Lockwood, L. B., and Coghill, R. D. (1944). *J. Amer. Chem. Soc.* **66,** 541-542.
Ward, G. E., Pettijohn, O. G., and Coghill, R. D. (1945). *Ind. Eng. Chem.* **37,** 1189-1194.
Wells, P. A., Moyer, A. J., and May, O. E. (1936). *J. Amer. Chem. Soc.* **58,** 555-558.
Wells, P. A., Moyer, A. J., Stubbs, J. J., Herrick, H. T., and May, O. E. (1937a). *Ind. Eng. Chem.* **29,** 653-656.
Wells, P. A., Stubbs, J. J., Lockwood, L. B., and Roe, E. T. (1937b). *Ind. Eng. Chem.* **29,** 1385-1388.
Wells, P. A., Lynch, D. F. J., Herrick, H. T., and May, O. E. (1937c). *Chem. Met. Eng.* **44,** 188-190.
Wells, P. A., Lockwood, L. B., Stubbs, J. J., Roe, E. T., Porges, N., and Gastrock, E. A. (1939). *Ind. Eng. Chem.* **31,** 1518-1521.
Wickerham, L. J. (1947). U.S. Patent No. 2,422,455.
Wickerham, L. J., and Dworschack, R. G. (1960). *Science* **131,** 985-986.
Wickerham, L. J., and Kuehner, C. C. (1956). U.S. Patent No. 2,764,487.
Wickerham, L. J., Flickinger, M. H., and Johnson, R. M. (1946). *Arch. Biochem. Biophys.* **9,** 95-98.

Microbiological Patents in International Litigation

JOHN V. WHITTENBURG

American Cyanamid Co.,
Stamford, Connecticut

A courtroom is hardly the forum scientists would choose to settle questions of identification and speciation of microorganisms and the metabolic products that they produce. Unfortunately, more and more patent matters raising such questions are being presented to courts for determination. Invariably the judge called upon to decide the matter, even if technically trained, is in a field of science largely unexplored in the courts and also one in which even the experts may sometimes disagree.

Courts have been presented with questions relating to patent infringements for many years. No infringement case is simple and some of the modern inventions in the fields of pure chemistry, physics, and electronics are, without question, extremely difficult to comprehend. However, in inventions in those fields, the elements involved are inanimate and their actions follow established rules of science. Courts have had experience in such fields and do have judicial and often scientific precedents upon which to rely when rendering their decisions. In the field of microbiology, as represented by fermentative production of a specified metabolic product, courts are faced with problems relating to living organisms whose actions are not always predictable.

The search for new microorganisms and the products produced thereby by means of fermentations have been concentrated in the genus *Streptomyces*. For that reason, this paper will deal principally with them—the problems associated with their proper identification, and their handling in legal proceedings in various countries.

Patent applications are customarily filed by scientists and/or groups of scientists when a new microorganism has been isolated which produces a new and useful antibiotic. Applications are also filed to cover the use of a known microorganism to produce a new metabolic product and, alternatively, to cover the use of a new microorganism to produce a known metabolic product. In order to protect his invention, an inventor in most countries of the world must have filed his patent application prior to any public disclosure or use thereof. Of course, there are exceptions to this rule, as for example, the provision in the United States Patent Statutes allowing the filing of a patent application within 1 year of a disclosure or use and the provision in the British Patents Act requiring that the publication be within the country. However, it is the better practice to file a patent application before any publication or use of the invention in order to avoid such bars to the grant of a subsequent patent.

A patent specification must give a full and complete disclosure so that a man skilled in the art can understand and reproduce the invention. The patent specification in total need not be as detailed or as theoretical as a doctoral thesis or a paper given at a symposium. However, it must present a detailed and complete scientific disclosure of the exact point of novelty, as for example, the microorganism and/or the metabolic product produced in microbiological inventions.

In accordance with established rules and practice, a scientist who first isolates and describes a new microorganism is entitled to name it. The name applied to the microorganism is binomial, i.e., given both the genus and species name. Bergey's (1957) defines a species of a microorganism as follows:

> A species of bacterium is the type culture of specimen together with all other culture or specimens regarded by an investigator as sufficiently like the type (or sufficiently closely related to it) to be grouped with it.

The first isolated and described microorganism becomes the type culture for the species. The fact that an alleged new microorganism has been described and named does not necessarily mean that the microorganism is new or that it is a new species. Before a new species of a microorganism becomes scientifically valid, it must be generally accepted as such by scientists working in the field (other than the person who isolated and named it).

The proposed establishment of a new species of microorganism may be based on an honest misinterpretation or erroneous observation of data or it may be the result of an obvious attempt to create a new species, regardless of the weight of scientific authority. While it is well known that many microorganisms may produce a plurality of metabolic products or that a particular metabolic product may be produced by more than one species of microorganism, uniform criteria must be applied when describing it. The patent and scientific literature discloses hundreds of species of *Streptomyces* and the metabolic products produced from them. Many texts, surveys, and review articles are available which contain disclosures from patent and scientific literature. However, the mere listing of a microorganism in the literature does not establish its validity. In fact, many authors indicate in survey articles that such compilations are based in whole or in part on the literature and that upon detailed study it would be possible to reduce many of the alleged different species to synonymy. Professor Waksman (1957) very clearly faced the problem of establishment of new species when he stated:

> Unfortunately, it has frequently been found much easier to assign undue importance to these variations and designate a freshly isolated culture as a new species. Some justification in this attitude has been found in the fact that the new culture possesses an important economic property, as that of producing a new antibiotic; this was frequently correlated with certain other morphological or cultural differences. This is largely the reason why within the last 10 years there have been more "new" species created than in all the previous 75 years since Ferdinand Cohn first described his *Streptothrix forsteri* .....
>
> It is commonly believed that to characterize a species on the basis of certain morphological and physiological properties, it is desirable to describe a large number of properties. This procedure is not always followed, as illustrated by the fact that many new species have been described recently, largely because it is easier to create a new species than to attempt to correlate the characteristics of a freshly isolated culture with those of known species already described in the literature. This problem has become particularly acute when Company A, for example, presents claims that to produce the same antibiotic or vitamins it is using a different species than that claimed in the patent granted to Company B. This is done, of course, to avoid patent infringement.

A clear distinction should be made between the criteria necessary to properly classify a microorganism and those recommended for describing same. A full and complete description of a microorganism will, of course, contain the information necessary for its proper classification. Inasmuch as the proper classification of a microorganism relies solely on those characteristics which have been shown to be most stable, many variable characteristics disclosed in a description of a microorganism are not used when the microorganism is classified. However, a competent taxonomist when classifying a microorganism will ordinarily carry out tasks other than those necessary for mere classification of the microorganism in order to show similarities and dissimilarities between it and another of established species rank.

Many keys and systems for classifications of microorganisms exist. One of the most widely used systems in the first steps employed in the proper classification of a *Streptomyces*, for example, is that described by Pridham *et al.* (1958). These authors suggest classification only according to selected groups. The genus *Streptomyces* is initially divided into seven morphological sections based on its sporophores determined by direct observation of Petri dish cultures at magnifications ranging from 100 to 500× on defined agars. The seven sections range from the simplest form (Rectus-Flexibilis) to the most complex (Biverticillus-Spira). Each section is then divided into six series based on spore color (color of sporulating aerial mycelium at maturity).

Primary emphasis in this grouping scheme is on morphology. After proper grouping is effected, these authors recognize that each series probably can be further delineated on the basis of physiological criteria into species.

A comprehensive system for classification of Streptomyces species has been advanced by Ettlinger *et al.* (1958). This system, while similar to the Pridham grouping system noted above, further provides for the ultimate classification of Streptomyces species. Ettlinger *et al.* as well as Pridham *et al.* recognize that there are many unreliable characteristics exhibited by *Streptomyces* and have discarded them as criteria for establishment of species status. These discarded characteristics include the production of soluble pigments, color of the vegetative mycelium, and antibiotic produced. In fact, it is generally accepted now by most taxonomists that antibiotic production is a strain rather than a species characteristic of a microorganism. Stable characteristics largely unaffected by media, conditions of culture, and age were selected by Ettlinger *et al.* to serve as criteria for species determination.

(1) The morphology of the spores;
(2) Color of the aerial mycelium;
(3) Morphology of the aerial mycelium;
(4) Chromogenicity.

All observations are made on mature, well-sporulated microorganisms.

It will be noted that two of the criteria used by Ettlinger *et al.* for identification of species, namely color of the aerial mycelium and morphology of the aerial mycelium, correspond to the criteria used by Pridham *et al.* in their system for grouping *Streptomyces*. There are, however, distinctions between the systems of Pridham *et al.* and Ettlinger *et al.* Whereas both suggest six groups or series for the color of the aerial mycelium, definition of the color range between the two is not identical. There is also a difference between the systems with regard to sporophore morphology. Pridham *et al.* have suggested seven groupings. On the other hand, Ettlinger recommends combining and/or eliminating certain of these groups so that only five groups will exist. However, the grouping suggested by Ettlinger *et al.* will also range from the simplest form to the most complex.

A further stable characteristic for differentiation of the species recognized by Ettlinger *et al.* makes use of the electron microscope in its determination. This is the morphology of the spores. The surface ornamentation of the individual spore which is not visible with the light microscope becomes clearly visible under the electron microscope and is found to be stable within the species. Spores are seen to have rough or smooth surfaces. Variations in the ornamentation of the spores having a rough surface and the shape of the spores having a smooth surface are shown to exist and grouping accordingly is suggested.

The final stable characteristic recognized by Ettlinger *et al.* is the ability of a *Streptomyces* to produce (or not produce) a melanin pigment or discoloration on certain tyrosine-containing nutrient media.

It appears that most taxonomists now follow, with minor variations, the system of classification recommended by Ettlinger *et al.* for differentiation of species within the genus *Streptomyces*. Reliance on highly variable characteristics as tools in species determination is to be avoided. Kuster (1963) states that the following morphological and physiological characters are important and commonly used criteria for classification of *Streptomyces:*

(1) Morphology of spore-bearing hyphae.
(2) Morphology of spore surface.
(3) Color of vegetative and aerial mycelium.
(4) Chromogenicity (melanin production).
(5) Utilization of carbohydrates.

Much confusion has arisen in the patent literature because of the various descriptions of microorganisms. Many of the descriptions have omitted characteristics which are now considered stable and critical for proper classification. Others have placed undue emphasis on unstable and variable characteristics. It will also be appreciated that many microorganisms now having valid specific rank and which were first disclosed in patent applications will perhaps have descriptions in the patent literature which differ in some respects from their descriptions in the scientific literature. The descriptions of microorganisms first disclosed around 1950, of course, reflect the systems then used to classify them. In many instances later descriptions of the microorganisms in the scientific literature will include characteristics not employed at the time of the writing of the original descriptions or developed later as, for example, the morphology of spore surface observed under the electron microscope.

In an attempt to avoid confusion and to obtain uniformity in description of microorganisms in patent specifications the Subcommittee on Taxonomy of the Actinomycetes of the International Committee on Bacteriological Nomenclature published Recommendations for Descriptions of Some Actinomycetales Appearing in Patent Applications (1962). The subcommittee recommends for the Actinomycetales (exclusive of the Mycobacteriaceae):

(1) That patent offices accept microorganisms for process patents

which fit any of the taxa in Rules 6, 7, or 8 of the International Codes of Nomenclature of Bacteria and Viruses;

(2) That the following minimum of criteria should be used in a description for a patent:

(a) Morphological observations of a well-sporulated culture if the organism being described can form spores
 1. Morphology of spore-bearing hyphae; simple or verticillate; whether straight, flexuous, loops (open spirals) or spirals (closed spirals). Included in the description of the sporophore should be a reproduction of a picture or drawing of these structures.
 2. Number of spores whether single, pairs, number of spores from 3 to 10, and more than 10 forming a chain.
 3. Presence of globular sporangia, as in Actinoplanaceae.
 4. Presence of flagellated spores, as in Actinoplanes.
 5. Ability to form aerial mycelium.
 6. Formation of conodiosphores and conidia on substrate and/or aerial mycelium.
 7. Tendency of the mycelium to fragment.
 8. Morphology of spore surface observed under the electron microscope.
 9. Occurrence of sclerotia.

(b) Color
Description of any significant color.
Chemically defined media on petri dishes should be used and age of culture, temperature, and medium should be stated. Record color of surface of well-sporulated aerial mycelium, also the reverse and surface of vegetative mycelium.
Record any diffusible pigment; other helpful observations would be the pH effect on color and general nature of the pigment.

(c) Physiological characters
 1. Melanin production studied on pepton-iron agar and/or on organic medium of gauze.
 2. Utilization of the following carbohydrates:
 Control without carbohydrates
 d-Glucose (as positive control)
 l-Arabinose i-Inositol
 Sucrose d-Mannitol
 d-Xylose d-Fructose

(d) Temperature

The ability to grow at 50°C should be determined.

(e) Microaerophilic growth

Following these recommendations would result in more uniformity in the drafting of descriptions of *Streptomyces* in patent applications.

It will be noted that the recommendations concern descriptions of members of the Actinomycetales and, therefore, encompass far more than the genus *Streptomyces*. In fact, a number of the recommendations have no bearing on *Streptomyces* at all but rather are directed exclusively to other genera of the order. All of the characteristics called for in the recommendations do not have equal value as criteria for species differentiation in the various genera of the order. However, the recommendations do include those criteria which are generally considered as most stable and therefore most useful for species differentiation. Certain criteria noted in the recommendations, being more variable from strain to strain in a given species, are less generally suitable for *Streptomyces* species differentiation, although in specific instances they may have substantial utility.

On the basis of negotiations between the Nordic Patent Authorities (Danish patent, 1962) special instructions were published as to the treatment of patent applications relating to microbiological processes in those countries. The instructions state that a known microorganism is sufficiently characterized by its systematic designation supplemented, if required, with a reference to such literature as describes the method for the systematic determination. Unknown microorganisms are required to be described in such detail as to avoid confusion with other microorganisms. Specifically, the instructions state that the description of unknown microorganisms shall include the following particulars:

(1) The name of the organism and, if so required, the designation in a public Collection of Cultures, and if possible also date and place of the isolation.

(2) Description of growth referred to a definite substrate, with a detailed description of the microscopic properties and characteristics and the microscopic morphology (including the shape and size of the spores, the morphology of the spore formation, the modus of ramification, and hyphal width of the mycelium).

(3) Growth properties (the morphology of the colonies, and information as to colors and, possibly, separated pigment) in at least ten standard substrates.

(4) Physiological properties referred to growth in substrates containing milk, nitrate, gelatine, starch, tyrosine, and, possibly, cellulose.

(5) The capability of the organism to produce hydrogen sulfide (melanin pigment) on organic or inorganic substrate.

(6) The capability of the organism to utilize a number of carbon sources.

(7) Reference to the most related species which is/are mentioned in Bergey's Manual of Determinative Bacteriology (1957) together with particulars as to how the said organism may be distinguished from the known organisms.

(8) Possible supplementary particulars regarding individual properties, for example, the production of antibiotics.

(9) In connection with the description of the physical and chemical properties of the antibiotic substance, a table on the quantitative special effects of the product shall be given, as well as a table wherein, by + or −, the activity of the antibiotic substance against gram-positive and gram-negative bacteria, fungi, and yeasts is stated and, if possible, also the activity against protozae, viruses, and rickettsiae, if such information is available.

Regardless of whether a description of the microorganism follows the strict requirements of the Nordic Patent Office or the more general recommendations of the International Committee on Bacteriological Nomenclature, it will be seen that much thought has been given to written descriptions of microorganisms appearing in patent applications. This is, of course, beneficial to all concerned in the very important field of *Streptomyces* fermentations. Not only will the stable characteristics of microorganisms be disclosed to allow proper classification of the species, but also sufficient other descriptive information will be given to assist in the characterization of strains within the species. It is hoped that uniformity of description of microorganisms will assist pure scientists in their work as well as investigators attempting to determine the true scope and coverage of a patent.

Differences exist around the world in the various patent offices and courts concerning the prosecution and interpretation of patent applications and patents relating to processes of producing antibiotics by fermentation of species of *Streptomyces*. Even in the absence of objections by the patent office to the written descriptions of microorganisms, the applicant still is faced with the problem concerning deposit and availability of the microorganism. In the United States of America the practice until recently required an applicant to have deposited his microorganism, if new or if a producer of a new metabolic product, in a recognized culture collection before he filed his application. If the microorganism had not been deposited prior to filing, the patent office rejected the application as being fatally defective in its disclosure and the defect could not be corrected by a later deposit of the microorganism.[1] Proof of the deposit had to be given at the time of filing. The United States Patent Office also required an applicant to release his microorganism to the public upon allowance of his application.

The decision by the Board of Appeals of the United States Patent Office in Ex parte Argoudelis *et al.* (1968) now requires that an applicant show that at the time of filing the application the organisms involved were known and available to persons skilled in the art. If this is done, subject to proof by evidence, the question of sufficiency of disclosure does not arise. The Argoudelis decision is now on appeal to the United States Court of Customs and Patent Appeals.

Currently United States patent applications involving new microorganisms which are not initially shown to be known and available at the time of the filing of the application are rejected as follows:

> All the claims are rejected as based upon an insufficient disclosure under 35 U.S.C. 112. The instant invention requires the use of a newly disclosed microorganism and therefore written description of the invention alone does not serve as a complete disclosure which would enable one skilled in the art to practice the invention at the time of filing the application. Ex parte Kropp, 143 USPQ 148. While the microorganism has been deposited in a public depository, it is not apparent from the record that the microorganism was freely available from the depository to anyone without applicants' permission at the time of filing the application. This rejection could be overcome by a statement that there was no agreement with the depository to limit the availability of the microorganism at the time of filing the application.

Applicants filing in the United States ordinarily deposit the microorganism at the American Type Culture Collection in Rockville, Maryland, or at the Northern Regional Research Laboratory in Peoria, Illinois. The American Type Culture Collection accepts deposit of a microorganism with respect to a pending patent application and, upon the request of the depositor, will not release it until the patent application concerned therewith has been allowed by the United States Patent Office. However, upon its release by the American Type Culture Collection, the microorganism is available throughout the world to investigators without restriction or by payment of a nominal sum. Microorganisms deposited at the American Type Culture Collection are listed in the ATCC catalog which is published from time to time so that investigators may be aware of the microorganisms deposited

---

[1] Ex parte Kropp, 143 USPQ 148.

at that culture collection. The American Type Culture Collection and other culture collections do not attempt to classify microorganisms or ascertain the correctness of the name and species applied to the microorganisms received for deposit, but rather list the microorganism in its collection under the name submitted with it by the depositor. The deposited microorganism is given a number by the culture collection and the microorganism is referred to by that number when investigators request subcultures.

The Northern Regional Research Laboratory will also accept a microorganism for deposit in its culture collection with respect to patent applications. It will accept the microorganism with restrictions regarding its release pending allowance of the patent application. The Northern Regional Research Laboratory does not publish a catalog of the microorganisms in its collection although microorganisms received for deposit are assigned collection numbers. Species of *Streptomyces* received for deposit by the Northern Regional Research Laboratory with respect to patent applications are referred to by the designation NRRL followed by a number. Species of *Streptomyces* received for deposit by this culture collection and not associated with a pending patent application are referred to by the designation NRRL followed by a letter, as for example B, and then a number.

Other well-known culture collections outside of the United States are the National Collection of Industrial Bacteria in Aberdeen, Scotland, and the Centraalbureau voor Schimmelcultures in Baarn, Holland. Both of these culture collections will receive microorganisms for deposit and will honor the request of the donor to restrict distribution of cultures until advised by the donor that release of the microorganism is permitted.

The Dutch Patent Office requires an applicant to have deposited and to make available to anyone having a reasonable interest therein cultures of microorganisms disclosed in patent applications. Release of the microorganism is not required when the application is published but is required upon grant of the patent. This ruling was handed down by the Appeal Department of the Dutch Patent Office on June 18, 1964, in Dutch patent Application 182,282 (Bijblad Industriele Ergendom, 1964). An opposition was lodged against the grant of Application 182,282 which claimed:

> Process for the preparation of antibiotics inclusive of oxytetracycline, by means of a *Streptomyces* species, characterized in that *Streptomyces vendargensis*, nov. spec., the properties of which are set forth in the specification, is cultivated under suitable conditions, the antibiotics formed, which are mentioned in the specification and one of which is called vengicide, being recovered from the culture liquor.

The opposition, among others, was based on the proposition that the microorganism S. *vendargensis* was not distinct from S. *rimosus* NRRL 2234 and, in any event, the applicant would not make a culture of the microorganism available to the opposer so that comparative studies could be carried out to determine if in fact the microorganisms in question were distinct species. The applicants stated that they were prepared to make a culture of S. *vendargensis* available to an independent expert so that he could carry out the tests desired by the opposer, but that they would release the culture to third parties only after grant of the patent.

In its decision the Appeal Department stated in part that there is no article in the Dutch Patents Act under which the applicant can be compelled to make his microorganism available to the public. However the Appeal Department continued and stated:

> ... that it stands to reason that for patents like the one under consideration the specification itself cannot be so drafted that only with help thereof the invention can be understood and applied, because more particularly for the aplication of invention it is indispensable to have the disposal of the actinomycete forming the starting point of the process according to the application;
>
> ... that accordingly the invention is only really made available to the public if the applicants are prepared upon the grant of the patent to place the organism found by them at the disposal of anyone having a reasonable interest in the availability of said organism;
>
> ... that the Appeal Department after considering the arguments and interests of applicants and opponents are of the opinion that upon the publication of the application the placing of the organism at the disposal of the public need not yet be made a requirement;
>
> ... that it cannot be maintained that the opponents can only oppose the grant of a patent if they have the disposal of the organism, because—for example in the present case—the very fact that the organism produces oxytetracycline already constitutes a sufficient ground to doubt whether a new species has been found and that the opposition can be based on said doubt;
>
> ... that it follows therefrom that as long as the patent has not been granted only the possibility need exist—in case an opposition is based on lack of novelty—to have a neutral expert verify whether applicants' pretensions are correct.

In this opposition proceeding the parties agreed to a neutral expert and the tests to be carried out by him. The report of the expert was that S. *vendargensis* and S. *rimosus* were separate species. Consequently, grant of the patent was ordered subject to the condition that the applicants irrevocably authorize the culture collection which had received a deposit of the microorganism to issue samples to interested parties and that the copy of the authorization and a copy of the acknowledgment of its receipt by the culture collection be sent to the Appeal Department.

This decision by the Dutch Patent Office followed the reasoning handed down by the District Court of The Hague in its decision (1961) of October 24, 1961 in the matter of American Cyanamid Company vs. Charles Pfizer & Co., Inc. wherein American Cyanamid had brought a nullity action against Dutch Patent No. 92,355. The patent claimed the production of tetracycline by fermentation of *Streptomyces* species ATCC Nos. 11652, 11653, and 11654. Pfizer had deposited the microorganisms at the American Type Culture Collection in conjunction with pending United States patent applications but had restricted distribution of these microorganisms until after the grant of the pending United States applications.

In its writ American Cyanamid had stated that it was not possible to determine whether the *Streptomyces* strains identified in the patent are identical with the *Streptomyces aureofaciens* disclosed in American Cyanamid's Dutch Patents Nos. 72,763 and 86,837 since it was not possible to obtain cultures of the microorganisms for comparative study. The District Court in holding Dutch Patent No. 92,355 null and void stated in part as follows:

> It is striking that, while the priority date was November 12, 1953, the species of strains had even on September 16, 1959 not been determined, which is also an indication that no species is concerned;
>
> Considering that, now that the cultures appear not to be accessible to the public, and the public therefore cannot know whether they would act in conflict with the attacked patent, there is all the more reason to assume that the specification contains insufficient data for a new method to be inferable.

In Great Britain the Patent Office[2] has also considered the question of release of a microorganism during an opposition proceeding. The parties involved were the same as those involved in the proceeding relating to Dutch Application No. 182,282 previously referred to. In the English case there was, in addition to the question of availability of the microorganism disclosed in the opposed application, also the question of its classification. The applicant filed evidence by independent Dutch experts referring to comparative tests made by them from which they came to the conclusion that *S. vendargensis* was a new species and not a variant of *S. rimosus*. The opponent argued that *S. vendargensis* was not a microorganism of different species from *S. rimosus* but was a strain or mutant thereof.

At the time of the opposition, samples of *S. rimosus* were available from the Northern Regional Research Laboratory under reference number NRRL 2234. Cultures of *S. vendargensis* were not available

---

[2] In the matter of an application for Patent No. 764,198 by Koninklijke Nederlandsche Gist-en Spiritusfabriek N. V. and in the matter of an opposition thereto by Charles Pfizer & Co., Inc.

from the applicant or a culture collection in Holland where the microorganism had been deposited by the applicant with a restriction that it not be released to third parties until a patent had been granted on a patent application made in Holland. The opponent argued that, since the applicant could submit evidence made on the comparative studies of the two microorganisms in issue, the Patent Office should make an order or adopt some procedure so that the microorganism of the applicant would be made available to them for study. Otherwise, the opposer argued that his evidence would be incomplete.

The Patent Office ruled that it could not require the applicant to release its microorganism during the opposition proceedings, stating in part:

> I can find no other provision in any section of the Act which might assist me in determining this matter and I am forced to conclude that I have no powers to compel the Applicant, when he is unwilling to do so, to make available a sample of the culture for examination by the Opponents.

Many jurisdictions grant patent protection to a chemical compound in addition to a process for preparing it. However, in such jurisdictions natural products, i.e., products of nature, are not patentable. This raises an interesting question since the natural habitat of most *Streptomyces* is the soil, a natural nutrient medium, and during its growth therein may produce metabolic products. If a *Streptomyces* does produce an antibiotic in its natural habitat, will such a product of nature negate patentability of the antibiotic when claimed as a chemical compound? This question was met directly and the product protection sustained by the District Court, Tel Aviv/Jaffa, Israel, in its decision of July 6, 1952 in the case of Parke, Davis & Company vs. Abic Chemical Laboratories Ltd. *et al*. This case involved an infringement of the compound chloramphenicol produced by fermentation of *S. venezuelae*. Infringement was not denied but the validity of the patent was attacked on the grounds that the antibiotic substance is a product of nature and no monopoly can be granted by a patent in respect of a product of nature. The Court in its decision stated in part as follows:

> Thus, it has not been proved to me that chloramphenicol does not exist in nature. The contrary is true. In this action, on the evidence before me and for the purpose of the action only, it has been proved that chloramphenicol exists in nature ... I find that chloramphenicol exists in nature, although in very small quantities only, and I understand that it is not stable there and cannot properly develop because it is destroyed as soon as it is produced ...
>
> On the strength of the evidence before me I think that there is no inventive step in the discovery of *S. venzuelae*. Of such step one could have spoken when one had found for the first time an antibiotic. But in the present case there was only a good organization and a great effort without any originality. The originality lies only in the sequence of the steps which led to the crystalline sub-

stance. The position in this case is, therefore, different from that in the above cases in that the inventive step was there in the discovery and the routine in in the method which brought the discovery to a practical use. Here there was no inventive step in the discovery but ingenuity in the practical method. This brings me to the question which more than any other problem of this case: if that is the position, have I to protect the inventive method only and to refuse the protection of the discovery, or have I to look on the discovery and on the practical method as an integral creation and to give protection to the substance if it is adopted for use by reason of finding such method . . .

I have come to the conclusion that there is subject matter for a patent if the applicant made a discovery without inventive ingenuity, if together therewith he found a practical process possessing inventive ingenuity, and that not only that process will have to be protected but also the product produced by the two stages of applicants' work.

An entirely different result was reached by the Canadian Supreme Court on February 20, 1968 in the case of Parke, Davis & Co. vs. Laboratoire Pentagone Ltee (1968), also involving the compound chloramphenicol produced by fermentation of S. *venezuelae* covered by Canadian Patent No. 479,333. The Canadian Law, Section 41 (1), of the Patent Act R.S.C. 1952, Chapter 203 as amended, states in part:

In the case of inventions relating to substances prepared or produced by chemical processes and intended for food or medicine, the specifications shall not include claims for the substance itself, except when prepared or produced by the methods or processes of manufacture particularly described and claimed or by their obvious chemical equivalents.

This section has been interpreted to mean that compound protection will be granted for products produced by fermentation which are useful as medicines but product protection will not be granted for compounds produced by a chemical reaction. One of the main points in issue was whether the substance chloramphenicol was prepared or produced by a chemical process.

It was proved by evidence that chloramphenicol is produced by fermentation of S. *venezuelae* but that as produced in the fermentation medium it is diluted, mixed with numerous impurities, and not usable in this impure state. It was further proved that an extraction process is indispensable for obtaining a usable substance for therapeutic purposes.

In arriving at its decision, the Canadian Supreme Court stated that it was not necessary to decide whether or not a fermentation process was a chemical process. The court rather decided that it was sufficient to examine the extraction process to determine if a chemical process was involved. The patentee argued that the extraction process was not a chemical process, whereas the defendent argued that it was. After considering the evidence, the court stated:

For these reasons, we must conclude that, according to the usual meaning of the expression 'chemical process,' the processes of extraction described in the patent in suit, are chemical processes according to paragraph 1 of Section 41 of the Patent Act, and consequently Claim 7 of the Canadian Patent No. 479,333 is invalid.

It is interesting to note that the Canadian Supreme Court referred to the Swiss patent law concerning the protection of processes for the preparation of fermentation products. In Switzerland patent protection is not granted to compounds per se. However, patent protection is granted on processes for the preparation of compounds. The Swiss Federal Act on Patents for Invention of June 25, 1954, Section 2, Paragraph 2, states:

> Patents shall be refused for inventions of medicines and inventions of processes for the manufacture of medicines by other than chemical methods.

The question as to whether a fermentation process for the production of an antibiotic was or was not a chemical process was directly in issue in the Swiss case of American Cyanamid Company vs. Lepetit S.p.A. (Experts report, 1965). This case involved the preparation of tetracycline by fermentation of S. *aureofaciens*. The expert committee appointed by the court was specifically asked if the fermentation process disclosed in Swiss Patent No. 324,085 was a chemical process or whether there is a question of a process for the manufacture of medicaments by other than chemical means which is not patentable. The experts in part stated:

> Fermentation is incontestably a chemical reaction. The fact that a microorganism participates in the fermentation changes nothing in this fact.

Thus, it will be seen that there is no uniformity in the interpretation of patent documents by courts of different jurisdictions, just as there has been no uniformity in the description of microorganisms in patent documents. There could be uniformity in the description of microorganisms in patent documents if the procedures previously discussed were followed in describing them. However, there is no simple solution whereby courts will uniformly interpret patents. There may always be different interpretations as long as the laws of countries are different. Since so many of the inventions in the field of industrial fermentations have made significant contributions to the art and the well-being of mankind, it is regrettable to see courts strike down these patents seemingly on a mere whim. While the laws of the countries are different, the patent should be interpreted in accordance with that law and not the law of other countries, even though there may seemingly be some contradictions between the laws.

To paraphrase a statement made by both the British and Canadian courts a patent should be interpreted by a mind willing to understand and uphold and not by a mind desirous of misunderstanding and striking down.

REFERENCES

"Bergey's Manual of Determinative Bacteriology," 7th ed., p. 21, Williams & Wilkins, Baltimore, Maryland, 1957.
Bijblad Industriele Eigendom, No. 12, 260, Dec. 15, 1964.
Canadian Supreme Court, Feb. 20, 1968, Laboratoires Pentagone Ltee. vs. Parke, Davis & Co.
Danish Patent J. **6** (1962).
District Court of The Hague, Oct. 24, 1961, American Cyanamid Company vs. Charles Pfizer & Co., Inc.
District Court, Tel Aviv/Jaffa, Israel, July 6, 1952, Parke, Davis & Co. vs. Abic Chemical Laboratories, Ltd.
Ettlinger, L. *et al.* (1958). *Arch. Mikrobiol.* **31**, 326–358.
Ex parte Argoudelis *et al.*, *Offic. Gaz.* **848**, 863, March 26, 1968.
Experts Report, April 14, 1965, Court of Justice of the Republic and Canton of Geneva; American Cyanamid Company vs. Lepetit S.p.A.
Int. Bull. Bacteriol. Nomenclature Taxonomy, **13** [No. 3], 169–170, July 15, 1962.
H. J. Kuster, (1963). *Microbiol. España*, **16**, 193–203.
Pridham, T. G. *et al.* (1958). *Appl. Microbiol.* **6**, 52–79.
Waksman, S. A. (1957). *Bacteriol. Rev.* **21**, 1–21.

Industrial Applications of Continuous Culture: Pharmaceutical Products* and Other Products and Processes

R. C. Righelato

*Glaxo Laboratories, Ltd.,
Ulverston, Lancashire, England*

AND

R. Elsworth[1]

*New Brunswick Scientific Co. Inc.,
New Brunswick, New Jersey*

I.	Introduction	399
II.	Pharmaceutical Products	401
	A. Introduction	401
	B. Large-Scale Processes	401
	C. Small-Scale Processes	403
	D. Conclusions	404
III.	Organic Chemicals	405
	Substances in Production or Development	405
IV.	Drink	407
	A. Brewing	407
	B. Continuous Wine Making	411
V.	Sewage and Trade Wastes	412
	A. Sewage Disposal	412
	B. Disposal of Trade Wastes	415
VI.	General Conclusions	415
	References	416

I. Introduction

Continuous industrial fermentations were discussed earlier in this series by Gerhardt and Bartlett (1959). More than half the review was given to an exposition of the technique, including the factors that control cell and product formation and the expected effects of contamination and mutation. It is interesting to refer to the applications which were reported then and to consider the predictions that were made at that time.

Two examples of cell production were included — bakers' and food yeast — processes which were developed before any theoretical basis was enunciated. In the case of bakers' yeast for which there is a sustained demand there is competition between the commercial merits of

---

*Dr. R. C. Righelato contributed the section on Pharmaceutical Products to this article.
[1] *Present address:* 25 Potters Way, Laverstock, Salisbury, Wilts., England.

the continuous and the batch processes. In the case of food yeast the cost of raw materials and competition from alternative sources of protein are the factors which determine whether the process—be it batch or continuous—can survive. In the studies of the continuous culture of cells, examples covering about forty organisms were given. Industrial potential was attributed to two—algae as a food source and mammalian cells for vaccines and enzymes.

Ethyl alcohol and vinegar were the examples of product formation. The growing competition of the chemical route to alcohol was mentioned, and the microbiological process was described as a salvage operation in waste disposal. Research on four antibiotic fermentations was reported—subtilin, chloramphenicol, penicillin, and streptomycin. The authors concluded that competition was forcing a reevaluation of continuous processing and that "an active period of research and development in this area was sure."

As for other chemical products, good prospects were seen for a continuous lactic acid process. Research by Iron Curtain workers on the acetone butanol process was discussed at length. No reference was made to the fact that in the West the death of the existing batch process was near, again because of chemical competition. The limited amount of work on product formation was referred to. It was forecast that because of the increased demand for sodium gluconate and the favorable growth and biosynthetic characteristics of the fermentation that "development of a . . . continuous gluconate process would appear likely of success." It was also concluded that "continuous citric acid fermentation offers a good likelihood of success, and increasing competition for the market will probably force development in this area."

Sewage disposal was discussed, particularly the work at Teddington. The use, in the digestive process, of sewage sludge enriched with $CaSO_4$, which then produces H_2S, was commended as an example of a continuous culture investigation which might well be profitable. The sole reference to the use of petroleum as a culture substrate was the example of a purifying process in which crude oil was reacted with an aqueous culture of bacteria to remove unwanted sulfur compounds by converting them into gaseous products.

This represented the state of the technique 10 years ago. The present article will describe some aspects of the present day situation, and should be regarded as an intelligence report, not a comprehensive record, since all the known facts about commercial processess are not published.

II. Pharmaceutical Products

A. INTRODUCTION

In the last decade continuous chemostat-type cultures have been described for most of the microbiologically produced pharmaceuticals (reviewed by Hospodka, 1966), and the application of continuous culture methods to industrial processes has been widely discussed (e.g., Lumb and Wilkins, 1961; Evans, 1965; Parker, 1967). Continuous fermentation offers high output rates, infrequent turnrounds, and steady-state processes amenable to simple automatic control. To the large-scale manufacturer this means efficient use of plant capacity and lower labor costs. Nevertheless, there are no reports of the large-scale production of pharmaceuticals by continuous fermentation.

To consider the industrial value of continuous culture the various processes can best be separated into the large-scale, long-residence time fermentations for the major antibiotics, vitamin B_{12}, and some extracellular enzymes, and the relatively small-scale processes for vaccine production and most bioconversions.

B. LARGE-SCALE PROCESSES

The products of greatest economic interest, which occupy the greater part of the industry's fermenter capacity, are the antibiotics. These are secondary metabolites of molds and actinomycetes for which sophisticated batch processes have been developed (Hockenhull and Mackenzie, 1968). The batch processes usually involve rapid growth of the organism to a high concentration at which time one substrate becomes exhausted or is fed at a growth-limiting rate. The control of the limited-growth phase, during which most product accretion occurs, is complex and sometimes breaks down. Nevertheless, prolongation of this phase, which has been a general trend in fermentation development, has had two important results, product concentration has increased and the yield per unit of substrate has increased due to the decreased importance of the quantity of substrate used in the initial growth phase in which little product formation occurs. It is perhaps mainly because the product concentration and substrate conversion efficiencies of continuous processes have not matched those of their batch counterparts that continuous fermentation is not used.

Approximate costs of a penicillin fermentation (Bungay, 1963) showed that raw materials contributed about 70% and running costs about 30% of the total. It would appear that increases in output rate, such as expected of continuous fermentations, will only produce significant savings if the yield per unit of substrate is not reduced. Extraction costs are small but to ensure high recoveries at the solvent extraction stage, a high concentration of penicillin in the broth is required.

Most antibiotics are, by a suitable choice of growth-limiting substrate or by a change from one set of conditions to another, produced at fairly high rates in chemostat culture. The question of whether a continuous fermentation can satisfactorily compete with a batch process is one of establishing, in the chemostat, the interrelationships of the variables that effect antibiotic production. Optimal conditions with respect to substrate utilization, output rate, and product concentration can then be attempted. Such an approach was used by Pirt and his co-workers with penicillin fermentation. Parameters such as temperature, pH, and medium composition were made optimal with respect to growth and penicillin production (Pirt and Callow, 1960). Relationships were then established between the rate of penicillin production and growth rate (Pirt and Righelato, 1967) and between growth rate and sugar utilization (Righelato et al., 1968). When the data was reviewed it was found that penicillin concentrations and yield per kilogram of sugar, similar to those obtained with batch cultures, can be obtained from a two-stage chemostat (Table I).

Some doubt has been cast on the applicability of these results to present industrial strains (Wright and Calam, 1968). Experiments with a high yielding strain of P. chrysogenum showed the steady-state production rate of penicillin to be about 40% of the maximum obtained in batch cultures, 6 units/mg dry w/hour. Although somewhat

TABLE I
PENICILLIN PRODUCTION IN GLUCOSE-LIMITED BATCH AND CHEMOSTAT CULTURES OF Penicillium chrysogenum WIS 54-1255 ON A SYNTHETIC MEDIUM

	Total residence time (hours)	Titer (units/ml)	Sugar utilization (MU/kg)	Output rate (MU/liter/hour)
Batch[a]	190	3600	24	.016
Chemostat[b]	120	3800	26	.032

[a]Glucose fed at 0.7 gm/liter/hour, initial concentration 1.5 gm/liter; ammonia added to control pH (Hosler and Johnson, 1953). Final dry weight 30 gm/liter.

[b]First stage dilution rate 0.014 per hour; second stage dilution rate 0.020 per hour. Dry weight 30 gm/liter throughout. Calculated after optimizing the data of Pirt and Righelato (1967).

higher than the steady-state production rate of the Wisconsin strain used by Pirt and Righelato (1967) it was clearly insufficient for an economic continuous process. The maximum penicillin production rate was maintained in neither batch nor continuous cultures and there is as yet no indication of the conditions necessary for its maintenance at that level. If steady-state production rates of 6 units/mg dry w/hour could be maintained, holding times of less than 100 hours would be sufficient for titers of up to 20,000 units/ml.

The discussion above has centered on the penicillin fermentation but similar arguments may be extended to other large-scale, long-residence time fermentations for other antibiotics and products such as vitamin B_{12}. It is perhaps also worth mentioning continuous fermentations other than the chemostat type, though there is little published information on their merits. A tower fermenter has been used for novobiocin production (Ross and Wilkin, 1968) and Dawson's phased continuous system has been used for antibiotics (Canadian Patents and Developments, 1968).

C. SMALL-SCALE PROCESSES

The pharmaceutical industry is also concerned with a number of other microbiological processes which occupy relatively little fermenter capacity. Among them are the bioconversions of steroids, the production of minor antibiotics, and the production of bacterial enzymes such as penicillin acylase and the vaccines.

Many of these processes have been done in continuous culture at high output rates. The 11α-hydroxylation of progesterone has been carried out in the second stage of a two-stage culture of *Rhizopus nigricans* giving a 50–60% yield with a 5-hour contact time (Reusser et al., 1961). High yields and production rates of tetanus and diphtheria toxin have been obtained in chemostats (Zaccharias and Björklund, 1968; Righelato and van Hemert, 1970). Several cell-bound antigens have been produced continuously at rates considerably in excess of the batch culture rates (Evans, 1965).

The microbiological steps of many of these processes often contribute only a small part of the total process cost. Moreover, the scale of the whole process is often insufficient to warrant the adoption, on the grounds of higher throughput, of a continuous step in what is otherwise a batchwise operation. In these processes conversion yields, product purity, or antigenic quality are more important than output rate. It is possible that the high degree of process control and the short residence times of chemostats will bring about improvements in product quality and thereby justify a continuous cultivation step.

Animal cells in tissue culture can be used for virus multiplication and the synthesis of hormones and other pharmaceutically interesting metabolites. The range of applications expands as medicine increasingly deals with aberrant metabolism on a molecular level. Chemostat techniques have been used in research but, for the present, the use of continuously growing cell lines for the production of material for human injection is prohibited. This prohibition does not apply to veterinary vaccines though for the reasons given above the small scale of the operation makes it unlikely that continuous processes would be developed for industry. However, the economic potential of cultures of cells of higher organisms, both animal and plant, has hardly been touched and it is possible that new culture methods will need to be developed to exploit their functions.

It is sometimes argued that degeneration (reduction in the productivity of an organism on prolonged cultivation), will preclude the adoption of continous processes for the synthesis of any product not essential to growth. Almost all of the pharmaceuticals fall into this category and indeed degeneration has proved a problem in antibiotic- and vaccine-producing cultures. Stable cultures are obviously desirable if maximum advantage is to be obtained from continuous processes and a variety of ways of avoiding the problem are suggested in the literature. Environment manipulation (Pirt et al., 1961; Sikita et al., 1964; Reusser, 1963) and techniques of reinoculation (Reusser, 1961) have proved successful. It is possible that some of these problems can be solved genetically (Ball, 1967).

D. Conclusions

Whether or not a process could, to advantage, be run continuously depends in part on the scale of operation. If a batch process fully occupies a plant, then, given a stable market, a continuous process would be worth adopting. The sphere in which continuous operation could prove most valuable is the large-scale antibiotic fermentations. Existing plants could fairly easily be adapted by providing a continuous sterile medium supply and sterile offtake. Most extraction trains already operate more or less continuously and little or no modification would be necessary. In such cases capital already invested in batch plants does not present a major obstacle to the use of continuous fermentation. Indeed the continuous processes may enable considerable expansion of output with only a little capital investment. However, continuous antibiotic processes will not be adopted until substrate conversions and product concentrations compare favorably with the batch fermentations.

Much of the published research on continuous fermentations has not been primarily concerned with developing economic processes but with systems suitable for biochemical or physiological study. If they wish to adopt continuous methods, industrial researchers must be prepared to carry out their own empirical research as they have done with batch cultures, perhaps developing anew both strains and media. Given such effort there appears to be no fundamental reasons why continuous methods should not be adopted for those processes which at present fully occupy tank capacity on a batchwise basis.

III. Organic Chemicals

Substances in Production or Development

Miall's recent review (1968) of organic chemical production does not confirm the earlier predictions of Gerhardt (1959) that lactic, gluconic, and citric acids were likely possibilities for production by continuous methods. Miall's list of continuous processes is limited to reports of the commercial production of vinegar, laboratory-scale production of lactic acid, and the fact that The British Petroleum Company is erecting two large-scale plants for edible protein production.

1. Vinegar

Fermentation vinegar is a two-stage process. First alcohol is made by fermenting a substrate such as barley malt or grape juices with yeast. The alcohol in the clarified mash is then oxidized by *Acetobacter aceti* to acetic acid. The acetification stage has been converted to a continuous process at Fardon's Vinegar Co., Birmingham, England. Some details of the conversion of wooden batch acetifiers have been given by White (1966). By regulating the flow of alcoholic mash to match the aeration capacity of the system, conversions in excess of 98% are obtained. Compared to a quick vinegar acetifier, output has been increased six-fold and raw material costs have been reduced by 40%.

2. Lactic Acid

Childs and Welsby (1966) have described a pH-controlled two-stage lactic acid fermentation, each stage of which used 2 liters of working volume. The culture was a mixture of strains of *Lactobacillus delbruekii* (as is used in large-scale batch production) operating at 50°C.

The medium was an acid hydrolyzate of a maize starch, barley meal mixture containing 9–10% reducing sugars, and supplemented with 0.05% ammonium sulfate. This is a fermentation in which it might be predicted should be successful in continuous culture; the results obtained supported this view. A run lasting 64 days, in which the best pH value was shown to be 5.8, was carried out. The optimum output rate was 1.5 volumes/day. It was concluded that if the results can be repeated on a large scale then a three-, or possibly only two-stage, process should suffice. This was said to be equivalent to a residence time of not more than 2 days, whereas, 5 days is required in the batch process. According to Miall (1969) there is no report that the process has been adapted in large scale.

In North America and France, chemical synthesis is now in competition with the fermentation process.

3. Edible Protein

Shacklady (1969) states that by the middle of 1971 The British Petroleum Company's new plant at Grangemouth, Scotland, for growing yeast on hydrocarbons, will be producing at a rate of 4000 tons/year. At a later date another plant producing at 17,000 tons/year will be operating at Lavera, near Marseilles. He also says that other organizations are working along similar lines — those in Japan, Central Europe, and the U.S.S.R. being the most active.

In the process developed at Grangemouth, yeast is grown on refined normal paraffins. The plant at Lavera will use gas oil, in which approximately 10% of the normal paraffins are selectively utilized. The procedures used in the development work have been outlined by Llewelyn (1968). These were based on continuous culture because this was considered to have economical and technical advantages. He refers to both aseptic and nonaseptic growth and it is implied that the Grangemouth unit will operate aseptically while that at Lavera will not. It is stated, that given a suitable strain of yeast, and appropriate culture pH value, temperature, and dilution rate, that the yeast remained the dominating organism, never falling below 99% by weight. In aseptic fermentations no difficulties were met with, and aseptic runs lasting many hundreds of hours were achieved.

In the same paper, Llewelyn gives information on product quality, toxicity testing, and feeding trials on rats, pigs, and poultry. The place of fermentation units in a refinery is discussed, including their role in dewaxing crude oil by assimilation of the normal paraffins into yeast. The economics of the process are thought of in relative terms.

It is concluded that "it is apparent that there are a number of areas where the product from either variant of the process could enter the rapidly expanding market for animal feedstuffs and could be sold at prices which would make the process viable." Finally there is speculation concerning commercial outlets, including incorporation into human food. The same field has been covered briefly by Shacklady (1969) and contains the additional information that there is no evidence of the material having any tendency to induce cancer, or to affect adversely growth, reproduction, or any other body function.

IV. Drink

A. Brewing

1. Introduction

There are now several published examples of commercial continuous brewing. Purssell and Smith (1968), of Arthur Guinness Son & Co., mention operational units in New Zealand, Canada, and Birmingham, England. A review by Hospodka (1966) which includes descriptions of various pilot-scale plants throughout the world indicates that the method is also being studied and applied in iron curtain countries. Evans (1965) briefly refers to study and application of the process in Australia and Japan. Purssell and Smith's critical paper is the most explicit of these three, because they have have included an assessment of the commercial value; it has been used as the basis of these notes which follow.

2. The Advantages of Continuous Brewing

After surveying the range of potential savings which apply to any continuous process, they see four advantages: smaller plant, reduced labor, less by-product yeast, and a savings in hops. They estimate the savings (per hectaliter of beer), as follows:

Smaller plant	15 U.S. cents
Reduced labor	1 U.S. cents
Saving on yeast and hops	3 U.S. cents
Total savings	19 U.S. cents

As disadvantages they mention the need for higher standards in equipment and improved control testing in order to combat the risk of infection. Commissioning a continuous plant, is, they say, a difficult matter. Owing to the general complexity and the dire consequences of incorrect operation, operatives need to be specially trained and a very disciplined procedure must be laid down. Those with experience of the supervision of process operations cannot but sympathize with such views.

Savings will result, because, as compared to a batch method, faster fermentation is a basic feature of any continuous process. Thus there is a consequent reduction in size of plant. In addition, unlike batch work, it is possible to work at artificially higher yeast concentrations in the fermenter. This is done by recycling some of the yeast cells separated from the effluent beer. In some circumstances it is possible to work at higher temperatures. Separately or together these two factors result in still faster fermentation and consequently a still smaller plant size. The other feature is that aerobic yeast reproduction is facilitated by the deliberate injection of air. This must effect an economy in the consumption of carbon substrates. As well the continuous process, as a consequence of recycling, generally produces less by-product yeast per unit of beer. This saving, in its turn, results in a saving of hops.

3. Examples of Continuous Processes

Two distinct continuous systems have been developed for brewing. One is a system of stirred vessels in series, with feedback of cells. The other uses a tubular reactor arranged for plug flow, with feedback of cells. This is contrived by either deliberate recycling, or by holdup of yeast. Feedback is essential, otherwise wash out of cells would occur.

Purssell and Smith give brief (but incomplete) descriptions of three systems which operate commercially:

1. The New Zealand Dominion Breweries plant.
2. The plant of J. Labatt Ltd., Ontario, Canada.
3. The A.P.V. tower fermenter which is operated commercially in Birmingham, England.

These processes shown diagrammatically in Fig. 1, are reproduced from Purssell and Smith. Table II gives some of the characteristics of each process. The Dominion system consists of two stirred tanks in series. Yeast propagation occurs in the first. Hospodka refers to a run of 12 months' duration without infection. He gives the controlled variables as temperature, yeast concentration, and dilution rate. The Labatt plant consists of three stirred tanks in series. Yeast propagation occurs in the first. Purssell and Smith comment that, although the beer has a high ester content, the taste panels found the beer to be "gratifyingly similar to the control batch fermented beer." Hospodka mentions, as well as the ones given above, additional variables including wort composition, aeration, stirring, and pressure. He quotes

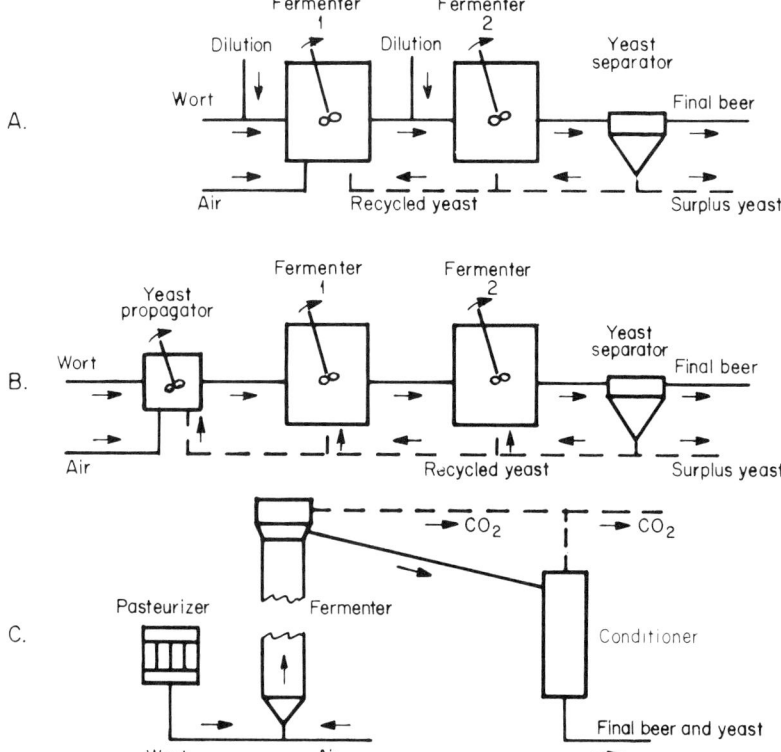

FIG. 1. Diagram of continuous beer fermentation processes (according to Purssell and Smith). (A) New Zealand Dominion Breweries. (B) J. Labatt Ltd., Ontario, Canada. (C) A.P.V. tower fermenter, operated at Mitchells and Butlers Ltd., Birmingham, England.

TABLE II
OPERATING VALUES FOR CONTINUOUS BEER FERMENTATION[a]

Process	Type	No. of stages	Yeast Concentration, (gm pressed) yeast/liter	Total production referred to batch	Total residence time (hours) for ale fermentation
Batch	—	—	22	1	240
Dominion	Stirred	2	Not given	Not given	35
Labatt	Stirred	3	30–50	¾	30–40
A.P.V.[b]	Plug flow	1[d]	250	⅛–⅓	4–8, 15[e]
A.P.V.[c]	Plug flow	1	200	1	4–6

[a] Values according to Purssell and Smith.
[b] Klopper et al., 1966.
[c] Ault et al., 1969.
[d] Followed by conditioning.
[e] For lager beer.

an example of pilot-plant operation of 4 months' duration in which all parameters, including yeast activity and freedom from contamination, were maintained.

In contrast to the above examples the A.P.V. plant is a tubular reactor. In the vertical tower the wort flows upward. Air is also introduced at the base to give a cocurrent system. At low flow rates yeast distribution in the tower is fairly uniform, at intermediate rates the yeast tends to concentrate toward the base of the tower as a porous plug. Further information on this and other details may be elicited from Klopper *et al.* (1966). The average yeast concentration is greater than in a batch fermentation, or than that quoted for the Labatt process. The residence time is less than that for the two other continuous methods — presumably a reflection of higher yeast concentration. A feature not mentioned in descriptions of the other methods is that to avoid bacterial contamination it was found necessary, in this system, to pasteurize the wort feed.

This was a preliminary report. In a very recent paper by members of the same team (Ault *et al.*, 1969) the earlier conclusions are modified. These stated that maximum efficiency resulted from using a high yeast concentration under conditions where little yeast growth occurred. In practice this condition of deficient aeration results in beer of high ester content and abnormal taste. This later work shows that if the wort is aerated to 70–75% of saturation, a tower fermenter will produce beers which can hardly be distinguished from batch beer, despite the very much shorter fermentation time. As a result there is an increase in yeast production which is now similar to that in batch fermentation. There is a further additional cost because the effluent beer must now undergo a yeast separation stage.

4. Conclusions

Purssell and Smith conclude:

(1) That in using a continuous process "beers can in most cases be produced having a flavor sufficiently close to the original not to be detected in the market."

(2) That application of the continuous process will have greater attraction in marketing a new brand of beer, where there is no obligation to match the flavor of an existing drink.

(3) That the process is best suited to regular production of one beer. As a rider they add that there will always be a need to retain some batch capacity in order to make special beer.

Finally they note: "However, the advantage to be gained from reduced costs by the use of continuous fermentation is not so large that other breweries brewing traditionally will be unable to compete with the new product. Thus it will always be the quality of the beer and the way that it is marketed that will mark the successful company."

This qualification echoes the earlier view of a fellow professional (Green, 1962), who questions and deplores what he terms the dubious chemical engineering concepts put forward by Dummett (1962). The latter claimed that the fermentation process "is a comparatively straightforward biochemical one." Green asserts that

> ... the brewer ... is concerned with the individuality of his product and its acceptance by the consumer. As a result this research will be directed towards the study of those phases of yeast metabolism, and its non-linear reaction to disturbances, which will continue to supply his particular product, either by batch or continuous processes.

Dummett has not left this rejoinder unanswered. In a press report of his presidential address to the Institution of Chemical Engineers (London) in May, 1969, he stated with reference to a continuous brewery operating in Spain that it was "the outcome of the application of a chemical engineering systems approach to a traditional biochemical operation." On the other hand, in Whitbread's new brewery at Luton, England, where a multiplicity of beers has to be produced, press reports state that batch is the chosen method. The deviation from tradition is that the batch fermentation process has been modified to make it faster. In the plant it is possible by consecutive batch brewing to complete up to a maximum of 24 brews each day. The associated processes, which are designed on a modular basis, can then be matched to the brewing output. In this way the overall process is effectively continuous.

B. Continuous Wine Making

An article by Peynaud (1967) refers to Argentinian, Italian, and French plants for continuous wine making from acid grapes. It is stated that the continuous process is unsuitable for mature, low-acid grapes. One hundred plants of the French design are said to be in operation mainly in the south of France.

All the fermenters described are upward-flow tower types in which the pressed grapes are introduced at the base. The wine is withdrawn at a point toward the top and the dregs which accumulate on the surface are withdrawn by mechanical means.

The meager details divulged indicate engineering ingenuity in mechanizing the existing process and so reducing labor costs and material losses. It is stated that the conversion of malic to lactic acid is enhanced. This is desirable for improving flavor and storage stability. Temperature control is also said to be improved. No other deliberate basic changes, affecting the microbiology or biochemistry of the process, appear to have been introduced.

V. Sewage and Trade Wastes

A. Sewage Disposal

1. The Activated Sludge Process

Innovation is not popular in an industry which is nonproductive. This is no doubt the reason why the use of sewage as a substrate in sulfur production, mentioned in the introduction, has not received serious attention, and why changes have been directed toward development of existing processes.

The activated sludge process is a complicated operation. It is a mixed aerobic culture of naturally occurring organisms ranging from bacteria to protozoa. The distribution of organisms (i.e., the ecology) varies between different plants, and a given unit is subject to diurnal and seasonal changes in temperature, and substrate composition and concentration. The purpose of sewage treatment is to remove or convert to stable compounds, the oxidizable components in the sewage, so that the receiving water course is not consequently depleted of oxygen to the detriment of its naturally occurring plant and fish life. This oxidizable capability is known as its biological oxygen demand (BOD), and is measured as the amount of dissolved oxygen absorbed by a sample when incubated at 20°C for 5 days.

The process has been described by Ainsworth (1966a) and is illustrated in Fig. 2. The bacterial floc, known as activated sludge, is prepared by aerating sewage to grow the naturally occurring organisms, separating the sludge by sedimentation, adding fresh sewage, and repeating the process several times. Besides biological activity, the other essential property of the sludge is that it should be easily separable by sedimentation. According to Ainsworth the concentrations usually encountered are 1–8 gm dry solids/liter in the aeration tank and 15–25 gm/liter in the settled sludge. The functions of the process are: (i) flocculation of suspended and colloidal matter in the sewage, (ii) oxidation of compounds of carbon and of ammonia, and (iii) autolysis of cell substance. British standards for an effluent, assumed to receive an eightfold dilution on discharge into the river, are 30 mg/liter of suspended solids and a BOD of 20 mg O_2/liter. Limitations are sometimes placed on ammonia content. Downing (1966) mentions an acceptable upper value, equivalent to 10 mg N/liter.

Downing (1966) states that the design of aeration systems is now

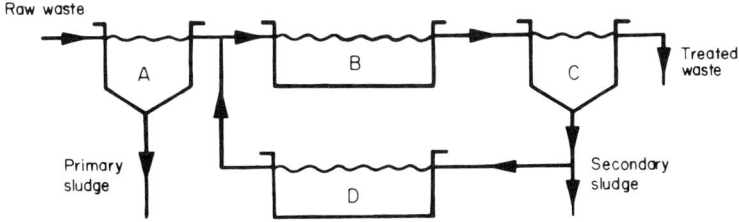

FIG. 2. Diagram of activated sludge processes. B and D, Aeration tanks; A and C, sludge settlers.

AERATION TANK CONTACT TIMES

Type	Tank B (hours)	Tank D (hours)
(a) Conventional	10–12	Not fitted
(b) High rate	1–4	Not fitted
(c) Extended aeration	40–60	Not fitted
(d) Contact stabilization[a]	1–4	3–12

[a] In type (d) the primary sludge separator (A) is omitted.

adequately understood. He then summarizes the variations on the basic method which are now in use or under development. Of the conventional type (10–12 hours contact time, Fig. 2a) he states that the process, which is also effective in removing ammonia, produces flocs which settle well and are amenable to subsequent anaerobic digestion, along with the primary sludge. The high rate process (1–4 hours contact time, Fig. 2b) yields sludge with poor settling properties and which is difficult to de-water. In addition, sludge yields are high, presumably because less autolysis occurs. Extended aeration (40–60 hours contact time, Fig. 2c) is said to be used increasingly for smaller communities. The plant is simple. There is no primary separation, and design of aeration is less critical. Ammonia removal is effective and sludge yields are lower. Contact stabilization (Fig. 2d) may be compared to a two-stage process. In the first stage (1–4 hours contact time) the oxidizable soluble components are disposed of accompanied by cell formation. The effluent from this stage is separated and the clarified water is discharged. The cell fraction is recycled via a second aeration system (3–12 hours contact time) in which further oxidation, autolysis, and partial ammonia removal occur. The result of rejecting clarified water after the brief first stage is an overall reduction in plant size. The dispersed growth method is sometimes used as a preliminary treatment of an industrial effluent before it is discharged for completion of treatment at a sewage works. It is a continuous culture in which there is no separation of solids. The contact time appropriate to a given effluent will depend on its composition and the growth rates of the organisms which develop. Two-stage processes are said to be used successfully for treating wastes of high BOD and would seem to be a method of exploiting any of the advantages of plug-flow which occurs in some single-stage systems.

In his conclusion Downing calls for further investigations on contact-stabilization and two-stage processes. In his Chairman's Address to the Water and Effluents Group of the Society of Chemical Industry (London, September, 25, 1969), Downing mentioned a long-term program of work being undertaken by the British Water Pollution Research Laboratory. This has the object of analyzing the role of the various organisms which are involved in the activated sludge process.

2. Percolating Bed Techniques

Packed beds, through which sewage trickles and creates microbial surface growth, carry out a similar process to that of the activated sludge process. They are appropriate for treating wastes from small communities. The traditional bed, packed with rock and stone, has disadvantages which have resulted in plastic packings becoming feasible alternatives. Askew (1966, 1967) has examined this process in detail and gives both technical and cost information for the treatment of wastes from distilleries, breweries, and fruit and vegetable processing factories. Pilot-plant studies on yeast, textile and resin wastes, and domestic sewage are also given. The saving in capital cost is estimated as ranging from 28 to 55% according to the waste being treated.

3. Anaerobic Sludge Digestion

Anaerobic digestion is frequently used to treat the primary and secondary sludges arising from the activated sludge process and also the solids which arise from percolating beds. The object is to produce an odorless product of reduced bulk. The process is usually two-stage, each stage having a holding time of about 30 days. The first stage in which most of the biological decomposition occurs, accompanied by CH_4 and CO_2 formation, is equipped for heating, mixing, and gas collection. The main function of the second stage is sedimentation. It is usually unheated and open to atmosphere. Mixed sludge is fed incrementally or continuously into the first unit. Supernatant from the second unit is recycled into the preceding aerobic stage. The sludge is disposed of in various ways which include dumping at sea and spreading on farmland. The methane-containing gas arising from the first stage is used for electricity generation. According to Kershaw (1969) detergent residues in sludge have serious effects on digestion and in some cases render the process inoperable. Randall (1967a,b) has reviewed the mechanism of the process and calls for research into the ecology of the system (cf. Downing above).

Anaerobic digestion is under competition from a chemical method – wet air oxidation in which the sludge is heated in the presence of air to 300–400°F under pressure (Teletzke, 1966). Kershaw (1969) in a review article puts forward the view that in the immediate future this may well be the approach to improved sludge processing.

B. Disposal of Trade Wastes

Trade wastes of animal and vegetable origin as well as spent gas liquor and those from organic chemical processes, including plastics manufacture, may all be treated by methods analogous to those used for sewage treatment. Ainsworth (1966b) has given examples of such processes in a recent review.

VI. General Conclusions

It is evident that continuous culture has not gained any foothold in pharmaceutical or organic chemical production. The development stage of yeast production from hydrocarbons has been completed, and the factor in determining the value of the process will be one of economics. In the brewing of beer the continuous method has been adopted, but there are clearly differences of opinion about its merits among the professionals. The aerobic process of sewage disposal is still under development. It seems to be agreed that research must turn from kinetics to ecology, and that further progress awaits a deeper understanding of the microbiology of the process.

In supplementing the speculations about the future made in previous sections, could another application be the production of cells as sources of their major chemical components? Herbert (1961) and Tempest (1970) have illustrated how, in continuous culture, bacterial cell composition can be manipulated by varying, for example, growth rate and the limiting substrate. Nucleic acid composition and distribution depends on growth rate. As an instance of substrate limitation

Sayer (1968) showed that *Escherichia coli* rich in alkaline phosphatase can be produced in phosphate-limited continuous culture. These are but two examples of the use of continuous culture in control of cell composition.

REFERENCES

I. INTRODUCTION

Gerhardt, P., and Bartlett, M. C. (1959). *Advan. Appl. Microbiol.* **1**, 215-260.

II. PHARMACEUTICAL PRODUCTS

Ball, C. (1967). *Genet. Res.* **10**, 173-183.
Bungay, H. R. (1963). *Biotechnol. Bioeng.* **5**, 1-7.
Canadian Patents and Developments Ltd. (1969). Brit. Patent 1,138,452.
Evans, C. G. T. (1965). *Lab. Pract.* **14**, 1168-1174.
Hockenhull, D. J. D., and Mackenzie, R. M. (1968). *Chem. Ind. (London)* **19**, 607-610.
Hosler, P., and Johnson, M. J. (1953). *Ind. Eng. Chem.* **45**, 871-874.
Hospodka, J. (1966). In "Continuous Culture of Microorganisms" (I. Malek and Z. Fencl, eds.), pp. 493-645. Academic Press, New York.
Lumb, M., and Wilkins, G. D. (1961). *Soc. Chem. Ind. (London) Monogr. No.* **12**, 254-264.
Parker, A. (1967). *Rep. Progr. Appl. Chem.* **52**, 483-491.
Pirt, S. J., and Callow, D. S. (1960). *J. Appl. Bacteriol.* **23**, 87-98.
Pirt, S. J., and Righelato, R. C. (1967). *Appl. Microbiol.* **15**, 1284-1290.
Pirt, S. J., Thackeray, J., and Harris-Smith, R. (1961). *J. Gen. Microbiol.* **25**, 119-130.
Reusser, F. (1961). *Appl. Microbiol.* **9**, 366-370.
Reusser, F. (1963). *Advan. Appl. Microbiol.* **5**, 189-215.
Reusser, F., Koepsell, H. J., and Savage, G. M. (1961). *Appl. Microbiol.* **9**, 346-348.
Righelato, R. C., and van Hemert, P. A. (1970). *Proc. 4th Int. Symp. Continuous Cultivation of Microorganisms, Prague, 1968.* Academic Press, New York.
Righelato, R. C., Trinici, A. P. J., Pirt, S. J., and Peat, A. (1968). *J. Gen. Microbiol.* **50**, 399-412.
Ross, N. G., and Wilkin, G. D. (1968). Brit. Patent 1,133,875.
Sikita, B., Slezák, J., and Herold, M. (1964). "Continuous Cultivation of Microorganisms." Proc. 2nd Symp., Prague, 1962, pp. 173-183. Publ. House Czechoslov. Acad. Sci., Prague.
Wright, D. G., and Calam, C. T. (1968). *Chem. Ind. (London)* pp. 1274-1275.
Zaccharias, B., and Björklund, M. (1968). *Appl. Microbiol.* **16**, 69-72.

III. ORGANIC CHEMICALS

Childs, G. G., and Welsby, B. (1966). *Process Biochem.* **1** (no. 8), 441-444.
Llewelyn, D. A. B. (1968). In "Microbiology" (P. Hepple, ed.), pp. 63-81. The Institute of Petroleum, London.
Miall, L. M. (1968). *Rep. Progr. Appl. Chem.* **53**, 403-412.
Miall, L. M. (1969). "Microbes in Industry," pp. 8-10. [*Suppl. to New Sci.* Sept. 25, 1969.]

Shacklady, C. A. (1969). "Microbes in Industry," pp. 5–7. [*Suppl. to New Sci.* Sept. 25, 1969.]
White, J. (1966). *Process Biochem.* 1 (No. 3), 139–144.

IV. DRINK

Ault, R. G., Hampton, A. N., Newton, R., and Roberts, R. H. (1969). *J. Inst. Brew. London* 75, 260–277.
Dummett, G. A. (1962). *Wallerstein Lab. Commun.* 25, 19–35.
Evans, C. G. T. (1965). *Lab. Pract.* 14, 1168–1174.
Green, S. R. (1962). *Wallerstein Lab. Commun.* 25, 337–348.
Hospodka, J. (1966). *In* "Continuous Culture of Microorganisms" (I. Malek and Z. Fencl, eds.), pp. 574–607. Academic Press, New York.
Klopper, W. J., Roberts, R. H., Royston, M. G., and Ault, R. G. (1966). *Proc. Eur. Brew. Conv. 1965*, pp. 238–259. Elsevier, Amsterdam.
Peynaud, E. (1967). *Process Biochem.* 2 (No. 12), 44–46.
Purssell, A. J. R., and Smith, M. J. (1968). *Proc. Eur. Brew. Conv., 1967*, pp. 155–165. Elsevier, Amsterdam.

V. SEWAGE AND TRADE WASTES

Ainsworth, G. (1966a). *Process Biochem.* 1 (No. 1), 15–22.
Ainsworth, G. (1966b). *Process Biochem.* 1 (No. 7), 385–390.
Askew, M. W. (1966). *Process Biochem.* 1 (No. 9), 483–492.
Askew, M. W. (1967). *Process Biochem.* 2 (No. 1), 31–34.
Downing, A. L. (1966). *Process Biochem.* 1, (No. 5), 257–293.
Kershaw, M. A. (1969). *Process Biochem.* 4 (No. 7), 46–48.
Randall, M. (1967a). *Process Biochem.* 2 (No. 11), 52–54.
Randall, M. (1967b). *Process Biochem.* 2 (No. 12), 17–20.
Teletzke, G. H. (1966). *Process Biochem.* 1 (No. 6), 329–333.

VI. GENERAL CONCLUSIONS

Herbert, D. (1961). *Symp. Soc. Gen. Microbiol. No.* 11, 391–416.
Sayer, P. D. (1968). *Appl. Microbiol.* 16, 326–329.
Tempest, D. W. (1970). *Advan. Microbial Physiol.* 4, 223–250.

Mathematical Models for Fermentation Processes*

A. G. Fredrickson, R. D. Megee, III, and H. M. Tsuchiya

*Chemical Engineering Department,
University of Minnesota,
Minneapolis, Minnesota*

I.	Introduction	419
II.	Formulation of Models for Fermentation Processes	423
III.	Some Basic Simplifying Assumptions	427
	A. Neglect of the Distribution of States	427
	B. Neglect of Segregation	429
	C. Neglect of Stochastic Phenomena	430
	D. Neglect of Biological "Structure"	430
IV.	Fermentors: Some Engineering Considerations	431
	A. Classification of Fermentations and Fermentors	431
	B. Properties of Ideal Fermentors	432
	C. Problems of Recycle	436
	D. Control of Continuous Fermentors	437
V.	Models for the Growth of a Pure Culture of Unicellular Microorganisms	438
	A. Unstructured Models	438
	B. A General Formulation of Structured Models	450
	C. Balanced and Unbalanced Growth	453
	D. Examples of Structured Models	457
VI.	Models for the Growth of Filamentous Organisms	459
VII.	Concluding Remarks	463
	References	464

I. Introduction

It seems appropriate to begin this review with a few remarks about mathematical models since there exists some confusion about their nature and some question concerning their utility. With regard to the latter, we can say that:

1. Models serve to correlate data and so provide a concise way of thinking about a system or process.

2. Models allow one—within limits—to predict quantitatively the performance of a system or process. Thus, they can reduce the amount of experimental labor necessary to design and/or optimize a process.

3. Models help to sharpen thinking about a system or process and can be used to guide one's reasoning in the design of experiments to isolate important parameters and elucidate the nature of the system or process. That is to say, the combination of mathematical modeling and experimental research often suggests new experiments that need to be done.

---

*During preparation of this review, the authors were supported in part by USDA Grant No. 12-14-100-9178 and USDPH Grant No. GM16692.

If the fermentation industry were confined to small-scale batch processing using pure culture techniques, the foregoing factors would be of little importance. However, the scope, the scale, and the complexity of fermentation processes are such that changes in technology will likely occur; indeed, there is already an emphasis on investigations on continous culture techniques, fermentation vessels are no longer always of the stirred-tank variety, and mixed cultures are starting to become common. We may therefore expect that the advantages of mathematical modeling will become more widely recognized in the fermentation industry and that the skills of the mathematically inclined bioengineer will take their place alongside those of the biochemist and microbiologist in this important field.

This sort of development has already taken place in much of the chemical industry, and mathematical modeling techniques are now applied with some confidence to very complex and large-scale chemical processes. The physical and chemical concepts used here comprise a major portion of modern chemical engineering science, and this science is also applicable to mathematical modeling in the fermentation industry. However, the fermentation industry is based on biological processes, and verified quantitative models for such processes are limited indeed. Most of the concepts used in models currently accepted in fermentation technology — in the United States, at any rate — were developed by the microbiologists of the late nineteenth and early twentieth centuries. Most published models of microbial growth are really only slight variations on themes that appeared in one or two classical papers more than two decades ago. This is not to say that new concepts and new models have not appeared. Rather, such concepts and models have not been appreciated and so have not undergone that process of testing and refinement which alone can establish their validity and utility. Hence, much of what is set forth here is admitted to be speculative, but hopefully it will stimulate testing of these tentative concepts.

Consider next the nature of mathematical models. By definition, a mathematical model of a system or process is a set of hypotheses concerning the mathematical relations between measured or measurable quantities associated with the system or process. Thus, mathematical models deal only with things that can be measured (assigned a number). A mathematical model may be simple or complex, depending on the system or process to be modeled, the skill and experience of the model maker, and the purpose for which the model is to be made. Most current models for fermentation processes are expressible as a system of nonlinear ordinary differential equations. Such models are mathematically tractable (at least by numerical means since the advent of electronic computers) but they are an order of magnitude more complex than the models originally used to describe the growth of biological populations; these were expressed as an algebraic equation—the "equation of the growth curve." More recently, models involving partial differential equations, or even nonlinear integro-differential equations have been proposed; most of these models are not yet mathematically tractable except in certain special cases.

Mathematical models of microbial growth are generated by a combination of three processes. First, well-established physical, chemical, or biological principles may provide certain parts of the model. Second, some hypotheses of the model may be plausible inferences from existing data. Finally, the remaining parts of the model are guesses, but one hopes, educated ones.

In order to test a mathematical model of a system or process, one must first of all be able to analyze the model mathematically to see what its quantitative predictions are for various feasible experimental conditions. The model is mathematically tractable if such analyses can be done. Models that are not mathematically tractable are of little use for practical purposes, but they should not be dismissed out of hand because they may provide important conceptual tools, and because advances in computing techniques may at any time render them tractable.

Any mathematical model of a fermentation process will contain parameters whose numerical value is unknown (although their physical or biological significance may specify what their orders-of-magnitude are) and cannot be measured directly. Hence, the second step in testing a model is to ascertain the values of model parameters by some sort of "curve-fitting" technique. That is, numerical values of the parameters are estimated by "fitting" the model's prediction to experimental data. Considerable difficulty of a mathematical or computational nature may arise here. If, for instance, the mathematical expression of the model is a set of nonlinear ordinary differential equations, it will rarely be possible to integrate these equations; that is, it will not in general be possible to find algebraic relations, clear of derivatives, between measured variables and model parameters. Hence, standard curve-fitting techniques cannot be applied, and numerical schemes must be employed. However, it is not possible to integrate the differential equations of the model numerically without knowing the numerical values of the model parameters beforehand. For this reason, some sort of trial-and-error parameter-estimating scheme must be adopted; see e.g., Heineken *et al.* (1967) or Bellman *et al.* (1967). At the present time, no generally applicable scheme for parameter estimation is available, and the success of a given scheme is found to be largely dependent on the specific problem considered.

The processes of model building, model testing, and curve fitting must not be confused. Curve fitting is part of model testing, but only part, and model building is surely not curve fitting.

Clearly, a mathematical model must be rejected or modified if it does not give a good fit of experimental data or if parameters estimated turn out to have unreasonable values. On the other hand, a good fit and reasonable parameter values do not "prove" that the model is valid. For one thing, goodness of fit is a rather subjective concept: some deviations between prediction and data are always present. How large can these become, and what kind must they be, before we will say that the model does not "fit"? In short, a good fit is a necessary but not sufficient condition for acceptance of a model.

Let us assume that the model has passed the first stage of the testing process; that is, reasonable values of parameters can be found such that the model fits experimental data in an acceptable fashion. The next step in the testing process involves doing a *different, independent* experiment to see if data so obtained are fitted by the model with the *same* set of parameters obtained in the previous stage of testing. If it does not, we must either reject the model (alter the basic hypotheses) or else restrict its use to those situations where a fit was obtained. If it does, we may go on to still other experimental tests. In this way, one can improve confidence in the generality of the model, though, of course, one can never prove that the model is *the* "right" one.

In the first paragraph of this introduction, some examples of *proper* usage of mathematical models were given. It is perhaps useful to close the introduction by giving some examples of *improper* usage of models. An obvious improper usage is to adhere to a model even when it is shown to violate some well-established chemical, physical, or biological principle, or when it is found to be grossly at variance with acceptable experimental data. A related example of improper usage is provided when unreasonable extrapolations of the model are made.

An example of a different kind of improper usage — perhaps improper evaluation would be a better word — of a mathematical model is rejection of a model that explains many experimental observations but says nothing about some others. No model can do everyting, and being hypercritical can be just as inhibiting for progress as being uncritical. Consider the utility of the Michaelis–Menten formulation of the kinetics of enzyme action; the equations have proved their value despite the uncovering of reactions which do not follow the predictions of the original model.

Finally, models of complex processes are sometimes rejected because they contain "too many constants" and "you can fit anything with a sufficient number of constants"; this is also an example of improper assessment of mathematical models. It might be possible, for example, to fit the time course of (say) optical density of a reaction mixture in a batch chemical reactor by a model containing one or two constants, even though five or six important reactions may be going on in that mixture. Such a model would surely fail, however, if the reaction system were changed (as by, e.g., adding recycle or changing the temperature). On the other hand, a mathematical model based on well-established physical, chemical, and biological principles would contain ten to twelve constants; these would be the frequency factors and activation energies of the important reactions of the system. Such a model would not be expected to fail so quickly if the conditions of the reaction were changed. To put it another way, one expects models to be robust in proportion to the degree to which they actually represent the processes that they are supposed to model.

II. Formulation of Models for Fermentation Processes

Mathematical models deal with the relations between measured or measurable *quantities,* or variables. Hence, it is important to see what variables must be considered in the modeling of fermentation processes. We shall divide these into two categories: biological and nonbiological variables.

The biological variables of significance depend on the ecological[1] complexity of the fermentation process. If the process involves mixed populations, then one class of variables important in the model is formed from those quantities that describe *community*[2] *structure.* Such quantities include the numbers and biomasses of the various populations of the community, and perhaps the distribution of physiological states within each population. If the process involves a *pure culture,* quantities of interest again include number of individuals (or population density), amount of biomass (or biomaterial concentration), and perhaps the distribution of physiological states within the population. The concept of physiological state is extremely important

---
[1]Used in the sense of interations between the biological forms and their environment.
[2]Used in the ecological sense.

for the understanding of fermentation processes; it is also quite difficult to quantify. More will be said concerning it.

Nonbiological variables of significance in the modeling of fermentation processes are those quantities that describe the *state* of the abiotic phase or phases of the fermentation system. In the case of the usual submerged culture technique, such variables include the concentrations of various chemical substances in the liquid phase, including dissolved oxygen and pH, the temperature, and pressure, etc. Another class of nonbiological variables that is important arises from the control exerted on the fermentation process. For instance, in continuous fermentation, dilution rate is an important variable. Still another class would be the mechanical shear which could alter the physiological state of the organisms.

As stated earlier, mathematical models are generated in part by application of well-established principles of physical and biological science. Physical principles of utility here may be divided into three categories: conservation, thermodynamic, and constitutive principles.

Conservation principles are essentially accounting principles. For any specified system, the general conservation principle may be expressed as the verbal equation: rate of accumulation of an entity in the system equals net rate of transfer of the entity into the system across its boundaries plus net rate of generation of the entity within the system. Entities subject to this principle include mass, chemical species, energy, and many others of lesser importance in model developing for fermentation processes. The system chosen may be small—a single cell or even a compartment within a cell—or it may be large—an entire fermentor or even an entire fermentation plant. The system may be an abstract one; in order to derive an equation for the distribution of ages in a bacterial population, we chose as a system the contents of the growth vessel *and* all cells whose age falls within specified limits. Depending on the system chosen and the nature of the transfer and generation processes involved, conservation principles will generate ordinary differential equations, partial differential equations, or even integro-differential equations.

Thermodynamic principles are, of course, expressions of the laws of thermodynamics as applied to appropriately chosen systems. The first law of thermodynamics is the principle of conservation of energy, and so is really a conservation principle. The second law, as is well known, places restrictions on the efficiency with which certain processes can occur, and on the extent to which they can occur. The second law is not much considered in models of fermentation processes, primarily because such models are not yet detailed enough to consider the extents and equilibria of intercellular processes. In ecology, however, this law finds important applications (see e.g., Odum, 1959, Chapter 3), particularly on the restrictions it places on the flow of energy between trophic levels in ecosystems. We may expect the second law of thermodynamics to find wider application in fermentation technology when processes involving mixed populations become more important.

Constitutive principles are those aspects of physiochemical law that govern the *rates* of physiochemical processes. Conservation principles and the laws of thermodynamics do not deal with mechanism, but constitutive principles do; they deal with mechanism as influenced by the constitution of matter. Hence, their generic title. Obvious examples of constitutive principles are the laws that govern the rates of transport of mass, charge, energy, and momentum (transport phenomena) and those that govern the rates of chemical reactions (reaction kinetics).

Since conservation principles do not depend on the constitution of matter, we may in general be confident that a mathematical equation expressing a conservation principle is "right" if we have been careful to include *all* inflows, outflows, and source and sink terms. Whether or not such an equation is useful depends primarily upon two things: the judgment exercised in the choice of the system and the availability of constitutive principles for the description of unknown quantities (such as reaction rates) which appear in the equation. Since biological mechanisms and biological constitutions are often poorly understood, it follows that constitutive relations present the greatest stumbling block to the maker of mathematical models of microbial growth.

It is not possible to codify biological principles of importance to model building so neatly. Hence, we shall simply state four principles, drawn from various areas of biology, that seem to be of most general importance in model building.

The first principle is ecological and physiological, and states that the activities of an organism, and the rates at which it carries on those activities, are dependent not only on the organism itself but also on the organism's environment. A corollary of this principle is that the constitution or state of the environment depends upon the activities of organisms inhibiting it. The principle and its corollary have long been recognized in fermentation technology (e.g., the concept of *limiting* substrate), and most models used therein *explicitly* incorporate the principle. However, a good deal of theoretical work on population dynamics *does not* incorporate the principle and its corollary explicitly (though it may do so *implicitly*) and so is suspect on *a priori* grounds.

The second principle is also ecological in part, and is in part drawn from genetics. It states that the current phenotype (constitution or state) of an organism depends not only on its genotype but also on the past history of environments seen by the organism. That is to say, by cultivating an organism in different environments, as one does in batch cultures, one can arrive at states with the same genotype but differing in physiologic function and morphologic structure. The principle is recognized in fermentation technology, of course, but often it is not incorporated in models or, if it is, it is usually done in an *ad hoc* and unsatisfactory manner. This bears upon the related questions of physiological state and structured models of which more will be said in subsequent pages. Here, suffice it to say that a model is *structured* if its formulation incorporates the foregoing principle; it is *unstructured* otherwise.

The third principle states that organisms may be divided into classes depending on their morphology, growth form, and their mode of reproduction. Thus Eubacteriales are unicellular, exhibit intercalary growth, and generally reproduce by binary fission, whereas, Actinomycetales and molds of industrial interest may be filamentous, multicellular or coenocytic, exhibit apical growth, and reproduce by a variety of means. These facts are familiar to every student of biology, of course, but their implications for modeling of microbial growth are not generally realized. Thus, fermentation technologists may apply the same mathematical model to the growth of both Eubacteriales and molds; the only apparent reason why such a process can have any success at all must be that the models used describe a highly specific aspect of growth that happens to be more or less similar in the two classes of organisms.

The fourth principle is the universally recognized fact of the mutability of organisms. The fact, if universally recognized, is also almost universally not considered in population dynamics, at least among those who construct models of microbial growth. True, Moser (1958) wrote a monograph on the dynamics of such changes in bacteria grown in a chemostat and some simple and highly specific models of the mutation process appear from time to time. Still, one feels that there are enormous challenges here, and that the theoretician has hardly scratched the surface. Involved are such questions as how does one formulate a mathematical theory of natural selection? Would such a theory take the form of an optimality principle, as suggested by Rosen (1967)? How would such a theory apply to problems of microbial technology?

Besides these rather general biological principles, there are many others, more or less specific, that are of importance to the present subject. However, we shall content ourselves with the four discussed above since these are recognized by everybody but not generally accounted for by model builders.

III. Some Basic Simplifying Assumptions

The growth of even a single population of microorganisms is of course a tremendously complicated process. To make any progress in the task of mathematically modeling such growth, simplifying assumptions must be made. These introduce inexactitudes into the models, in the sense that we are thereby certain that the model will not describe *all* of the various facets of growth that we know do occur. Hopefully, the models will describe some facets of growth that we consider important, but even that is by no means certain; only constant recourse to experimental testing can keep the maker of models on the right track.

Many published models of microbial growth make certain simplifying assumptions without stating such assumptions explicitly. It seems important that the more common hypotheses of this sort be discussed. Hence, we shall consider four such in this section.

A. Neglect of the Distribution of States

Biological populations are composed of individual organisms, and the individuals are by no means physiologically, morphologically, or even genetically identical. As the beginning student of microbiology learns by a glance through the microscope, individuals differ in size, shape, staining properties, and perhaps in other characteristics, such as motility. Examination for a period of time teaches that individuals differ in "age," if we define a cell's age as the chronological time since it was formed by fission of its parent, since fissions are never fully synchronous events. More refined analyses would perhaps reveal even more fundamental, if less obvious, distinctions between individuals. To sum up, it is evident that the individuals of a population do not all exist in the same state—however we might define that term—but rather represent a distribution of states.

The first simplifying assumptions made by most workers trying to model microbial growth are that the foregoing distribution of states can be ignored and that the properties of the culture can be adequately described in terms of a "typical" individual whose behavior represents some sort of average over the distribution of states. Such assumptions at once lead to uncertainty about the validity of the model, since they imply that a whole host of parameters of the population—the moments of the distribution of states higher than the first—are not important in determining the properties or activities of the population.

There are three reasons for making these assumptions; one involves a conceptual difficulty and the other two involve practical difficulties. The conceptual difficulty is: What do we mean by the "state" of an individual organism anyway? Can we use some obvious parameter associated with a cell as a measure or index of its state? Would the age of a cell (as defined above) serve this purpose? If so, Von Foerster (1959), Trucco (1965), Fredrickson and Tsuchiya (1963), Koževšník (1964), and others have supplied mathematical apparatus that can be used to model such things as changes in the age distribution of bacteria brought about by changes in environmental conditions. Or can we use the size of a bacterial cell as an index of its state? This possibility was considered by Koch and Schaechter (1962), Eakman *et al.* (1966), and Subramanian and Ramkrishna (1970) showed how to construct a model of population growth using size as an index of state. Or is some more general notion of state, such as the biochemical constitution of the cell as proposed by Fredrickson *et al.* (1967) needed? How can recent work in molecular biology, such as that summarized by Maaløe and Kjeldgaard (1966) be translated into a mathematical description of "state"?

An obvious practical difficulty involved in the use of any model that recognizes a distribution of states is: How does one *measure* the states of individual organisms so that some idea of the distribution can be obtained? The idea of bacterial size as an index of state is attractive here, since electronic devices can now measure size distributions quickly (and one hopes, accurately). But many workers might prefer to use age or some other cell feature as an index of state, and in addition, the theoretical work of Fredrickson *et al.* (1967) has shown that any *single* index of state is going to be inadequate in all but the most rigidly controlled growth conditions.

The other practical difficulty involved here is the fact that the equations resulting from models recognizing a distribution of states quickly became mathematically intractable. Von Foerster's equation, which results if age is used as an index of state, is handled rather readily, but the equation of Eakman *et al.*, which is based on size as an index of state, requires extensive numerical computation for its solution. More complicated indexes of state present even more formidable mathematical difficulties.

For these reasons, in the rest of this review we shall consider only those models where the distribution of states among individual organisms is neglected.

B. Neglect of Segregation

In unicellular microorganisms, life is *segregated* into structurally and functionally discrete units—cells. Hence, the *number* of individual organisms present in a population must be an important parameter for the description of the population. It is clearly not the only parameter so important; such quantities as the biomass of the population must also be involved. Nevertheless, number must be a quantity of prime importance, since, for instance, the biological potentialities of a population composed of $2n$ organisms and having total biomass m are not the same as those of a population composed of n organisms but having total biomass m. ($2n$ and n as used here refer to *number* of organisms, not to ploidy!)

In spite of the foregoing arguments, many models currently in favor make the assumption that the segregation of life into discrete units can be ignored. With such an assumption, number of organisms is not admitted as a parameter to be described by the model, and in effect, the model views the population as biomass *distributed continuously* throughout the culture. Models based on such neglect of segregation will be called here *nonsegregated*. [In an earlier review (Tsuchiya *et al.*, 1966), we called these models *distributed*, but nonsegregated seems to be a better word.]

One may well wonder how a nonsegregated model can have any success at all in the description of unicellular growth. True, there may be no *practical* need for knowing numbers of organisms present (biomass may be the quantity of practical importance), but in general, increase in number (proliferation) and increase in biomass (growth) are coupled processes, so that one cannot really omit the one from a model purporting to describe the course of the other. A possible explanation for the success of some nonsegregated models is that they have been applied to situations that are *balanced growth*, or nearly balanced growth. In this situation (discussed in detail below), growth and proliferation are *proportional*, so that knowledge of amount of biomass is tantamount to knowledge of number of organisms.

C. Neglect of Stochastic Phenomena

It is not possible to predict the behavior of individual microbial cells with certainty. Thus, Kelly and Rahn (1932), Powell (1955, 1958), and later workers found that the generation times of individual bacterial cells were not all the same, but rather showed random deviations about mean values. Schaechter et al. (1962) found that the size (length) of rodlike bacteria at fission also showed random fluctuations about a mean. The coefficient of variation of the size at fission was considerably smaller than the coefficient of variation of the generation time, thus implying that cell size is a better index of cell state than is cell age (Koch and Schaechter, 1962).

Models for microbial propagation, at least those that are used in fermentation technology, generally make the simplifying assumptions that the foregoing stochastic phenomena can be neglected, and that growth can be treated as a deterministic process. If this assumption cannot be made, then nonsegregated models cannot be used, either; if cells divide at random times, the number of cells present must be a variable of the model.

Stochastic population models (so-called "birth-and-death processes") very quickly lead to formidable mathematical difficulties, even when one attempts to model only very simple biological phenomena. For some examples, see Bartlett (1960) or Bharucha-Reid (1960). Hence, it is desirable to avoid such models whenever possible.

Fortunately, this is usually permissible in fermentation technology. The circumstance that permits one to neglect randomness here is that one is dealing with such an enormous number of cells that random deviations from a mean average out. There are situations where random deviations are important, however, and these always deal with cases where the *total number* of cells involved is small. Sterilization is one such situation (Fredrickson, 1966) and various transient growth situations, in which the population size for one reason or another becomes quite small (such as near the critical dilution rate in continuous culture) are others.

In the following discussion, we shall ignore stochastic phenomena.

D. Neglect of Biological "Structure"

Two microorganisms having the same biomass and inhabiting the same environment may nevertheless have widely different properties and activities. This is the problem of *state* again; we would say that the two organisms have different states. If our model recognizes the existence of a distribution of states, it should also recognize that that distribution may change in response to changes in the environment. Or, if our model does not recognize the existence of a distribution of states, it should at least recognize the possibility that the state of the average or typical organism (which is all such models consider) can change in response to changes in the environment. This means that parameters *in addition* to population number and population biomass must in general be important for the description of population behavior.

Many models currently used in fermentation technology do not recognize this; in most of these models, population biomass is the sole variable employed for describing the population. Since this procedure regards organisms and population biomass as featureless, structureless entities, we shall call such models *unstructured*. In the following, we shall consider both unstructured and structured models.

IV. Fermentors: Some Engineering Considerations

A. Classification of Fermentations and Fermentors

Fermentations may be classified in many different ways. Those primarily interested in the biology of a fermentation might classify it on the basis of the number of biological populations involved (pure culture, mixed culture) or on the basis of the kind or kinds of biological populations involved (yeasts, molds, bacteria, algae). Those primarily interested in physical aspects of the fermentation might classify it on the basis of the number of phases involved or on the basis of the kinds of phases involved. For instance, in the usual submerged culture fermentation, three phases are involved: a gas, a liquid, and a "solids" phase (the *biophase*). In fermentations employing a hydrocarbon substrate, a second liquid phase must be added to the foregoing list. In the trickling filter, a second solid phase, the substratum on which organisms grow, must be added to the list. Finally, those primarily interested in the biochemistry of the fermentation might classify it on the basis of the principal substrates consumed or the principal products formed or on the basis of the stage of the fermentation at which substrate is consumed and/or product is formed.

In this review, we shall concern ourselves almost exclusively with pure culture fermentations in which there is a gas phase, one liquid phase, and a biophase. Unless otherwise stated, the biophase will be composed of organisms reproducing by binary fission. However, the biochemistry of the fermentation is assumed not to be so rigidly fixed; mathematical models for different fermentation patterns are thus a subject of this review.

Fermentors themselves are classified most directly on the nature of *flow* of the phases in and through them. The utility of this classification is that flow patterns in most fermentors actually used either approximate to one of two ideal types of flow situations or else they approximate to simple combinations of the two ideal types. The ideal types are of course the continuous flow, stirred vessel (C^* for short) and the continuous, plug flow, tubular fermentor (or simply tubular fermentor). The batch fermentor is generally a special case of the C^* obtained by letting the flow-through rate approach zero.

An exhaustive review of the patterns of flow in and through processing vessels is given by Levenspiel and Bischoff (1963). In this section, we shall summarize those properties of the two ideal types of fermentor that are most important in the construction of fermentation models. Three basic assumptions will be made in all cases unless otherwise stated. These are : (1) there is no "wall growth"; (2) temperature, pressure, and other thermodynamic variables are time-independent and spatially uniform; and (3) mass transfer between gas and liquid phases is not a rate-limiting process, neither is mass transfer across cellular membranes as e.g., in fermentations of steroids, hydrocarbons, etc. Obviously, in many industrial fermentations, one or more of these assumptions may have to be relaxed; indeed, such a relaxation may sometimes be the essence of understanding the fermentation.

B. Properties of Ideal Fermentors

1. The C^* Fermentor

In this type of fermentor, culture medium or culture itself is continuously fed to the fermentor, where growth and its attendent processes occur. The culture within the fermentor is agitated somehow; generally a mechanical agitator is provided for this purpose. Culture is removed from the vessel at the same volumetric rate as feed is admitted, so that (barring volume changes) the volume of the culture in the fermentor remains constant.

In the *ideal* C^*, agitation is assumed to be so vigorous that mixing is *complete*. The criteria of complete or perfect mixing may be stated in different, but equivalent ways. The usual criterion is simply that the *composition* of the culture in the vessel is *uniform*, so that the composition of the stream leaving the fermentor is the same as that of any sample drawn from the interior of the fermentor. From a statistical point of view, mixing is perfect if (1) the probability that a "particle" (organism, molecule) will be in a given subvolume in the culture is the same as the probability that it will be in any other subvolume of equal size — regardless of location in the vessel — of the culture; and (2) "particles" move independently through the vessel. The statistical criteria are the more general, and the first criterion given can be derived from them.

No real vessel can be perfectly mixed, but in practice, the behavior of a real vessel can be made to approach that of an ideal C^* very closely.

Let c be the concentration (amount per unit volume of *culture*) of some substance, biotic or abiotic, in a culture. Then application of the principle of conservation of mass to the system composed of the culture (liquid plus biophase) in the C^* yields the differential equation

$$V(dc/dt) = q(c_f - c) + Vr \tag{1}$$

relating the rate of change of c to the rate of flow through the vessel (q) and the rate of production of the substance per unit volume (r). In the equation, V is the volume of culture in the vessel (a constant), and c_f is the concentration of substance in the feed to the C^*.

From this equation follows the first point of importance concerning the C^*. Suppose that a steady state $(dc/dt = 0)$ has been attained and that a substance is consumed or destroyed by reactions within the vessel $(r \neq 0)$. Then we see that

$$c_f - c = -(1/D)\,r$$

where $D = V/q$ is the *dilution rate*. This equation shows that the composition of the feed stream undergoes a *discontinuous change* when the stream enters the C^*. Thus, if the feed to the C^* contains organisms, these will experience a discontinuous change, in other words, an *environmental shock*, upon entering the C^*.

A second point of importance follows from Eq. (1) although we shall give no proof of the fact here (see e.g., Aris, 1965). This is that the *residence time* of a particle in a C^* is not fixed but is subject to statistical fluctuations. In particular, the density of the distribution of residence times is negative exponential, and is

$$De^{-Dt}$$

where D is the dilution rate. The most probable residence time is zero. The mean residence time is then D^{-1}, and this is also the standard deviation of the distribution of residence times. If a cascade (series) arrangement of equal volume C^*'s is used, say N of them, then the residence time follows a gamma distribution, with density function

$$(ND) \cdot [(NDT)^{N-1} e^{-NDt}]/(N-1)!$$

The mean is again D^{-1} (V is the *total* volume of all N tanks) but the standard deviation is smaller than in the one-vessel case, and is $D^{-1} N^{-1/2}$. As the number of tanks in the cascade becomes larger and larger, while D is held constant (thus implying that the individual tanks become smaller and smaller), the standard deviation becomes smaller and smaller. In the limit of very large N, the residence time is no longer subject to statistical fluctuations, and is always D^{-1}.

A third point of importance concerning the C^* is that the probability that a particle in the vessel at some time will be "washed out" in some subsequent time interval is independent of such factors as the size of the particle or its residence time in the C^*. Since washout is equivalent to death so far as the population in the C^* is concerned, one sees that in effect, the flow through the vessel imposes a *nonselective* death rate on the population. This fact should always be borne in mind, particularly when two or more populations are growing symbiotically in a C^*.

2. The Tubular Fermentor

In this type of fermentor, culture medium, or culture itself, are continuously fed to the fermentor, where growth and its attendent processes occur. The culture within the fermentor is not stirred, since the object is to have elements of culture move progressively through the fermentor without mixing.

In the ideal or plug-flow tubular fermentor, adjacent elements of culture are assumed to move progressively through the tube, without exchange of material between such elements. In addition, the composition of the culture is assumed to be uniform over any cross section, though the composition obviously changes from section to section.

Evidently, no real fermentor can be an ideal tubular system, but in some cases, the behavior of real apparatus can be made to approach the ideal.

As before, let c be the concentration of some substance in the culture. Application of the principle of conservation of mass to a system of infinitesimal length moving with the velocity of flow through the tube (v) then yields the partial differential equation

$$(\partial c/\partial t) + v(\partial c/\partial z) = r \qquad (2)$$

where z is the axial distance from the inlet of the tube, and r is the production rate per unit volume, as before.

In this case, the residence time of a particle is not subject to statistical fluctuations; a particle at position z has been in the vessel for a time z/v, and if the tube has length L, the transit time will be L/v.

In many cases, mixing in the direction of flow may be important. This is accounted for most easily (though only approximately) by introducing an effective axial diffusion coefficient, $\mathscr{D}$. Material balance in this case leads to the more complicated equation

$$(\partial c/\partial t) + v(\partial c/\partial z) = \mathscr{D}\,(\partial^2 c/\partial z^2) + r \qquad (3)$$

where the mixing term involves the second derivative of the concentration. The diffusivity $\mathscr{D}$ is not the usual diffusion coefficient that is used in studies of transport phenomena. It is dependent on the flow situation in the fermentor and independent of the usual (molecular) diffusion coefficient, at least if flow is turbulent; see e.g., Levenspiel and Bischoff (1963).

If axial mixing does occur, then particle residence times are again subject to statistical fluctuations. No simple expression for the density of the distribution of residence times emerges in this case [though it can be found by solving Eq. (3) subject to appropriate boundary and initial conditions]. However, one can say that the mean residence time will be L/v and the standard deviation of the residence time will increase as $\mathscr{D}$ increases; the distribution will be skewed, with the most probable residence time being smaller than the mean. If the dimensionless parameter $\mathscr{D}/vL$ is small compared to unity, then the distribution of residence times is approximately Gaussian, with mean L/v and standard deviation $(L/v)(2\mathscr{D}/vL)^{1/2}$.

Biologically speaking the most important feature of the tubular fermentor is the progressive *change* in environmental conditions seen by an organism traversing the tube. This is in marked contrast to the constant conditions seen by an organism traversing an ideal C*. In fact, the situation in the ideal tubular fermentor is entirely similar to that in the batch fermentor, with residence time, z/v, replacing batch time, t. Hence, the ideal tubular fermentor admits the possibility of accomplishing on a continuous, steady-state basis that which is done discontinuously in batch situations.

The simple model of the tubular fermentor given above becomes considerably more complicated if we try to account for the presence of a flowing gas phase, in addition to the liquid and biophases, in the fermentor. A review of these complications is given by Cichy et al. (1969).

C. Problems of Recycle

In certain applications, it may be desirable or necessary to return a portion of the effluent culture from a fermentor to the inlet of the fermentor. Thus, in a tubular fermentor with no axial mixing and fed with sterile medium, no steady-state growth is possible unless recycle is employed.

The formal mathematical treatment of recycle is generally quite simple; material balances of the type familiar in elementary chemical engineering are usually all that is involved. Figure 1 shows a fairly

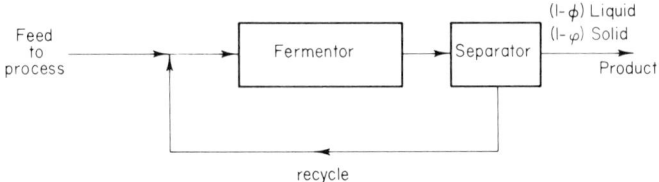

Fig. 1. Recycle arrangements for a continuous fermentor.

general scheme. In the figure, a fraction ϕ of the *liquid* effluent from the fermentor is returned as recycle, and a fraction φ of the solid (biological) effluent is returned as recycle. These two fractions need not be equal. If solids are centrifuged out and all liquid removed, then $\phi = 0$ but $\varphi > 0$; if culture is simply returned without separation of solid and liquid, then $\phi = \varphi > 0$. In the situation shown, the concentration of biomaterial in the stream entering the fermentor is x_f; this will be related to the concentration in the stream leaving the fermentor by

$$x_f = (1 - \phi)x_0 + \varphi x$$

where x_0 is the concentration in the feed stream to the process. The concentration in the stream leaving the process is just

$$\frac{(1 - \varphi)x}{(1 - \phi) + (1 - \varphi)x}$$

Both of these relations are established by simple material balances.

On the other hand, recycle does lead to a number of other problems that are not handled so readily. Consider a recycle employed with either a tubular fermentor or with a cascade (more than one) of C^*'s in series. Suppose that solids and liquids in the effluent stream are not separated ($\phi = \varphi$) so that a fraction of the effluent stream is simply returned to the inlet of the fermentor. Then organisms so recycled are suddenly subjected to a new environment when they reenter the fermentor, since obviously, conditions change along the length of a tubular reactor or from tank-to-tank in a C^* cascade. Hence, organisms receive an *environmental shock* as they reenter the fermentor from the recycle stream, and something very like (if not exactly like) the lag phase following inoculation of a batch culture must ensue. Hence, adequate modeling of systems employing recycle seems to require models that can predict lag phases when they are applied to batch growth situations. Many models currently in use *cannot* predict a lag phase (in particular all but one of the so-called "unstructured" models treated in Section V,A below cannot predict a lag phase) for batch growth, and so their use for systems employing recycle is suspect, and on an *a priori* basis, likely to be incorrect.

D. Control of Continuous Fermentors

Efficient operation of fermentation equipment generally requires that control be exercised on a fairly large number of operating variables, such as pH, temperature, level of dissolved gases. Here, however, we consider briefly only the control exerted on the flow rate to a continuous fermentor. In the so-called *chemostat*, the flow rate is maintained constant so that no feedback control is used. On the other hand, feedback control of the flow rate is sometimes used. This is done by monitoring some quantity that is a measure of population density, biomass concentration, or product concentration in the fermentor, and this is then used to alter the flow rate. The object is of course to maintain a desired level of the measured quantity in the fermentor. Systems operating with such control are often called *turbidostats* if the controlled quantity is the turbidity of the culture.

Evidently, use of feedback control introduces further elements of dynamical behavior that must be built into the model. In the C^*, the dilution rate now becomes a function of (say) turbidity, rate of change of turbidity, etc., whereas in the tubular fermentor, the velocity becomes a function of these quantities. These facts complicate the dynamical behavior of biological reactors, but this has not yet been recognized in the literature. It would seem worthwhile to do a rather thorough simulation study of the dynamics of various fermentations, to see how they are affected by the addition of feedback control on flow rate.

V. Models for the Growth of a Pure Culture of Unicellular Microorganisms

A. Unstructured Models

Physicochemical conservation principles may be applied to appropriately chosen systems to derive the basic equations for stirred tank and tubular fermentors. The simplifying assumptions discussed previously will be made in all cases.

Let x be the concentration of biomass in the culture; x is measured in mass of biomaterial (usually on a dry basis) per unit volume of culture (solids and liquid). Then one obtains from Eq. (1)

$$dx/dt = D(x_f - x) + \mu x \tag{4}$$

in the stirred tank fermentor, and Eq. (3) yields the partial differential equation

$$(\partial x/\partial t) + v(\partial x/\partial z) = \mathscr{D} \, (\partial^2 x/\partial z^2) + \mu x \tag{5}$$

for the tubular fermentor. In Eq. (4), it is assumed that the inflow stream contains biomass at concentration x_f; this term may arise, for instance, if recycle is used or if the vessel is one of a series of a cascade. The quantity D is the usual *dilution rate* (volumetric flow rate divided by the volume of culture in the vessel).

In effect, Eqs. (4) and (5) serve to *define* the *specific growth rate*, μ, for the two fermentation systems considered.

The balance principle may also be applied to numbers of organisms, and leads to the equation

$$dn/dt = D(n_f - n) + \nu n \tag{6}$$

for the stirred fermentor, but to the equation

$$(\partial n/\partial t) + v(\partial n/\partial z) = \mathscr{D} \, (\partial^2 n/\partial z^2) + \nu n \tag{7}$$

for the tubular fermentor. Here, n is the population density, number of organisms per unit volume of culture. In these equations, it is assumed that the stirred fermentor may be fed with organisms at population density n_f and that the axial diffusivity of biomaterial and individual organisms is the same in the tubular fermentor.

In effect, Eqs. (6) and (7) serve to *define* the *specific multiplication rate*, ν, for the two fermentation systems considered.

According to the biological principles stated in Section II above, the environment of a population affects its growth and multiplication, which in turn effects the environment. Hence, equations describing these effects must be part of the model. Some of the required equations are obtained by applying the balance principle to the individual chemical species in the environment. Specifically, if the principle be applied to the ith chemical species in the environment, there results

$$ds_i/dt = D(s_{if} - s_i) - F_i \qquad (8)$$

for the stirred fermentor fed by a stream containing the ith species at concentration s_{if}, and

$$(\partial s_i/\partial t) + v(\partial s_i/\partial z) = \mathscr{D}\ (\partial^2 s_i/\partial z^2) - F_i \qquad (9)$$

for a tubular fermentor, in which s_i is the *concentration* of the ith chemical substance in the culture. Again, we assume that the axial diffusivities for all species in the environment are the same, and the same as those of biomaterial and individual cells. In Eqs. (8) and (9), F_i is the rate (per unit volume of culture) at which the ith environmental substance is *consumed* by growth and multiplication; F_i is *negative* if that substance is *produced* by growth and multiplication. In effect, Eqs. (8) and (9) serves to *define* the *consumption rates* F_i.

In general, there will be one equation of the form of Eq. (8) or (9) for *each* substance present in the environment, whether it is a substrate, an inhibitor, a toxin, or a by-product. Suppose there are A such substances. Then we have to write A equations of the form of Eqs. (8) or (9). This can be done most economically by introducing vector-matrix notation:

$$\mathbf{s} = [s_1\ s_2\ \ldots\ s_A]$$

$$\mathbf{F} = [F_1\ F_2\ \ldots\ F_A]$$

Then the A equations (8) become

$$d\mathbf{s}/dt = D(\mathbf{s}_f - \mathbf{s}) - \mathbf{F} \qquad (10)$$

and the A equations (9) become

$$(\partial \mathbf{s}/\partial t) + v(\partial \mathbf{s}/\partial z) = \mathcal{D}\,(\partial^2 \mathbf{s}/\partial z^2) - \mathbf{F} \tag{11}$$

In writing the foregoing equations for environmental substances, it was assumed that there is no *transfer* of such substances between the culture and any other phase (such as a gas phase) present in the reactor. Obviously, this is not true for substances such as oxygen and carbon dioxide, and so Eqs. (8) and (9) or their vector forms need to be generalized to account for transfer. We shall consider this briefly below.

The equations given above, while useful, do not allow us to attack many problems because they are *incomplete;* in other words, there are more variables than there are equations. No further equations can be obtained by application of balance principles, so we seek constitutive equations. In this case, the constitutive equations must describe the dependence of the unknown quantities μ, ν, F_1, F_2, ..., F_A on the constitution or *state* of the fermentation system. In unstructured models, such as we are using here, the only variables of state are x, n, and $s_1, s_2, \ldots, s_A$ (plus such things as temperature and pressure, which we assume to be constant). Hence, the constitutive relations will, in general, be of the functional form

$$\mu = \mu(x, n, \mathbf{s})$$
$$\nu = \nu(x, n, \mathbf{s}) \tag{12}$$
$$\mathbf{F} = \mathbf{F}(x, n, \mathbf{s})$$

Specification of these constitutive equations gives a number of equations equal to the number of unknowns.

Before considering specific examples of unstructured models, it is necessary to say a few words about the mathematical problems posed by the foregoing equations. Consider first the problem of the stirred fermentor. If this operates at steady state, the pertinent equations reduce to the set of algebraic equations

$$0 = D(x_f - x) + \mu x$$
$$0 = D(n_f - n) + \nu x \tag{13}$$
$$\mathbf{0} = D(\mathbf{s}_f - \mathbf{s}) - \mathbf{F}$$

together with Eqs. (12), which must still be used. These form a well-defined mathematical problem. On the other hand, if we consider the unsteady state, then Eqs. (4), (6), and (10), together with Eqs. (12), do not form a well-defined mathematical problem; we have to specify in addition *initial conditions*, that is, the state of the system at some arbitrarily chosen instant of time.

Consider next the tubular fermentor. In the steady state, the pertinent equations reduce to

$$v(dx/dz) = \mathcal{D}\,(d^2x/dz^2) + \mu x \tag{14}$$

$$v(dn/dz) = \mathcal{D}\,(d^2x/dz^2) + \nu n \tag{15}$$

$$v(ds/dz) = \mathcal{D}\,(d^2s/dz^2) - \mathbf{F} \tag{16}$$

together with Eqs. (12) which must still be used. This set of differential equations does not form a completely defined problem; we must specify in addition *boundary conditions*. The appropriate conditions appear to be those of Danckwerts (1953):

$$\lim_{z \downarrow 0}\;[x_f - x + (\mathcal{D}/v)(dx/dz)] = 0$$

$$\lim_{z \downarrow 0}\;[n_f - n + (\mathcal{D}/v)(dn/dz)] = 0 \tag{17}$$

$$\lim_{z \downarrow 0}\;[\mathbf{s}_f - \mathbf{s} + (\mathcal{D}/v)(d\mathbf{s}/dz)] = \mathbf{0}$$

$$\lim_{z \uparrow L}\;(dx/dz) = 0$$

$$\lim_{z \uparrow L}\;(dn/dz) = 0 \tag{18}$$

$$\lim_{z \uparrow L}\;(d\mathbf{s}/dz) = \mathbf{0}$$

where the subscript f denotes feed conditions, and L is the length of the fermentor. Conditions (17) say in effect that there is no diffusion *out* of the fermentor at its entrance; the significance of conditions (18) is not so easily stated. A discussion of the Danckwerts boundary condition is given by e.g., Aris (1965). If there is no axial dispersion ($\mathcal{D}$

vanishes), Eqs. (15) and (16) become identical to those of an unsteady state batch fermentor; the "time" is then the residence time, z/v. In this case, the boundary conditions at $z = L$ do not apply, and those at $z = 0$ reduce to $x = x_f$, $n = n_f$, $s = s_f$.

In the unsteady state, the Danckwerts boundary conditions still hold (but with the ordinary derivatives replaced by partial derivatives), but one must specify in addition the *initial distribution* of x, n, and s along the fermentor length.

We shall now consider some specific examples that have appeared in the literature; these arise by specifying the functions appearing in Eqs. (12).

1. Monod's Model and Its Modifications

Monond (1950) assumed that the rate of growth was limited by the availability of a single substrate. For constitutive equations, he assumed that μ showed a Briggs–Haldane (Michaelis–Menten) form and that the consumption rate of the limiting substrate was proportional to the specific growth rate. Hence, the equations of his model are two in number, and are

$$dx/dt = D(x_f - x) + [\mu_m x s/(K + s)] \qquad (19)$$

$$ds/dt = D(s_f - s) - (1/Y) \cdot [\mu_m x s/(K + s)] \qquad (20)$$

for a stirred fermentor, or

$$(\partial x/\partial t) + [v(\partial x/\partial z)] = \mathscr{D}\ (\partial^2 x/\partial z^2) + [\mu_m x s/(K + s)] \qquad (21)$$

$$(\partial s/\partial T) + [v(\partial s/\partial z)] = \mathscr{D}\ (\partial^2 s/\partial z^2) - (1/Y) \cdot [\mu_m x s/(K + s)] \qquad (22)$$

for the tubular fermentor. This model contains three biologically significant constants: μ_m, the maximum specific growth rate; K, the so-called *Michaelis constant;* and Y, the *yield coefficient*. In Monod's model, it is assumed that population density (n) does not affect the growth rate; the model says nothing about multiplication rate. The model is mathematically complete, however, since there are two unknowns $(x$ and $s)$ and two equations.

Equations (19) and (20) apply to a single vessel without recycle and with no organisms in the feed if one sets $x_f = 0$. They apply to the nth vessel of a cascade if one takes x, s, D to be the values for the nth vessel, and x_f, s_f to be the values for the $(n - 1)$st tank. The latter problem has been treated by many authors; see, e.g., Herbert (1964), Fencl

(1964), Powell and Lowe (1964), and Aiba *et al.* (1965). Finally, the equations apply to a single vessel with recycle. In this case, x_f and s_f will depend on x and s, respectively; the exact dependence is determined by the specific way that recycle is returned. Treatments of this problem are given by Herbert (1961) and Fencl (1966).

In the rest of this review, we shall consider only the single stirred vessel, without recycle. It should be fairly obvious how any specific model is to be adopted to other cases.

Monod's model was developed from data on steady-state continuous propagation in stirred vessels, and in many cases, the model "fits" data for that specific situation. That it has its shortcomings is evident, however, if one applies it to a batch growth situation ($D = 0$). One finds then that the model predicts no lag phase nor can it predict a phase of decline; basically, the model predicts a phase of exponential growth followed by a stationary phase when the substrate is exhausted. When the model is applied to transient phenomena in continuous flow stirred tanks, it predicts either washout [if $D > \mu_m s_f/(K + s_f)$] or a stable steady state [if $D < \mu_m s_f/(K + s_f)$], but the approach to either washout or a steady state is predicted to take place *without oscillations* (Tsuchiya et al., 1966).

It is often found that Monod's model with a constant yield coefficient will not fit experimental data. Herbert (1958) and others showed that this could be due to the occurrence of *maintenance reactions;* that is, not all substrate consumed is used for growth, as is assumed in Monod's model. Herbert proposed to include a negative term, independent of substrate concentration, in Monod's expression for μ; his equations for a stirred vessel were thus

$$dx/dt = D(x_f - x) + [\mu_m xs/(K + s)] - \mu_c x \qquad (23)$$

$$ds/dt = D(s_f - s) - (1/Y)[\mu_m xs/(K + s)] \qquad (24)$$

where μ_c is the specific *maintenance rate.* It is assumed that the four quantities μ_m, μ_c, K, and Y are constants. If we define the *apparent yield coefficient* as that value Y_a which makes Monod's equation fit data, then according to Herbert's model, Y_a will not be constant but will satisfy

$$Y_a = Y[D/(D + \mu_c)] \qquad (25)$$

for the steady state of the stirred vessel (no organisms in the feed). Thus, the apparent yield coefficient will increase toward a maximum value of Y as the dilution rate is increased. The foregoing relation was confirmed experimentally by Marr et al. (1963).

According to Herbert's model of maintenance, the organism uses up part of its own biomass in order to maintain its viability and ability to function; it is assumed that no exogeneous substrate is consumed for maintenance. Various authors have pointed out that maintenance may in fact consume exogenous substrate. Variations of Herbert's model that account for this have the general form

$$dx/dt = D(x_f - x) + [\mu_m xs/(K + s)] - \mu_c x \tag{26}$$

$$ds/dt = D(s_f - s) - (1/Y)[\mu_m xs/(K + s)] - k_s x \tag{27}$$

Unfortunately, these equations contain the undesirable feature that they predict that substrate will be consumed even when none is present; this is seen simply by setting s and D equal to zero in Eq. (27). Hence, these equations should not be used.

An alternate model of maintenance in which exogenous substrate is consumed was advanced by Ramkrishna et al. (1966). The equations of this model are

$$dx/dt = D(x_f - x) + [\mu_m xs/(K + s)] - [K' \mu_c x/(K' + s)] \tag{28}$$

$$ds/dt = D(s_f - s) - (1/Y)[\mu_m xs/(K + s) - a[K' \mu_c xs/(K' + s)] \tag{29}$$

where K' and a are additional constants. If $s \ll K'$, these equations reduce to Herbert's model of maintenance. For large values of s, they reduce to the expression postulated by Marr et al. (1963). The rationalization for Eqs. (28) and (29) is given in Ramkrishna et al. (1966).

Andrews (1968) noted that Monod's model sometimes fails at moderate to high substrate concentrations because the substrate actually begins to inhibit growth. He proposed to model this by using an expression for substrate inhibition from enzyme kinetics; his modification of Monod's model for a stirred vessel can be written as

$$dx/dt = D(x_f - x) + \{\mu_m s/[K + s + (s^2/K')]\} \tag{30}$$

$$ds/dt = D(s_f - s) - (1/Y)\{\mu_m s/[K + s + (s^2/K')]\} \tag{31}$$

One sees that at large values of s, the specific growth rate will *decrease* as s^{-1}, whereas at small values of s, it increases in proportion to s.

In some fermentations, different substrates may be limiting at different stages of the fermentation. Megee (1970) found such a situation in his study of the growth of *Lactobacillus casei*. He modeled this situation by replacing the single-substrate Briggs–Haldane expression in Monod's model with a double-substrate expression; he had also to write material balance equations for both substrates. His model was

$$dx/dt = D(x_f - x) + \{\mu_m x s_1 s_2/[(K_1 + s_1)(K_2 + s_2)]\} \tag{32}$$

$$ds_1/dt = D(s_{1f} - s_1) - (1/Y_1)\{\mu_m x s_1 s_2/[(K_1 + s_1)(K_2 + s_2)]\} \tag{33}$$

$$ds_2/dt = D(s_{2f} - s_2) - (1/Y_2)\{\mu_m x s_1 s_2/[(K_1 + s_1)(K_2 + s_2)]\} \tag{34}$$

where the subscripts 1 and 2 denote substrates 1 and 2, respectively. In enzyme kinetics, one cannot, in general, factor the denominator of the Briggs–Haldane expression as has been done in Eqs. (32)–(34), since the general expression for the specific growth rate contains three *independent* constants in the denominator (see e.g., Mahler and Cordes, 1966):

$$\mu = \frac{\mu_m s_1 s_2}{K_2 A + K_2 s_1 + K_1 s_2 + s_1 s_2}$$

where A is a constant not in general equal to K_1. However, if K_1 and K_2 are approximately equal, as they must be if there is to be a possibility for either substrate to be limiting, then $A \approx K_1$, and the denominator factors as assumed in Eq. (32).

An interesting case of double-substrate kinetics arises if one of the limiting substrates is a dissolved gas that is transferred to the abiotic phase of a culture from a gas phase. In this case, Eqs. (32)–(34) must be modified for mass transfer; they become

$$dx/dt = D(x_f - x) + \{\mu_m x s_1 s_2/[(K_1 + s_1)(K_2 + s_2)]\} \tag{35}$$

$$ds_1/dt = D(s_{1f} - s_1) - (1/Y_1)\{\mu_m x s_1 s_2/[(K_1 + s_1)(K_2 + s_2)]\} \\ + K_L a(s_1^* - s_1) \tag{36}$$

$$ds_2/dt = D(s_{2f} - s_2) - (1/Y_2)\{\mu_m x s_1 s_2/[(K_1 + s_1)(K_2 + s_2)]\} \tag{37}$$

In these equations, s_1 is the concentration of dissolved gas, s_1^* is the concentration that would be in equilibrium with the gas phase, and $K_L a$ is the volumetric mass transfer coefficient (Finn, 1954; Aiba et al., 1965). In batch culture $(D = 0)$ with the second substrate not rate-limiting $(s_2 \gg K_2)$ and $s_1 \ll s_1^*$, we see that growth will be *linear* (i.e., growth rate independent of biomaterial concentration):

$$dx/dt \approx Y_1 K_L a s_1^* \qquad (38)$$

This situation is fairly common in situations where oxygen is the limiting substrate.

Monod's model and the various modifications of it that we have described above focus attention on increase of biomaterial and consumption of substrates. In many cases, of course, one is interested in the formation of products other than biomass during growth. Leudeking and Piret (1959) advanced a model for product formation which, when Monod's model of growth is assumed, can be written as

$$dx/dt = D(x_f - x) + [\mu_m x s/(K + s)] \qquad (39)$$

$$ds/dt = D(s_f - s) - (1/Y)[\mu_m x s/(K + s)] \qquad (40)$$

$$dp/dt = D(p_f - p) + \alpha[\mu_m x s/(K + s)] + \beta x \qquad (41)$$

where p is the concentration of the product in the culture. The first two equations are Monod's model, of course; the third equation is due to Luedeking and Piret. The term involving the constant α is the *growth-associated* rate of product formation; the term involving the constant β is the *nongrowth-associated* rate of product formation. This term involves certain difficulties and raises questions about the source of material that is eventually converted into product. If that source is the cellular mass itself—that is, product is formed as a by-product of maintenance—then maintenance terms similar to that used by Herbert should be added to Eq. (39). If this is not done, the model would predict that product formation would go on indefinitely in a batch culture. On the other hand, if the source for the product is the growth rate-limiting substrate, a term should be added to Eq. (40) to express this. If such is not done, the model would predict that product formation would go on even when all substrate for it is exhausted.

Finally, consider the case where product is formed from an exogenous substrate not limiting for growth as well as by a growth associated process. A model for this could be written as

$$dx/dt = D(x_f - x) + [\mu_m x s_1/(K + s_1)] \tag{42}$$

$$ds_1/dt = D(s_{1f} - s_1) - (1/Y_1) [\mu_m x s_1/(K + s_1)] \tag{43}$$

$$ds_2/dt = D(s_{2f} - s_2) - [kxs_2/(k_2 + s_2)] - (1/Y_2) [\mu_m x s_1/(K + s_1)] \tag{44}$$

$$dp/dt = D(p_f - p) + \alpha[\mu_m x s_1/(K + s_1)] + \beta[kxs_2/(K_2 + s_2)] \tag{45}$$

One sees that in this case product formation would stop when both substrates are exhausted. An interesting feature of this model is that, depending on conditions, it would exhibit either the Type 1 pattern or the Type 3 pattern of Gaden's "typical" industrial fermentations (1959).

2. Extensions of Monod's Model: Rate Processes Dependent on Products of Growth

In the foregoing models, all substances affecting growth were assumed to be supplied in some feed stream entering the fermentor. We shall now consider models for cases where products of growth and metabolism affect growth and its attendent processes.

A fairly typical example of such a case is that where some product of growth inhibits growth. There are many plausible ways to model such inhibition, and we shall consider three. First, suppose that inhibition is similar to the noncompetitive inhibition of enzyme kinetics (see e.g., Mahler and Cordes, 1966). In this case, we can write the model as

$$dx/dt = D(x_f - x) + \{\mu_m x s/[(K + s)(K_i + i)]\} \tag{46}$$

$$ds/dt = D(s_f - s) - (1/Y) \{\mu_m x s/[(K + s)(K_i + i)]\} \tag{47}$$

$$di/dt = D(i_f - i) + \alpha\{\mu_m x s/[(K + s)(K_i + i)]\} + \beta x \tag{48}$$

in which i is the concentration of the inhibitor in the culture. Inspection of the third of these equations shows that the inhibitor is assumed to form both by a growth-associated and a nongrowth-associated process.

A second class of models of inhibition can be found by assuming that a product of growth actually causes destruction of biomaterial. In the absence of specific kinetic data, the simplest plausible models for this would seem to involve a second-order reaction between the toxin and the biomass. Ramkrishna et al. (1967) considered a number of models of this class; the simplest is expressible as

$$dx/dt = D(x_f - x) + [\mu_m xs/(K + s)] - kxi \qquad (49)$$

$$ds/dt = D(s_f - s) - (1/Y)[\mu_m xs/(K + s)] \qquad (50)$$

$$di/dt = D(i_f - i) + \alpha[\mu_m xs/(K + s)] \qquad (51)$$

where i is the concentration of the toxin in the culture. Special cases of these equations appeared earlier; these were proposed for describing the ethanol fermentation (Rahn, 1929; Klem, 1933). The third of the foregoing equations assumes that toxin production is growth associated. Other assumptions are possible and are discussed in detail in the paper cited. Perhaps Eqs. (49)–(51) should not be said to comprise an inhibitor model; "toxin model" would be a better description. The principal difference between the two models just given is their predictions relating to long-term behavior in batch culture. The competitive inhibition model predicts a stationary phase of indefinite duration; the toxin model predicts a phase of decline, exhibiting a logarithmic order of death, upon exhaustion of substrate. It should be noted that the inhibitor model, Eqs. (49)–(51) can be used even if the identity of the inhibitor is not known. This is because the inhibitor concentration can be eliminated from the equations.

As a third example of inhibition phenomena, consider the case where some product of growth accelerates degradative processes that would eventually lead to death of cells. Assume, however, that the cell has a mechanism involving diversion of substrate from the growth process which can arrest the degradative processes. Megee (1970) conceived these ideas in order to explain his data on the effect of pH on the growth of *Lactobacillus casei;* the substance that accelerated degradative processes was, in his case, the hydrogen ion. The equations of the model are

$$dx/dt = D(x_f - x) + [\mu_m xs/(K + s)] - [kxi^2/(K'i + s)] \qquad (52)$$

$$ds/dt = D(s_f - s) - (1/Y)[\mu_m xs/(K + s)] - \beta[kxis/(K'i + s)] \quad (53)$$

$$di/dt = D(i_f - i) + \alpha[\mu_m xs/(K + s)] \quad (54)$$

These equations are similar to the maintenance model of Ramkrishna et al., 1966; they differ in that the rate of consumption of substrate for maintenance depends on the concentration, i, of the substance that accelerates degradation.

An entirely different class of models having some interesting properties arises if one considers that organisms must somehow alter or "condition" substances present in their abiotic environment before they can assimilate them. In the simplest case, we can assume that the organisms first break down the substrate precursor into a substance, e.g., starch into glucose, by some enzymic reaction, and then utilize the product in the usual way. If s' is the concentration of the substrate in the culture, the simplest model of this process assumes the form

$$dx/dt = D(x_f - x) + [\mu_m xs'/(K + s')] \quad (55)$$

$$ds/dt = D(s_f - s) - [kxs/(K' + s)] \quad (56)$$

$$ds'/dt = D(s_f' - s') - (1/Y)[\mu_m xs'/(K + s')] + \alpha[kxs/(K' + s)] \quad (57)$$

where α is the amount of substrate formed per unit amount of substrate precursor consumed. In batch systems, the foregoing model will predict the existence of a lag phase, because the substrate precursor must be broken down into the substrate before growth can commence at maximal rate.

3. Critique of Unstructured Models

Clearly, the list of unstructured models given above does not exhaust the possibilities for forming such models. One can easily imagine cases where various elements of the models must be combined to form new models. One can also imagine cases where entirely new phenomena come into play. It seems unlikely that there is any general model of microbial growth that will cover all the possibilities, so that no further examples of unstructured models will be given here.

How "good" are the unstructured models we have been discussing? Many of those described above have in fact been used to fit experimental data, but none of them have been subjected to the full testing process described in Section I. Hence, we can say that such models are useful in the sense that they will describe specific situations, but that the full range of their utility is unknown.

At this point, it is helpful to recall the biological principles stated in Section II to be of importance in the formulation of models. According to the second of those principles, the phenotypic expression of a genotype depends not only on the genotypic constitution, but also on the history of environments seen by the organism. If that history has been changing, we expect the phenotypic expression of the genotype to be changing.

Now unstructured models have no way of accounting for changes in the phenotype. The variables x and n are measures of the *quantity* of biomaterial in a culture, but they tell us nothing about the *quality*, or *physiological state* of that biomaterial. Hence, we can expect on *a priori* grounds that unstructured models will be inadequate to handle situations in which the state of a population's environment is changing at an appreciable rate. In order to handle such situations, a quantitative measure of the physiological state of a population is needed. Models using such a concept are called *structured* (Tsuchiya *et al.*, 1966) because they regard biomaterial as a structured, changeable substance.

B. A General Formulation of Structured Models

We do not know the "right" way to formulate the concept of structure of a population of living organisms. We can, however, proceed in a way that seems capable of accommodating the current views of molecular biology. Models so formulated, while not perfectly rigorous, will nevertheless be much more flexible and plausible than the unstructured models discussed before.

Hence, let us assume that the physiological state of a microbial population is specified to a sufficient degree of accuracy by the concentrations $y_1, y_2, \ldots, y_B$ of the various materials that comprise cells. The vector **y**

$$\mathbf{y} = [y_1 \ y_2 \ \ldots \ y_B]$$

will then be called the *physiological state vector* of the population. The elements y_j of **y** will be the concentration of various proteins, nucleic acids, structural material, etc., in cells. It may be possible in many cases to lump substances together in this regard; thus y_j might represent the concentration of cell wall material, y_2 the concentration of RNA, etc., even though it is recognized that cell walls are not a homogeneous chemical substance and that there are various kinds of RNA.

The foregoing definition of the physiological state vector is due to Fredrickson *et al.* (1967), although their definition was given for an individual organism rather than for a population. It is implicit in the much earlier work of Hinshelwood (1946) and in the recent book of Dean and Hinshelwood (1966). It is surely not the most general definition of physiological state, since among other things, it does not allow one to account for the geometrical structure of cells, nor for the dependence of diffusional processes upon such structure.

Nevertheless, we shall assume that **y**, the physiological state vector of the biotic part of a culture, and **s**, the state vector of the abiotic part of a culture, determine the rates of all biological processes occurring in that culture. Evidently, such an assumption is consistent with biological principles stated above: a changing **s** vector will cause the **y** vector to change, and that by definition is what is meant by a changing phenotypic expression of the genotype.

The only rate processes that we can model by the definitions of states adapted are homogeneous chemical reactions. Of course, intracellular biochemical reactions are rarely homogeneous, but in the view that neglects the segregation of life structure and function into discrete units, heterogeneous reactions are modeled as homogeneous ones.

Suppose that R biochemical reactions occur among the A components of the abiotic phase of the culture and the B components of the biotic phase of the culture. Let r_k be the rate of the kth such reaction, per unit volume of culture. If A_{ki} is the amount of the ith component of the abiotic phase produced by one unit of the kth reaction, then the volumetric rate of production of that component must be

$$\sum_{k=1}^{R} A_{ki}\, r_k$$

Similarly, if B_{kj} be the amount of the jth component of the biotic phase produced by one unit of the kth reaction, then the volumetric rate of production of that component must be

$$\sum_{k=1}^{R} B_{kj}\, r_k$$

The A_{ki} and the B_{kj} are the *stoichiometric coefficients* of the biochemical reactions (Aris, 1965).

The general differential equations of structured models can now be obtained by applying the principle of conservation of biochemical species to each of the components of the culture. For a stirred-tank fermentor, this leads to

$$dy_j/dt = D(y_{jf} - y_j) + \sum_{k=1}^{R} B_{kj} r_k \qquad j = 1, 2, \ldots, B \qquad (58)$$

for the components of the biotic phase, and to

$$ds_i/dt = D(s_{if} - s_i) + \sum_{k=1}^{R} A_{ki} r_k \qquad i = 1, 2, \ldots, A \qquad (59)$$

for the components of the abiotic phase. If we define $R \times A$ and $R \times B$ matrices **A** and **B** by

$$\mathbf{A} = \begin{bmatrix} A_{11} & A_{12} & \ldots & A_{1A} \\ A_{21} & A_{22} & \ldots & A_{2A} \\ \cdot & \cdot & & \cdot \\ \cdot & \cdot & & \cdot \\ \cdot & \cdot & & \cdot \\ A_{R1} & A_{R2} & \ldots & A_{RA} \end{bmatrix}, \quad \mathbf{B} = \begin{bmatrix} B_{11} & B_{12} & \ldots & B_{1B} \\ B_{21} & B_{22} & \ldots & B_{2B} \\ \cdot & \cdot & & \cdot \\ \cdot & \cdot & & \cdot \\ \cdot & \cdot & & \cdot \\ B_{R1} & B_{R2} & \ldots & B_{RB} \end{bmatrix} \qquad (60)$$

then Eqs. (58) and (59) are more concisely written as

$$d\mathbf{y}/dt = D(\mathbf{y}_f - \mathbf{y}) + \mathbf{rB} \qquad (61)$$

$$d\mathbf{s}/dt = D(\mathbf{s}_f - \mathbf{s}) + \mathbf{rA} \qquad (62)$$

where $\mathbf{r} = [r_1 \, r_2 \, \ldots \, r_k]$ is the vector of reaction rates. In addition, we still have the number balance equation

$$dn/dt = D(n_f - n) + \nu n \qquad (63)$$

The set of Eqs. (61)–(63) serves to *define* the vector of reaction rates, **r**, the stoichiometric matrices, **A** and **B**, and the specific multiplication rate, ν.

As in the case of unstructured models, the model has more variables than the number of equations so far available. The missing equations are constitutive equations for the reaction rate vector and the specific multiplication rate. According to the hypothesis stated above, these two quantities are determined by the state of the biotic material present and by the state of the abiotic environment. In addition, the population density may be involved, so that in general, the requisite constitutive equations will be of the form

$$\mathbf{r} = \mathbf{r}(\mathbf{y}, \mathbf{s}, n) \tag{64}$$

$$\nu = \nu(\mathbf{y}, \mathbf{s}, n) \tag{65}$$

The stoichiometric matrices are assumed to be constant.

It is important to note that the vector of reaction rates and the specific multiplication rate at a given time are assumed to depend on states (**y** and **s**) and population density *at that time*, and *not* on states at earlier times. This does not contradict the principle that phenotypic expression of a genotype depends on the past history of environments, because the current state of the biomaterial is determined by the past history of the environment.

We shall shortly consider some structured models which have appeared in the literature. Before doing so, however, it is necessary to see how unstructured models are related to structured models. We shall also consider the concepts of balanced and unbalanced growth, since these have a bearing on the validity of unstructured models.

C. Balanced and Unbalanced Growth

In order to introduce the concept of balanced growth, it is helpful to put the balance equations for the components of the biotic material into a somewhat different form. It will be recalled that the rate of production of the *j*th component was given by

$$\sum_{k=1}^{R} B_{kj} r_k$$

per unit volume of culture. Hence, μ_j, the rate per unit amount of the *j*th substance, will be

$$\mu_j = \frac{1}{y_j} \sum_{k=1}^{R} B_{kj} r_k \tag{66}$$

and we call μ_j the *specific synthesis rate* for the jth component of the biotic phase. Equation (58) can then be written as

$$dy_j/dt = D(y_{jf} - y_j) + \mu_j y_j \qquad (67)$$

which parallels the expression [Eq. (4)] given earlier for the biomaterial concentration. The set of quantities μ_j ($j = 1, 2, \ldots, B$) depends on **y**, **s**, and n, of course.

The concept of *balanced growth* was first given explicit statement by Campbell (1957); he says that "growth is *balanced* over a time interval if, during that interval, every extensive property of the growing system increases by the same factor." Thus, growth is balanced during time t to $t + dt$ if during that interval

$$\mu_1 = \mu_2 = \ldots = \mu_B = \nu \qquad (68)$$

since only if the foregoing equation is true will the extensive properties of a growing system ($y_1, y_2, \ldots, y_B, n$) all increase by the same factor.

If growth is not balanced it is *unbalanced*.

Let us examine briefly a number of common growth conditions to see if they are cases of balanced growth. Consider first a single stirred vessel, operating at steady state, with no biomaterial in the feed. According to the latter hypothesis, $y_{1f}, y_{2f}, \ldots, y_{Bf}$, and n_f all vanish. If operation is steady then

$$dy_1/dt = dy_2/dt = \ldots = dy_B/dt = 0$$

and

$$dy_n/dt = 0$$

Hence, Eq. (67) requires that

$$D = \mu_1 = \mu_2 = \ldots = \mu_B$$

whereas Eq. (63) requires that $D = \nu$. One sees that the conditions given guarantee that Eq. (68) will be satisfied, so that growth is balanced.

Consider next the case of a single vessel operating at steady state, but with a constant fraction (ϕ) of the effluent stream recycled. In this case

$$y_{jf} = \phi\, y_j$$

$$n_f = \phi\, n$$

and so application of the condition that operation be steady again leads to Eq. (68), and growth is balanced.

Suppose, however, that the vessel under consideration is the Nth one in a series of equal-volume vessels; in this case, y_{jf} and n_f are the values of y_j and n in the $(N-1)$st vessel. In the steady state, we have

$$\mu_j = D[1 - (y_{jf}/y_j)]$$

$$\nu = D[1 - (n_f/n)]$$

Inasmuch as the ratios y_{jf}/y_j are not necessarily the same for all components of the biotic phase, nor are they necessarily equal to n_f/n, we see that such growth is not balanced.

Finally, consider batch growth $(D = 0)$. We see that such growth will be balanced if

$$\frac{d\ln y_1}{dt} = \frac{d\ln y_2}{dt} = \cdots = \frac{d\ln y_B}{dt} = \frac{d\ln n}{dt}$$

If the common value of these derivatives is constant, then growth is exponential as well as balanced. Fredrickson et al. (1967) showed that if batch growth is balanced, it *must* be exponential; this contradicts Campbell's statement that balanced growth and exponential ("logarithmic") growth are "independent." Perret (1960) emphasizes the point that exponential batch growth need *not* be balanced, and that is of course quite true.

Qualitatively, one expects to obtain balanced growth when the environment of a population has been constant for a sufficient length of time for cellular adaption processes to complete themselves. In the first two cases cited above, one sees that cells and their ancestors have seen the same environment for an indefinite time in the past, so that it is not surprising that these are balanced growth situations. In the third case, however, cells receive an environmental "shock" when they are transferred from one vessel to the next, because of the discontinuity in concentration prevailing between a feed stream and the contents of a perfectly stirred vessel. Hence, such growth cannot be balanced. In batch growth, the environment is continually changing, so that batch growth can never be *exactly* balanced. However, the relative change in environmental state during the early stages of growth may be slight indeed, so that the population may adapt to the environment (and growth be essentially balanced), provided that no rate-limiting factor comes into play before such adaptation is complete.

The concepts of balanced and unbalanced growth can be applied to the problem raised earlier; vis., how good are unstructured models of microbial growth? First, however, one needs to see the relation between structured and unstructured models of microbial growth.

Fredrickson *et al.* (1967) point out that the biomaterial concentration, x, must be a *linear and homogeneous* function of the elements y_j of the physiological state vector. Thus,

$$x = \mathbf{ya} \qquad (69)$$

where **a** is the column vector *(B × 1 matrix)* whose elements a_j are the weight factors for the y_j in the definition of x. The differential Eq. (4) for biomaterial concentration then follows from Eq. (61) by post-multiplying it by **a** and defining μ, the specific growth rate, by

$$\mu = [(\mathbf{rB}) \ \mathbf{a}]/\mathbf{ya} \qquad (70)$$

Similarly, the differential Eq. (10) for the vector **s** follows upon putting

$$\mathbf{F} \equiv -\mathbf{rA} \qquad (71)$$

Fredrickson *et al.* (1967) pointed out that these two equations pose a dilemma for *any* unstructured model. On the one hand, the equations show that μ and **F** are functions not only of **s** and n, but *also* of the physiological state vector, **y**. On the other hand, unstructured models contain only enough equations to cover dependence of μ and **F** on **s**, n, and x. Only in the case of balanced growth do unstructured models contain a number of equations equal to the number of variables. Hence, in any situation other than balanced growth, unstructured models are *incomplete*. From this, we conclude that unstructured models cannot be a generally valid representation of microbial growth except in the case of balanced growth.

At this point, one encounters the objection that unstructured models have in fact been used to fit data from situations that are not balanced growth, so that the mathematical deduction just made must be based on faulty premises. To this, it may be replied that the situations where unstructured models fit data are probably not far from balanced growth. For satisfactory modeling of situations in which cells are responding to a considerable environmental shock, we maintain that it will be necessary to introduce some measure of physiological state (and hence of "structure").

D. EXAMPLES OF STRUCTURED MODELS

Here, we can examine in detail only one structured model, that of Williams (1967); this is fairly typical of the kind of models that have been proposed, and illustrates the kind of results that one obtains from such models.

Williams begins by assuming that the biomass of a population is composed of two portions: a synthetic portion and a structural/genetic portion. The synthetic portion collects substrate from the surroundings, assimilates it, and supplies the assimilate to the structural/genetic portion. This latter portion in turn uses the assimilate to form new structural and genetic material. It is assumed that a viable cell has a certain basic amount of structural/synthetic portion always; when this is doubled, the cell divides. Thus, the population density (in a population dividing asynchronously) will be proportional to the concentration of structural/genetic biomass in the culture.

Let y_1 and y_2 be concentrations of the synthetic portion of the biomass and the structural/genetic portion of the biomass, respectively. The concentration of total biomass is simply

$$x = y_1 + y_2$$

whereas the population density is

$$n = \gamma^{-1} y_2 \qquad (72)$$

where γ is the minimal structural/genetic biomass in a viable cell. Thus, there are two biotic substances ($B = 2$). However, it is assumed that there is only one limiting substrate, so that one need consider only one abiotic substance ($A = 1$).

Insofar as reactions are concerned, the model assumes that two occur. The assimilation reaction is taken to be second order: first order in (environmental) substrate and first order in *total*[3] biomass:

$$r_1 = k_1 \, s(y_1 + y_2) = k_1 \, sx \qquad (73)$$

---

[3]Possibly, the model where the rate is $r_1 = k_1 \, sy_1$, would be of interest, as well.

where s is as usual the substrate concentration in the culture. The other reaction is the replication of structural/genetic material, and this is also taken to be second order: first order in synthetic material and first order in structural/genetic material:

$$r_2 = k_2\, y_1\, y_2 \tag{74}$$

As Williams points out, assumption of Briggs–Haldane kinetic expressions would probably produce a quantitatively better model.

With concentrations in mass units, the matrices **A** and **B** become

$$\mathbf{A} = \begin{bmatrix} -1 \\ 0 \end{bmatrix}, \quad \mathbf{B} = \begin{bmatrix} 1 & 0 \\ -1 & 1 \end{bmatrix}$$

so that the equations for a chemostat culture can be written as

$$dy_1/dt = -Dy_1 + k_1\, s(y_1 + y_2) - k_2\, y_1\, y_2 \tag{75}$$

$$dy_2/dt = -Dy_2 + k_2\, y_1\, y_2 \tag{76}$$

$$ds/dt = D(s_f - s) - k_1\, s(y_1 + y_2) \tag{77}$$

The equation for population density is not independent, but follows from Eq. (76) and is

$$dn/dt = \gamma^{-1}(dy_2/dt) = \gamma^{-1}[-Dy_2 + k_2\, y_1\, y_2] = -Dn + k_2\, y_1\, n \tag{78}$$

Williams solved these equations on an analog computer for both batch and continuous operation. For the batch case, he found that his model exhibited the following features, all of them typical of batch growth of most unicellular organisms: (i) A lag phase in which there is growth in size but not much multiplication occurs; (ii) the duration of the lag phase is short if the inoculum is drawn from a culture in exponential growth; it is long if the inoculum is drawn from a stationary phase culture; (iii) exponential growth follows the lag phase (for appropriate initial values of s, y_1, and y_2); (iv) cells are largest during the exponential phase; (v) the composition of cells changes during both the lag and exponential phases, so growth is *not balanced;* (vi) the biomass concentration reaches its asymptotic value before the population density does; and (vii) a stationary phase occurs when nutrient is exhausted, and cells return to a minimum size.

In his discussion of the shortcomings of his model, Williams notes that one of these is the difficulty experienced trying to relate y_1 or y_2 to actual measured components of biomass. Is the synthetic portion of the biomass the pool of small metabolites in the cell or is it these plus ribosomal material? Is the structural/genetic material the total of all cellular macromolecules or is it these minus ribosomal material? Ramkrishna et al. (1966, 1967) considered a number of structured models similar to that of Williams, in that biomaterial was divided into two categories. But in the models of Ramkrishna and his co-workers, different categories were assumed: one was nucleic acids, the other was the rest of the biomass. This type of model also exhibits quite a number of features actually observed in microbial growth.

In retrospect, it appears that while the models of Williams and of Ramkrishna et al. might give a good qualitative correlation of many of the phenomena observed in microbial growth, they probably will not give a good quantitative correlation. The reason is not so much the fact that there are uncertainties about what are the proper constitutive relations to use (e.g., Briggs–Haldane kinetics, second-order reaction, or something else), but that not enough control mechanism has been built into the models. Recent experimental work in molecular biology, as summarized, for instance, by Maaløe and Kjeldgaard (1966) or Watson (1965), suggests that mechanisms more involved and more flexible than those assumed in the foregoing models need to be incorporated in newer models. We suggest that attempts to bring such mechanisms into population models might be most fruitful.

VI. Models for the Growth of Filamentous Organisms

In the discussion of biological principles important in modeling of fermentation processes (Section II, above), it was emphasized that the differing morphology of molds from that of bacteria probably indicates the need for different models. Different models may also be needed for the economically important Actinomycetales; these exhibit growth forms somewhat intermediate to those of bacteria and molds. Evidently, Emerson (1950) was the first worker to recognize this, and he predicted that in batch growth, a submerged mold culture ought not to exhibit exponential growth, but instead, should grow somewhat slower than exponential. In particular, he predicted that biomass concentration should increase in proportion to the *cube* of the time elapsed, instead of in proportion to the exponential of the time. His data, and that of Marshall and Alexander (1960) on oxygen consumption by molds, show that this so-called "cube-root law" does indeed "fit" data better than does the logarithmic relation.

Emerson's prediction seems to be based on the following observations regarding mold growth. Imagine first a mold culture constrained to grow in a single direction (as in, e.g., the tube method of Ryan et al., 1943). The experimental observation here is that, after an initial lag period, the linear rate of advance of the fungal hyphae is a constant. If one assumes that the amount of biomass is proportional to the *length* of the mold culture, then clearly,

$$dx/dt = \text{constant} \tag{79}$$

in this case. Now consider a mold culture growing on a surface, and exhibiting radial growth. The experimental observation here (Fawcett, 1921) is that, after an initial lag phase, the rate of increase of the colony radius is a constant. If one assumes that the amount of biomass is proportional to the *square* of the colony radius (that is, to the *area* of the colony), then clearly,

$$dx/dt = (\text{constant})\, x^{1/2} \tag{80}$$

Finally, consider the case where growth is three-dimensional (i.e., is unconstrained). Then one ought to observe (by extrapolation from the one- and two-dimensional cases) that the rate of increase of colony radius is constant, so that if biomass is proportional to the cube of the colony radius, then

$$dx/dt = (\text{constant})\, x^{2/3} \tag{81}$$

Integration of this equation then yields Emerson's equation

$$x = (x_0^{1/3} + at)^3 \tag{82}$$

where x_0 is the value of x at $t = 0$ (end of the lag phase), and a is a constant expressing the rate of growth.

It is possible to criticize this procedure on two grounds. First, one might question the extrapolation from the one- and two-dimensional cases to the three-dimensional case. It is well known that molds in submerged culture may exhibit "pulpy" or "pelletlike" growth [Kluyver and Perequin (1933), Burkholder and Sinnott (1945), and many others], and one certainly expects Emerson's extrapolation to be valid in the latter case. But what about the former case? Does the concept of a three-dimensional colony expanding at a constant rate apply here? Is not exponential growth to be expected in this case? Second, it is assumed in the three cases cited above that the *concentration* of biomass is uniform in the colony, otherwise, the biomass would not be proportional to length (one-dimension), radius squared (two-dimensions), or radius cubed (three-dimensions). But Plomley's studies (1959) on two-dimensional growth show that the biomass concentration is not uniform throughout the colony, but is greater in the "older" portions of the colony.

These criticisms of Emerson's model should not obscure the basic point of his paper, which was that the special morphology and mode of growth of molds probably require models differing from those used for bacteria. Nor should they cast any doubt on experimental data suggesting the validity of the cube-root law [although that law does not always obtain in mold growth; see e.g., Meyrath (1963)]. Hence, we enquire as to whether there are other models which can lead to the cube-root law.

One such model follows from some work by Yano *et al.* (1961), on pelletlike growth of *Aspergillus niger*. These authors considered the diffusion of oxygen into mold pellets. If it is assumed that pellets are spherical, of uniform biomass density (probably not a good assumption), and consume oxygen at a constant rate Q per unit mass if oxygen is present, then they showed that there is a critical radius r_c such that if r, the pellet radius, exceeds r_c, there will be a zone in the interior of the pellet which is anaerobic. The critical radius is given by

$$r_c = (6 \mathcal{D} s / \rho_m Q)^{1/2} \tag{83}$$

where $\mathcal{D}$ is the effective diffusion coefficient of oxygen within the mold pellet, s the concentration of oxygen in the liquid surrounding the pellet, and ρ_m the bulk density of mycelial biomass in the pellet. Values of $\mathcal{D}$ measured by Yano *et al.* are about an order of magnitude smaller than the molecular diffusivity of oxygen in water; this is as one would expect.

If it is assumed that oxygen diffusion through the pellet is the rate-limiting step in mold growth, one can derive an expression for the rate of increase of biomass in batch culture. Further assumptions made are that biomass synthesis rate is proportional to oxygen uptake rate (yield coefficient $= Y$), that growth is as pellets, and that the population density (n) of pellets does not change. Then clearly

$$dx/dt = QY(1-f)x \qquad (84)$$

where f is the fraction mycelial volume that is anaerobic.

Suppose initially, pellets are so small that $r \ll r_c$. Then none of the mycelium will be anaerobic, $f = 0$, and growth will be exponential for a time. But suppose the pellets become so large that $r \gg r_c$. In this case, the aerobic part of a pellet will be confined to a thin layer, of thickness δ, around the outside of the pellet. The thickness δ is easily calculated from the laws of diffusion, since curvature of the pellet can be neglected; the thickness turns out to be

$$\delta = (1/\sqrt{3})\, r_c \qquad (r \gg r_c) \qquad (85)$$

Hence,

$$f = [(4/3)\pi(r-\delta)^3]/[(4/3)\pi\, n^3] = [1 - (\delta/r)^3]$$

$$\approx 1 - 3(\delta/r) = 1 - \sqrt{3}\,(r_c/r) \qquad (86)$$

Now if all pellets have the same radius, then

$$x = (4/3)\pi r^3 \rho_m n$$

so that

$$1 - f \approx \sqrt{3}\, r_c (4\pi\rho n/3x)^{1/3}$$

and hence

$$dx/dt = \sqrt{3}\, QY r_c\, (4\pi\rho n/3)^{1/3}\, x^{2/3} \qquad (87)$$

This, of course, is the same differential equation that leads to Emerson's cube-root law. Hence, diffusion limitation of growth rate can lead to the cube-root law.

The foregoing consideration, while perhaps correct in certain cases of pelletlike growth, does not come to grips with the basic biological problem; viz., given the observed morphology and mode of growth of molds, do the quantitative aspects of mold growth differ from those of bacteria or other unicellular organisms? What sort of models should one use for mold growth?

It appears that the most basic difference between mold growth and bacterial growth — and a difference that should be reflected in models —

is the occurrence of *differentiation in molds*. Zalokar (1959), Yanagita and his co-workers (1966), Stine (1968a,b), and others have studied this phenomenon in the case of molds grown on a solid substratum. Distinct differences in morphology can be observed in different parts of the mycelium, and cytological differences are correlated with physiological differences. For instance, the appearance of certain enzymes, or the production of certain metabolic products, can be associated with the prevalence of certain morphological characteristics. In batchwise, submerged culture of mold, the appearance of various fermentation products at different times of the culture cycle suggests that differentiation is occuring there, too. Shu (1961) and Terui *et al.* (1967) have proposed models to account for product formation in submerged mold culture, but these models do not introduce the notion of differentiation *explicitly*.

Megee *et al.* (1969) have advanced a tentative model of mold growth in which differentiation is an explicit part of the model. The basic ideas are that mold biomass exists in a number of discrete states (different degrees of differentiation), that these states pass into one another, generally in one direction only, and that the rates of passage are mediated by the availability of substrates. When plausible hypotheses for the various mechanisms involved are made, the resulting model exhibits a number of features unique to the growth of molds (such as the cube-root law of Emerson for batch growth). A great deal of work remains to be done on the model, however, before its conceptual basis is regarded as established. For instance, is the concept of *discrete* states of differentiation appropriate or should we use a *continuous spectrum* of degrees of differentiation. Operationally, how does one recognize a certain state of differentiation? By cytological characteristics? By kinetic or physiological behavior? Is the foregoing concept of differentiation, which is based mostly on observations made on growth on solids, valid for submerged mold culture? These are difficult questions to answer, but it is upon the availability of answers to them that further progress in quantitative modeling of mold growth probably depends.

VII. Concluding Remarks

In the foregoing pages, we have made no attempt to give a comprehensive review of the literature on mathematical models for fermentation processes. An attempt was made to cite selected examples which, one hopes, are representative of models that have been and are being used in fermentation technology. However, we considered it most important to try to state the general principles upon which *any* model must be built, and to suggest directions of research which appear to us to hold the most promise for future progress.

REFERENCES

Aiba, S., Humphrey, A. E., and Millis, N. (1965). "Biochemical Engineering," Chaps. 5 and 6. Academic Press, New York.
Andrews, J. F. (1968). *Biotechnol. Bioeng.* **10**, 707–723.
Aris, R. (1965). "Introduction to the Analysis of Chemical Reactors," Chaps. 2, 7, and 9. Prentice-Hall, Englewood Cliffs, New Jersey.
Bartlett, M. S. (1960). "Stochastic Population Models." Methuen, London.
Bellman, R., Jacquez, J., Kabala, R., and Schwimmer, S. (1967). *Math. Biosci.* **1**, 71–76.
Bharucha-Reid, A. T. (1960). "Elements of the Theory of Markov Processes and Their Applications," Chap. 4. McGraw-Hill, New York.
Burkholder, P. R., and Sinnott, E. W. (1945). *Amer. J. Bot.* **32**, 424–431.
Campbell, A. (1957). *Bacteriol. Rev.* **21**, 263–272.
Cichy, P. T., Ultman, J. S., and Russell, T. W. F. (1969). *Ind. Eng. Chem.* **61** [8], 6–14.
Danckwerts, P. V. (1953). *Chem. Eng. Sci.* **2**, 1–13.
Dean, A. C. R., and Hinshelwood, C. N. (1966). "Growth, Function and Regulation in Bacterial Cells." Oxford Univ. Press, London.
Eakman, J. M., Fredrickson, A. G., and Tsuchiya, H. M. (1966). *Chem. Eng. Prog. Symp. Ser. No. 69*, **62**, 37–49.
Emerson, S. (1950). *J. Bacteriol.* **60**, 221–223.
Fawcett, H. S. (1921). *Univ. Calif. Berkeley Publ. Agr. Sci.* **4**, 183–232.
Fencl, Z. (1964). *In* "Continuous Cultivation of Microorganisms" (I. Málek, K. Beran, and J. Hospodka, eds.) pp. 109–117. Academic Press, New York.
Fencl, Z. (1966). *In* "Theoretical and Methodological Basis of Continuous Culture of Microorganisms" (I. Málek and Z. Fencl, eds.) pp. 69–153. Academic Press, New York.
Finn, R. K. (1954). *Bacteriol. Rev.* **18**, 254–274.
Fredrickson, A. G. (1966). *Biotechnol. Bioeng.* **8**, 167–182.
Fredrickson, A. G., Ramkrishna, D., and Tsuchiya, H. M. (1967). *Math. Biosci.* **1**, 327–374.
Fredrickson, A. G., and Tsuchiya, H. M. (1963). *AIChE J.* **9**, 459–468.
Gaden, E. L., Jr. (1959). *J. Biochem. Microbiol. Technol. Eng.* **1**, 413–429.
Heineken, F. G., Tsuchiya, H. M., and Aris, R. (1967). *Math. Biosci.* **1**, 115–141.
Herbert, D. (1958). *In* "Continuous Cultivation of Microorganisms. A Symposium." pp. 45–52. Publ. House Czech. Acad. Sci., Prague.
Herbert, D. (1961). *In* "Continuous Culture of Micro-Organisms." pp. 21–53. Soc. Chem. Ind., London.
Herbert, D. (1964). *In* "Continuous Cultivation of Microorganisms" (I. Málek, K. Beran, and J. Hospodka, eds.) pp. 23–44. Academic Press, New York.
Hinshelwood, C. N. (1946). "The Chemical Kinetics of the Bacterial Cell." Oxford Univ. Press, London.
Kelly, C. D., and Rahn, O. (1932). *J. Bacteriol.* **23**, 147–153.
Klem, A. (1933). *Hvalrådets Skr.* **7**, 55–91.
Kluyver, A. J., and Perequin, L. H. C. (1933). *Biochem. Z.* **266**, 68–81.
Koch, A. L. and Schaechter, M. (1962). *J. Gen. Microbiol.* **29**, 435–454.
Koževnik, J. (1964). *In* "Continuous Cultivation of Microorganisms" (I. Málek, K. Beran, and J. Hospodka, eds.) pp. 59–68. Academic Press, New York.

Leudeking, R., and Piret, E. L. (1959). *J. Biochem. Microbiol. Technol. Eng.* **1**, 393–412; 431–459.
Levenspiel, O., and Bischoff, K. B. (1963). *Advan. Chem. Eng.* **4**, 95–198.
Maaløe, O., and Kjeldgaard, N. O. (1966). "Control of Macromolecular Synthesis." Benjamin, New York.
Mahler, H. R., and Cordes, E. H. (1966). "Biological Chemistry," Chap. 6. Harper and Row, New York.
Marr, A. G., Nilson, E. H., and Clark, D. J. (1963). *Ann. N. Y. Acad. Sci.* **102**, 536–548.
Marshall, K. C., and Alexander, M. (1960). *J. Bacteriol.* **80**, 412–416.
Megee, R. D., III. (1970). Ph.D. Thesis, University of Minnesota, Minneapolis, Minnesota.
Megee, R. D., III, Kinoshita, S., Fredrickson, A. G., and Tsuchiya, H. M. (1969). Paper presented at 62nd Ann. Meeting, Amer. Inst. Chem. Eng., Washington, D. C., Nov., 1969.
Meyrath, J. (1963). *Antonie van Leeuwenhoek J. Microbiol. Serol.* **29**, 57–78.
Monod, J. (1950). *Ann. Inst. Pasteur Paris* **79**, 390–410.
Moser, H. (1958). "The Dynamics of Bacterial Populations Maintained in the Chemostat." Carnegie Inst. Washington, Washington, D. C.
Odum, E. P. (1959). "Fundamentals of Ecology," 2nd ed., Chap. 3. Saunders, Philadelphia, Pennsylvania.
Perret, C. J. (1960). *J. Gen. Microbiol.* **22**, 589–617.
Plomley, N. J. B. (1959). *Aust. J. Biol. Sci.* **12**, 53–64.
Powell, E. O. (1955). *Biometrika* **42**, 16–44.
Powell, E. O. (1958). *J. Gen. Microbiol.* **15**, 492–511.
Powell, E. O., and Lowe, J. P. (1964). *In* "Continuous Cultivation of Microorganisms" (I. Málek, K. Beran, and J. Hospodka, eds.) pp. 45–53, Academic Press, New York.
Rahn, O. (1929). *J. Bacteriol.* **18**, 207–226.
Ramkrishna, D., Fredrickson, A. G., and Tsuchiya, H. M. (1966). *J. Gen. Appl. Microbiol.* **12**, 311–327.
Ramkrishna, D., Fredrickson, A. G., and Tsuchiya, H. M. (1967). *Biotechnol. Bioeng.* **9**, 129–170.
Rosen, R. (1967). "Optimality Principles in Biology." Plenum, New York.
Ryan, F. J., Beadle, G. W., and Tatum, E. L. (1943). *Amer. J. Bot.* **30**, 784–799.
Schaecter, M., Williamson, J. P., Hood, J. R., Jr., and Koch, A. L. (1962). *J. Gen. Microbiol.* **29**, 421–434.
Shu, P. (1961). *J. Biochem. Microbiol. Technol. Eng.* **3**, 95–109.
Stine, G. J. (1968a). *Can. J. Microbiol.* **13**, 1203–1210.
Stine, G. J. (1968b). *J. Cell Biol.* **37**, 81–88.
Subramanian, G., and Ramkrishna, D. (1970). *Chem. Eng. Sci.*, in press.
Terui, G., Okazaki, M., and Kinoshita, S. (1967). *J. Ferment. Technol.* **45**, 497–503.
Trucco, E. (1965). *Bull. Math. Biophys.* **27**, 285–304, 449–471.
Tsuchiya, H. M., Fredrickson, A. G., and Aris, R. (1966). *Advan. Chem. Eng.* **6**, 125–206.
Van Foerster, H. (1959). *In* "The Kinetics of Cellular Proliferation" (F. Stohlman, ed.) pp. 382–407. Grune & Stratton, New York.
Watson, J. D. (1965). "Molecular Biology of the Gene." Benjamin, New York.
Williams, F. M. (1967). *J. Theoret. Biol.* **15**, 190–207.
Yanagita, T. (1966). *J. Ferment. Technol.* **44**, 313–320.
Yano, T., Kodama, T., and Yamada, K. (1961). *Agr. Biol. Chem.* **25**, 580–584.
Zalokar, M. (1959). *Amer. J. Bot.* **46**, 602–610.

AUTHOR INDEX

Numbers in italics refer to the pages on which the complete references are listed.

A

Abbott, D. D., 344, *360*
Abelson, J., 122, *137*
Abercrombie, M. J., 34, *87*
Abou-El-Seould, M., 150, *160*
Abraham, E. P., 163, 164, 165, 188, 193, 220, 222, 224, *230, 231, 234, 235*
Adamiec, A., 296, *310*
Adams, G-A., 289, *314, 315*
Adams, G. L., 378, *380*
Adams, M. H., 121, 122, 123, *136*
Agnistikova, V. N., 286, 303, *310, 314*
Ahern, L. K., 214, 223, 224, 230, *236*
Aiba, S., 114, *116*, 443, 446, *464*
Ainsworth, G., 412, 415, *417*
Alexander, M., 460, *465*
Alexopoulos, C. J., 25, 26, 27, 86, *87*
Allen, S. E., 318, *340*
Altschul, A. M., 142, *159*
Amsterdam, D., 269, *282*
Anchel, M., 19, *21*
Anderson, R. C., 226, *234*
Anderson, R. E., 103, *117*
Anderson, R. F., 378, *379, 381*
Anderson, R. J., 352, 353, *360*
Andrews, J. F., 444, *464*
Andrews, S. L., 195, 196, 197, 198, 199, 200, 218, 219, 228, 230, *234, 236*
Anonymous, 318, *339*
Anzai, K., 272, 273, *281*
Aposhian, H. V., 133, *136*
Ararashi, S., 170, 171, *236*
Arcos, J. C., *339*
Arcos, M., *339*
Aris, R., 421, 422, 428, 429, 433, 441, 443, 450, 452, *464, 465*
Arison, B. H., 292, *310*
Arnold, M., 378, *379*
Aschner, M., 34, *87*
Asensio, C., 290, *312*
Ash, M. M., 345, 346, *361*
Askew, M. W., 414, *417*
Asseva, I. V., 292, 309, *311, 313*
Atashari, S., 174, *234*

Aube, C., 309, *311*
Ault, R. G., 409, 410, *417*

B

Babayan, G. S., 309, *314*
Bachrach, H. L., 104, *116*
Backer-Dirks, O., 347, *359*
Backus, R. C., 123, *136*
Bailey, F., 292, *312*
Bak, A. L., 32, *87, 89*
Ball, C., 404, *416*
Ballou, C. E., 38, 43, 44, 47, 53, 55, *88*
Banno, I., 27, *87*
Barber, M., 195, *230*
Barbu, E., 31, *88*
Barei, S., 112, *118*
Barlin, G. B., 261, *265*
Barnett, J. A., 30, 31, *87*
Barrett, G. C., 239, *265*
Bartha, R., 319, 320, 324, 325, 327, 329, 330, 332, 337, *339, 340*
Bartholomew, W. H., 114, *116*
Bartlett, M. C., 399, 405, *416*
Bartlett, M. S., 430, *464*
Bauer, S., 43, *88*
Baumgartner, W. E., 292, *311*
Bautista, E. M., 324, *340*
Beadle, G. W., 460, *465*
Beale, A. J., 96, *116*
Beaumont, P. C., 14, *21*
Beereboom, J. J., 243, 245, 247, 251, 252, 255, *265, 266*
Behki, R. M., 278, *281*
Belasco, I. J., 329, *339*
Bellman, R., 422, *464*
Belozersky, A. N., *87*
Benedict, R. G., 1, 5, 6, 9, 10, 11, 12, 13, 18, *21, 23*, 378, *379, 380, 381*
Benham, R. W., 33, 34, *87*
Benjamin, C. R., 379, *380*
Benner, E. J., 222, *230*
Bennet, J. V., 222, *230*
Bentley, R., 287, *311*

Bentley, T., 290, *313*
Berliner, D. L., 108, *118*
Berman, D. S., 352, 353, 355, *361*
Berman, L., 96, *118*
Bharucha-Reid, A. T., 430, *464*
Bhattacharjee, S. S., 34, 37, 44, 47, 49, 51, *87, 88*
Bianco, E. J., 242, *266*
Bibby, B. G., 350, *359*
Bickel, H., 165, *232*
Biesele, J. J., 94, *117*
Bilai, V. I., 287, 288, 290, *311*
Billeter, M. A., 134, 135, *136*
Billings, R. E., 227, *236*
Birch, J. R., 99, 100, *116*
Bird, H. L., Jr., 286, 292, *311, 312*
Bischoff, K. B., 432, 435, *465*
Bishop, C. T., 83, *87*
Bitler, B. A., *265*
Bixler, D., 351, 352, *359*
Björklund, M., 403, *416*
Black, H. R., 228, *234*
Black, P. H., 96, *116*
Blackwood, A. C., 271, 272, 273, 279, *280, 281*
Blackwood, R. K., 239, 243, 245, 247, 251, 252, 253, 255, *265, 266*
Blank, F., 83, *87*
Bloch, D. E., 195, 196, 197, 198, 199, 200, 228, *236*
Blom, R. H., 378, *379, 380*
Bogard, M. O., 378, *381*
Bolden, T. E., 345, 346, *359*
Bolgarev, P. T., 291, *311*
Bondarchuk, A. O., 287, *311*
Boniece, W. S., 229, *230, 236*
Bonvicino, G. E., 247, *265*
Boothe, J. H., 243, 245, 247, 248, 249, 251, 255, 261, *265, 266*
Bordeleau, L. M., 321, 325, 327, 329, 330, *339, 341*
Borgstrom, G., 158, *159*
Borrow, A., 284, 285, 289, 290, 292, 293, 295, 296, 297, 298, 299, 300, 301, 302, 304, 305, 306, 307, *311*
Borst, P., 133, *137*
Borud, A. M., 378, *381*
Bowersox, O. C., 103, *117*
Bowling, C. C., 322, *339*

Bowman, F. W., 220, *230*
Boyd, P. F., 163, *231*
Bradbury, E. M., 37, *87*
Bradle, B., 378, 379, *380*
Bradley, S. G., 150, *159*
Brady, L. R., 1, 5, 6, 9, 10, 11, 12, 13, 16, 17, 19, *21, 22, 23*
Braun, P., 228, *230*
Braun, W., 154, *159*
Bressani, R., 155, *159*
Brian, P. W., 283, 284, 285, 290, 293, 296, 309, *311, 315*
Briggs, D. E., 309, *311*
Brindle, S. A., 95, *117*
Brodie, B. B., *340*
Brodie, J. L., 222, *230*
Bronfenbrenner, J., 122, *136*
Broschard, R. W., 243, *266*
Broshard, R. B., 247, *265*
Brotzu, G., 163, *231*
Brown, C. N., 198, 199, 201, 202, 204, 205, *236*
Brown, F., 102, *116*
Brown, J. C., 286, *311*
Brown, M. E., 309, *311*
Brown, R. E., 156, *160*
Brown, Sheila, 292, 296, 297, 298, 299, 300, 301, 302, 304, 305, 306, 307, *311*
Brudevold, F., 349, 350, 353, 356, 357, *359, 360, 361*
Brunings, K. J., 244, 247, *265, 266*
Bryant, G., 378, *381*
Bryant, J. C., 94, 99, 100, *116*
Buck, R. W., 18, *21*
Buckley, H., 73, 77, 79, *89*
Bullock, F. A., 348, *360*
Bu'Lock, J. D., 18, *21*, 298, *311*
Bunker, H. J., 150, 152, 154, *159*
Bunn, P. A., 171, *231*
Buonocore, M. G., 358, *360*
Burkholder, P. R., 460, *464*
Burlingham, S. K., 309, *311*
Burnett, G. W., 347, *360*
Burns, A. A., 107, 108, 112, 113, *117*
Burrows, T. M., 107, *118*
Burt, A. M., 94, 95, *116*
Burton, H. S., 164, *231*
Burton, K. A., 29, 36, *89*

C

Burton, M. O., 268, *280*
Butler, K., 242, *265*

Cadmus, M. C., 36, *89,* 378, *379, 381*
Cahn, R. D., 102, *116*
Calam, C. T., 402, *416*
Callow, D. S., 107, 109, *117,* 402, *416*
Caltrider, P. G., 267, *280*
Cameron, J., 113, *116*
Campbell, A., 454, *464*
Campbell, J. B., 218, 228, *231*
Campbell, J. J. R., 268, 269, 270, 279, *280, 280, 281*
Canvin, D. T., 287, *312*
Capell, G. H., 113, *116*
Cappellini, R. A., 290, *311*
Capstick, P. B., 94, 95, 108, *116*
Caras, G. J., 317, *340*
Cardinal, E. V., 292, *311*
Carito, S. L., 288, *311*
Carski, R. R., 96, *116*
Carter, H. E., 154, *159*
Carter, R. E., 271, *280*
Cartwright, B., 102, *116*
Cavell, B. D., 284, 286, 287, 289, 292, 295, *311*
Cerecedo, L. R., 150, *161*
Chailakhyan, M. K., 309, *313*
Chamberlain, J. W., 218, 228, *231*
Chang, P. C., 271, 272, 273, 279, *280*
Chang, R. S., 109, *116*
Chang, T.-W., 220, *231*
Chaphery, J. A., 287, *311*
Chapman, W. G., 94, 95, *116*
Chauvette, R. R., 168, 169, 170, 171, 174, 176, 177, 178, 180, 181, 183, 187, 188, 189, 194, 195, 196, 197, 198, 199, 200, 214, 218, 228, 229, *231, 236*
Cheminais, L., 309, *314*
Cheney, L. C., 245, *265*
Cherry, W. R., 94, 107, *116*
Chester, V. E., 284, 285, 290, 293, 296, *311*
Childs, G. G., 405, *416*
Chilton, N. W., 350, 353, 356, 357, *359, 360*
Christofinis, G., 96, *116*
Ciba, S. A., 174, 180, *231*

Cichy, P. T., 436, *464*
Ciegler, A., 378, *379*
Claisse, M. L., 32, *87*
Clare, B. G., 32, *87*
Claridge, C. A., 165, *231*
Clark, D. J., 444, *465*
Clark, G. R., 17, *21*
Cleland, R. E., 309, *311*
Clement, G., 147, *159*
Clemo, G. R., 274, *280*
Clive, D. L., 239, *265*
Closs, G. L., 4, *22*
Cocker, J. D., 188, 208, 214, 219, *231*
Coghill, R. D., 373, 374, 375, 376, *380, 381, 382*
Cohen, A., 346, *360*
Cohen, E. P., 107, *116*
Cohen, J. S., 86, *87*
Cohen, S., 225, *234*
Cohen, S. S., 125, *137*
Coker, J. N., 17, *21*
Cole, S. E., 309, *313*
Collins, J. F., 208, *231*
Conover, L. H., 242, 244, 247, *265, 266*
Convit, J., 144, 148, *160*
Conway, H. F., 272, 273, *281,* 284, 293, 297, 302, 304, 307, *315,* 378, *379, 381*
Cook, F. D., 278, *280, 281*
Coombs, R. R. A., 96, *116*
Cooper, F. P., 83, *87*
Cooper, P. D., 94, 95, *116,* 208, *231*
Cooper, R. C., 146, *160*
Cordes, E. H., 445, 447, *465*
Corke, C. T., 337, *341*
Corman, J., 378, *380, 381*
Cowley, B. R., 188, *231*
Cox, J. S. G., 188, *231*
Cragg, G., 295, 305, *312, 315*
Crane-Robinson, C., 37, *87*
Crast, L. B., Jr., 171, 174, 180, 229, *231*
Crawford, K., 163, *231*
Cremer, H. D., 155, *159*
Crocker, C. K., 378, *379*
Cronquist, A., 1, *21*
Cross, B. E., 283, 286, 287, 289, 295, 296, 300, 302, 305, 309, *311, 312*
Cruikshank, C. N. D., 95, *116*
Culp, H. W., 225, *231*
Currie, J. N., 364, *380, 381*

Curtis, P. J., 283, 284, 285, 286, 287, 289, 290, 293, 296, 300, 305, *311, 312*
Curtis, R. W., 309, *312*

D

Dal Prato, A., 104, 112, *118*
Dalziel, A. M., 292, *311*
Danckwerts, P. V., 441, *464*
Daniels, D. G. H., 330, *339*
Darken, Marjorie, A., 284, 303, 306, *312*
Davies, J. R., 93, *118*
Davis, E. V., 94, 107, *119*
Davis, F. E., 107, 109, 112, *117*
Davis, G. E., 198, *231*
Davis, J. B., 157, *159*
Davis, J. G., 274, *281*
Davison, P. F., 125, 126, *136*
Dean, A. C. R., 451, *464*
Deijs, A., 14, *22*
Delahunt, C. S., 239, 258, 260, 264, *266*
Demain, A. L., 164, 198, *231, 232*
De Moss, R. D., 269, 271, 273, 275, 279, *281*
Dendze-Pletman, B. B., 294, 305, *314*
Dennis, R. W. G., 8, 11, *21*
De Rose, A. F., 294, *312*
Devay, J. E., 307, *316*
DeZeeuw, J. R., 147, *159*
Diekmann, H., 289, *312*
Dietrich, K. R., 303, *312*
Dimler, R. J., 36, *88*
Divekar, P. V., 16, *21*
Djien, K. S., 378, *380*
do Carmo-Sousa, L., 77, 78, *88*
Doerschuk, A. P., 248, 251, *265, 266*
Dolan, M. M., 180, *236*
Donzanti, A., 346, *360*
Doty, P., 31, 32, *88, 89*
Dougherty, T. F., 108, *118*
Downing, A. L., 412, *417*
Dubovaya, L. P., 286, 303, 304, *310, 314*
Duke, W. B., 319, *339*
Dulbecco, R., 96, *119*
Dummett, G. A., 411, *417*
Dworschack, R. G., 378, *380, 382*
Dymovich, V. O., 287, *311*

E

Eachus, A. H., 177, *234*

Eagle, H., 99, 107, *116*
Eagles, B. A., 268, *280*
Eakman, J. M., 428, *464*
Eardley, S., 188, 208, 214, 219, *231*
Earle, W. R., 94, 100, *116*
Edge, W. J., 292, *312*
Edwards, O. E., 278, *281*
Edwards, R. L., 14, 17, 18, *21*
Edwards, T. E., 42, *88*
Eggers, S. H., 220, *232*
Ehrlich, P. R., 139, *159*
Einbinder, J. M., 34, *87*
Elinov, N. P., 37, *87*
Ellis, J. J., 378, 379, *380*
Elsworth, R., 99, 101, 103, 107, 108, 112, 113, 114, 115, *116, 117, 118,* 124, 126, *136*
Elsworthy, G. C., 14, 17, *21*
Emerson, S., 459, *464*
Emmons, L. R., 96, *117*
Emslie, R. D., 347, 353, 354, 357, *360*
English, A. R., 239, 264, *265*
Erokhina, L. I., 291, 303, *312, 315*
Ertola, R. J., 157, *159*
Esse, R. C., 245, 246, *265, 266*
Essery, J. M., 174, *231*
Estabrook, R. W., 32, *87*
Ettlinger, L., 386, *398*
Euler, K. L., 17, *21, 22*
Evans, C. G. T., 401, 403, 407, *416, 417*
Evans, V. J., 94, 100, *116*
Eveleigh, D. E., 43, 47, 49, 55, *88*
Ezhov, V. A., 293, *312*

F

Falcone, G., 42, *87*
Falkow, S., 31, 32, *87, 88*
Fanning, E. A., 353, 354, 357, *360*
Farison, R. E., 378, *379*
Farrant, J., 97, *116*
Farrar, W. E., Jr., 224, *232*
Fawcett, H. S., 460, *464*
Fechtig, B. W., 165, *232*
Fedchenko, A. N., 304, *314*
Feger, V. H., 378, *381*
Felton, D. G. I., 270, 271, *281, 282*
Fencl, Z., 442, 443, *464*
Fennell, Dorothy, I., 284, 293, 297, 302, 304, 307, *315,* 378, *381*

Ferrero, E., 245, *265*
Filipello, S., 12, *22*
Fink, H., 146, *159*
Finland, M., 228, *230*
Finn, R. K., 446, *464*
Finn, S. B., 348, 351, 356, 357, *360*
Fisher, A. W., Jr., 146, *159*
Fleming, P. C., 219, *232, 234*
Flickinger, M. H., 378, *382*
Flippin, R. S., 304, *312*
Flodin, N. W., 141, *160*
Flynn, E. H., 165, 167, 168, 169, 170, 171, 174, 176, 177, 178, 180, 181, 183, 187, 188, 189, 194, 195, 196, 197, 198, 199, 200, 214, 218, 228, 229, *231, 232, 234, 236*
Focke, I., 286, *314*
Foda, M. S., 34, *87*
Fogh, J., 96, *116*
Foord, R. D., 228, 229, *235*
Forbes, R. M., 156, *160*
Ford, J. W. S., 103, *116*
Forrest, J. O., 344, 346, *360*
Fosdick, L. S., 347, *361*
Fosket, R. R., 347, *361*
Fouts, J. R., *340*
Frank, L. H., 269, 271, 273, 275, 279, *281*
Frank, R. L., 17, *21*
Frasher, L. A., 347, *360*
Frazer, D., 125, *136*
Frechet, J., 42, *87*
Fredrickson, A. G., 421, 422, 428, 430, 443, 444, 448, 449, 450, 451, 455, 456, 459, 463, *464, 465*
Freifelder, D., 125, 126, *136*
Frer, D. S., 322, 323, *340*
Friedkin, M., 372, *380*
Friedland, W. C., 292, *315*
Fujisawa, K. I., 180, 228, *235*
Fukanbara, 292, 300, 302, 308, *313, 315*
Fukazawa, Y., 33, 37, *87, 89*
Furtner, W., 12, 14, 15, *22*
Fuska, J., 284, 293, 302, 304, 305, *312, 313, 314*

G

Gabel, N. W., 4, *22*

Gabriel, M., 12, 15, *21*
Gaden, E. L., Jr., 102, *118*, 447, *464*
Galsky, A. G., 309, *312*
Galstyan, A. S., 330, *340*
Galt, R. H. B. 286, 287, 289, 295, 300, 302, 305, 309, *311, 312*
Gancendo, C., 290, *312*
Gancendo, J. M., 290, *312*
Garber, E. D., 32, *87*
Gardner, D. E., 349, *359*
Garibaldi, J. A., 379, *380*
Garland, A. J., 94, *116*
Gary, N. D., 103, *117*
Gasdorf, H., 378, *379*
Gastrock, E. A., 368, 380, *382*
Geissbühler, H., 319, 325, *340*
Geissman, T. A., 295, 305, *312, 315*
Gerber, N. N., 274, 275, 279, 280, *281*
Gerber, P., 96, *116*
Gerhardt, P., 399, 405, *416*
Gerner, R. E., 107, 108, 112, 113, *117*
Gerzon, K., 286, *312*
Gessard, C., 268, *281*
Gey, G. O., 107, *117*
Gey, M. K., 107, *117*
Geyer, R. P., 100, 109, *116, 118*
Giardinello, F. E., 107, 109, 112, *117*
Gillespie, D., 32, *87*
Gillespie, D. C., 278, *280, 281*
Gish, C. W., 353, *360*
Glass, C. A., 378, *380*
Glass, D. G., 219, *232, 234*
Glass, R. L., 347, *360*
Glauert, R. H., *266*
Globus, G. A., 303, *315*
Godfrey, E. I., 113, *116*
Godfrey, J. C., 174, 229, *234*
Goldman, A. A., 243, *266*
Goldner, M., 219, *232, 234*
Goldstein, I. J., 45, *87*
Goljamanos, T., 353, 354, 357, *360*
Golueke, C. G., 144, 145, 146, *160*
Gordon, P. N., 245, 247, 251, 255, *265, 266*
Gordon, W. L., 290, *312*
Gorin, P. A. J., 34, 36, 37, 39, 40, 42, 43, 44, 45, 46, 47, 48, 49, 51, 53, 55, 72, 73, 75, 76, 78, 79, 82, 84, 85, *87, 88, 89*
Gotaas, H. B., 144, *160*
Gottstein, W. J., 177, *234*, 245, *265*

Gradnik, B., 245, *265*
Graff, S., 94, *116*
Graham, A. F., 133, *136*
Graham, H. D., 291, *312*
Gray, W. D., 150, 151, *160*
Greco, A. E., 123, 124, 129, *136*
Green, A., 243, *265*
Green, H., 94, *118*
Green, S. R., 411, *417*
Gregory, G. I., 188, 208, 214, 219, *231*
Griffith, R. S., 228, *234*
Griffiths, J. B., 99, 100, 108, *116*
Gröger, D., 5, 11, *22*
Groenewegen, H. J. 271, 280, *281*
Gron, P., 350, *360*
Gronlund, A. F., 269, 279, *281*
Grosser, B. I., 108, *118*
Grossowicz, N., 269, *281*
Grove, John Frederick, 283, 286, 287, 289, 300, 305, *312*
Growich, J. A., 248, *266*
Guardia, F. S., 325, 326, 327, *340*
Guess, W. L., 17, *22*
Guiffre, N. A., 93, 95, 99, 107, *117*
Guignard, 270, *281*
Gupta, A. K. S., 332, *340*
Gurner, B. W., 96, *116*
Gurvich, I. A., 294, *313*
Gwinnett, A. J., 358, *360*

H

Hacker, C., 96, *116*
Hale, C. W., 163, 193, *230, 231, 234*
Hall, H. H., 378, *379, 380*
Hall, M. E., 208, 214, 219, *231*
Halpern, Y. S., 269, *281*
Hamilton, D., 298, *311*
Hamilton-Miller, J. M. T., 222, 224, *234*
Hampton, A. N., 409, 410, *417*
Hannibal, D. V., 378, *379*
Hanson, A. W., 278, *281*
Hanson, J. R., 286, 287, 289, 295, 300, 302, 305, 309, 319, *311, 312, 339*
Hardcastle, G. A., Jr., 171, *234*
Hardie, J. M., 352, 353, *361*
Hargrove, R. E., 379, *380*
Harhash, A. W., 287, 288, 290, 304, *312*
Harman, R. E., 198, *231*

Harris-Smith, R., 126, *136*, 404, *416*
Harrison, D. M., 309, *312*
Harrison, J. W. E., 344, *360*
Hartmann, G., 242, *265*
Hasenpusch, P., 107, 108, 112, 113, *117*
Hassan, T. K., 332, *340*
Hatfield, G. M., 19, *21*
Hattori, K., 170, 171, 174, 225, *234, 235, 236*
Haworth, W. N., 44, 49, *88*
Hay, G. W., 45, *87*
Hayano, S., 277, *282*
Hayashi, K., 290, *314*
Hayashi, T., 283, 285, 287, 292, *315*
Hayat, P., 269, *281*
Hayden, J., 347, *360*
Hayes, W. J., 317, *340*
Hayflick, L., 93, 94, 95, 101, 103, *116, 117*
Haynes, W. C., 272, 273, *281*, 378, *379, 380*
Head, W. F., 291, *312*
Heath, R. L., 49, *88*
Heatly, N. G., 163, *231*
Hedén, C. G., 91, 105, *117*
Heger, E. N., 372, 378, *381*
Heim, R., 4, 5, 6, *21, 22*
Hein, J. W., 347, *360*
Heineken, F. G., 421, 422, *464*
Heinen, W., 290, *312*
Hellbach, R., 366, 367, *380*
Hellman, N. N., 378, *380, 381*
Hemming, H. G., 284, 285, 290, 293, 296, *311*
Henderson, J. H. M., 291, *312*
Henehan, Catherine, 284, 285, 290, 293, 296, *311*
Henry, E. G., 13, *22*
Hensley, D. E., 378, *379, 381*
Henson, W., 307, 308, *314*
Herbert, D., 103, *117*, 415, *417*, 442, 443, *464*
Herbert, R. B., 277, *281*
Herd, J. K., 356, *360*
Herold, E., 146, *159*
Herold, M., 404, *416*
Herr, E. B., Jr., 226, *234*
Herrett, R. A., 319, *340*
Herrick, H. T., 366, 367, 368, *380, 381, 382*
Hershey, A. D., 122, *136*
Hesler, L. R., 18, *22*
Hess, W., 246, *361*

Hesseltine, C. W., 378, 379, *380, 381*
Higgins, H. M., 194, 195, 196, 197, 198, 199, 200, 228, *236*
Hill, K. L., 319, *340*
Hill, R. D., 287, *312*
Hill, T. J., 347, *360*
Hilleman, H., 268, *281*
Hilleman, M. R., 96, *118*
Hinshelwood, C. N., 451, *464*
Hitzman, D. O., 304, *312*
Hlavka, J. J., 255, *265, 266*
Hochstein, F. A., 244, 247, *265, 266*
Hockenhull, D. J. D., 401, *416*
Hodgkin, D. C., 164, *234*
Hoffman, C. A., 378, *381*
Hoffman, R. K., 101, *118*
Hofmann, A., 4, 5, *22*
Hofstra, G., 319, *340*
Hogan, L. B., Jr., 228, *234*
Holbrook, A. H., 292, *312*
Holdsworth, H., 113, *119*
Holliman, F. G., 277, *281*
Holloway, W. J., 228, *234*
Holme, T., 308, *312*
Holmes, D. H., 229, *230*
Holmes-Siedle, A. G., 330, 332, *340*
Holmström, B., 100, 107, 112, 113, *117*
Hoover, J. R. E., 180, 191, 192, *236*
Hora, F. B., 8, 11, *21*
Horak, E., 5, *22*
Horibe, S., 170, 171, *236*
Horitsu, K., 34, 43, 46, 49, *88*
Horne, R. W., 40, *88*
Horowitz, H. S., 353, *360*
Hoskins, S. W., 345, 346, *360*
Hosler, P., 402, *416*
Hospodka, J., 401, 407, *417, 416*
Hou, C. I., 269, 279, *281*
House, W., 102, *117*
Hranisovljevic, M., 83, *87*
Huang, H. T., 165, *234*
Hubbe, I., 155, *159*
Hudgins, H. R., 322, *339*
Hudson, C. S., 49, *88*
Huffman, C. W., 318, *340*
Hull, R. N., 94, 96, 107, *116, 117*
Hulme, M. A., 298, *311*
Humphrey, A. E., 114, *116,* 152, *160,* 443, 446, *464*
Hunt, E. E., *360*

Hutchings, B. L., 243, *266*

I

Ijichi, K., 379, *380*
Ikekawa, N., 292, *312*
Imshenetsky, A. A., 290, 291, 293, *313*
Ingle, T. R., 48, *88*
Ingledew, W. M., 269, 270, 279, 280, *281*
Ingram, J., 279, *281*
Iofo, R. N., 305, *314*
Ionova, I. V., 294, *314*
Ionova, N. B., 405, *314*
Irie, T., 275, *281*
Isaac, P. C. G., 146, *160*
Isenberg, H., 129, *136*
Isono, K., 272, 273, *281*
Izaki, K., 230, *236*

J

Jackson, B. G., 165, 168, 169, 170, 171, 174, 176, 177, 178, 180, 181, 183, 187, 189, 194, 195, 196, 197, 198, 199, 200, 214, 218, 219, 228, 230, *231, 236*
Jackson, D., 352, 353, *360*
Jackson, I., 114, *117*
Jackson, R. W., 272, 273, *281,* 284, 293, 297, 302, 304, 307, *315,* 378, 379, *380, 381*
Jacobs, P., 101, 103, *117*
Jacobson, K. H., 324, 337, *341*
Jacoby, W. B., 320, *340*
Jacquez, J., 422, *464*
Jago, M., 220, 224, *235*
Jakovljevic, M., 83, *87*
Jakubowitch, R. A., 228, *234*
James, P. M. C., 352, 353, *360*
Jamison, H. C., 348, 351, 352, 354, 356, 357, *360, 361*
Jaworski, E. G., 319, 323, *340*
Jayko, L. G., 378, *379*
Jeanes, A. R., 36, *88*
Jeffery, J. D. A., 188, *234*
Jefferys, E. G., 284, 285, 290, 292, 293, 296, 297, 298, 299, 300, 301, 302, 304, 305, 306, 307, *311*
Jenkins, G. N., 348, 349, *360*
Jensen, A. L., 155, *160,* 284, 303, 306, *312*
Jensen, E. R., 251, *266*
Jensen, F. C., 108, *118*

Jerrel, E. A., 125, *136*
Johnson, A. W., 19, *22*
Johnson, D. A., 171, *234*
Johnson, E. L., 155, *161*
Johnson, J. D., 242, *265*
Johnson, J. E., 104, *117*
Johnson, M. J., 115, *117,* 402, *416*
Johnson, R. M., 378, *382*
Johnson, R. W., 96, *117*
Jones, G. H., 43, 47, *88*
Jones, J. K. N., 34, *87*
Jordan, E. O., 268, *281*
Jordan, W. A., 352, *360*
Jordan, W. S., 96, *117*
Jorgensen, J., 144, 148, *160*
Julian, G., 378, *380*

K

Kabala, R., 422, *464*
Kabins, S. A., 225, *234*
Käärik, A., 3, *22*
Kaesberg, P., 129, *136*
Kagawa, H., 299, *315*
Kagawa, T., 292, 300, 302, 308, *313, 315*
Kahn, R. H., 93, *117*
Kale, N., 17, *21*
Kalmanson, G., 122, *136*
Kambe, K., 285, 287, *315*
Kamm, J. J., *340*
Kane, V. V., 220, *232*
Kaplan, K., 228, *235*
Katayama, E., 283, 292, *316*
Kathirvelu, R., 290, *313*
Katznelson, H., 309, *313*
Kaufman, B., 156, *160*
Kaufman, D. D., 321, *340*
Kavanagh, F., 292, *313*
Kawabata, Y., 153, 157, *161*
Kawagita, S., 33, 37, *89*
Kawaguchi, H., 180, 228, *235*
Kawarada, A., 283, 284, 287, 296, 303, 307, *313, 315*
Kearney, P. C., 320, 325, 326, 327, *340*
Keil, J. G., 287, *311*
Kelley, S. E., 372, *381*
Kelly, B. K., 163, *231*
Kelly, C. D., 430, *464*
Kenny, G. E., 95, 96, *118*
Kerr, D. W., 346, *360*

Kershaw, M. A., 415, *417*
Kersten, H., 242, *265*
Kersten, W., 242, *265*
Kesel, R. G., 346, *360*
Kessell, R. H. J., 292, 296, 297, 298, 299, 300, 301, 302, 304, 305, 306, 307, *311*
Kestle, D. G., *234*
Ketchum, B. H., 144, *160*
Khanna, J. M., 12, 17, *22*
Khlopenkova, L. P., 309, *313*
Kilburn, D. G., 99, 100, 105, 106, 107, 109, 110, 112, *117, 118*
Kimmelmann, B. B., 345, 346, *360*
Kind, A. C., *234*
King, R. J., 345, 346, *360*
King, W. J., 345, 346, *359*
Kinoshita, S., 291, *315,* 463, *465*
Kirby, S. M., 222, 224, 228, 229, *235*
Kirby, W. M. M., 222, *230, 234*
Kitamura, H., 283, 284, 287, 296, 307, 308, *313, 315*
Kitching, D., 115, *116*
Kjeldgaard, N. O., 428, 459, *465*
Kleiner, G. I., 294, 405, *314*
Klem, A., 448, *464*
Klöker, W., 86, *88*
Klopper, W. J., 409, 410, *417*
Kluyver, A. J., 272, 273, *281,* 460, *464*
Klyne, W., 286, *312*
Knight, S. G., 290, *313*
Knoll, E. W., 220, *230*
Knowles, C. O., 332, *340*
Knox, R., 222, *234*
Knox, W. E., 323, *340*
Kobayashi, K., 157, *161*
Koch, A. L., 428, 430, *464*
Kodama, T., 461, *465*
Kögl, F., 14, 17, *22,* 270, 271, 280, *281*
Koepsell, H. J., 378, *380, 381, *403, *416*
Komagata, K., 31, 32, 75, *88*
Kon, S. K., 155, *160*
Korst, J. J., 242, *265*
Koschel, K., 242, *265*
Koval, E. Z., 304, *316*
Kožešnik, J., 428, *464*
Krasilnikov, N. A., 305, 309, *313*
Krause, J. M., 224, *232*
Krauss, R. W., 143, *160*
Kravchenko, B. F., 305, *314*
Krazinsky, H., 255, *265*

Kreger-van Rij, N. J. W., 26, 29, 71, 72, 73, 75, 77, 84, 85, *88*
Krotkov, O., 272, 273, 276, *282*
Kruse, P. F., Jr., 104, *117*
Krutova, R. L., 294, 305, *314*
Kuceram, C. J., 107, 109, 112, *117*
Kuchaeva, A. G., 305, *313*
Kucherov, V. F., 286, 289, 294, 303, *310, 313, 314*
Kuchler, R. J., 107, *117*
Kuehner, C. C., 378, *382*
Kuhn, R., 44, *88*
Kuhr, I., 284, 286, 293, 302, 304, 305, *312, 313, 314*
Kuhr, K., 302, *313*
Kukolja, S., 188, 189, *234*
Kulik, U., 155, *159*
Kumar, S. A., 290, *313*
Kunin, C. M., 96, *117*
Kurachi, M., 269, *281*
Kurita, M., 170, 171, 174, *234, 236*
Kurosawa, E., 275, *281*, 283, *313*
Kuster, H. J., 387, *398*
Kuwabara, S., 220, *235*
Kuzel, N. R., 292, *313*
Kuzirian, O., 322, *341*
Kwant, G. W., 347, *359*
Kyes, F. M., 350, 351, *360*

L

Labeyrie, S., 268, *282*
Lachance, P. A., 149, *160*
Lacoste, A. M., 268, *282*
Lagoda, A. A., 378, *379*, 381
Lambert, G. F., 141, *161*
Lamberts, B. L., 378, *380*
Lanen, J. M., 378, *380*
Lang, K., 155, *159*
Langford, C. T., 284, 293, 297, 302, 304, 307, *315, 378, 381*
Langlykke, A. F., 378, *380*
Lanzilotta, R. P., 319, 320, 321, 324, 337, *339, 340*
Lasseur, P., 271, *281*
Lavagnino, E. R., 168, 169, 170, 171, 174, 176, 177, 178, 180, 181, 183, 187, 189, 214, 218, 219, 228, 230, *231, 234*
Lavate, W. V., 290, *313*

Lawburg, E. J. L., 95, *116*
Lazenby, J. K., 188, *231*
Lazer, L. S., 292, *311*
Lechevalier, M. P., 274, 275, *281*
Ledingham, G. A., 86, *88*
Lee, C.-C., 226, *234*
Lee, K. Y., 31, *88*
Lee, Y.-C., 38, 44, 53, 55, *88*
Leibowitz, J., 34, *87*
Leimgruber, W., 278, *282*
Lein, J., 165, *231*
Lekareva, T. A., 286, 303, *310, 314*
LeMense, E. H., 378, *380*
Lemieux, R. U., 229, *235*
Leonard, M. J., 94, *119*
Lesley, S. M. 278, *281*
Lethbridge, J. H., 123, 125, 127, 134, *136*
Leudeking, R., 446, *465*
Leung, A., 4, 5, *22*
Levan, A., 94, *117*
Leveille, G. A., 144, 146, *160*
Levenspiel, O., 432, 435, *465*
Levitch, M. E., 273, 276, 279, *281*, 379, *380*
Lewis, A., 167, 168, 169, 170, 171, 174, 175, 176, 177, 178, 180, 181, 182, 183, 184, 186, 187, 228, 229, *234, 235*
Lewis, B. A., 45, *87*
Lewis, D. G., 18, *21*
Lewis, J. C., 379, *380*
Lieb, H. B., 332, *340*
Liener, I. E., 155, *161*
Liepins, H., 109, *116*
Liepins, P., 378, *381*
Lillehoj, E. B., 378, *379*
Lillick, L., 144, *160*
Lilly, M. D., 110, *117, 118,* 157, *159*
Lincoln, R. E., 98, 107, *119*
Lindenfelser, L. A., 378, *379*
Linder, F., 240, 245, *266*
Linke, H. A. B., 325, 327, 329, 330, 332, *339, 340*
Linkens, H. F., 290, *312*
Lin Wang, E., 277, *281*
Lippincott, J. A., 309, *312*
Litchfield, J. H., 151, *160*
Llewelyn, D. A. B., 406, *416*
Lloyd, Eithne C., 292, 296, 297, 298, 299, 300, 301, 302, 304, 306, 307, *311*
301, 302, 304, 306, 307, *311*

Lloyd, P. B., 284, 285, 290, 292, 293, 296, 297, 298, 299, 300, 301, 302, 304, 306, 307, *311*
Lock, M. V., 34, *87*
Locke, J. M., 272, 273, *281*
Lockhart, H. B., 141, *161*
Lockwood, L. B., 368, 369, 372, 373, *380, 381, 382*
Lodder, J., 26, 29, 71, 73, *88*
Loder, B., 165, *234*
Lodge, M., 146, *160*
Loeb, T., 128, *136*
Löw, I., 44, *88*
Loewenburg, J. R., 290, *313*
Long, A. G., 208, 214, 219, *231*
Lorenzo, J., 345, 346, *360*
Lowe, G., 220, *232*
Lowe, J. P., 443, *465*
Lowery, J. A., 246, *265*
Lumb, M., 401, *416*
Lunan, K. D., 125, 126, 127, *136*
Lunin, M., 347, *360*
Lunn, J. S., 171, *231*
Lupova, L. M., 286, 303, *310, 314*
Luria, S. E., 123, *136*
Luttinger, J. R., 165, *231*
Lynch, D. F. J., 368, *382*
Lynch, J. E., 264, *265*

M

Maaløe, O., 428, 459, *465*
McCann, H. G., 348, *360*
McCarty, K. S., 94, *116*
McComb, A. J., 308, *313*
McCormick, J. R. D., 239, 248, 251, *265, 266*
MacDonald, J. C., 267, 268, 269, 270, *281*
McDowell, M. E., 144, 146, *160*
McGinnis, J., 379, *380*
McIlwain, H., 274, *280, 281*
Mackareth, F. J. H., 110, *117*
McKean, T. W., 350, 351, *360*
Mackenzie, R. M., 401, *416*
MacKenzie, S. L., 34, *88*
McLimans, W. F., 94, 107, 109, 112, *117, 119*
McMahon, R. E., 225, 226, 227, *231, 236*
MacMillan, J., 284, 286, 287, 289, 292, 294, 295, 309, *311, 312, 314*

Macpherson, I. A., 93, 94, 95, 96, 99, *117, 118*
McRae, D. H., 322, 323, 331, 335, 336, *340, 341*
Macrae, I. C., 324, *340*
Maeda, K., 277, *281, 282*
Mager, J., 34, *87*
Magus, R. J., 39, 43, 44, 49, 53, 55, 73, 78, 79, *88*
Mahadevan, A., 290, *313*
Mahler, H. R., 445, 447, *465*
Maidment, B. J., 103, *117*
Majer, M., 93, *117*
Malashenko, Yu. R., 290, *311*
Malone, M. H., 12, 17, *21, 22*
Mandel, I., 347, *360*
Mandel, M., 31, 32, *88*
Mansager, E. R., 331, *341*
Margoliash, E., 32, 33, *88, 147, 160*
Margolish, M., 109, *116*
Markuse, Z., 155, *160*
Marmur, J., 31, 32, *88, 89*
Marr, A. G., 444, *465*
Marshall, F. J., 225, *231*
Marshall, K. C., 460, *465*
Martell, M. J., 245, 248, 249, 255, 261, *266*
Marthaler, T. M., 350, *360*
Martin, S. M., 289, *314*
Martin, W. J., 352, 353, 355, *361*
Mascitelli-Coriandoli, E., 339, *340*
Mascoli, C. C., 96, *117*
Maskell, M. A., 101, *118*
Maslen, E. N., 164, *234*
Masler, L., 43, *88*
Masters, R. C., 94, *116*
Mathern, R. O., 148, *160*
Mathsubara, T., 224, *235*
Matsuhashi, M., 230, *236*
Matsunaka, S., 322, *340*
May, O. E., 366, 367, 368, *380, 381, 382*
Mazurek, M., 34, 37, 44, 47, 49, 51, *88*
Mead, S. W., 153, *160*
Meffert, R. M., 345, 346, *360*
Megee, R. D., III, 445, 448, 463, *465*
Mehler, A. H., 323, *340*
Merchant, D. F., 93, 107, *117*
Mercier, L., 271, *281*
Mertz, D., 307, 308, *314*
Meselson, M., 31, *88*

Meyrath, J., 461, *465*
Miall, L. M., 405, 406, *416*
Micetich, R. G., 180, 229, *235*
Mickelson, M. N., 304, *312*
Miedema, E., 104, *117*
Milicich, S., 171, *231*
Mill, P. J., 37, *88*
Miller, D. E., 321, *340*
Miller, G. A., 115, *116*
Miller, L. M., 164, *232*
Miller, L. S., 180, *236*
Miller, M. W., 71, 84, 85, *88,* 244, *266*
Miller, P. A., 248, 251, *266*
Miller, S. A., 141, *160*
Millican, R. C., 269, *281*
Millis, N. F., 114, *116,* 443, 446, *464*
Mills, A. M., 304, *312*
Milner, H. W., 144, 145, *160, 161*
Milshtein, I. M., 294, *313*
Mine, Y., 224, *235*
Minner, J. R., 96, *117*
Minor, W. F., 245, *265*
Minter, P. C., 317, *340*
Mitsuda, H., 148, *160*
Miyata, J. T., 13, *22*
Miyazaki, T., 34, *88*
Molander, L. T., 99, 103, *118*
Molin, O., 91, 105, *117*
Molle, W. H., 344, *360*
Monod, J., 442, *465*
Montuelle, B., 309, *314*
Moore, G. E., 93, 100, 101, 103, 107, 108, 112, 113, *117, 118, 119*
Morehead, P. S., 93, 94, *116*
Moreland, D. E., 319, *340*
Moreland, W. T., 244, *266*
Morin, R. B., 165, 168, 169, 170, 171, 174, 176, 177, 178, 180, 181, 183, 187, 189, 214, 218, 219, 228, 230, *231, 234*
Morrison, A., 286, 287, 289, 300, 305, *312*
Morse, M. D., 318, *340*
Mortara, M., 12, *22*
Morton, J., 243, *265*
Moser, H., 426, *465*
Moser, M., 5, *22*
Motoyama, H., 283, 292, *316*
Moyer, A. J., 366, 367, 368, 374, 375, 376, 378, 379, *380, 381, 382*
Mrak, E. M., 71, 84, *88,* 158, *160*

Mueller, R. A., 168, 169, 170, 171, 174, 176, 177, 178, 180, 181, 183, 187, 189, 214, 218, 219, 228, 230, *231, 234*
Muggleton, P. W., 97, *117,* 222, 223, 224, 226, 228, 229, *235*
Muhler, J. C., 351, 352, 353, *359,* 360
Mull, R. P., 150, *161*
Munyon, W. H., 107, *117*
Murai, K., 244, 245, 247, 251, 255, *265, 266*
Murase, M., 277, *281*
Murayama, N., 289, *316*
Muromtsev, G. S., 286, 289, 291, 294, 303, 304, 305, *310, 314*
Murphy, W. H., 93, *117*
Murray, J., 18, *22*
Musilek, V., 302, 305, *312, 314*
Myers, J., 144, 146, *160, 161*
Myers, P. L., 296, *311*
Myhr, B. C., 104, *117*

N

Nabors, C. J., 108, *118*
Nagaoka, I., 275, *281*
Nagata, Y., 290, *314*
Nagle, S. C., Jr., 100, 103, *117*
Naito, T., 180, 228, *235*
Nakagawa, S., 180, 228, *235*
Nakajima, T., *22*
Nakamura, H., 148, *160*
Nakamura, S., 277, *281*
Nakamura, Y., 287, 289, *314*
Nakanishi, I., 271, 273, *282*
Nakase, T., 31, 32, 75, *88*
Nardelli, L., 104, 112, *118*
Nathorst-Westfelt, L., 165, 225, *235*
Naylor, M. N., 353, 354, 357, *360*
Nelson, C. D., 272, 273, 276, *282*
Nelson, C. T., 34, *87*
Nelson, G. E. N., 284, 294, *315,* 372, 373, 378, *379, 380, 381*
Nelson, W. O., 156, *160*
Nestyuk, M. N., 291, 294, 405, *311, 314*
Netien, G., 309, *314*
Neuzil, E., 268, *282*
Newkirk, J. F., 164, 198, *231, 232*
Newman, J. F. E., 102, *116*
Newton, G. G. F., 164, 165, 188, 193, 220, 230, *234, 235*

Newton, R., 409, 410, *417*
Nickerson, W. J., 42, *87*
Nilson, E. H., 444, *465*
Nimmo-Smith, R. H., 323, *340*
Nishida, M., 224, 225, *235*
Nita, L., 309, *316*
Nixon, I. S., 284, 285, 290, 293, 296, *311*
Nord, F. F., 150, *161*
Norris, G. L. F., 284, 285, 290, 293, 296, *311*
Northcote, D. H., 40, *88, 89*
Norton, K., 295, 302, 309, *311, 312*

O

Oach, F. F., 150, *160*
O'Callaghan, C. H., 222, 223, 224, 226, 228, 229, *235*
Ochoa, S., 133, *137*
Oddoux, L., 309, *314*
O'Dell, N. M., 224, *232*
Odum, E. P., 425, *465*
Oelke, E. A., 318, *340*
Ogata, Y., 277, *282*
Ogur, M., 156, *160*
Okada, T., 153, 157, *161*
Okami, Y., 277, *281, 282*
Okazaki, M., 463, *465*
Okui, M., 224, 225, *235*
Ola'h, G. M., 5, *22*
Onihara, T., 153, 157, *161*
Ono, J., 287, 289, *314*
Orlando, M. D., 96, 103, *117*
Ortengren, B., 165, 225, *235*
Orton, P. D., 8, 11, *21*
Ostrom, C. A., 351, *361*
Oswald, W. J., 144, 145, 146, *160*
Otsuki, T., 283, 284, 296, 303, *313*
Overell, B. G., 356, *360*
Overton, N. F., 350, 351, *360*
Owens, O. V. H., 107, *117*

P

Packman, E. W., 344, *360*
Page, Z., 96, *116*
Pameijer, J. H. N., 349, *360*
Panetta, C. A., 171, *234*
Panina, G., 104, 112, *118*
Panosyan, A. K., 309, *314*
Paranchych, W., 133, *136*
Parker, A., 401, *416*
Parker, R. F., 109, *118*
Pasternack, R., 244, 247, *265, 266*
Patchett, A. A., 174, *235*
Pathak, S. G., 150, *160*
Paul, A. G., 4, 5, *22*
Paul, J., 93, *117*
Pease, H. L., 329, *339*
Peat, A., 401, *416*
Peat, S., 42, 49, *88*
Pedrazzoli, A., 245, *265*
Peffley, G. E., 352, *360*
Pennington, F. C., 247, *266*
Peppler, H. J., 153, *160*
Perequin, L. H. C., 460, *464*
Perez-Llano, G. A., 155, *160*
Perkins, F., 101, 103, *117*
Perlin, A. S., 44, 45, 86, *87, 88*
Perlman, D., 93, 95, 96, 99, 107, *117*
Perlman, E., 269, *282*
Perret, C. J., 455, *465*
Perrin, D. D., 261, *265*
Perry, M. B., 34, 83, *87*
Peter, H. H., 165, *232*
Peterson, E. A., 278, *280, 281*
Peterson, J. K., 352, 357, *360*
Peterson, J. L., 290, *311*
Peterson, M. H., 292, *315*
Peterson, R. E., 378, *379*
Petisi, J., 255, *266*
Petisi, J. P., 243, 247, *265*
Peto, S., 103, *117*
Pettijohn, O. G., 373, *382*
Peynaud, E., 411, *417*
Pfeifer, V. F., 372, 378, *379, 381*
Phaff, H. J., 26, 27, 28, 34, 71, 77, 78, 84, 85, *87, 88*
Phillips, C. R., 101, *118*
Phillips, G. E., 154, *159*
Phillips, I., 222, *235*
Phinney, B. O., 295, 305, *312, 314, 315*
Pilgrim, F. J., 242, 244, 247, *265, 266*
Pinto, P. V. C., 150, *160*
Pioch, R. P., 168, 169, 170, 171, 174, 176, 177, 178, 180, 181, 183, 187, 189, 214, 218, 228, *231*
Piret, E. L., 446, *465*
Pirt, S. J., 99, 100, 107, 108, 109, *116, 117*, 401, 402, 403, 404, *416*

AUTHOR INDEX

Pisano, M. A., 288, *311*
Pitt, J., 228, *235*
Pittsley, J. E., 36, *88*
Plimmer, J. R., 325, 326, 327, *340*
Plumley, N. J. B., 461, *465*
Podojil, M., 275, 280, *281*, 284, 292, 302, 304, 305, *312*, *313*, *314*
Polatnick, J., 104, *116*
Pollock, M. E., 95, 96, *118*
Pollock, M. R., 220, *235*
Porges, N., 368, *380*, *382*
Portner, D. M., 101, *118*
Postowsky, J. J., 17, *22*, 270, *281*
Powell, A. J., 298, *311*
Powell, E. O., 430, 443, *465*
Pramer, D., 319, 320, 324, 325, 327, 329, 330, 332, 337, *339*, *340*
Prasad, I., 337, *340*
Price, R. T., 108, *118*
Pridham, T. G., 272, 273, *281*, 378, *379*, *381*, 385, *398*
Probst, G. W., 294, *314*
Prox, A., 12, 14, 15, *22*
Pryce, R. J., 284, 292, *311*
Pugh, C. T., 292, *311*
Pulvertaft, R. J. V., 93, *118*
Pumper, R. W., 99, 103, *118*
Pursell, A. J. R., 407, *417*

Q

R

Radike, A. W., 350, 353, *360*
Radlett, P. J., 99, 100, 109, 110, 111, *118*
Radley, Margaret, 284, 285, 290, 293, 296, *311*
Rahman, S. B., 96, *117*
Rahn, O., 430, 449, *465*
Rake, G. W., 107, 109, 112, *117*
Rakovskii, Y. S., 304, *314*
Ramkrishna, D., 428, 444, 448, 449, 451, 455, 456, 459, *465*
Randall, M., 415, *417*
Raper, K. B., 284, 293, 297, 302, 304, 307, *315*, 375, 378, *381*
Rapp, R., 180, 229, *235*
Raymond, R. L., 153, *160*
Read, G., 16, *21*

Redemann, C. T., 295, *314*
Redfield, A. C., 144, *160*
Redin, G. S., 255, *266*
Redman, C. E., 229, *230*
Reese, E. T., 290, *313*
Reeves, M. D., 376, *381*
Regna, P. P., 244, 247, *265*, *266*
Rennhard, H. H., 245, 247, 251, 252, 255, *265*, *266*
Renz, J., 287, *315*
Reusser, F., 157, *160*, 403, 404, *416*
Rhodes, L. J., 378, *380*
Rhodes, R. A., 378, *381*
Richards, J. H., 271, *280*
Richards, J. W., 103, *118*
Richmond, M. H., 208, *231*
Ricicova, A., 302, 304, *314*
Ridley, M., 222, *235*
Rietz, P., 276, *281*
Righelato, R. C., 401, 402, 403, *416*
Riley, A. C., Jr., 378, *381*
Riley, J. M., 103, *117*
Rippon, J. W., 32, *87*
Robbers, J. E., 1, 10, 11, *22*, *23*
Roberts, L. P., 156, *160*
Roberts, R. H., 409, 410, *417*
Robertson, J. A., 379, *380*
Robinson, D. M., 97, *118*
Robinson, L. B., 95, *118*
Roe, E. T., 368, 369, 381, *382*
Roeske, R. W., 165, 168, 169, 170, 171, 174, 176, 177, 178, 180, 181, 183, 187, 189, 214, 218, 222, 223, 228, 229, *231*, *234*, *235*
Rogerson, C. T., 19, *21*
Rogerson, D. L., Jr., 123, 124, 129, *136*
Rogovin, S. P., 378, *380*, *381*
Roizman, B., 95, *118*
Rolfe, R., 31, *88*
Romagnesi, M. H., 18, *22*
Rose, W. C., 141, *161*
Rosen, J. D., 326, 332, *340*
Rosen, R., 427, *465*
Ross, A. S., 245, 248, 249, *266*
Ross, M., 345, 346, *361*
Ross, N. G., 403, *416*
Rothwell, B., 292, 296, 297, 298, 299, 300, 301, 302, 304, 306, 307, *311*
Royston, M. G., 409, 410, *417*
Ruddle, C. H., 94, *118*

Runyon, W. S., 100, *118*
Rusanova, N. V., 305, *314*
Rushizky, G. W., 123, 124, 129, *136*
Russell, T. W. F., 436, *464*
Ryan, C. W., 168, 169, 170, 171, 174, 176, 177, 178, 180, 181, 183, 187, 189, 191, 192, 193, 214, 218, 228, 229, *231, 235*
Ryan, D. M., 222, 224, 228, 229, *235*
Ryan, F. J., 460, *465*
Ryman, I. R., *87*

S

Sabath, 220, 222, 224, *235*
Sackston, W. E., 309, *311*
Salisbury, G. B., 344, *360*
Sallans, H. R., 157, *160*
Sandhu, R. S., 287, 288, *314*
Santero, G., 104, *118*
Sargeant, K., 96, *118,* 123, 124, 125, 127, 129, 130, 133, 134, 135, 136, *136*
Sassiver, M. L., 166, 167, 168, 169, 170, 171, 174, 175, 176, 177, 178, 180, 181, 182, 183, 184, 186, 187, 228, 229, *234, 235*
Saunders, B. C., 330, 332, *339, 340*
Sauvageau, 270, *281*
Savage, G. M., 403, *416*
Sawada, M., 17, *22*
Sayer, P. D., 115, *116,* 416, *417*
Scanlon, W. B., 218, 219, 230, *234*
Schach von Wittenau, M., 239, 243, 244, 245, 247, 248, 251, 252, 255, 258, 260, 263, 264, *265, 266*
Schaechter, M., 428, 430, *464, 465*
Schejter, A., 32, 33, *88,* 147, *160*
Scheske, F. A., 292, *315*
Schidkraut, C. L., 32, *89*
Schilling, E. L., 94, 100, *116*
Schleicher, J. B., 91, 105, *118*
Schlie, I., 146, *159*
Schmid, W. E., 143, 147, *161*
Schmidt, W. H., 376, *381*
Schneider, G., 286, *314*
Schneierson, S. S., 269, *282*
Schneller, A., 255, *265*
Schoental, R., 276, *282*
Schreiber, K., 286, *314*
Schuerch, C., 42, *87*
Schultes, R. E., 5, *22*
Schwimmer, S., 422, *464*

Scola, F. P., 351, *361*
Scott, A. I., 171, *234*
Scrimshaw, N. S., 139, *161*
Segal, A. H., 355, *361*
Seidel, W., 240, 245, *266*
Self, D. A., 110, *117, 118*
Sembdner, G., 286, *314*
Sentandreu, R., 40, *89*
Senti, F. R., 378, *380, 381*
Serebryakov, E. P., 303, *310,* 286, 289, *314*
Sermar, J. B., 96, *117*
Serzedello, A., 304, *315*
Seta, Y., 283, 284, 287, 296, 303, 308, 307, *313, 315*
Seto, T. A., 165, *234*
Sevcik, V., 284, 292, 293, 302, 305, *312, 313, 314*
Shacklady, C. A., 406, 407, *417*
Shannon, G. M., 378, *381*
Shapiro, E., 108, *118*
Sharabi, N. E., 321, 337, *340, 341*
Sheda, R., 30, 31, *88*
Shedden, W. I. H., 98, *118*
Shepard, C. C., 96, *116*
Shepherd, D., 298, *311*
Shepherd, R. G., 166, 167, 168, 169, 170, 171, 174, 175, 176, 177, 178, 180, 181, 182, 183, 184, 186, 187, 228, 229, *234, 235*
Sheppard, A. C., 284, 292, *311*
Shibasaki, K., 378, *380, 381*
Shiere, F. R., 347, *361*
Shimomura, T., 287, 289, *314*
Shirokov, O. G., 305, *313*
Shklyar, M. Z., 303, *315*
Shooter, K. V., 123, 125, 127, 134, *136*
Shotwell, O. L., 378, 379, *380, 381*
Shu, P., 155, *160,* 284, 303, 306, *312,* 463, *465*
Shull, G. M., 165, *234*
Siasoco, R., 228, *235*
Siddiqui, I. R., 289, *315*
Sieger, G. M., 246, *265*
Siewierski, M., 332, *340*
Sigal, M. V., 195, 196, 197, 198, 199, 200, 228, *236*
Sikita, B., 404, *416*
Sikl, D., 43, *88*
Silverman, W. B., 19, *21*
Simolin, A. V., 294, *313*

Simon, R. L., 191, 192, 193, 228, 229, *235*
Simpson, W. F., 96, *118*
Singer, R., 1, 2, 3, 4, 5, 6, 7, 9, 10, 11, 12, 15, 20, 21, *22*
Sinnott, E. W., 460, *464*
Sinsheimer, R. L., 125, 126, 127, 128, *136*
Siu, F. Y., 174, 194, 195, 196, 197, 198, 199, 200, 228, 229, *236*
Sjöberg, B., 165, 225, *235*
Sjolander, N. O., 248, *266*
Slack, G. L., 352, 353, 355, *361*
Slezak, J., 404, *416*
Slife, F. W., 319, *339*
Slodki, M. E., 34, 36, *89*, 378, *379*
Sloneker, J. H., 36, *88*
Sloof, W. Ch., 26, 71, *88*
Sly, J. C. P., 188, *231*
Smalley, H. M., 298, *311*
Smart, R. S., 347, *361*
Smiley, K. L., 378, *379, 381*
Smith, A. H., 5, 6, 12, *21, 22*
Smith, B. A., 345, 346, *361*
Smith, C., 304, *312*
Smith, C. E., 378, *380*
Smith, F., 45, *87*
Smith, G. A., 163, *231*
Smith, G. N., 298, *311*
Smith, H. G., 18, *21*
Smith, H. M., 107, *118*
Smith, I., 6, *22*
Smith, J. T., 222, *234*
Smith, M., 378, *380*
Smith, M. J., 407, *417*
Smith, N., 163, *231*
Smith, R. J., 318, *341*
Smith, R. J., Jr., 326, *340*
Snell, N. S., 379, *380*
Soder, A., 240, 245, *266*
Sohns, V. E., 272, 273, *281*, 284, 293, 297, 302, 304, 307, *315*, 378, *379, 380, 381*
Sokolova, E. V., 291, *312, 315*
Solomon, M., 345, 346, *360*
Somerfield, G. A., 188, *231*
Sorensen, N. A., 19, *22*
Sorokin, C., 144, *161*
Spector, C., 295, *314, 315*
Speirs, R., 349, *359*
Spence, D. J., 284, 294, *315*, 378, *381*
Spencer, J. F. T., 26, 27, 28, 34, 36, 37, 39, 40, 42, 43, 44, 45, 46, 47, 48, 49, 51, 53, 55, 72, 73, 75, 76, 78, 79, 82, 84, 85, *87, 88, 89*, 157, *160*
Spencer, J. L., 168, 169, 170, 171, 174, 176, 177, 178, 180, 181, 183, 187, 189, 194, 195, 196, 197, 198, 199, 200, 214, 218, 228, 229, *231, 236*, 255, *266*
Spiegelman, S., 32, *87*
Spinelli, M., 349, *359*
Spirin, A. S., *87*
Spoehr, H. A., 144, *161*
Stadtman, E. R., 273, 279, *281*
Stahly, E. A., 310, *315*
Standiford, H. C., *234*
Stark, B. P., 330, 332, *340*
Starmons, J. L. E. M., 347, *359*
Stecher, P. G., 337, *341*
Stedman, R. J., 180, 191, 192, *236*
Steglich, W., 12, 14, 15, *22*
Stein, S., 4, *22*
Stein, W. J., 243, *266*
Stenderup, A., 32, *87, 89*
Stent, G. S., 121, *136*
Stephens, C. R., 242, 243, 244, 245, 247, 248, 251, 252, 255, *265, 266*
Sternberg, M., 287, 289, 291, *315*
Stevens, W. K., 226, 229, *235*
Stewart, D. L., 94, 95, *116*
Stewart, J. C., 295, *312*
Still, C. C., 322, 332, *340, 341*
Still, G. G., 322, 323, 330, 331, 333, *340, 341*
Stine, G. J., 463, *465*
Stockwell, F. T. E., 108, 113, *116*
Stoker, M. G. P., 93, 94, 99, *117, 118*
Stoll, C., 287, 289, 299, 300, *315*
Stone, C. J., 101, 107, 109, 112, *118*
Stoodley, R. J., 34, *87*
Storck, R., 31, *89*
Stowe, B. B., 283, *315*
Strack, E., 268, *282*
Strandberg, G. W., 378, *381*
Strange, C. G., 349, *361*
Strauss, J. H., Jr., 129, *136*
Stringer, C. S., 378, *380, 381*
Strominger, J. L., 230, *236*
Stubblefield, R. D., 378, *380, 381*
Stubbs, J. J., 367, 368, 369, *381, 382*
Studola, F. H., 272, 273, *281*, 283, 284, 291, 293, 294, 302, 304, 307, *315*, 378, *381*

Stulberg, C. S., 96, *118*
Stuntz, D. E., 6, 9, 18, *21, 23*
Subramanian, G., 428, *465*
Sueoka, N., 31, *89*
Sullivan, G., 13, 16, 17, *22*
Sullivan, H. R., 226, 227, *236*
Sulser, G. F., 347, *361*
Sumiki, Y., 283, 284, 292, 296, 300, 303, 307, 308, *312, 313, 315, 316*
Sure, B., 155, *161*
Sutcliffe, P., 352, 353, *360*
Suter, P. J., 292, *314*
Sutherland, S. A., 171, *234*
Suzuki, S., 272, 273, *281*
Swait, J. C., 292, 296, 297, 298, 299, 300, 301, 302, 304, 306, 307, *311*
Swan, D. G., 270, *281*
Swan, G. A., 271, 274, *282*
Swanstrom, M., 122, *136*
Sweat, M. L., 108, *118*
Sweeney, M. S., 225, *234*
Sweet, B. H., 96, *118*
Swered, K., 191, 192, *236*
Swift, A. C., 180, *236*
Swim, H. E., 108, 109, *118*
Switzer, C. M., 319, *340*
Sylvester, J. C., 292, *315*
Syrbu, G. A., 318, *341*
Syverton, J. T., 96, *118*

T

Tabenkin, B., 368, 369, *380, 381, 382*
Takahashi, J., 153, 157, *161*
Takahashi, K., 180, 228, *235*
Takahashi, N., 283, 284, 287, 296, 307, *313, 315*
Takai, M., 287, *313*
Takano, T., 170, 171, 174, *234, 236*
Takeda, R., 271, 272, 273, *282*
Takemoto, T., *22*
Tamiya, H., 144, *161*
Tamorria, C. R., 245, 246, *265, 266*
Tamura, M., 307, *315*
Tamura, S., 287, *313*
Tamura, T., 289, *316*
Tanner, F. W., Jr., 244, *265*, 378, *381*
Tashjian, A. H., 108, *118*
Tatum, E. L., 460, *465*
Taylor, A. B., 194, *236*

Teletzke, G. H., 415, *417*
Telling, R. C., 94, 95, 99, 100, 101, 103, 107, 108, 109, 110, 111, 112, 113, *116, 117*
Temnikova, T. V., 304, *314*
Tempest, D. W., 415, *417*
Teneth, M., 269, *281*
Terui, G., 463, *465*
Terui, M., 299, *315*
Thackeray, J., 404, *416*
Theriault, R. J., 292, *315*
Thom, C., 364, *381*
Thomas, A. E., 352, 354, *361*
Thomas, C. A., Jr., 122, *137*
Thomas, W. H., 143, *160*
Thomas, W. J., 94, 107, *119*
Thompson, F. R., 337, *341*
Thompson, K. W., 108, *118*
Thoms, H., 13, *22*
Tillotson, J. R., 228, *230*
Tipper, D. J., 230, *236*
Tobin, R., 378, *380*
Todaro, G., 94, *118*
Tönnis, B., 271, 280, *281*
Toohey, J. I., 272, 273, 276, *282*
Toplin, I., 102, *118*
Traufler, D. H., 372, 378, 379, *381*
Treccani, V., 157, *161*
Tredwell, P. E., 95, *118*
Tribble, H. R., Jr., 103, *117*
Trinici, A. P. J., 401, *416*
Trischmann, H., 44, *88*
Tritsch, G. L., 100, 103, *118*
Trucco, E., 428, *465*
Tscherter, H., 5, *22*
Tsien, W. S., 155, *161*
Tsuchiya, H. M., 378, *380, 381,* 421, 422, 428, 429, 443, 444, 448, 449, 450, 451, 455, 456, 459, 463, *465*
Tsuchiya, T., 33, 37, *87, 89*
Tyler, V. E., 5, 6, 13, *21*
Tyler, V. E., Jr., 1, 4, 5, 6, 9, 10, 11, 12, 13, 16, 17, 18, *21, 22, 23*

U

Ubertini, B., 104, 112, *118*
Udagawa, K., 291, *315*
Ulman, J. S., 436, *464*
Ulrich, K., 113, *118*
Ulyanova, O. M., 290, 291, *313*

Umezawa, H., 277, *281, 282*
Unger, V. H., 322, *341*
Uno, S., 283, 292, *315*
Unrau, A. M., 287, *312*
Ursprung, J. J., 255, *265*

V

Valanju, N., 19, *21*
Vallette, J. P., 268, *282*
Vancura, V., 309, *315*
van der Walt, J. P., 71, 72, 84, 85, *89*
van Hemert, P. A., 105, 106, *118*, 403, *416*
Van Heyningen, E., 163, 168, 169, 170, 171, 174, 176, 177, 178, 180, 181, 183, 187, 188, 189, 190, 198, 199, 201, 202, 204, 205, 207, 214, 218, 223, 224, 228, 230, *231, 236*
Van Heyningen, E. M., 191, 192, 193, 228, 229, *235*
van Kerken, A. E., 72, 85, *89*
Van Lanen, J. M., 378, *380, 381*
van Uden, N., 73, 77, 79, *89*
van Wezel, A. L., 92, 105, 106, *118, 119*
Varner, E. L., 292, *311*
Verbiscar, A. J., 295, 305, *312, 315*
Verduin, J., 143, 147, *161*
Verner, D. O., 287, *311*
Vickers, T. G., 135, 136, *136*
Vidal-Lieria, M., *89*
Vincent, M. M., 108, *118*
Vining, L. C., 16, *21*
Vinson, L. J., 150, *161*
Vischer, E., 165, *232*
Vitovskaya, G. A., 37, *87*
Vogel, P., 346, *361*
Vogelsang, J., 13, *22*
Voinescu, R., 291, *315*
Vojnovich, C., 372, 378, *381*
Volpe, A. R., 345, 346, *359*
Von Foerster, H., 428, *465*
Vosseller, G. V., 101, 107, 113, *119*
Vowles, N. J., 353, 354, 357, *360*
V'yun, A. A., 287, *311*

W

Wachtel, L. W., 349, *361*
Wagner, R. L., 247, *266*
Wahl, R., 31, *88*
Wakagi, S., 283, *315*
Waksman, S. A., 385, *398*

Walker, J. A. H., 113, *119*
Walker, J. S., 98, 107, *119*
Walker, P. M. B., 32, *89*
Wallace, D., 350, *361*
Waller, C. W., 243, 247, *265, 266*
Walsh, J. P., 347, *361*
Walters, M. B., 18, *23*
Walton, R. B., 164, *232*
Wang, H. L., 378, *380, 381*
Ward, G. E., 368, 369, 372, 373, 376, *380, 381, 382*
Washburn, W. H., 292, *315*
Washington, O., *87*
Wasserman, E. A., 153, *161*
Waterworth, P. M., 195, *230*
Watkins, J. F., 96, *119*
Watling, R., 2, 3, 4, 5, 6, *21, 23*
Watson, J. D., 459, *465*
Watson, P. R., 36, *88*
Webb, F. C., 99, 100, 107, 109, 110, 112, *117*, 157, *159*
Weibull, C., 254, *266*
Weigele, M., 278, *282*
Weinstein, L., 220, 228, *231, 235*
Weir, J. K., 4, *23*
Weirether, F. J., 98, 107, *119*
Weisburger, E. K., 338, *341*
Weisburger, J. H., 338, *341*
Weiss, L., 93, *118*
Weiss, R. E., 91, 105, *118, 119*
Weissman, C., 133, 134, 135, *136, 137*
Wellock, W. D., 349, *360, 361*
Wells, P. A., 367, 368, *380, 381, 382*
Welsby, B., 405, *416*
Whelan, W. J., 42, *88*
Whistler, R. L., 48, *88*
Whitaker, A. R., 115, *116*
Whitaker, N., 304, *315*
White, J., 405, *417*
White, M., 220, *230*
White, P. B., 104, *117*
Wichelhausen, R. H., 95, *118*
Wick, W. E., 226, 229, *230, 236*
Wickerham, L. J., 29, 34, 36, 74, 75, 79, 84, *89*, 378, 379, *380, 382*
Wierzchowska, Z.,
Wierzchowski, P., 286, *315*
Wiesen, C. F., 378, *380*
Wilcox, C., 228, *230*
Wildy, P., 98, *118*

Wilkins, G. D., 401, 403, *416*
Wilkinson, J. A., 93, *118*
Wilkinson, R. G., 243, 247, *265*
Willcox, O. W., 155, *161*
Williams, C. H., 324, 337, *341*
Williams, F. M., 457, *465*
Williams, J. H., 243, *265*, *266*
Williams, M. W., 310, *315*
Williams, R. C., 123, *136*
Williams, V., 126, *136*
Williamson, L., 357, *360*
Wilson, D. V., 18, *21*
Wilson, H. F., 322, 323, 331, 335, 336, *340*, *341*
Wilson, J. N., 94, 95, *116*
Winnett, G., 326, 332, *340*
Wishart, D. R., 222, *235*
Wixom, R. L., 141, *161*
Wolf, C. F., 243, *266*
Wolf, F. A., 29, *89*
Wolf, F. T., 29, *89*
Woodard, R. B., 242, 244, 247, *265*, *266*
Woolf, D. O., 286, *312*
Wrede, F., 268, *282*
Wright, D. G., 402, *416*
Wyatt, G. R., 125, *137*

Y

Yabuta, T., 283, 285, 287, 289, 292, *315*, *316*
Yagashita, K., 277, *282*
Yamada, K., 153, 157, *161*, 461, *465*
Yamaki, T., 283, *315*
Yamashiroya, H. M., 99, 103, *118*
Yanagita, T., 463, *465*
Yano, T., 461, *465*
Yanofsky, C., *161*
Yarrow, D., 30, 31, 79, *88*, *89*
Yasumota, K., 148, *160*
Yeary, R., 239, 263, 264, *266*
Yeo, R. G., 123, 125, 127, 134, 135, 136, *136*
Yergaman, G., 94, *119*
Yih, R. Y., 322, 323, 331, 335, 336, *340*, *341*
Yokobe, T., *22*
Yokota, Y., 224, *235*
Young, J., 355, *361*

Z

Zaccharias, B., 403, *416*
Zacharias, B., 308, *312*
Zacherl, W., 353, *361*
Zafran, J. N., 345, 346, *360*
Zaichenko, A. M., 304, *316*
Zajieck, I., 304, *312*
Zakordonets, L. A., 288, *311*
Zalokar, M., 463, *465*
Zander, H. A., 347, *361*
Zanevich, V. E., 290, *311*
Zarnescu, A., 309, *316*
Zayon, G. M., 345, 346, *360*
Ziegler, D. W., 94, 107, *119*
Zimmerman, S. B., 133, 134, *137*
Zinder, N. D., 128, *136*
Zobell, C. E., 157, *161*
Zweig, G., 307, *316*

SUBJECT INDEX

A

Acylaminocephalosporanic acid, unsaturated, 178, 182
7-Acylaminocephalosporanic acids
  cyclic analogs of thiouronium derivatives, 201, 207
  effect of chain length, 169
  thiouronium, 201, 209
7-Acyl-7-aminocephalosporanic acid dimethyldithio carbamates, 199, 203
7-Acyl-aminocephalosporin
  -4'-carboxamido-pyridinium derivatives, 196, 200
  pyridinium derivatives, 196, 199
Aeruginosin A and B, 277
Aminocephalosporanic acids
  heterocyclic thiol derivatives, 201, 208
7-Aminocephalosporanic acids
  aminoacyl-, 171–174
  azi- and cyano-alkylacyl-, 174, 178
  cycloaliphatic acyl-, 178, 184
  fused ring acyl-, 171, 175
  haloacyl-, 171, 176
  heteroacyl-, 174, 178–181
  N-substituted aminoacyl-, 181, 186
  phenacetyl-, alpha-substituted, 170, 171
  phenylated acyl-, 171, 176
  ring substituted, 167
    phenoxyacetyl-, 168
    phenyl thioacetyl-, 168
  substituted thioalkylacyl-, 174, 177
  substituted urea acylamino-, 181, 185
Arlington Farms Laboratories of USDA, 365

B

Bacteriophage production, 121
  calculation of yields, 123
  contamination prevention, 124
  DNA bacteriophages, 125
  ØX174 bacteriophage, 128
  T7 bacteriophage, 125–127
  Growth methods, 122
  RNA bacteriophages, 128
  MS2 bacteriophage, 129
  R17 bacteriophage, 133
  μ2 bacteriophage, 130–133
Basidiomycete chemotaxonomy, 1
  agaric acid, 13
  anthraquinones, 14
  biosynthesis of indolyl hallucinogens, 7
  diatretynes, 20
  indolyl derivatives, 4, 5
  fungal products (miscellaneous), 9–13
  involution, 17
  polyacetylenic products, 19
  styryl pyrones, 18
Terphenyl quinones, 15–17
  tetronic acids, 13, 15
  utility of macrochemical tests, 3
  urea, 6, 8
2,3-Butanediol, by fermentation, 373

C

Cephaloglycin, 166
Cephaloridines
  alkyl-substituted, 194, 195
  amide-substituted, 195, 196
  halogen-substituted, 195
Cephalosporanic acid, 164, 165, 167
Cephalosporin(s)
  3-alkoxymethyl derivatives, 205, 216
  3-aminomethyl, 205, 214
  3-azidomethyl, 201, 209
  3-azidomethyl-7-glyoxalyl, 205, 213
  biological resistance to lactamase, 219
  competitive inhibitors, 223
  in vivo activity, 228, 229
  β-lactam cleavage, 220, 221
  metabolism, 225, 226
  3-methoxymethyl, 205, 217
  mode of action, 225, 230
  pharmacology, 225, 227
  phenoxymethyl penicillin conversion to, 218
  structure/activity relationships, 163
  xanthate derivatives of, 198, 201

485

Cephalosporin C, 163, 164
Cephalosporin N, 164
Cephalothin, 166
  dithio carbamates of, 199, 202
  piperazinodithio carbamates, 199, 204
  quaternized piperazinodithio carbamates, 201, 205
  thio acid derivates, 201, 209
  zwitterionic dithio carbamates, 199, 201, 204
Chlororaphine, 270–272
Citric acid, fermentation, 364, 368
Continuous culture, industrial applications of, 399
  brewing, 407–410
  wine, 411
  organic chemicals, 405
    acetic acid, 405
    edible proteins, 406
    lactic acid, 405
  pharmaceutical products, 401
    antibiotics, 402
    enzymes and vaccines, 403
    steroid hydroxylation, 403
  sewage and trade wastes, 412–415
Currie, J. N., 364

D

Dentafrices, therapeutic, 343
  caries-inhibition, 345
    ammonia and urea compounds in, 345
    antibiotics in, 346
    antienzyme in, 347
    fluoride in, 348–351
    Crest®, 352–353
    Cue®, 354
  desensitizing, 344
  phosphate fluorides, 355–357
  plaque and calculus-inhibition, 344
  Super Stripe®, 343
Desacetoxycephalosporanic acid
  D-phenylglycyl derivatives of, 192, 193
Desacetoxycephalosporins, 191
  7-acylamino-, 191, 192
Desacetylcephalosporins, 188, 189
  acyl derivatives of, 189, 190
  O-carbamyl derivatives of, 191

F

Fermentation
  classification of, 431
Fermentation processes
  mathematical model for, 419
    for growth of filamentous organisms, 459
Fermentor(s)
  continuous flow, stirred vessel (C*), 432
  continuous plug flow, tubular, 434
  control of, 437
  design of, 367, 370
  recycling, 436

G

Gibberellin(s)
  assay methods, 291–293
  biosynthesis of, 294–297
  fermentation, 283
    enzymes found, 290
    other compounds detected, 287–289
  structures, 284, 285
  uses of, 309–310
Gibberellin A_1, 307
Gibberellin $A_{\overline{2}}$, 307
Gibberellin A_3, 307, 308
Gibberellin A_4, A_7, A_9, A_{12}, A_{14}, A_{16}, A_{24}, 308
Gibberellin A_{10}, A_{13}, A_{15}, 309
Gibberellin-producing organisms, 284, 285
Gibberellin-related compounds,
  from fermentations, 286
Gluconic acid, by fermentation, 366, 367
Griseolutein A and B, 277, 278

H

Herbert's model of maintenance, 444
Herbicides, acylanilide, 317–319
  biological activity of metabolites, 336, 337
  metabolism
    by cultures and enzymes, 320, 321, 329
    in coil, 319, 324–327
    in plants, 322, 323, 330
  pathways of, 333–335
Herrick, H. T., 364
1-Hydroxyphenazine, 276
2-Hydroxyphenazine, 276

SUBJECT INDEX

2-Hydroxyphenazine-1-carboxylic acid, 276

I

Iodinin, 274, 275
Itaconic acid, by fermentation, 372

K

Ketogluconic acid, by fermentation, 369
Kojic acid, by fermentation, 368

L

Lactic acid, by fermentation, 368

M

Mammalian cells, large-scale culture, *in vitro*, 91
  adaptation to suspension culture, 94
  contamination
    by bacteria, 95
    by fungi, 95
    by viruses, 96
    by other cells, 96
  maintenance of cultures, 92, 94
  mass cultivation of cells, 104
    microcarriers, 105, 106
    monolayers, 104, 105
    submerged culture, 107
      effect of environment, 108–111
      equipment, 112–114
  media
    composition of, 99, 100
    sterility testing of, 103
    sterilization of, 101–103
  selection of, 92, 96
  spontaneous variation of, 93
  storage of, 92, 97
Mathematical models
  nonsegregated, 429
  segregated, 429
  structured
    general formulation of, 450
    examples of, 457
  unstructured, 438
    critique of, 449
  variables
    biological, 423
    nonbiological, 424
May, O. E., 364

Microorganisms, as food, 139
  algae, 142–149
  bacteria, 156, 157
  fungi, 149–152
  lichens, 155
  yeast, 152
    fermentation of hydrocarbons, 153, 154
    fermentation of industrial wastes, 153
Monod's model, 442–443
  extensions of, 447
  modifications of, 443–447
Myxin, 278, 279

N

NRRL, 371

P

Patents, microbiological, 383
  definition of species, 384–389
  deposition of cultures, 390–392
  distinction of strains, 393–395
  occurrence of antibiotics in nature, 395–397
Penicillin fermentation, 373–377
Phenazine-1-carboxylic acid, 272, 273
7-Phenylacetylamino cephalosporanic acid, 3-acyl-aminomethyl, 205, 215, 216
7-Phenylamino cephalosporanic acid
  carboxamide substituted, pyridinium derivatives, 196, 197
Pyocyanine, 267–270

S

Sorbose, by fermentation, 369

T

Tetracyclines
  activity of
    basic requirements for, 239–242
    electronic and lipolytic effects on, 258–264
  modification of
    at C-2 position, 242–245
    at C-4 position, 245–247
    at C-6 position, 248–253
    at C-7 and C-9 positions, 253–256
    at C-12A position (esters), 258–259

structure/activity relationships, 237
5A(11A)-dehydro- and 5A-epi-, 246, 247
11A-substituted, 256–257

Y

Yeast(s)
  classification of
    methods of, 29, 30
    by PMR spectra, 54–83
    genera and species, 26–28
  mannose-containing polysaccharides of
    isolation of, 40–42
    methods of determining structure of, 43–48
    PMR spectra of, 48–54
  nucleic acids, 31
  polysaccharides, 33–36
    PMR studies of, 37–39
  proteins, 32
    serological properties of, 33

QR
1
A38
v.13
1970

JAN 5 1971